# Effect of Processing and Cooking on Physicochemical, Sensory, and Functional Properties of Food

# Effect of Processing and Cooking on Physicochemical, Sensory, and Functional Properties of Food

Guest Editor

**Sheng-Dun Lin**

MDPI

Basel • Beijing • Wuhan • Barcelona • Belgrade • Novi Sad • Cluj • Manchester

*Guest Editor*
Sheng-Dun Lin
Department of Food Science
and Technology
Hungkuang University
Taichung
Taiwan

*Editorial Office*
MDPI AG
Grosspeteranlage 5
4052 Basel, Switzerland

This is a reprint of the Special Issue, published open access by the journal *Foods* (ISSN 2304-8158), freely accessible at: https://www.mdpi.com/journal/foods/special_issues/J4HAT4ZGMP.

For citation purposes, cite each article independently as indicated on the article page online and as indicated below:

| Lastname, A.A.; Lastname, B.B. Article Title. *Journal Name* **Year**, *Volume Number*, Page Range. |
|---|

**ISBN 978-3-7258-4553-8 (Hbk)**
**ISBN 978-3-7258-4554-5 (PDF)**
**https://doi.org/10.3390/books978-3-7258-4554-5**

# Contents

# About the Editor

**Sheng-Dun Lin**

Dr. Sheng-Dun Lin is currently a Professor in the Department of Food Technology at Hungkuang University. He received his Ph.D. in Agronomy from the Institute of Food Science at National Chung Hsing University in January 1995 and worked as a Postdoctoral Researcher at National Chung Hsing University for half a year. He joined the military in July 1995, and after being discharged from the military in May 1997, he joined Chen Hsiang Food Industrial Co., Ltd., where he worked in food research and development and quality control. In August 1998, he switched to the education sector to engage in teaching and research, and has often assisted private enterprises in developing new products, improving processes and food quality, and helping to establish quality control systems. Dr. Lin's expertise includes food processing, functional food development, food quality analysis and management, and food industry management. From September 1998 to May 2025, he implemented 103 government and private enterprise projects and transferred 20 technologies. He has also won many awards, such as the Taiwan National Science and Technology Council's subsidy for outstanding talents in colleges and universities, and the Taiwan Association for Food Science and Technology's "Outstanding Food Promotion Service Personnel Award".

# Preface

Food ingredients can be processed and cooked using physical, chemical and microbiological methods, with cooks able to deploy either a single technique or a combination of two or more methods to change the appearance, nutritional content and phytochemical make-up of their ingredients, thereby increasing the final product's shelf life, palatability and functionality. Additionally, the use of innovative processing and cooking techniques can increase the reuse of natural resources or processing by-products. This is the main focus of this Special Issue, "The effects of processing and cooking on the physicochemical, sensory and functional properties of foods". Thanks to the efforts of many scholars and MDPI editorial colleagues, a total of 16 high-quality original research articles are published here across four parts: 1. Enhancing the nutritional value and functionality of traditional foods. 2. The reuse of processing by-products. 3. The effects of different processing and cooking methods on food quality characteristics. 4. The application of new food processing technologies. The contents of these manuscripts are diverse and valuable, helping researchers to conduct follow-up studies and provide pathways for the food industry to develop new products.

**Sheng-Dun Lin**
*Guest Editor*

foods

MDPI

*Editorial*

# Effect of Processing and Cooking on Physicochemical, Sensory, and Functional Properties of Food

**Sheng-Dun Lin**

Department of Food Science and Technology, Hungkuang University, 1018, Sec. 6, Taiwan Boulevard, Shalu District, Taichung 433304, Taiwan; lin54@hk.edu.tw; Tel.: +886-4-26318652 (ext. 5038); Fax: +886-4-26319176

Food ingredients come from a wide range of sources, and different processing and cooking methods can change their appearance, nutrients and phytochemicals, thereby increasing the product's shelf life, palatability and functionality. In addition, the reuse of natural resources can be improved through innovative processing or cooking techniques. The main focus of this Special Issue is the "Effect of Processing and Cooking on Physicochemical, Sensory, and Functional Properties of Food", which is linked to the maintenance or enhancement of the nutritional quality, bioactive components and functionality of food through processing or cooking techniques or product formulation. This Special Issue contains 16 high-quality original research articles on enhancing the nutritional value and functionality of traditional foods, the reuse of processing by-products, the effects of different processing and cooking methods on food quality characteristics, and the application of new food processing technologies.

## 1. Enhancing the Nutritional Value and Functionality of Traditional Foods

Many research reports have pointed out that plant foods, in addition to providing essential nutrients, are also rich in a variety of bioactive compounds [1,2]. These bioactive compounds are directly related to human health and various conditions [3,4], such as cardiovascular disease and metabolic syndrome. Therefore, if foods containing bioactive ingredients can be added to traditional foods using processing technology, their nutritional value, antioxidant properties, and health benefits can be enhanced [4,5]. Traditional wheat products, such as wheat noodles, baked bread and steamed bread, are popular among consumers, but are often criticized for lacking bioactive compounds and dietary fiber.

The flowers of *Clitoria ternata* (CT) can be used as a natural pigment in the food industry. The flowers and extracts of CT are rich in bioactive compounds and possess antioxidant, anti-inflammatory, antibacterial, antidiabetic and antitumor activities [6]. Shiau et al. (contribution 1) investigated the effects of CT flower extract (CTFE) on the quality characteristics of dry and boiled noodles. The results showed that cooking the noodles significantly decreased their anthocyanin content and blue color, but increased the green intensity of the noodles. The addition of 20–30% CTFE produced blue noodles with rich bioactive substances, antioxidant and good sensory attributes compared to the control.

Tamarillo (*Solanum betaceum* Cav.) has a unique flavor and is rich in nutrients, bioactive compounds, and health benefits [4,7]. Syu et al. (contribution 3) investigated the effects of freeze-dried tamarillo powder (TP) on the physicochemical, sensory, and functional properties of mantou. The results showed that TP was rich in phytochemicals and had good antioxidant properties. Taking wheat bread (a high GI food) as a reference, T20 (manto made by replacing 20% of wheat flour with TP) was classified as a medium GI food. Adding an additional 15–20% water to the T20 formulation increased its volume and specific volume, and the overall liking score (7.4–7.5 on a 9-point hedonic test) was not significantly different from the control ($p > 0.05$). A new medium GI manto with more bioactive compounds and good antioxidant properties was successfully developed.

Frozen dough is increasingly used in bread and bakery products, but the freezing and thawing process during production destroys the gluten structure and prevents the bread from expanding, resulting in a smaller volume and a tougher crumb [8]. Casein hydrolysate can not only be used as a protein supplement, but can also improve the quality characteristics of various foods [9]. The effects of casein savinase hydrolysate (CSH) (1–2 g/100 g flour) on the physicochemical quality properties of wheat dough and corresponding bread were studied by Bekiroglu et al. (contribution 9). This study confirmed that adding CSH to frozen dough can reduce the quality decline of frozen bread.

## 2. Reuse of Processing By-Products

Nutrient-rich and safe agricultural processing by-products can be sustainably used to produce value-added products with economic and environmental benefits [10]. Cocoa bean shells (CBSs) are a by-product of chocolate processing (accounting for approximately 12–20% of cocoa beans) and their nutritional composition is not significantly different from cocoa beans and should not be discarded [11]. Wang et al. (contribution 16) investigated the effects of replacing 5–25% of wheat flour (WF) with Taiwanese cocoa bean shells (CBSs) on the quality properties of bread. The heavy metal content of CBS is lower than the Codex Alimentarius limit for cocoa powder, and ochratoxin A and aflatoxin (B1, B2, G1 and G2) were not detected. CBS is not only rich in nutrients but also has good antioxidant properties. This study also successfully developed a new medium GI bread containing total dietary fiber and phytochemicals by partially replacing wheat flour with CBS.

During the process of making mushrooms into vacuum fried crisps, the blanching broth (BB) and centrifuged broth (CB) generated are often discarded, increasing the amount of wastewater and treatment costs. Chen et al. (contribution 5) determined the quality characteristics of BB and CB and formulated them into instant beverages using functional indigestible dextrin. The BB and CB are rich in bioactive compounds, especially polysaccharides, ergothioneine, and total phenols. This study successfully developed BB and CB into functional foods in the form of beverages.

## 3. Effects of Different Processing and Cooking Methods on Food Quality Characteristics

Appropriate processing or cooking techniques are an important research topic because they can maintain or improve the nutrients, phytochemicals, sensory and functional properties of foods. Glucosinolates can be hydrolyzed into isothiocyanates, antithyroid substances, and goitrogens. Research has shown that washing, soaking, and boiling (destroy myrosinase) can help reduce the amount of goitrogens in food and prevent the release of goitrogens from thyrotropin [12]. However, heating also destroys glucosinolates and isothiocyanates [13]. Panduang et al. (contribution 7) investigated the optimization of cooking conditions to reduce goitrogens while retaining isothiocyanates in cruciferous vegetables. The results showed that home cooking methods to preserve isothiocyanates and reduce goitrogens include steaming cabbage at 80–100 °C for 4 min and stir-frying Chinese kale at 60–100 °C for 2 min. These methods can preserve phytochemicals and reduce food hazards.

Vegetables can be pickled through anaerobic fermentation with lactic acid bacteria, acetic acidification with or without added salt, or a combination of these processes [14]. Radman et al. (contribution 8) prepared unfermented sea fennel leaf preservatives using different pickle juices made from apple cider vinegar, wine vinegar, and alcohol vinegar and compared their quality properties. The results showed that sea fennel pickled in alcohol vinegar scored the highest in all sensory parameters, while sea fennel samples pickled in wine vinegar scored the lowest. A strong odor of wine vinegar was described as an off-flavor of this sample.

Parboiling and polishing are common rice processing methods in Bangladesh. Akhter et al. (contribution 10) studied the effects of parboiling and polishing on the nutritional composition of five high-yielding rice varieties. The results suggest that consumers should be encouraged to consume unpolished rice in order to reduce the loss of nutrients in rice.

The cooking process can significantly change the quality characteristics of meat. Wang et al. (contribution 11) studied the effects of different frying temperatures and times on the physicochemical properties and volatile components of pork tenderloin, and their correlations. The results show that b* value and moisture content have the greatest impact on volatile substances. The b* value was positively correlated with the aroma components, but negatively correlated with the moisture content. This study suggests that the b* value and moisture content can be detected to predict the changes in the types and contents of volatile substances in fried pork.

Frankfurters, bratwursts, and wieners are becoming less popular among consumers due to their lack of fiber or high salt, nitrite, and fat content [15]. Pérez et al. (contribution 12) evaluated the nutritional quality of artisanal sausages produced in 10 factories in Ecuador. The results showed that the proximate composition contents of handmade sausages were within the recommended range for commercial sausages. This suggests that "artisan" is just a name for a product.

Peach palm fruit contains a large amount of oil and protein, as well as a large amount of carbohydrates and insoluble dietary fiber, and is therefore considered to be a fruit with high nutritional value. However, peach palm fruits need to be cooked in water before consumption to inhibit anti-nutritional factors, eliminate oxalate crystals in the peel [16], and inactivate peroxidase in the pulp to avoid irritation of the throat mucosa [17]. Soares et al. (contribution 13) first reported the changes in the quality characteristics of albino peach palm before and after the cooking and drying processes. The results showed that the albino peach palm pulp and powder could be utilized by the food, pharmaceutical and biotechnology industries.

*Agaricus bisporus* has been extensively studied due to its unique quality properties, such as flavor and high amino acid score [18]. However, there are few studies on the changes in its flavor compounds during cooking. Xie et al. (contributor 14) used gas chromatography–mass spectrometry and electronic nose to study the effects of different cooking methods on the volatile aroma components of mushrooms. The results showed that the electronic nose could effectively distinguish samples prepared by different cooking methods. Cluster analysis of differential aroma substances showed that the aroma of the steamed group and the raw group was more similar. 1-Octen-3-one was identified as the main aroma component of cooked mushrooms.

## 4. Application of New Food Processing Technologies

In recent years, many scholars have applied new food processing technologies to product development and quality research. The processing technology of forming three-dimensional structures by stacking edible materials is called three-dimensional food printing (3DFP) technology [19]. The fruits of *Aronia melanocarpa* (AM) are rich in polyphenols with functional properties such as antioxidant, lipid-lowering, anti-inflammatory, and anti-cancer effects [20]. Zhou et al. (contributor 2) aimed to investigate the printability and storage properties of AM gels in 3DFP. The results showed that the AM polyphenol gel loading system most suitable for 3DFP processing and printing requirements was AM slurry:methyl cellulose:pea albumin:hyaluronic acid = 100:14:1:1. This study provides a theoretical basis for 3D printing applications using fruit pulp as raw material.

Ultrasound is a non-thermal radiation technology. Its interaction with the medium will produce mechanical effects, cavitation effects and thermal effects [21]. Ultrasonic pretreatment can effectively enhance heat transfer, improve drying efficiency and inhibit the deterioration of material quality. Shang et al. (contributor 4) studied the effects of different ultrasonic pretreatment processes on the quality properties of licorice. Compared with the control, ultrasonic pretreatment combined with far-infrared drying significantly shortened the drying time of the samples, reduced the moisture content, and improved the quality of the samples. The results of this study can be used as a reference for the industrialization of licorice drying.

The traditional frying drying method is to heat the oil to 160–180 °C under atmospheric pressure to dry out the moisture in the raw materials through boiling and evaporation in contact with air [22]. However, the physical and chemical properties of food will change during the frying process, causing defects such as rancid odor, dark brown color, and shrinkage in shape. Compared with traditional frying, vacuum frying is a process in which the frying and heating process is carried out in a vacuum and sealed container [23]. Since the boiling point of water decreases under reduced pressure, the water in the vacuum system evaporates quickly and is removed even at low oil temperatures. The resulting dehydrated food becomes crispy while retaining its original color and flavor. Vacuum frying is rarely used in the production of aquatic products (especially fish). Chien et al. (contributor 6) studied the application of vacuum frying technology in the preparation of dried silver herring products. The results showed that this technology was suitable for preparing dried herring products. Vacuum frying temperature significantly affects the water activity, oil content and color of the samples. This study can provide seafood processors with optimal processing conditions to achieve better quality.

Cold plasma (CP) technology can reduce the use of chemicals, so its introduction into the food industry is of great significance [24]. Plasma-reactive substances can reduce microbial contamination, modify specific food structures, and extract nutrients from samples [25]. Cao et al. (contributor 15) evaluated the quality characteristics of rice grains treated repeatedly with low-pressure radio frequency (RF) helium CP. The results showed that low-pressure RF 120W helium CP treatment for 20 s had a lasting effect on rice grains, and intermittent CP treatment had the additional benefit of improving the cooking and gelatinization properties of rice and maintaining the hygroscopic properties of rice.

In conclusion, this Special Issue provides some essential data on the effects of food processing and cooking techniques on food nutritional content, phytochemicals, sensory qualities, antioxidant properties and functionality. It also provides several effective strategies for the innovation and improvement of various types of food in the future, which is of great significance to researchers and food manufacturers. The papers included in this Special Issue have made positive contributions to the research progress in the field of food processing and cooking technology.

**Funding:** This research received no external funding.

**Acknowledgments:** The Guest Editors of the Special Issue would like to sincerely thank each and every author whose important work was included in this Special Issue and helped make it a success.

**Conflicts of Interest:** The author declares no conflicts of interest.

## List of Contributions

1. Shiau, S.-Y.; Yu, Y.; Li, J.; Huang, W.; Feng, H. Phytochemical-Rich Colored Noodles Fortified with an Aqueous Extract of Clitoria ternatea Flowers. *Foods* **2023**, *12*, 1686. https://doi.org/10.3390/foods12081686.
2. Zhou, Q.; Nan, X.; Zhang, S.; Zhang, L.; Chen, J.; Li, J.; Wang, H.; Ruan, Z. Effect of 3D Food Printing Processing on Polyphenol System of Loaded *Aronia melanocarpa* and Post-Processing Evaluation of 3D Printing Products. *Foods* **2023**, *12*, 2068. https://doi.org/10.3390/foods12102068.
3. Syu, P.-C.; Zhang, Q.-F.; Lin, S.-D. Physicochemical, Antioxidant, Sensory, and Starch Digestibility Properties of Steamed Bread Fortified with Tamarillo Powder. *Foods* **2023**, *12*, 2306. https://doi.org/10.3390/foods12122306.
4. Shang, J.; Zhang, Q.; Wang, T.; Xu, Y.; Zang, Z.; Wan, F.; Yue, Y.; Huang, X. Effect of Ultrasonic Pretreatment on the Far-Infrared Drying Process and Quality Characteristics of Licorice. *Foods* **2023**, *12*, 2414. https://doi.org/10.3390/foods12122414.
5. Chen, S.-Y.; Tseng, J.; Wu, C.-R.; Lin, S.-D. Quality Evaluation of Shiitake Blanched and Centrifuged Broths as Functional Instant Drinks. *Foods* **2023**, *12*, 2925. https://doi.org/10.3390/foods12152925.
6. Chien, H.-I.; Hwang, C.-C.; Lee, Y.-C.; Huang, C.-Y.; Chen, S.-C.; Kuo, C.-H.; Tsai, Y.-H. Determining the Optimal Vacuum Frying Conditions for Silver Herring (*Spratelloides gracilis*) Using the Response Surface Methodology. *Foods* **2023**, *12*, 3533. https://doi.org/10.3390/foods12193533.
7. Panduang, T.; Phucharoenrak, P.; Karnpanit, W.; Trachootham, D. Cooking Methods for Preserving Isothiocyanates and Reducing Goitrin in Brassica Vegetables. *Foods* **2023**, *12*, 3647. https://doi.org/10.3390/foods12193647.
8. Radman, S.; Brzovi'c, P.; Raduni'c, M.; Rako, A.; Šaroli'c, M.; Runjić, T.N.; Urlić, B.; Mekinić, I.G. Vinegar-Preserved Sea Fennel: Chemistry, Color, Texture, Aroma, and Taste. *Foods* **2023**, *12*, 3812. https://doi.org/10.3390/foods12203812.
9. Bekiroglu, H.; Ozulku, G.; Sagdic, O. Effects of Casein Hydrolysate Prepared with Savinase on the Quality of Bread Made by Frozen Dough. *Foods* **2023**, *12*, 3845. https://doi.org/10.3390/foods12203845.
10. Akhter, K.T.; Shozib, H.B.; Islam, M.H.; Sarwar, S.; Islam, M.M.; Akanda, M.R.; Siddiquee, M.A.; Mohiduzzaman, M.; Rahim, A.T.M.A.; Shaheen, N. Variations in the Major Nutrient Composition of Dominant High-Yield Varieties (HYVs) in Parboiled and Polished Rice of Bangladesh. *Foods* **2023**, *12*, 3997. https://doi.org/10.3390/foods12213997.
11. Wang, Z.; Nie, T.; Zhang, H.; Wang, W.; Chen, H.; Wang, S.; Sun, B. Correlation Analysis between Volatile Compounds and Quality Attributes in Pork Tenderloin in Response to Different Stir-Frying Processes. *Foods* **2023**, *12*, 4299. https://doi.org/10.3390/foods12234299.
12. Pérez, L.; Pincay, R.; Salazar, D.; Flores, N.; Escolastico, C. Evaluation of the Quality and Lipid Content of Artisan Sausages Produced in Tungurahua, Ecuador. *Foods* **2023**, *12*, 4288. https://doi.org/10.3390/foods12234288.
13. Soares, S.D.; dos Santos, O.V.; da Conceição, L.R.V.; Costi, H.T.; Silva Júnior, J.O.C.; Nascimento, F.d.C.A.d.; Pena, R.d.S. Nutritional and Technological Properties of Albino Peach Palm (*Bactris gasipaes*) from the Amazon: Influence of Cooking and Drying. *Foods* **2023**, *12*, 4344. https://doi.org/10.3390/foods12234344.
14. Xie, L.; Guo, S.; Rao, H.; Lan, B.; Zheng, B.; Zhang, N. Characterization of Volatile Flavor Compounds and Aroma Active Components in Button Mushroom (*Agaricus bisporus*) across Various Cooking Methods. *Foods* **2024**, *13*, 685. https://doi.org/10.3390/foods13050685.

15. Cao, Z.; Li, X.; Song, H.; Jie, Y.; Liu, C. Effect of Intermittent Low-Pressure Radiofrequency Helium Cold Plasma Treatments on Rice Gelatinization, Fatty Acid, and Hygroscopicity. *Foods* **2024**, *13*, 1056. https://doi.org/10.3390/foods13071056.
16. Wang, C.-W.; Shen, H.-S.; Yang, C.-W.; Syu, P.-C.; Lin, S.-D. Physicochemical, Antioxidant, Starch Digestibility, and Sensory Properties of Wheat Bread Fortified with Taiwanese Cocoa Bean Shells. *Foods* **2024**, *13*, 2854. https://doi.org/10.3390/foods13172854.

## References


1. Guo, Y.; Zhang, H.; Shao, S.; Sun, S.; Yang, D.; Lv, S. Anthocyanin: A review of plant sources, extraction, stability, content determination and modifications. *Int. J. Food Sci. Technol.* **2022**, *57*, 7573–7591. [CrossRef]
2. Nolden, A.A.; Forde, C.G. The nutritional quality of plant-based foods. *Sustainability* **2023**, *15*, 3324. [CrossRef]
3. Hemler, E.C.; Hu, F.B. Plant-based diets for cardiovascular disease prevention: All plant foods are not created equal. *Curr. Atheroscler. Rep.* **2019**, *21*, 18. [CrossRef] [PubMed]
4. Diep, T.T.; Rush, E.C.; Yoo, M.J.Y. Tamarillo (*Solanum betaceum* Cav.): A review of physicochemical and bioactive properties and potential applications. *Food Rev. Int.* **2022**, *38*, 1343–1367. [CrossRef]
5. McClements, D.J.; Grossmann, L. Next-generation plant-based foods: Challenges and opportunities. *Annu. Rev. Food Sci. Technol.* **2024**, *15*, 79–101. [CrossRef] [PubMed]
6. Li, C.; Tang, W.; Chen, S.; He, J.; Li, X.; Zhu, X.; Li, H.; Peng, Y. Phytochemical Properties and In Vitro Biological Activities of Phenolic Compounds from Flower of *Clitoria ternatea* L. *Molecules* **2022**, *27*, 6336. [CrossRef] [PubMed]
7. Chen, S.-Y.; Zhang, Q.-F.; Lin, S.-D. Nutritional and phytochemical composition of the red tamarillo grown in Taiwan. *J. Food Compos. Anal.* **2024**, *131*, 106258. [CrossRef]
8. Naito, S.; Fukami, S.; Mizokami, Y.; Ishida, N.; Takano, H.; Koizumi, M.; Kano, H. Effect of freeze-thaw cycles on the gluten fibrils and crumb grain structures of breads made from frozen doughs. *Cereal Chem.* **2004**, *81*, 80–86. [CrossRef]
9. Bekiroglu, H.; Bozkurt, F.; Karadag, A.; Ahhmed, A.M.; Sagdic, O. The effects of different protease treatments on the techno-functional, structural, and bioactive properties of bovine casein. *Prep. Biochem. Biotechnol.* **2022**, *52*, 1097–1108. [CrossRef] [PubMed]
10. Galali, Y.; Omar, Z.A.; Sajadi, S.M. Biologically active components in by-products of food processing. *Food Sci. Nutr.* **2020**, *8*, 3004–3022. [CrossRef] [PubMed]
11. Rojo-Poveda, O.; Barbosa-Pereira, L.; Zeppa, G.; Stévigny, C. Cocoa bean shell—A by-product with nutritional properties and biofunctional potential. *Nutrients* **2020**, *12*, 1123. [CrossRef] [PubMed]
12. Oloyede, O.O.; Wagstaff, C.; Methven, L. The Impact of Domestic Cooking Methods on Myrosinase Stability, Glucosinolates and Their Hydrolysis Products in Different Cabbage (*Brassica oleracea*) Accessions. *Foods* **2021**, *10*, 2908. [CrossRef] [PubMed]
13. Baenas, N.; Marhuenda, J.; García-Viguera, C.; Zafrilla, P.; Moreno, D.A. Influence of Cooking Methods on Glucosinolates and Isothiocyanates Content in Novel Cruciferous Foods. *Foods* **2019**, *8*, 257. [CrossRef] [PubMed]
14. Pérez-Díaz, I.M.; Breidt, F.; Buescher, R.W.; Arroyo-López, F.N.; Jiménez-Díaz, R.; Garrido-Fernández, A.; Bautista-Gallego, J.; Yoon, S.S.; Johanningsmeier, S.D. Fermented and Acidified Vegetables. In *Compendium of Methods for the Microbiological Examination of Foods*; American Journal of Public Health: Washington, DC, USA, 2015. Available online: https://ajph.aphapublications.org/doi/10.2105/MBEF.0222.056 (accessed on 2 March 2025).
15. Méndez-Zamora, G.; García-Macías, J.A.; Santellano-Estrada, E.; Chávez-Martínez, A.; Durán-Meléndez, L.A.; Silva-Vázquez, R.; Quintero-Ramos, A. Fat reduction in the formulation of frankfurter sausages using inulin and pectin. *Food Sci. Technol.* **2015**, *35*, 25–31. [CrossRef]
16. Pinheiro, R.C.; Ballesteros, L.F.; Cerqueira, M.A.; Rodrigues, A.M.C.; Teixeira, J.A.; Silva, L.H.M. Peach palm (*Bactris gasipaes* Kunth) and mammee apple (*Mammea americana* L.) seeds: Properties and potential of application in industry. *LWT* **2022**, *170*, 114089. [CrossRef]
17. dos Santos, M.A.S.; Protázio, D.C.; da Costa, G.P.; Rebello, F.K.; Martins, C.M.; Bezerra, A.S.; da Silva Nogueira, A. Profile of peach palm fruit consumers in the metropolitan region of Belém, Pará, Brazilian Amazon. *Int. J. Innov. Educ. Res.* **2021**, *9*, 550–560. [CrossRef]
18. Krishnamoorthi, R.; Srinivash, M.; Mahalingam, P.U.; Malaikozhundan, B. Dietary nutrients in edible mushroom, *Agaricus bisporus* and their radical scavenging, antibacterial, and antifungal effects. *Process Biochem.* **2022**, *121*, 10–17. [CrossRef]
19. Montoya, J.; Medina, J.; Molina, A.; Gutiérrez, J.; Rodríguez, B.; Marín, R. Impact of viscoelastic and structural properties from starch-mango and starch-arabinoxylans hydrocolloids in 3D food printing. *Addit. Manuf.* **2021**, *39*, 101891. [CrossRef]
20. Sun, W.T.; Wang, S.; Wang, M.; Xu, H.T.; Zhou, Q.C. Research Progress of Extraction, Purification and Function of Overseas *Aronia melanocarpa* in Recent Years. *Food Ind.* **2019**, *40*, 262–265. [CrossRef]

21. Yuan, L.; He, X.; Lin, R.; Cheng, S. Effect of ultrasonic pretreatment on moisture state and hot air drying characteristics of kiwifruit. *J. Agric. Eng.* **2021**, *37*, 263–272. [CrossRef]
22. Sharanabasava, M.R. Vacuum processing of food—A mini review. *MOJ Food Process. Technol.* **2018**, *6*, 283–290. [CrossRef]
23. Diamante, L.M.; Shi, S.; Hellmann, A.; Busch, J. Vacuum frying foods: Products, process and optimization. *Int. Food Res. J.* **2015**, *22*, 15–22. Available online: http://www.ifrj.upm.edu.my (accessed on 7 March 2025).
24. Zadeh, J.H.; Pazir, F. Invesigation of the potential application of cold plasma technology in food safety. *GIDA* **2023**, *48*, 614–626. [CrossRef]
25. Warne, G.R.; Williams, P.M.; Pho, H.Q.; Tran, N.N.; Hessel, V.; Fisk, I.D. Impact of cold plasma on the biomolecules and organoleptic properties of foods: A review. *J. Food Sci.* **2021**, *86*, 3762–3777. [CrossRef]

foods

MDPI

Article

# Physicochemical, Antioxidant, Starch Digestibility, and Sensory Properties of Wheat Bread Fortified with Taiwanese Cocoa Bean Shells

**Chun-Wei Wang, Hui-Shan Shen, Chih-Wei Yang, Pei-Ci Syu and Sheng-Dun Lin ***

Department of Food Science and Technology, Hungkuang University, Taichung 433304, Taiwan; w506016@gmail.com (C.-W.W.); roxenshen33@gmail.com (H.-S.S.); n91768oris@sunrise.hk.edu.tw (C.-W.Y.); spc679163@gmail.com (P.-C.S.)

* Correspondence: lin54@hk.edu.tw; Tel.: +886-4-26318652 (ext. 5038); Fax: +886-4-26319176

**Abstract:** The effects of replacing 5–25% of wheat flour (WF) with Taiwanese cocoa bean shells (CBSs) on the physicochemical, antioxidant, starch digestion, and sensory properties of the bread were studied. The lead (0.18) and cadmium (0.77) contents (mg/kg) of the CBSs were below the Codex Alimentarius specifications for cocoa powder. Ochratoxin A and aflatoxins ($B_1$, $B_2$, $G_1$, and $G_2$) were not detected in the CBSs. The CBSs were rich in dietary fiber (42.9%) and bioactive components and showed good antioxidant capacity. The ash, fat, protein, dietary fiber, crumb $a^*$ and $c^*$, hardness, chewiness, total phenols, and antioxidant activities of the bread increased with an increasing CBSs level. The starch hydrolysis rate (45.1–36.49%) of the CBS breads at 180 min was lower than that of the control (49.6%). The predicted glycemic index of the bread (CBS20 and CBS25) with 20–25% of the WF replaced with CBSs was classified as a medium-GI food using white bread as a reference. In the nine-point hedonic test, the overall preference scores were highest for control (6.8) and CBS breads, where CBSs replaced 5–10% of WF, with scores of 7.2 and 6.7. CBS20 supplemented with an additional 20–30% water improved its volume, specific volume, and staling rate, but the overall liking score (6.5–7.2) was not significantly different from the control ($p > 0.05$). Overall, partially replacing wheat flour with CBSs in the production of baked bread can result in a new medium-GI value food containing more dietary fiber, bioactive compounds, and enhanced antioxidant capacity.

**Keywords:** cocoa bean shells; functional bread; physicochemical quality; antioxidant property; starch hydrolysis; predicted glycemic index; sensory evaluation

## 1. Introduction

Cocoa beans are the main raw material used to make chocolate. The total global cocoa bean production in 2021 was approximately 5.242 million tons, with approximately 77.3% of cocoa beans being produced in Africa. The top five cocoa producing countries are Côte d'Ivoire, Ghana, Ecuador, Cameroon, and Nigeria [1]. The chocolate products from cocoa fruits generate a large amount of waste. Only 10% of the total weight of cocoa fruits is used for commercialization, whereas the remaining 90% is discarded as waste or by-products [2]. The cocoa bean shells (CBSs) is a by-product of cocoa bean processing for making chocolate, accounting for approximately 12–20% of cocoa beans [3]. Most of the CBSs, after separation from cocoa beans, is considered a by-product or waste, but its nutritional content is not much different from that of cocoa beans [4]. Value-added strategies for CBSs have emerged in different areas. Its new uses in the food industry, livestock feed, or industrial applications, such as biofuels, absorbents, or composites are considered the most common [4]. CBSs is not only rich in dietary fiber (39.25–66.33%) [4], but it also contains a variety of bioactive compounds [5,6], such as polyphenols (quercetin, quercetin-3-O-glucoside, catechin, and epicatechin), methylxanthine alkaloids (caffeine, theobromine, and theophylline), and fatty acids (i.e., oleic and linoleic acids).

Wheat bread is one of the oldest and most popular food products among consumers. Traditional white wheat bread is rich in carbohydrates but lacks bioactive ingredients and dietary fiber, so it may be detrimental to human health and may lead to metabolic-syndrome-related diseases [7]. Therefore, adding functional ingredients to bakery products has become increasingly popular. Adding dietary fiber can reduce the risk of chronic diseases, as well as improve the basic nutritional function of the raw materials (i.e., providing calories) [8]. Food manufacturers are trying to produce baked goods that are rich in dietary fiber and bioactive ingredients that have higher nutritional and health value, with attractive palatability. Raw materials or by-products rich in bioactive ingredients and dietary fiber can be used to functionally enhance food, so these raw materials and by-products are often combined [9].

Food waste and food processing by-products represent a serious global problem, endangering the long-term sustainability of the food supply chain [10]. Approximately one-third of the total food produced globally each year is wasted: of the 1.5 billion tons of fruits and vegetables produced each year, an estimated 500 million tons are wasted during processing or are converted into by-products such as peels, seeds, shells, pods, and pomace [11]. These by-products can be used and converted into valuable products, such as dietary fibers, polyphenols, polysaccharides, essential oils, resins, flavor compounds, and pigments [12]. Agricultural food waste and processing by-products can be sustainably used to produce value-added products that are economically and environmentally beneficial and contribute to the sustainability and circular economy of the food industry [13].

Taiwan introduced cocoa from Indonesia in 1922 for trial cultivation. Cocoa requires a high-temperature and humid environment for growth, and cocoa can only be grown at latitudes between 20 degrees north and south. Therefore, the climate in Pingtung County is the most suitable for cocoa cultivation. In the early days, the cocoa processing method was unfamiliar, and cocoa pods cannot be eaten fresh, so cocoa remained an unpopular crop fir a long time. In recent years, Taiwan's cocoa bean cultivation technology has improved, and cocoa beans can be grown in many areas. Therefore, cocoa yield and quality have increased each year. Although the government spares no effort in promoting the development of the cocoa industry, CBSs application development is not yet either effective or diversified. Currently, many foreign researchers have used CBSs or its extracts to study functional foods, such as CBS chocolate cake [14], CBS tea [15], and CBS-extract-enhanced extra virgin olive oil jam [16], finding that CBSs extract enhances the functionality of biscuits [17] and that CBSs and walnut oil can be used as animal fat substitutes and as healthy bioactive sources for beef burgers [18].

The cocoa planting area and production in Taiwan are increasing every year, as is the amount of CBSs produced. However, few studies have focused on products using CBSs or its extracts. As such, we used Taiwanese cocoa bean shells grown without pesticides as test materials to examine their heavy metal and mycotoxins, as well as their physicochemical quality characteristics. We replaced 5%, 10%, 15%, 20%, and 25% of the wheat flour (13.5% protein) with CBSs powder to make bread and analyzed its physicochemical quality characteristics, starch digestibility, and sensory evaluation. The bioactive components and antioxidant properties of the freeze-dried CBSs extract, extraction yield, total phenols content, and the antioxidant properties of the freeze-dried bread extracts were analyzed using an aqueous solution with 50% ($v/v$) ethanol as the extraction solvent. In addition, the effects of adding different amounts of water on the quality characteristics of CBS bread were studied.

## 2. Materials and Methods

### 2.1. Materials and Reagents

Raw cocoa beans (Forastero variety) from plants grown without pesticides were fermented and sun-dried. We divided 32 kg of dried cocoa beans into 4 batches, and we roasted each batch in a far-infrared roller roasting machine (NTM-610, Feli Technology Co., Ltd., New Taipei, Taiwan) at 130 °C for 15 min (50 rpm). The cooled cocoa beans were poured into a cocoa bean cracker and winnower (NCM-102, Feli Technology Co., Ltd., New Taipei, Taiwan) for shelling, and the shelling yield was 12.32%. The cocoa bean shells (CBSs) (Figure 1) were ground into powder with a mill (RT-30HS, Rong Tsong Precision Technology Co., Taichung, Taiwan) and sieved (<0.7 mm). The obtained CBS powder was sealed in an aluminum foil-laminated bag (PET/Al/PE) and stored in a −25 °C freezer. Wheat flour (13.5 g protein/100 g) was purchased from Chia Fha Enterprise Co., Ltd. (Taichung, Taiwan). Sucrose (Taiwan Sugar Co., Ltd., Tainan, Taiwan), sodium chloride (Taiyen Biotech Co., Ltd., Tainan, Taiwan), Lesaffer saf-instant yeast gold (Lesaffer, Marcq-en-Baroeul, France), and unsalted butter (Fonterra brands (New Young) Pte., Ltd., Taiwan Branch, Taipei, Taiwan) were purchased from a local market.

**Figure 1.** The appearance and color of cocoa bean shells and its powder.

Mycotoxin standards (aflatoxin $B_1$, $B_2$, $G_1$, $G_2$, ochratoxin A (OTA)), theobromine, caffeine, caffeic acid, catechin, epicatechin, quercetin, protocatechuic acid, gallic acid, Folin–Ciocalteu phenol reagent, ammonium formate, formic acid, 1,1-diphenyl-2-picrylhydrazyl (DPPH), potassium ferricyanide, ferric chloride hexahydrate, ferrozine, pepsin, α-amylase (heat stable), α-amylase (from porcine pancreas type VI-B), α-amylase (from human saliva, type XIII-A), 3,5-dinitrosalicylic acid, maltose, protease (from *Bacillus licheniformis*), amyloglucosidase, celite filter aid, and acetic acid were purchased from Sigma-Aldrich Chemical (St. Louis, MO, USA). Lead and cadmium were purchased from AccuStandard, Inc. (New Haven, CT, USA). Ferrous chloride tetrahydrate was purchased from Alfa Aesar (Ward Hill, MA, USA). Hexane was purchased from Tedia (Fairfield, OH, USA). Trichloroacetic acid was purchased from Thermo Scientific Chemicals Inc. (Waltham, MA, USA). Acetonitrile, methanol, and acetone were purchased from Avantor Performance Materials (Center Valley, PA, USA). Ethanol was purchased from Taiwan Tobacco & Liquor (Tainan, Taiwan). Dialysis membranes (molecular-weight cut-off of 6000–8000) were purchased from Membrane Filtration Products (Seguin, TX, USA). Cupric sulfate, anhydrous sodium carbonate, and sodium hydroxide were purchased from Shimakyu's Pure Chemicals (Osaka, Japan); potassium chloride was purchased from Showa Chemical (Tokyo, Japan). Methylene blue and anhydrous sodium acetate were purchased from Katayama Chemical Industries (Osaka, Japan), and methyl red pure was purchased from Koch Light Research Laboratories (Gauteng, South Africa). Potassium sodium (+)-tartrate tetrahydrate was purchased from Wako Pure Chemical Industries (Osaka, Japan), and potassium sulfite was purchased from Nihon Shiyaku Reagent (Tokyo, Japan). Nitric acid (ultrapure grade, 60–70%), hydrochloric acid, sulfuric acid, sodium dihydrogen phosphate (monobasic), and sodium phosphate (dibasic, 12 hydrates) were purchased from Union Chemical (Hsinchu, Taiwan).

### *2.2. Bread Preparation*

The formula (baking percentage, Table 1) and production of the control bread were based on those used in a previous study [19]. The control bread recipe involved the use of wheat flour (700 g), sugar (70 g), salt (10.5 g), yeast (7 g), unsalted butter (42 g), and water (434 g). The CBSs powder replaced 5%, 10%, 15%, 20%, and 25% (*w*/*w*) of the wheat flour in the recipe to make various breads, labeled as CBS5, CBS10, CBS15, CBS20, and CBS25, respectively. The dough was manufactured using a straight dough method. Yeast was dissolved in water at room temperature and mixed with flour, CBSs powder, sugar, and salt, using a mixer (HL-11007, San Neng Bake Ware Co., Ltd., Taichung, Taiwan) at low speed (150 rpm) for 4 min, then mixed at medium speed (316 rpm) mix for 6 min. Melted unsalted butter was added to the mixer and mixed on low speed for 4 min. After the dough was completely mixed, the dough was folded and rolled and then placed in an incubator (Chung Pu Baking Machinery, Taichung, Taiwan) and incubated at 27 °C and 75% RH for 90 min. Each piece of dough was shaped and poured into a loaf pan (196 × 106 × 110 mm, SN2052, San Neng Bakeware, Taichung, Taiwan), which was placed in an incubator at 38 °C and 85% RH for 45 min. The dough was baked in a preheated oven (K35E, Chung Pu Baking Machinery, Taichung, Taiwan) with the upper heat set at 170 °C and the bottom heat set at 230 °C for 40 min. After this, the baked bread was removed from the oven, allowed to cool to room temperature for 2 h, then removed from the baking pan and weighed. Baked breads were packed in polyethylene bags before analyzing their quality characteristics. Seven batches (two loaves of 560 g per batch) were prepared for each recipe. All treatments were randomly generated.

**Table 1.** Formulae of cocoa bean shells breads.

| Ingredient (%) | Control [1] | CBS5 | CBS10 | CBS15 | CBS20 | CBS25 | CBS20W5 | CBS20W10 | CBS20W15 | CBS20W20 | CBS20W25 | CBS20W30 |
|---|---|---|---|---|---|---|---|---|---|---|---|---|
| Wheat flour | 100 | 95 | 90 | 85 | 80 | 75 | 80 | 80 | 80 | 80 | 80 | 80 |
| CBS powder [1] | 0 | 5 | 10 | 15 | 20 | 25 | 20 | 20 | 20 | 20 | 20 | 20 |
| Sugar | 10 | 10 | 10 | 10 | 10 | 10 | 10 | 10 | 10 | 10 | 10 | 10 |
| Sodium chloride | 1.5 | 1.5 | 1.5 | 1.5 | 1.5 | 1.5 | 1.5 | 1.5 | 1.5 | 1.5 | 1.5 | 1.5 |
| Yeast | 1 | 1 | 1 | 1 | 1 | 1 | 1 | 1 | 1 | 1 | 1 | 1 |
| Unsalted butter | 6 | 6 | 6 | 6 | 6 | 6 | 6 | 6 | 6 | 6 | 6 | 6 |
| Water | 62 | 62 | 62 | 62 | 62 | 62 | 67 | 72 | 77 | 82 | 87 | 92 |
| Total | 180.5 | 180.5 | 180.5 | 180.5 | 180.5 | 180.5 | 185.5 | 190.5 | 195.5 | 200.5 | 205.5 | 210.5 |

[1] Control, CBS5, CBS10, CBS15, CBS20, and CBS25: bread prepared with 0%, 5%, 10%, 15%, 20%, and 25% ($w/w$) replacement of wheat flour with cocoa bean shells powder, respectively. CBS20W5, CBS20W10, CBS20W15, CBS20W20, CBS20W25, and CBS20W30: formula of CBS20 with an additional 5%, 10%, 15%, 20%, 25%, and 30% ($w/w$) water, respectively, relative to the amount of wheat flour in the control, which was set at 100%.

The effect of the use of different amounts of water on the quality characteristics of CBS20 bread was studied. The recipe for the CBS20 bread involved the use of flour (560 g), CBSs powder (140 g), sugar (70 g), salt (10.5 g), yeast (7 g), unsalted butter (42 g), and water (434 g) (Table 1). Water amounts of 5%, 10%, 15%, 20%, 25%, and 30% ($w/w$) were added, based on a control wheat flour amount setmat 100% (Table 1). Therefore, the total amounts of water in the CBS20W5, CBS20W10, CBS20W15, CBS20W20, CBS20W25, and CBS20W30 bread were 469 g, 504 g, 539 g, 574 g, 609 g, and 644 g, respectively.

### 2.3. Determination of Minerals

The contents of lead and cadmium in the CBSs powder were determined according to the method described by the Taiwan Food and Drug Administration [20]. Briefly, a sample (0.5 g) was placed in a microwave digestion flask, and 0.5 mL of internal standard solution (rhodium) and 6 mL of nitric acid (60–70%, ultrapure grade) were added. Then, we placed the sample in a microwave digester (SINEO Microwave Chemistry Technology, Shanghai, China) for digestion, according to the following program: 1. 1000 W, heating time 5 min, temperature 100 °C; 2. 1000 W, heating time 15 min, duration 5 min, temperature 220 °C; 3. 1800 W, heating time 10 min, duration 10 min, temperature 240 °C. After digestion and cooling, we poured the solution into a volumetric flask, washed the microwave digestion flask with 5 mL of deionized water each time, placed the washing liquid into the volumetric flask, diluted the liquid to 50 mL with deionized water, and filtered the resulting liquid through a 0.45 μm PTFE syringe filter to create the test solution. We added 0.5 mL of internal standard solution and 6 mL of nitric acid (ultrapure grade) to another blank microwave digestion flask. The subsequent steps were the same as those for the test solution to create a blank test solution. All the test solutions were analyzed using high-resolution ICP-MS

(Thermo Fisher Scientific, Waltham, MA, USA). Each mineral was identified and quantified on the basis of the calibration curve of the pure compound.

## 2.4. Determination of Mycotoxins

The aflatoxin $B_1$, $B_2$, $G_1$, $G_2$, and OTA contents in the CBSs powder were measured following the method described by the Taiwan Food and Drug Administration [21]. Briefly, a sample (5 g) was placed in a centrifuge tube, and 5 mL of a phosphate buffer solution was added; the two were mixed until even. Then, 20 mL of methanol solution containing 70% acetonitrile was added, the sample was shaken for 30 min, and then centrifuged at 4300× *g* for 5 min. The supernatant of the solution (5 mL) was placed in a test tube and dried with nitrogen in a 50 °C water bath. The residue was dissolved in 20% acetonitrile solution and diluted to 1 mL, filtered through a 0.22 μm PTFE filter membrane, and used as a test solution. The test solution was analyzed with a liquid chromatography/tandem mass spectrometer (Waters Xevo TQ-MS, Milford, MA, USA) equipped with an ACQUITY BEH C18 column (2.1 mm × 10 mm, 1.7 μm; Waters Co., Milford, MA, USA). Solvent A of the mobile phase, consisting of ammonium formate (0.315 g) and formic acid (1 mL), was dissolved in deionized water to create 1000 mL, which was then filtered through a 0.22 um filter membrane. Solvent B, ammonium formate (0.315 g) and formic acid (1 mL), was dissolved in methanol to make 1000 mL, which was then filtered through a 0.22 μm PTFE filter membrane. Elution was performed with the following linear gradient: 5–85% B from 0 to 5.5 min, 85–100% B from 5.5 to 5.8 min, 100% B from 5.8 to 6.9 min, 100–5% B from 6.9 to 7.0 min, and 5–5% B from 7.0 to 9.0 min. The mobile phase flow rate was 0.3 mL/min. The sample injection volume was 10 μL, with the injection performed using an autosampler. The capillary voltage was 2.0 KV, the ion source temperature was 150 °C, $ESI^+$ ionization mode was used, the desolvation temperature was 500 °C, and the desolvation flow was 1000 L/h.

## 2.5. Determination of Physical Characteristics

The water-holding capacity of the sample (g $H_2O$ absorbed/g sample) was measured by following a previously reported method [19]. A scale (GG4002-S, Mettler-Toledo, Langacher Greifense, Switzerland) was used to measure the weight of the bread. The bread volume was determined using the rapeseed displacement method [22]. The specific bread volume was calculated by dividing the bread volume by its weight.

The reflective color of each sample was measured with a spectrophotometer (CM-5, Konica Minolta Sensing, Inc., Tokyo, Japan), setting the CIE $L^*$, $a^*$, and $b^*$ values, using a D65 light source at 10° [19]. The crust color was measured on the top of the bread. The crumb was measured by cutting the midsection of the bread into cubes (3.5 cm × 3.5 cm × 3.5 cm) with a knife. The hue angle ($h^*$) and chroma ($c^*$) of the samples were calculated as $\arctan(b^*/a^*)$ and $(a^{*2} + b^{*2})^{1/2}$, respectively. The total color difference ($\Delta E$) between the sample and the control was calculated as $\Delta E = [(L^*_c - L^*_s)^2 + (a^*_c - a^*_s)^2 + (a^*_c - a^*_s)^2]^{1/2}$, where $L^*$, $a^*$, and $b^*$ were the color coordinates in the control (c) and CBS bread (s).

A texture analyzer (TA-XT2, 25 kg model; Stable Micro Systems, Godalming, Surrey, UK) was used to measure the texture of the middle part (2.5 cm × 2.5 cm × 2.5 cm) of the bread, according to a previous study [19]. The studied textural qualities included hardness, cohesiveness, springiness, chewiness, and resilience. Additionally, the bread hardness after being stored at room temperature for 24 h was measured, and the staling rate was calculated using the following formula [23]:

$$\text{Staling rate} = \left(\frac{\text{crumb hardness after 24 h [N]} - \text{crumb hardness after 2 h [N]}}{\text{hardness after 2 h [N]}}\right) \quad (1)$$

## 2.6. Determination of Proximate Composition and Total Dietary Fiber

The moisture, protein, fat, and ash contents of the samples were analyzed following the methods of the American Association of Cereal Chemists [24]. The nitrogen conversion factors for the CBS powder, wheat flour, and baked bread used for crude protein calculation were 6.25, 5.7, and 5.7, respectively. The carbohydrate content (g/100 g) was 100%, minus those of moisture, protein, fat, and ash. The total dietary fiber of the sample was determined according to AOAC method 985.29 [25].

### 2.7. Determination of Bioactive Components in the Extract

We placed 15 g of powder sample into a centrifuge bottle and added 150 mL of 50% (*v*/*v*) ethanol. The mixture was then shaken in a 75 °C water bath at 150 rpm for extraction for 30 min and centrifuged (4450× *g*, 30 min). The supernatant was filtered through Advantec No. 1 filter paper. The residue was extracted again following the above steps. The two filtrates were combined, concentrated under reduced pressure (45 °C), and freeze-dried under vacuum.

The total phenols contents of the freeze-dried extracts were measured according to the Folin–Ciocalteu method [26]. To prepare the extract solution, 0.2 g of lyophilized CBSs extract or 0.5 g of lyophilized wheat flour and bread extract was weighed into a centrifuge tube, with the volume increased to 10 mL with methanol. A centrifuge tube was placed in an ultrasonic oscillator (53 kHz) for 15 min and centrifuged (1462× *g*, 10 min). The obtained supernatant was the extract solution. The total phenolic content of the extract solution were calculated following the standard curve [760 nm absorbance = 0.0009 $C_{\text{gallic acid}}$ (μg/mL) + 0.00106, $R^2$ = 0.9995]. The results are expressed as milligrams of gallic acid equivalents (GAE) per gram of lyophilized extract.

The contents of theobromine, caffeine, caffeic acid, protocatechuic acid, catechin, epicatechin, and quercetin in the lyophilized extracts were analyzed according to the method of Rojo-Poveda et al. [27]. The extract solution was prepared following the same methods used for the total phenols. The supernatant after centrifugation was filtered through a PVDF syringe filter (13 mm × 0.45 μm) and then analyzed using HPLC. The HPLC system consisted of a pump (Hitachi 5110, Hitachi High-Tech Co., Tokyo, Japan), a diode array detector (Hitachi 5430), and a Luna C18(2) column (250 mm × 4.6 mm, 5 μm particle size; Phenomenex, Torrance, CA, USA). The column temperature was maintained at 35 °C. The mobile phase consisted of water containing formic acid at 0.1% *v*/*v* (solvent A) and 100% methanol (solvent B) at a flow rate of 1.0 mL/min. Elution was performed with the following linear gradient: 10% B from 0 min to 2 min, 10–50% B from 2 to 18 min, 80% B from 18 to 40 min, 90% B from 40 to 42 min, and 10% B from 42 to 45 min. The sample injection volume was 10 μL. The bioactive components in the samples were identified by comparing their relative retention times and UV spectra with those of the verified compounds. The contents of the various bioactive components were calculated based on the calibration curves corresponding to the pure compounds.

### 2.8. Determination of Antioxidant Property of the Extract

Various concentrations of each extract were prepared in methanol (0 to 0.5 mg extract/mL for CBSs and wheat flour; 0 to 3.6 mg extract/mL for bread). The DPPH radical scavenging ability of these solutions was then determined according to the method of Shimada et al. [28]. Each extract (4 mL) was mixed with 1 mL of methanolic solution containing DPPH radicals, resulting in a final concentration of 0.2 mmol/L DPPH. The mixture was shaken vigorously and left to stand for 30 min in the dark, and the absorbance of the mixture was then measured at 517 nm against a blank.

$$\text{Scavenging ability (\%)} = \left\{1 - \left[\frac{\text{Absorbance value of sample added with DPPH} - \text{Absorbance value of sample added with methanol)}}{\text{(Absorbance value of control added with DPPH} - \text{Absorbance value of control added with methanol)}}\right]\right\} \times 100\%$$

The reducing power of the extract solutions prepared with methanol at various concentrations (0–0.6 mg extract/mL for CBSs and wheat flour; 0–15 mg extract/mL for bread) was measured according to the method of Oyaizu [29]. Each extract (2.5 mL) was mixed with 2.5 mL of 200 mmol/L sodium phosphate buffer (pH 6.6) and 2.5 mL of 1% (*w*/*v*) potassium ferricyanide, and the mixture was incubated at 50 °C for 20 min. After adding 2.5 mL of 10% (*w*/*v*) trichloroacetic acid, the mixture was centrifuged at 200× *g* for 10 min.

The upper layer (5 mL) was mixed with 5 mL of deionized water and 1 mL of 0.1% ($w/v$) ferric chloride, and the absorbance of the mixture was measured at 700 nm against a blank to determine the amount of ferric ferrocyanide (Prussian blue) formed. A higher absorbance indicates a higher reducing power. Reducing power = absorbance value of the sample with $FeCl_3{\cdot}6H_2O$ added-absorbance value of the sample without $FeCl_3{\cdot}6H_2O$ added.

The ferrous ion chelation ability of the extract solutions prepared with methanol at various concentrations (0–0.6 mg extract/mL for CBSs flour and wheat flour; 0–6 mg extract/mL for bread) was analyzed following the method of Dinis et al. [30]. Each extract (1 mL) was mixed with 3.7 mL of methanol and 0.1 mL of 2 mmol/L ferrous chloride. After standing for 30 s, 0.2 mL of 5 mmol/L ferrozine was added. After 10 min at room temperature, the absorbance of the mixture was determined at 562 nm against a blank. A lower absorbance indicates a higher chelating ability.

$$\text{Chelating ability (\%)} = \left\{1 - \left[\frac{\text{Absorbance value of sample added with reagent} - \text{Absorbance value of sample added with water)}}{\text{Absorbance value of control added with methanol}}\right]\right\} \times 100\%$$

The estimated $EC_{50}$ value (mg extract/mL) represents the effective concentration at which 50% of the DPPH free radicals are scavenged, the absorbance is 0.5 as a measure of reducing power, and 50% of the ferrous ions are chelated. The estimated $EC_{50}$ values of the CBS and CBS bread extracts were obtained via the interpolation of the linear regression analysis, while the $EC_{50}$ values of the wheat flour and control bread were obtained via the extrapolation of the linear regression analysis.

### 2.9. Determination of Starch Hydrolysis and Predicted Glycemic Index (pGI)

The starch hydrolysis of the baked bread (room temperature) was tested using a restricted (dialysis) system method [31]. The hydrolysis index (HI, %) was calculated by dividing the area under the hydrolysis curve of the sample by that of white bread, and the resulting value × 100%. The pGI of the starch hydrolysis value of bread at 180 min was calculated according to the following formula: pGI = 39.71 + (0.549 × HI) [32].

### 2.10. Sensory Evaluation

Sensory evaluation of the baked bread was performed using the hedonic test. Fifty untrained consumers (17 men and 33 women, aged 21–54 years) interested in the CBS breads were recruited from Hungkuang University students, staffs, and teachers. Each CBS sample (2.5 cm × 2.5 cm × 2.5 cm) was placed on a white plate, identified by a random three-digit number, and then presented to the consumer in a counterbalanced and randomized order. Consumers were asked to rinse their mouths with warm water before tasting the next sample to reduce the impact of sample residue. A nine-point hedonic scale questionnaire was used to assess the sensory quality characteristics of the bread, including appearance, crumb color, aroma, flavor, texture, and overall liking (1 = dislike very much, 5 = neither like nor dislike, 9 = like very much).

### 2.11. Statistical Analysis

Each measurement was performed four times, except for the hedonic test ($n$ = 50). The variance of the experimental data was analyzed using the Statistical Analysis System (SAS Institute, Cary, NC, USA). Duncan's multiple range test was used to determine the significance of differences among the means at the 0.05 level.

## 3. Results and Discussion

### 3.1. Heavy Metal and Mycotoxin Contents in the CBS

The heavy metal (lead and cadmium) contamination in cocoa raw materials and products has been widely studied [33,34]. The possible factors affecting the heavy metal contamination of cocoa include the use of chemical fertilizers, pesticides, and insecticides, as well as exposure to metal equipment during fermentation and drying [35]. The International Agency for Research on Cancer (IARC) lists inorganic lead compounds as Group 2A (probably carcinogenic to humans) and cadmium as Group 1 (carcinogenic to humans) [36]. However, tetraethyl lead is still commonly used in several cocoa-producing countries. During cocoa bean fermentation, drying, and transportation, emissions from the combustion of leaded gasoline may directly contact CBSs [37]. Lead poisoning involves symptoms such as anemia, weight loss, and cognitive impairment [38]. Assa et al. [39] found that the lead concentrations in CBSs range from 5.810 to 11.146 mg/kg, which are 58~111-fold higher than the lead concentrations observed in cocoa beans (0.1 mg/kg) and substantially higher than the European Union-permitted lead contamination levels of cocoa powder (1.00 mg/kg). However, these values vary widely in different studies [4]. Cadmium is easily absorbed by crops. However, even when the cumulative cadmium concentration is high, plants do not show signs of poisoning, but the concentration is enough to threaten human and animal health [40]. Cadmium affects the body's physiological metabolism, which may be related to the development of various cancers (such as breast, kidney, pancreatic, prostate, and bladder cancers) or chronic diseases (such as Alzheimer's disease) [41]. Lewis et al. [42] found that the cadmium content of CBSs was 0.05–0.10 mg/kg, which is lower than the Codex Alimentarius standard of 1.5 mg/kg for cocoa powder [39,43]. The Taiwan Food and Drug Administration has not yet announced the heavy metal standards for CBSs. The lead and cadmium contents (mg/kg, dry basis) in the CBSs in this study were 0.19 and 0.80 (Table 2), respectively, which are both lower than the Codex Alimentarius specification for cocoa powder (Pb < 1.0 mg/kg and Cd < 1.5 mg/kg).

**Table 2.** Heavy metals, mycotoxins, and physicochemical quality characteristics of cocoa bean shells and wheat flour.

| | CBSs [1] | WF |
|---|---|---|
| | Heavy metals (mg/kg, dry basis) | |
| Pb | 0.19 ± 0.01 [2] | <0.01 |
| Cd | 0.80 ± 0.02 | <0.01 |
| | Mycotoxins (μg/kg, dry basis) | |
| Aflatoxin $B_1$ | <0.2 | <0.2 |
| Aflatoxin $B_2$ | <0.1 | <0.1 |
| Aflatoxin $G_1$ | <0.2 | <0.2 |
| Aflatoxin $G_2$ | <0.1 | <0.1 |
| Ochratoxin A | <0.5 | <0.5 |
| | Proximate composition (g/100 g) | |
| Moisture | 4.31 ± 0.30 b | 13.55 ± 0.07 a |
| Crude protein | 16.22 ± 0.50 a | 13.51 ± 0.07 b |
| Crude fat | 10.33 ± 0.33 a | 1.22 ± 0.03 b |
| Crude ash | 7.40 ± 0.33 a | 0.47 ± 0.02 b |
| Carbohydrate | 61.74 ± 0.29 b | 71.25 ± 0.07 a |
| Total dietary fiber | 42.90 ± 0.76 a | 1.96 ± 0.04 b |
| | Water holding capacity (g water absorbed/g sample) | |
| | 3.47 ± 0.04 a | 0.83 ± 0.01 b |
| | Color property | |
| $L^*$ | 40.35 ± 0.24 b | 89.98 ± 0.10 a |
| $a^*$ | 15.21 ± 0.20 a | 0.16 ± 0.04 b |
| $b^*$ | 23.02 ± 0.38 a | 9.95 ± 0.17 b |
| $\Delta E$ | | 53.49 ± 0.11 |

[1] CBSs: Cocoa bean shells. WF: wheat flour. [2] Each value is expressed as mean ± standard deviation ($n$ = 4). Means with different lowercase letters within a row differ significantly ($p < 0.05$).

The mycotoxins produced by members of the *Aspergillus* and *Penicillium* genera are also major safety concerns regarding CBSs. Cocoa beans can become contaminated during fermentation, drying, and storage. Notably, mycotoxins are heat-resistant and cannot be completely eliminated by roasting [5,44]. OTA is associated with nephrotoxic, teratogenic, and immunosuppressive activities and is classified as a 2B carcinogen (possible human carcinogen) [36,45]. The main OTA content in total cocoa beans is found in CBSs, usually accounting for approximately 50–95%, with a concentration of 0.13–2.01 g/kg, which is within the acceptable range for cocoa beans, as determined by the European Commission (<2 g/kg) [44–47]. However, the OTA of CBSs can be as high as 23.1 g/kg [48]. Aflatoxin is hepatotoxic, teratogenic, mutagenic, and carcinogenic in humans [49]. The concentrations (μg/kg) of aflatoxin $B_1$, $B_2$, $G_1$, and $G_2$ detected in the CBSs were 0.01–0.84, <0.003–0.02, <0.003–0.44, and <0.002–0.06, respectively, which are considered safe values [50]. Considering the extensive use of cocoa products by children, the Brazil National Health Surveillance Agency has set the limits for the OTA and total aflatoxin contents of cocoa beans to less than 10 and 5 μg/kg, respectively. The EU Rapid Alert System for Food and Feed (RASFF) found that the aflatoxin levels in organic cocoa powder produced in Ghana and sold in the Netherlands have increased [51]. The levels of aflatoxin $B_1$, $B_2$, $G_1$ and $G_2$, and OTA were below the detection limits in CBSs (Table 2), and the detection limits were 0.2, 0.1, 0.2, 0.1, and 0.5 μg/kg, respectively.

In summary, the CBSs used in this study as a raw food material for wheat bread should not pose any risk of food poisoning caused by lead, cadmium, and mycotoxins.

### *3.2. Physicochemical Properties of CBSs and Wheat Flour*

The crude protein, fat, ash, and total dietary fiber contents of the CBSs were significantly higher than those of the WF, whereas the opposite was found for moisture and carbohydrates (Table 2). The moisture, protein, fat, ash, and dietary fiber contents (g/100 g) of the CBSs from São Tomé cocoa beans (Forastero variety) were 5.9, 20.9, 2.3, 7.9, and 55.1, respectively [27]. The moisture, protein, ash, and dietary fiber contents were higher than those in this study (Table 2), whereas the fat content was lower than that in the samples in this study. The moisture, protein, fat, ash, and total dietary fiber contents (g/100 g) of CBSs (Forastero variety) from the Republic of Ghana were 6.79–7.05, 15.59–17.13, 3.00–5.60, 7.00–7.34, and 61.18–63.14, respectively [3]. The water and dietary fiber contents were higher than those in the samples in this study, but its fat content was lower. The protein and ash contents were comparable to those of the samples in this study (Table 2). The possible reasons for the differences in the contents of proximate components of CBSs include shelling rate, production location, and cultivation and drying methods.

The water-holding capacity of the CBSs was significantly higher than that of the WF (Table 2) because the dietary fiber content of the CBSs was higher than that of the WF. The polysaccharides in dietary fiber are highly hydrophilic, and water can be retained in the hydrophilic sites of the fiber or in the interstitial spaces in the molecular structure [52].

The $L^*$ value of the WF was significantly higher than that of the CBSs, indicating that the WF was brighter than the CBSs (Table 2). The $a^*$ and $b^*$ values of CBSs are redder and yellower than those of the WF. The $\Delta E$ value showing the color difference between CBSs and WF was 53.49, which means that the observers would notice the differences in color between the two samples [53].

### *3.3. Bioactive Components and Antioxidant Properties of CBSs and Wheat Flour Extracts*

We used 50% aqueous ethanol as the extraction solvent. The extraction yield and total phenol concentrations of the CBSs extract were significantly higher than those of the WF extract (Table 3). When the total phenolic content of the extract was converted into CBSs powder, the total phenolic content (9.06 mg GAE/g powder) was also significantly higher than that of the WF (0.70 mg GAE/g powder). The main phenolic compounds in CBSs include epicatechin, catechins, and proanthocyanidins [27,54]. The total phenolic content of CBSs from different regions of Venezuela is 5.88–7.52 g GAE/kg powder (Criollo and Trinitario) [55], which is lower than that found in this study. The possible reasons for the difference in the total phenolic content of CBSs include differences in variety, cultivation practices, drying, shelling rate, solvent, and extraction method [27,56].

**Table 3.** The yield and bioactive components of cocoa bean shells powder and wheat flour extracts.

| | Content (mg/g Lyophilized Extract) | |
|---|---|---|
| | CBSs [1] | WF |
| Yield (%) [2] | 27.35 ± 0.05 a [4] | 9.99 ± 0.06 b |
| Total phenols [3] | 33.12 ± 0.18 a | 7.04 ± 0.22 b |
| Theobromine | 119.1 ± 0.1 | $<10^{-5}$ |
| Caffeine | 3.44 ± 0.08 | $<10^{-5}$ |
| Epicatechin | 10.05 ± 0.14 | $<10^{-4}$ |
| Catechin | 15.12 ± 1.07 | $<10^{-4}$ |
| Caffeic acid | 0.31 ± 0.04 | $<10^{-5}$ |
| Protocatechuic acid | $<10^{-5}$ | $<10^{-5}$ |
| Quercetin | 0.04 ± 0.01 | $<10^{-4}$ |

[1] CBSs: cocoa beans shells. WF: wheat flour. [2] Yield (%) = (weight of lyophilized extract/weight of sample) × 100%. [3] Total phenols unit: mg gallic acid equivalent/g lyophilized extract. [4] Each value is expressed as mean ± standard deviation ($n$ = 4). Means with different lowercase letters within a row differ significantly ($p < 0.05$).

Among the bioactive components in CBSs in this study, the theobromine content was the highest, followed by those of catechin > epicatechin > caffeine > caffeic acid > quercetin, whereas protocatechuic acid was not detected (Table 3). The methylxanthines in cocoa, including theobromine and caffeine, can positively affect our mood and alertness. The theobromine content of the CBSs in this study was higher than that of caffeine. This result is similar to those of Rojo-Poveda et al. [27] and Martins et al. [57]. In terms of phenolic compounds, the contents of catechin and epicatechin in the CBSs were the highest (Table 3). Cocoa polyphenol extract and epicatechin can protect human endothelial cells from oxidative damage by regulating the production of oxygen free radicals, antioxidant enzymes, and nonenzymatic defenses [57]. After the fresh cocoa beans were dried and defatted, the samples contained epicatechin concentrations of 21.89–43.27 mg/g. Fermentation reduces the epicatechin concentration of cocoa beans [58]. Rojo-Poveda et al. [27] used six home coffee-brewing techniques to extract CBSs. Methylxanthines and phenolic compounds, which are antioxidants and inhibit α-glucosidase, were detected in the extracted extracts [27].

The progression of various human diseases, such as diabetes and atherosclerosis, is related to free radicals [59]. Naturally occurring antioxidants can delay the progression of many chronic human diseases by scavenging free radicals [60]. The antioxidant capacity of the CBSs extract in this study, including its ability to scavenge DPPH radicals, reduce power, and chelate ferrous ions, was significantly enhanced with the increase in extract concentration (Figure 2). When the extract concentrations were 0.1, 0.2, and 0.4 mg extract/mL, the DPPH radical scavenging ability of the CBSs extracts was 37.89%, 61.73%, and 87.82%, respectively, whereas that of the WF extracts was only 2.88%, 3.54%, and 4.71%, respectively (Figure 2A). At a concentration of 0.1, 0.3, and 0.6 mg extract/mL, the reducing power of the CBSs extracts (0.191 AU, 0.538 AU, and 0.989 AU, respectively) was better than that of the WF extract (0.040 AU, 0.043 AU, and 0.043 AU, respectively) (Figure 2B). When the concentration (mg extract/mL) of the extract was 0.1 and 0.5, the CBSs extract had a better ability to chelate ferrous ions (14.642% and 52.273%) than the WF extract (4.566% and 5.853%) (Figure 2C). Ferrous ions can be chelated by intermolecular hydroxyl groups or adjacent hydroxyl and carbonyl groups in the molecule to form complexes [61]. Therefore, the specific functional groups in the polyphenolic structure of CBSs extracts may be one of the reasons for their ability to chelate ferrous ions. The estimated $EC_{50}$ values (mg extract/mL) of the CBSs extract for scavenging DPPH radicals, reducing power, and chelating ferrous ions were 0.15, 0.28 and 0.47, respectively, which were substantially higher than those of the WF extract. This result is related to the fact that the CBSs extract is rich in bioactive components (Table 3).

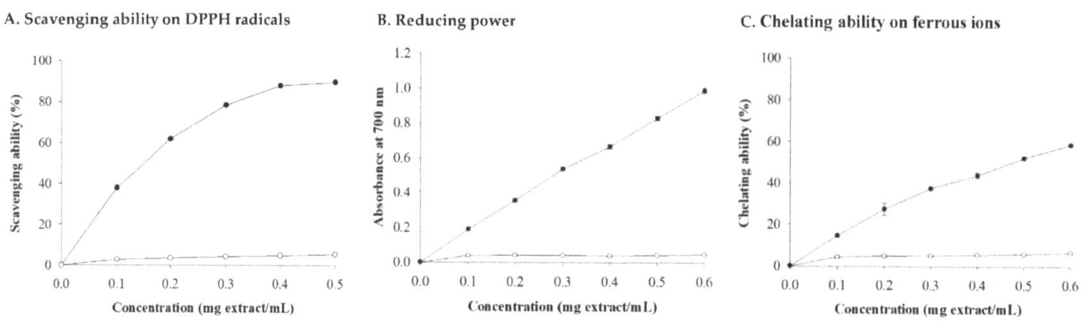


**Figure 2.** Antioxidant property of cocoa bean shells and wheat flour extracts. Each value is expressed as mean ± standard deviation ($n$ = 4). CBS (•), WF (○). CBS: cocoa beans shell. WF: wheat flour.

The above results show that CBSs is a rich source of dietary fiber and bioactive components, and replacing the wheat flour in traditional wheat bread recipes with CBSs can produce functional products.

### 3.4. Color Characteristics of the Breads

Color affects the consumer acceptance of bread. Figure 3 shows the appearance, as well as the crust and crumb color of the bread made with CBSs replacing part of WF. The $L^*$, $b^*$, $h^*$, and $c^*$ values of all bread crusts decreased as the amount of CBSs increased (Table 4). That is, as the amount of CBSs replacing WF increased, the color of the bread crust became darker ($L^*$), less yellow ($b^*$), and less saturated ($c^*$). The $h^*$ value shows that the color of the bread crust changed from yellow to orange (59.73–47.95°) [62,63]. The $\Delta E$ values for CBS5, CBS10, CBS15, CBS20, and CBS25, compared to the control, were all greater than 5; the $\Delta E$ values for CBS10, CBS15, CBS20, and CBS25, compared to those for CBS5, were 7.72, 7.69, 8.04, and 10.5, respectively, which were also greater than 5. When the $\Delta E$ of two samples was greater than 5, an observer can notice two different colors [53]. The $\Delta E$ values for CBS15 and CBS20, compared to CBS10, were 0.03 and 0.32, respectively; the $\Delta E$ value for CBS20 and CBS15 was 0.35, indicating that the color difference would be unrecognizable by a standard observer. The $\Delta E$ values for CBS10, CBS15, and CBS20, compared to CBS25, were 2.75, 2.81, and 2.46, respectively, indicating that inexperienced observers could spot the difference [53].

**Figure 3.** The appearance and color of cocoa bean shells breads.

**Table 4.** Color property of cocoa bean shells breads.

| | Control [1] | CBS5 | CBS10 | CBS15 | CBS20 | CBS25 |
|---|---|---|---|---|---|---|
| | | | Crust color property | | | |
| $L^*$ | 52.31 ± 2.52 a [2] | 43.14 ± 0.79 b | 34.51 ± 0.79 c | 34.41 ± 0.47 c | 34.30 ± 0.97 c | 32.33 ± 0.51 d |
| $a^*$ | 13.02 ± 0.75 b | 13.78 ± 0.18 a | 13.82 ± 0.37 a | 13.93 ± 0.32 a | 13.48 ± 0.33 ab | 12.87 ± 0.72 b |
| $b^*$ | 22.31 ± 0.62 a | 16.89 ± 0.59 b | 16.48 ± 0.93 b | 16.92 ± 0.74 b | 16.09 ± 0.73 b | 14.27 ± 0.78 c |
| $h$ (°) | 59.73 ± 1.62 a | 50.78 ± 1.25 b | 49.99 ± 0.89 b | 50.52 ± 0.81 b | 50.04 ± 0.73 b | 47.95 ± 1.16 c |
| $c^*$ | 25.84 ± 0.65 a | 21.80 ± 0.40 b | 21.51 ± 0.94 b | 21.92 ± 0.74 b | 20.99 ± 0.76 b | 19.22 ± 0.98 c |
| $\Delta E$ | | 10.77 ± 2.07 c | 18.78 ± 2.69 b | 18.75 ± 2.31 b | 19.10 ± 2.55 b | 21.56 ± 2.91 a |
| | | | Crumb color property | | | |
| $L^*$ | 78.42 ± 2.35 a | 56.88 ± 2.67 b | 48.55 ± 2.16 c | 43.07 ± 2.13 d | 39.53 ± 1.40 e | 34.27 ± 1.46 f |
| $a^*$ | 0.39 ± 0.06 f | 6.23 ± 0.34 e | 8.64 ± 0.30 d | 9.65 ± 0.34 c | 10.39 ± 0.28 b | 10.67 ± 0.33 a |
| $b^*$ | 16.95 ± 0.75 d | 18.05 ± 0.61 c | 18.53 ± 0.52 b | 18.78 ± 0.49 ab | 18.88 ± 0.45 a | 18.96 ± 0.43 a |
| $h$ (°) | 88.69 ± 0.17 a | 70.99 ± 0.49 b | 64.99 ± 0.80 c | 62.80 ± 0.59 d | 61.17 ± 1.02 e | 60.61 ± 0.86 f |
| $c^*$ | 16.95 ± 0.75 e | 19.10 ± 0.68 d | 20.45 ± 0.53 c | 21.12 ± 0.55 b | 21.55 ± 0.37 a | 21.76 ± 0.43 a |
| $\Delta E$ | | 22.37 ± 3.41 e | 31.05 ± 3.48 d | 36.60 ± 2.82 c | 40.21 ± 2.23 b | 45.38 ± 2.50 a |

[1] Control, CBS05, CBS10, CBS15, CBS20, and CBS5: bread prepared with 0%, 5%, 10%, 15%, 20%, and 25%. ($w/w$) replacement of wheat flour with cocoa bean shells powder, respectively. [2] Each value is expressed as mean ± standard deviation ($n$ = 4). Means with different lowercase letters within a row differ significantly ($p < 0.05$).

The $a^*$, $b^*$, $c^*$, and $\Delta E$ values of the crumb increased as the amount of CBSs replacing WF increased, whereas the trends in the $L^*$ and $h^*$ value were the opposite (Table 4). This shows that the redness, yellowness, and chroma of the crumb strengthened and darkened as the amount of CBSs replacing WF increased. The $h^*$ value showed that as the amount of CBSs replacing WF increased, the color of the bread crumbs changed from yellow-green to yellow (88.69–60.61°) [62,63]. Based on the $\Delta E$ value of the crumbs, an observer could notice two different colors between breads. The $\Delta E$ between CBS15 and CBS20 was 3.61, which means any observer could easily notice the difference [53]. The change in crust color of the baked bread may be related to the Maillard reaction, caramelization reaction, and the pigment of CBSs [64]. The internal temperature of the bread during the baking process does not exceed 100 °C, so the color of the crumb is usually affected by the color of the flour and the raw materials. Therefore, the crumb is less prone to the Maillard and caramelization reactions, which creates color changes [65].

### 3.5. Weight, Volume, Texture, and Proximate Composition of the Breads

The weight of the bread increased as the amount of CBSs in the bread increased (Table 5). This may have been due to the total dietary fiber content and water-holding capacity of CBSs being significantly higher than those of the WF (Table 2), making the moisture in the bread with a higher CBSs content less likely to evaporate during the baking process, as the weight of the dough put into the mold is the same (560 g). The moisture content of CBSs (4.31%) was significantly lower than that of wheat flour (13.55%) (Table 2); therefore, bread with higher CBSs content retains higher solid content. The volume and specific volume of the bread decreased as the amount of CBSs replacing the WF increased (Table 5). The reduction in bread volume was due to the effect of gluten dilution. Although the crude protein content of the bread increased as the amount of CBSs replacing the WF increased (Table 5), the protein structure of CBSs was different from that of WF (glutenin and gliadin). Therefore, as the amount of CBSs replacing WF increased, the gluten content of the bread also decreased, and its volume decreased. In addition, dietary fiber competes with gluten for water, thereby hindering the normal development of a dough's gluten network. The dilution of the dough's gluten changes the continuity of the gluten network, causing the gluten network structure to weaken and the dough air cells to rupture during the early fermentation process. This results in less dough expansion [66].

**Table 5.** Weight, volume, texture, and proximate composition of cocoa bean shells breads.

| | Control [1] | CBS5 | CBS10 | CBS15 | CBS20 | CBS25 |
|---|---|---|---|---|---|---|
| Weight (g) | 496 ± 2 e | 497 ± 2 e | 508 ± 1 d | 510 ± 2 c | 516 ± 1 b | 520 ± 1 a |
| Volume (mL) | 2343 ± 13 a | 2204 ± 17 b | 2096 ± 15 c | 1788 ± 10 d | 1551 ± 8 e | 1213 ± 14 f |
| Specific volume (mL/g) | 4.72 ± 0.03 a | 4.43 ± 0.05 b | 4.13 ± 0.04 c | 3.50 ± 0.03 d | 3.01 ± 0.01 e | 2.33 ± 0.03 f |
| Texture profile analysis | | | | | | |
| Hardness (N) | 0.98 ± 0.09 d [2] | 1.04 ± 0.09 cd | 1.21 ± 0.19 bc | 1.30 ± 0.16 b | 1.35 ± 0.16 b | 4.01 ± 0.48 a |
| Cohesiveness | 0.83 ± 0.04 a | 0.82 ± 0.06 a | 0.81 ± 0.05 a | 0.82 ± 0.04 a | 0.80 ± 0.02 ab | 0.76 ± 0.05 b |
| Springiness | 1.00 ± 0.01 a | 1.00 ± 0.02 a | 1.00 ± 0.02 a | 0.99 ± 0.03 ab | 0.98 ± 0.02 b | 0.96 ± 0.02 c |
| Chewiness (N) | 0.81 ± 0.08 c | 0.85 ± 0.10 c | 0.97 ± 0.15 bc | 1.05 ± 0.14 b | 1.06 ± 0.12 b | 2.93 ± 0.40 a |
| Resilience | 0.37 ± 0.02 a | 0.37 ± 0.07 a | 0.36 ± 0.03 a | 0.34 ± 0.03 ab | 0.32 ± 0.02 b | 0.31 ± 0.02 b |
| Proximate composition (g/100 g) | | | | | | |
| Moisture | 35.81 ± 0.43 a | 35.55 ± 0.99 a | 35.36 ± 0.43 a | 35.26 ± 0.03 a | 34.42 ± 0.84 b | 34.23 ± 0.65 b |
| Crude ash | 1.21 ± 0.03 f | 1.45 ± 0.05 e | 1.51 ± 0.08 d | 1.69 ± 0.02 c | 1.90 ± 0.02 b | 2.09 ± 0.06 a |
| Crude fat | 2.66 ± 0.05 f | 3.55 ± 0.09 e | 4.39 ± 0.23 d | 5.15 ± 0.08 c | 5.93 ± 0.17 b | 6.63 ± 0.24 a |
| Crude protein | 7.47 ± 0.08 f | 7.67 ± 0.10 e | 7.94 ± 0.05 d | 8.11 ± 0.14 c | 8.47 ± 0.06 b | 8.70 ± 0.08 a |
| Carbohydrates | 52.85 ± 0.42 a | 51.78 ± 1.03 b | 50.80 ± 0.37 c | 49.79 ± 0.09 d | 49.28 ± 0.76 d | 48.35 ± 0.74 e |
| Total dietary fiber | 0.70 ± 0.01 f | 2.08 ± 0.12 e | 3.60 ± 0.11 d | 5.15 ± 0.10 c | 6.70 ± 0.19 b | 8.12 ± 0.12 a |

[1] Control, CBS05, CBS10, CBS15, CBS20, and CBS5: bread prepared with 0%, 5%, 10%, 15%, 20%, and 25% (*w*/*w*) replacement of wheat flour with cocoa bean shells powder, respectively. [2] Each value is expressed as mean ± standard deviation ($n$ = 4). Means with different lowercase letters within a row differ significantly ($p < 0.05$).

Hardness is often used as an indicator of bread quality. As the amount of CBSs replacing WF increased, the hardness of the bread also increased (Table 5). Increases in bread hardness are related to its volume and specific volume. The volume of bread decreased with the increase in CBSs addition (Table 5), and the hardness increased. Cohesiveness refers to the degree to which food can withstand deformation before being destroyed. This means that bread with high cohesion forms clumps rather than disintegrating during chewing [67]. The results of this study showed that when CBSs replaced up to 25% of the WF, the cohesiveness of the bread significantly decreased, which may have been caused by the decrease in gluten content. Springiness is measured as the degree of recovery between the first and second compressions. Our findings showed that when the amount of CBSs replacing WF was 20% and 25%, the springiness of bread significantly decreased. This may have been due to the gradual increase in the proportion of CBSs replacing WF in the bread formula, resulting in a lower gluten content, smaller volume, and higher density. Chewiness is calculated as hardness × cohesiveness x springiness. Chewiness represents the force required to chew food, which involves the chewing characteristics of teeth, such as the compression, shearing, cutting, tearing, and crushing of food. The results showed that when the amount of CBSs replacing WF was more than 15%, the chewiness of bread significantly increased. Resilience is the ratio of energy that can be recovered after the first compression. The results showed that when the amount of CBSs replacing WF was 20% and 25%, the resilience of bread significantly decreased ($p < 0.05$). In summary, replacing WF with CBSs increases the hardness, gumminess, and chewiness of bread but also reduces springiness, cohesiveness, and resilience. These changes are primarily attributed to gluten dilution. If the CBSs is used to replace the WF in bread, the number of air cells may be considerably reduced, resulting in a reduction in specific volume, thereby reducing springiness and increasing hardness.

The ash, fat, protein, and total dietary fiber contents of the bread increased as the amount of CBSs replacing WF increased, but the moisture and carbohydrate contents gradually decreased (Table 5). This could be attributed to the differences in proximate compositions between CBSs and WF (Table 2).

### 3.6. Starch Hydrolysis and pGI of the Breads

Measuring the GI value of foods in vivo is a resource-intensive and relatively time-consuming experiment. The in vitro enzymatic method is used to determine the rates of starch digestion and glucose absorption in the small intestine, which is particularly suitable for screening starch-rich products during product development prior to in vivo gastrointestinal testing. Therefore, the in vitro enzymatic method is a faster and more cost-effective method [32,68]. The pGI of baked bread in this study was estimated using the in vitro enzyme hydrolysis of starch. Using white bread as a reference, the starch hydrolysis was 49.6% at 180 min in the restricted system (Figure 4). This result is similar to that in previous studies (42–50%) [31,68–70]. Figure 4 depicts the starch hydrolysis curve of the baked bread in this study. At the end of the 180 min incubation period, the starch hydrolysis percentages of CBS5 (45.1%), CBS10 (43.6%), CBS15 (40.7%), CBS20 (38.9%), and CBS25 (36.5%) were significantly lower than that of the control (50.0%). This result indicates that bread made by replacing part of the WF with CBSs may be more beneficial to the health of consumers and those with type 2 diabetes. When using white bread as the reference food, the pGI values of the control, CBS5, CBS10, CBS15, CBS20, and CBS25 ranged from 95.99 to 79.44 (Table 6). Among the breads tested, CBS20 and CBS25 were medium-GI foods [71], possibly due to their total dietary fiber and bioactive ingredients.

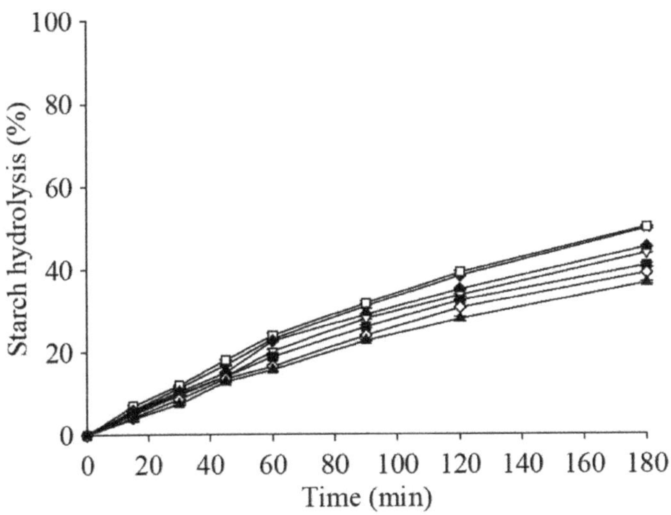


**Figure 4.** Starch hydrolysis of control and cocoa bean shells breads. Each value is expressed as mean ± standard deviation ($n$ = 4). White bread (★), control (□), CBS5 (◆), CBS10 (▽), CBS15 (■), CBS20 (◇), CBS25 (▲).

**Table 6.** Predicted glycemic index (pGI) of cocoa bean shells breads in the in vitro test.

| Sample | HI (%) | pGI [2] | Category |
|---|---|---|---|
| Control [1] | 102.52 ± 0.40 A [3] | 95.99 ± 0.22 A | High |
| CBS5 | 92.28 ± 0.67 B | 90.37 ± 0.37 B | High |
| CBS10 | 88.01 ± 0.31 C | 88.03 ± 0.17 C | High |
| CBS15 | 83.27 ± 0.35 D | 85.42 ± 0.19 D | High |
| CBS20 | 77.79 ± 0.74 E | 82.42 ± 0.41 E | Medium |
| CBS25 | 72.37 ± 0.46 F | 79.44 ± 0.25 F | Medium |

[1] Control, CBS05, CBS10, CBS15, CBS20, and CBS5: bread prepared with 0%, 5%, 10%, 15%, 20%, and 25% ($w/w$) replacement of wheat flour with cocoa bean shells powder, respectively. [2] GI categories, with the white bread used as reference food as follows: high GI value, GI > 85; medium-GI value, 85 > GI > 60; and low GI value, GI < 60. [3] Each value is expressed as mean ± SD ($n$ = 4). Means with different capital letters within a column differ significantly ($p < 0.05$).

### 3.7. Hedonic Quality Characteristics of the Breads

The appearance and crumb color scores of control (8.0 and 7.7), CBS5 (7.7 and 7.0), and CBS10 (7.5 and 7.2) were not significantly different ($p > 0.05$) (Figure 5), and were higher than those of other samples. The control (6.5 and 6.8), CBS5 (7.2 and 6.8), CBS10 (7.5 and 7.3), and CBS15 (7.6 and 6.7) exhibited no significant differences in aroma and flavor scores, which were higher than those of other breads. The texture score of control (6.8), CBS5 (7.2), and CBS10 (7.2) was not significantly different and was higher than that of the other samples. There was no significant difference in the overall liking scores of the control (6.8), CBS5 (7.2), and CBS10 (6.7), while CBS25 (5.0) had the lowest score. Based on the data obtained, high levels of CBSs will reduce the hedonic quality characteristic scores of the bread. Changes in these hedonic quality characteristics are positively correlated with the volume and specific volume of the CBS bread and negatively correlated with hardness and chewiness. Among all the samples, the bread made with CBSs replacing 25% of WF had the lowest score.

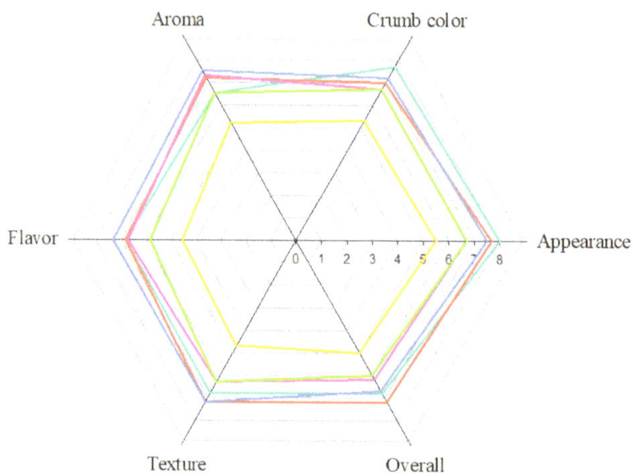


**Figure 5.** The radar plot of hedonic quality characteristic results of cocoa bean shells breads. Each value is expressed as mean ($n$ = 50). Control (———), CBS5 (———), CBS10 (———), CBS15 (———), CBS20 (———), CBS25 (———). Nine-point hedonic scale with 1, 5, and 9 representing extremely dislike, neither like nor dislike, and extremely like, respectively.

### 3.8. Total Phenols and Antioxidant Properties of the CBS Bread Extracts

In this study, 50% aqueous ethanol was used as the extraction solvent for the bread. The extraction yield (20.15–28.27%) and total phenolic content phenolic content (3.64–11.91 mg GAE/g lyophilized extract) of the obtained extract increased with the increase in the amount of CBS replacing WF (Table 7). This was due to the total phenolic content of the CBSs being significantly higher than that in the WF (Table 2). The antioxidant ability increased in a dose-dependent manner with the increase in bread extract concentrations (Figure 6). At a dose of 0.6 mg extract/mL, the CBS25 extract (49.05%) had the strongest ability to scavenge DPPH radicals, followed by CBS20 (38.24%), CBS15 (29.36%), CBS10 (12.29%), CBS5 (6.69%), and the control (4.71%) (Figure 6A). When the extract concentration increased to 2.4 mg extract/mL, the order of the extract's ability to scavenge DPPH radicals was CBS25 (88.12%) > CBS20 (82.88%) > CBS15 (66.93%) > CBS10 (54.95%) > CBS5 (49.67%) > control (9.45%). At 3.6 mg extract/mL, the scavenging ability of the CBS25 and CBS20 extracts was more than 90%. The estimated $EC_{50}$ value (mg extract/mL) showed that the ability of the bread extract to scavenge DPPH radicals in the descending order was CBS25 (0.64) > CBS20 (0.93) > CBS15 (1.50) > CBS10 (2.19) > CBS5 (2.41) > control (15.07).

**Table 7.** Yield and total phenolic content of cocoa bean shells bread extracts.

| | Control [1] | CBS5 | CBS10 | CBS15 | CBS20 | CBS25 |
|---|---|---|---|---|---|---|
| Yield (%) [2] | 20.15 ± 0.33 d [4] | 20.87 ± 0.06 f | 23.23 ± 0.39 c | 25.49 ± 0.16 b | 26.13 ± 0.01 b | 28.27 ± 0.56 a |
| Total phenols (mg GAE [3]/g lyophilized extract) | 3.64 ± 0.05 f | 5.52 ± 0.08 e | 7.35 ± 0.06 d | 8.87 ± 0.23 c | 10.67 ± 0.30 b | 11.91 ± 0.11 a |

[1] Control, CBS5, CBS10, CBS15, CBS20, and CBS25: bread prepared with 0%, 5%, 10%, 15%, 20%, and 25% (*w/w*) replacement of wheat flour with cocoa bean shells powder, respectively. [2] Yield (%) = (weight of lyophilized extract/weight of sample) × 100%. [3] GAE: gallic acid equivalent. [4] Each value is expressed as mean ± standard deviation ($n$ = 4). Means with different lowercase letters within a row differ significantly ($p < 0.05$).

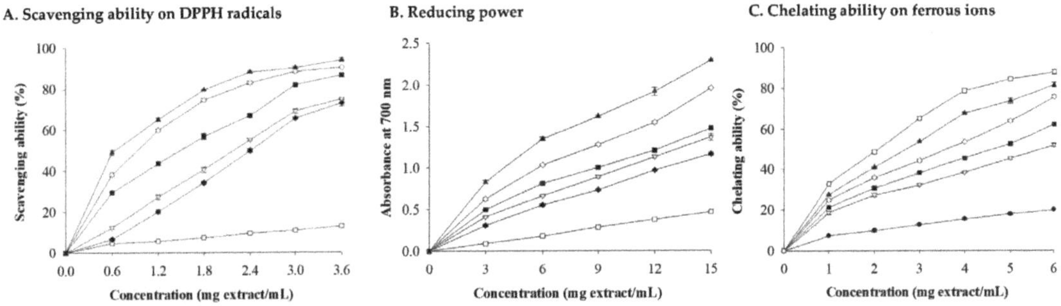


**Figure 6.** Antioxidant property of cocoa bean shells bread extracts. Each value is expressed as mean ± standard deviation ($n$ = 4). Control (□), CBS5 (◆), CBS10 (▽), CBS15 (■), CBS20 (◇), CBS25 (▲).

When the concentration was 3 mg extract/mL, the reducing power of the extract followed the order of CBS25 (0.83AU) > CBS20 (0.63 AU) > CBS15 (0.50 AU) > CBS10 (0.41 AU) > CBS5 (0.31 AU) > control (0.09 AU) (Figure 6B). When the extract concentration was increased to 15 mg extract/mL, the reducing power of the extracts, in descending order, was CBS25 (2.29 AU) > CBS20 (1.95 AU) > CBS15 (1.47 AU) > CBS10 (1.36 AU) > CBS5 (1.16 AU) > control (0.47 AU). The estimated $EC_{50}$ value (mg extract/mL) showed that CBS25 (1.67) had the strongest reducing power, followed by CBS20 (2.50), CBS15 (3.21), CBS10 (4.15), CBS5 (5.34), and the control (16.79).

When the dose was 1 mg extract/mL, CBS25 (32.78%) had the strongest ability to chelate ferrous ions, followed by CBS20 (27.52%) > CBS15 (24.87%) > CBS10 (21.28%) > CBS5 (18.63%) > control (7.30%) (Figure 6C). When the dose was increased to 6 mg extract/mL, the ability of the extracts to chelate ferrous ions in was, descending order, CBS25 (87.80%) > CBS20 (81.51%) > CBS15 (75.66%) > CBS10 (62.16%) > CBS5 (51.97%) > control group (19.92%). The results of the estimated $EC_{50}$ value (mg extract/mL) showed that CBS25 (2.09) had the strongest ability to chelate ferrous ions, followed by CBS20 (2.71), CBS15 (3.63), CBS10 (4.64), CBS5 (5.69), and the control (17.54).

Based on the above results, we found a significant positive correlation between the total phenols content and the antioxidant capacity of the extract.

### *3.9. The Impact of Adding Additional Water on the Quality Characteristics of CBS20*

CBS25 had the highest content of total dietary fiber (8.12 g/100 g) (Table 5) and total phenols (11.91 mg GAE/g lyophilized extract) (Table 7), the lowest starch hydrolysis rate (Figure 4), and the strongest antioxidant ability (Figure 6). Therefore, we first explored the feasibility of adding additional water to make CBS25 bread. The results showed that the skins of all breads had large tear openings, the tissue was too solid and rough (Figure 7), and the purpose of improvement was not achieved. Because the gluten concentration of the dough was diluted, the gluten network structure weakened, and the air cells of the dough ruptured during the early fermentation process, resulting in less dough expansion [66].

**Figure 7.** The appearance and color of control and CBS25 breads with various amounts of water added.

As such, we explored the impact of adding additional water on the quality characteristics of CBS20 bread, which also has medium GI. Adding an appropriate amount of water to CBS20 improved its appearance (Figure 8). We found no significant difference in the weights (488–498 g) of CBS20W5, CBS20W10, CBS20W15, CBS20W20, CBS20W25, and CBS20W30 (Table 8), but the weights of these breads were all significantly lower than those of the control and CBS20. This is because the weight of the dough placed in the mold was 560 g. Theoretically, the bread with less water added before baking (CB20) contained more solids than the bread with more water added before baking. Furthermore, the dough was baked at a high temperature in the oven, and the moisture in the sample with additional water added easily evaporated before the crust developed a hardened surface. The volume of the bread significantly increased with increased added water amount (Table 8), with the volume of CBS20W30 (2283 mL) bread being closest to that of the control (2342 mL). This may be caused by the yeast in the dough having enough moisture to ferment. In addition, when water turns to steam, it expands to 1100 its original volume. Steam is one of the important leavening agents [72]. Thus, the moisture in the dough will form steam during the baking process, which will also help increase the volume of the bread.

**Figure 8.** The appearance and color of control and CBS20 breads with various amounts of water added.

The WF in bread does not fully absorb water, which leads to a low gluten yield and restricted expansion [70]. Additionally, if dough lacks free water to maintain the normal metabolic activity of the yeast, carbon dioxide production is insufficient. In this study, by increasing the amount of water, the gluten in CBS20 bread not only fully expanded but also contributed to an increased yeast proliferation rate and gas production, so the volume of the bread also significantly increased. Specific volume is one of the most important parameters in bread making, which indicates the final gas retention of bread, affecting consumer preference [73]. We found no significant difference in the weight of bread made by adding additional water to CBS20, and we found a positive correlation between specific volume and volume. The increase in bread hardness is often used as an indicator of bread staling, which reflects the deterioration of bread quality and reduces consumer acceptability [74]. The staling rates of the samples were determined from the hardness changes 2 h and 24 h after baking (Table 8). Among all samples, CBS20W30 and CBS20W25 had the slowest staling rate, whereas CBS20 (1.67) had the fastest staling rate. This result indicated that increasing the moisture content of CBS20 slowed its staling rate. This may have occurred because the higher humidity of the more hydrated dough increased the number of disulfide bonds during mixing [75]. The above results show that adding additional water can improve the physical properties of CBS20.

**Table 8.** Weight, volume, staling rate, and sensory evaluation of control and CBS20 breads with various amounts of water added.

| | Control [1] | CBS20 | CBS20W5 | CBS20W10 | CBS20W15 | CBS20W20 | CBS20W25 | CBS20W30 |
|---|---|---|---|---|---|---|---|---|
| Weight (g) | 496 ± 2 b [2] | 516 ± 1 a | 488 ± 2 c | 498 ± 3 c | 487 ± 4 c | 487 ± 3 c | 486 ± 2 c | 487 ± 3 c |
| Volume (mL) | 2342 ± 13 a | 1551 ± 8 h | 1888 ± 35 g | 2022 ± 38 f | 2121 ± 13 e | 2179 ± 11 d | 2229 ± 18 c | 2283 ± 18 b |
| Specific volume (mL/g) | 4.72 ± 0.03 a | 3.01 ± 0.01 g | 3.87 ± 0.07 f | 4.15 ± 0.07 e | 4.35 ± 0.05 d | 4.48 ±< 0.01 c | 4.58 ± 0.02 b | 4.71 ± 0.03 a |
| Texture profile analysis 2 h after baking | | | | | | | | |
| Hardness (N) | 0.98 ± 0.09 b | 1.35 ± 0.16 a | 0.96 ± 0.12 b | 0.87 ± 0.10 c | 0.83 ± 0.15 d | 0.75 ± 0.16 e | 0.69 ± 0.08 f | 0.56 ± 0.10 g |
| Texture profile analysis 24 h after baking | | | | | | | | |
| Hardness (N) | 1.33 ± 0.12 c | 3.57 ± 0.18 a | 2.01 ± 0.16 b | 1.42 ± 0.15 c | 1.17 ± 0.15 d | 0.96 ± 0.16 e | 0.84 ± 0.09 f | 0.66 ± 0.10 g |
| Staling rate | 0.37 ± 0.19 e | 1.67 ± 0.28 a | 1.10 ± 0.20 b | 0.63 ± 0.09 c | 0.41 ± 0.09 d | 0.29 ± 0.07 f | 0.22 ± 0.06 g | 0.18 ± 0.04 h |
| Hedonic quality characteristics | | | | | | | | |
| Flavor | 6.8 ± 0.8 a | 5.8 ± 0.8 b | 4.7 ± 0.5 c | 5.0 ± 0.6 c | 5.8 ± 0.4 b | 6.3 ± 0.5 ab | 6.5 ± 0.5 ab | 6.3 ± 0.5 ab |
| Texture | 6.8 ± 1.0 ab | 6.3 ± 1.2 b | 6.3 ± 0.5 b | 6.3 ± 0.5 b | 6.5 ± 0.5 ab | 6.8 ± 0.4 ab | 7.3 ± 0.5 a | 6.5 ± 0.5 ab |
| Overall | 6.8 ± 1.2 ab | 6.0 ± 0.6 b | 6.2 ± 0.4 b | 6.2 ± 0.8 b | 6.3 ± 0.8 ab | 6.5 ± 0.5 ab | 7.2 ± 0.4 a | 6.5 ± 0.8 ab |

[1] Control and CBS20: bread prepared with 0% and 20% ($w/w$) replacement of wheat flour with cocoa bean shells powder, respectively. CBS20W5, CBS20W10, CBS20W15, CBS20W20, CBS20W25, and CBS20W30 refer to the CBS20 formula with an additional 5%, 10%, 15%, 20%, 25%, and 30% ($w/w$) water, respectively. [2] Each value is expressed as mean ± standard deviation ($n$ = 4; $n$ = 50 for hedonic quality characteristics). Means with different lowercase letters within a row differ significantly ($p < 0.05$). Nine-point hedonic scale with 1, 5, and 9 representing extremely dislike, neither like nor dislike, and extremely like, respectively.

The CBS20 formula with an additional 20–30% water produced the bread with the highest flavor score, with no significant difference (Table 8). There was no significant difference in texture and overall liking scores among CBS20W15, CBS20W20, CBS20W25, and CBS20W30. Among them, CBS20W25 scored the highest, significantly improving the hedonic quality characteristics of CBS20.

## 4. Conclusions

We used CBSs obtained from cocoa grown without pesticides in Taiwan as the test material. The heavy metal content of the CBSs was lower compared to the EU limits, and no mycotoxins were detected. The CBSs was found to be rich in dietary fiber and bioactive components, and the CBSs showed strong antioxidant properties. We successfully developed a functional and novel type of bread. Compared with traditional bread made with 100% WF, CBSs bread contains more dietary fiber and bioactive components, as more CBSs is used to replace WF. The incorporation of CBSs can reduce the GI value of traditional bread and provide a richer cocoa flavor. In addition, adding an additional 20–30% water to the CBS20 formula can improve its appearance, color, volume, staling rate, and hedonic quality characteristics. We think that the results of this study should serve as a reference for governments, farmers, and food processors to promote the recycling of CBSs in the cocoa bean industry and achieve the goal of zero waste.

**Author Contributions:** Conceptualization, S.-D.L.; data curation, C.-W.W., H.-S.S., C.-W.Y., P.-C.S. and S.-D.L.; formal analysis, C.-W.W., H.-S.S., P.-C.S. and S.-D.L.; funding acquisition, S.-D.L.; investigation, C.-W.W., H.-S.S., C.-W.Y., P.-C.S. and S.-D.L.; methodology, C.-W.W., H.-S.S., C.-W.Y. and S.-D.L.; project administration, S.-D.L.; resources, S.-D.L.; software, C.-W.W., H.-S.S. and S.-D.L.; supervision, S.-D.L.; validation, C.-W.W., H.-S.S. and S.-D.L.; visualization, C.-W.W., H.-S.S. and S.-D.L.; writing—original draft, C.-W.W., H.-S.S. and S.-D.L.; writing—review and editing, S.-D.L. All authors have read and agreed to the published version of the manuscript.

**Funding:** This research received no external funding.

**Data Availability Statement:** The original contributions presented in the study are included in the article, further inquiries can be directed to the corresponding author.

**Acknowledgments:** We are grateful to Sy Young Forestry Co., Ltd. (Taichung, Taiwan) for providing cocoa beans grown without pesticides as experimental materials for this study.

**Conflicts of Interest:** The authors declare no conflicts of interest.

## References

1. International Cocoa Organization. Production of Cocoa Beans (Thousand Tonnes). ICCO Quarterly Bulletin of Cocoa Statistics, 2023, Vol. XLIX, No.1, Cocoa Year 2022/23. Available online: https://www.icco.org/wp-content/uploads/Production_QBCS-XLIX-No.-1.pdf (accessed on 12 May 2024).
2. Battegazzore, D.; Bocchini, S.; Alongi, J.; Frache, A. Plasticizers, antioxidants and reinforcement fillers from hazelnut skin and cocoa by-products: Extraction and use in PLA and PP. *Polym. Degrad. Stab.* **2014**, *108*, 297–306. [CrossRef]
3. Botella-Martínez, C.; Lucas-Gonzalez, R.; Ballester-Costa, C.; Pérez-Álvarez, J.Á.; Fernández-López, J.; Delgado-Ospina, J.; Chaves-López, C.; Viuda-Martos, M. Ghanaian cocoa (*Theobroma cacao* L.) bean shells coproducts: Effect of particle size on chemical composition, bioactive compound content and antioxidant activity. *Agronomy* **2021**, *11*, 401. [CrossRef]
4. Rojo-Poveda, O.; Barbosa-Pereira, L.; Zeppa, G.; Stévigny, C. Cocoa bean shell—A by-product with nutritional properties and biofunctional potential. *Nutrients* **2020**, *12*, 1123. [CrossRef]
5. Soares, I.D.; Okiyama, D.C.G.; da Costa Rodrigues, C.E. Simultaneous green extraction of fat and bioactive compounds of cocoa shell and protein fraction functionalities evaluation. *Food Res. Int.* **2020**, *137*, 109622. [CrossRef]
6. Mellinas, A.C.; Jiménez, A.; Garrigós, M.C. Optimization of microwave-assisted extraction of cocoa bean shell waste and evaluation of its antioxidant, physicochemical and functional properties. *LWT-Food Sci. Technol.* **2020**, *127*, 109361. [CrossRef]
7. McKeown, N.M.; Meigs, J.B.; Liu, S.; Saltzman, E.; Wilson, P.W.; Jacques, P.F. Carbohydrate nutrition, insulin resistance, and the prevalence of the metabolic syndrome in the Framingham Offspring Cohort. *Diabetes Care* **2004**, *27*, 538–546. [CrossRef]
8. Eswaran, S.; Muir, J.; Chey, W.D. Fiber and functional gastrointestinal disorders. *Am. J. Gastroenterol.* **2013**, *108*, 718–727. [CrossRef]
9. Lin, S. Dietary fiber in bakery products: Source, processing, and function. Adv. *Food Nutr. Res.* **2022**, *99*, 37–100. [CrossRef]
10. Kumar, S.; Konwar, J.; Purkayastha, M.D.; Kalita, S.; Mukherjee, A.; Dutta, J. Current progress in valorization of food processing waste and by-products for pectin extraction. *Int. J. Biol. Macromol.* **2023**, *239*, 124332. [CrossRef]
11. FAO. *The State of Food and Agriculture 2019: Moving Forward on Food Loss and Waste Reduction*; FAO: Rome, Italy, 2019. Available online: https://openknowledge.fao.org/server/api/core/bitstreams/11f9288f-dc78-4171-8d02-92235b8d7dc7/content (accessed on 7 May 2024).
12. Marić, M.; Grassino, A.N.; Zhu, Z.; Barba, F.J.; Brnčić, M.; Brnčić, S.R. An overview of the traditional and innovative approaches for pectin extraction from plant food wastes and by-products: Ultrasound-, microwaves-, and enzyme-assisted extraction. *Trends Food Sci. Techol.* **2018**, *76*, 28–37. [CrossRef]
13. Galali, Y.; Omar, Z.A.; Sajadi, S.M. Biologically active components in by-products of food processing. *Food Sci. Nutr.* **2020**, *8*, 3004–3022. [CrossRef]
14. Souza, F.N.S.; Vieira, S.R.; Campidelli, M.L.L.; Rocha, R.A.R.; Rodrigues, L.M.A.; Santos, P.H.; de Deus Souza Carneiro, J.; de Carvalho Tavares, I.M.; de Oliveira, C.P. Impact of using cocoa bean shell powder as a substitute for wheat flour on some of chocolate cake properties. *Food Chem.* **2022**, *381*, 132215. [CrossRef]
15. Siow, C.S.; Chan, E.W.C.; Wong, C.W.; Ng, C.W. Antioxidant and sensory evaluation of cocoa (*Theobroma cacao* L.) tea formulated with cocoa bean hull of different origins. *Future Foods* **2022**, *5*, 100108. [CrossRef]
16. Hernández-Hernández, C.; Morales-Sillero, A.; Fernández-Prior, M.Á.; Fernández-Bolaños, J.; de la Paz Aguilera-Herrera, M.; Rodríguez-Gutiérrez, G. Extra virgin olive oil jam enriched with cocoa bean husk extract rich in theobromine and phenols. *LWT-Food Sci. Technol.* **2019**, *111*, 278–283. [CrossRef]
17. Papillo, V.A.; Locatelli, M.; Travaglia, F.; Bordiga, M.; Garino, C.; Coïsson, J.D.; Arlorio, M. Cocoa hulls polyphenols stabilized by microencapsulation as functional ingredient for bakery applications. *Food Res. Int.* **2019**, *115*, 511–518. [CrossRef]
18. Botella-Martínez, C.; Lucas-González, R.; Lorenzo, J.M.; Santos, E.M.; Rosmini, M.; Sepúlveda, N.; Teixeira, A.; Sayas-Barberá, E.; Pérez-Alvarez, J.A.; Fernandez-Lopez, J.; et al. Cocoa coproducts-based and walnut oil gelled emulsion as animal fat replacer and healthy bioactive source in beef burgers. *Foods* **2021**, *10*, 2706. [CrossRef]
19. Mau, J.L.; Lee, C.C.; Yang, C.W.; Chen, R.W.; Zhang, Q.F.; Lin, S.D. Physicochemical, antioxidant and sensory characteristics of bread partially substituted with aerial parts of sweet potato. *LWT-Food Sci. Technol.* **2020**, *117*, 108602. [CrossRef]
20. Taiwan Food and Drug Administration. Method of Test for Heavy Metals in Vegetables, Fruits, Jams and Jellies (MOHWH0024.00). In *Inspection Methods Announced by the Food and Drug Administration of the Republic of China*; Taiwan Food and Drug Administration: Taipei, Taiwan, 2020. Available online: https://www.fda.gov.tw/Tc/siteListContent.aspx?sid=103&id=33847 (accessed on 6 January 2023).

21. Taiwan Food and Drug Administration. Method of Test for Mycotoxins in Foods—Test of Multimycotoxin (MOHWT0010.02). In *Inspection Methods Announced by the Food and Drug Administration of the Republic of China*; Taiwan Food and Drug Administration: Taipei, Taiwan, 2017. Available online: https://www.fda.gov.tw/tc/siteListContent.aspx?sid=103&id=24258 (accessed on 6 January 2023).
22. Lu, T.M.; Lee, C.C.; Mau, J.L.; Lin, S.D. Quality and antioxidant property of green tea sponge cake. *Food Chem.* **2010**, *119*, 1090–1095. [CrossRef]
23. Sahin, A.W.; Wiertz, J.; Arendt, E.K. Evaluation of a new method to determine the water addition level in gluten free bread systems. *J. Cereal Sci.* **2020**, *93*, 102971. [CrossRef]
24. AACC International. *Approved Methods of the AACC*, 10th ed.; American Association of Cereal Chemists: St. Paul, MN, USA, 2000.
25. AOAC. *Official Methods of Analysis. Method 985.29. Total Dietary Fiber in Foods, Enzymatic-Gravimetric Method*, 17th ed.; Association of Official Analytical Chemists International: Gaithersburg, MD, USA, 2003.
26. Singleton, V.L.; Rossi, J.A. Colorimetry of total phenolics with phosphomolybdic-phosphotungstic acid reagents. *Am. J. Enol. Vitic.* **1965**, *16*, 144–158. [CrossRef]
27. Rojo-Poveda, O.; Barbosa-Pereira, L.; Mateus-Reguengo, L.; Bertolino, M.; Stévigny, C.; Zeppa, G. Effects of particle size and extraction methods on cocoa bean shell functional beverage. *Nutrients* **2019**, *11*, 867. [CrossRef]
28. Shimada, K.; Fujikawa, K.; Yahara, K.; Nakamura, T. Antioxidative properties of xanthan on the autoxidation of soybean oil in cyclodextrin emulsion. *J. Agric. Food Chem.* **1992**, *40*, 945–948. [CrossRef]
29. Oyaizu, M. Studies on products of browning reactions: Antioxidative activities of products of browning reaction prepared from glucosamine. *Jpn. J. Nutr.* **1986**, *44*, 307–315. [CrossRef]
30. 30 Dinis, T.C.P.; Madeira, V.M.C.; Almeida, L.M. Action of Phenolic Derivatives (Acetaminophen, Salicylate, and 5-Aminosalicylate) as Inhibitors of Membrane Lipid Peroxidation and as Peroxyl Radical Scavengers. *Arch. Biochem. Biophys.* **1994**, *315*, 161–169. [CrossRef]
31. Syu, P.C.; Zhang, Q.F.; Lin, S.D. Physicochemical, Antioxidant, Sensory, and Starch Digestibility Properties of Steamed Bread Fortified with Tamarillo Powder. *Foods* **2023**, *12*, 2306. [CrossRef] [PubMed]
32. Goñi, I.; Garcia-Alonso, A.; Saura-Calixto, F. A starch hydrolysis procedure to estimate glycemic index. *Nutr. Res.* **1997**, *17*, 427–437. [CrossRef]
33. Kruszewski, B.; Obiedziński, M.W.; Kowalska, J. Nickel, cadmium and lead levels in raw cocoa and processed chocolate mass materials from three different manufacturers. *J. Food Compos. Anal.* **2018**, *66*, 127–135. [CrossRef]
34. Thyssen, G.M.; Keil, C.; Wolff, M.; Sperling, M.; Kadow, D.; Haase, H.; Karst, U. Bioimaging of the elemental distribution in cocoa beans by means of LA-ICP-TQMS. *J. Anal. At. Spectrom.* **2018**, *33*, 187–194. [CrossRef]
35. Barišić, V.; Jozinović, A.; Flanjak, I.; Šubarić, D.; Babić, J.; Miličević, B.; Doko, K.; Ačkar, Đ. Difficulties with use of cocoa bean shell in food production and high voltage electrical discharge as a possible solution. *Sustainability* **2020**, *12*, 3981. [CrossRef]
36. International Agency for Research on Cancer. IARC Monographs on the Identification of Carcinogenic Hazards to Humans. *Agents Classified by the IARC Monographs*. 2024, Volume 1–136. Available online: https://monographs.iarc.who.int/list-of-classifications (accessed on 8 June 2024).
37. Rankin, C.W.; Nriagu, J.O.; Aggarwal, J.K.; Arowolo, T.A.; Adebayo, K.; Flegal, A.R. Lead contamination in cocoa and cocoa products: Isotopic evidence of global contamination. *Environ. Health Perspect.* **2005**, *113*, 1344–1348. [CrossRef]
38. Ahamed, M.; Siddiqui, M.K.J. Environmental Lead Toxicity and Nutritional Factors. *Clin. Nutr.* **2007**, *26*, 400–408. [CrossRef] [PubMed]
39. Assa, A.; Noor, A.; Yunus, M.R.; Misnawi; Djide, M.N. Heavy metal concentrations in cocoa beans (*Theobroma cacao* L.) originating from East *Luwu*, South Sulawesi, Indonesia. *J. Phys. Conf. Ser.* **2018**, *979*, 012011. [CrossRef]
40. Genchi, G.; Sinicropi, M.S.; Lauria, G.; Carocci, A.; Catalano, A. The effects of cadmium toxicity. *Int. J. Environ. Res. Public Health* **2020**, *17*, 3782. [CrossRef]
41. Charkiewicz, A.E.; Omeljaniuk, W.J.; Nowak, K.; Garley, M.; Nikliński, J. Cadmium Toxicity and Health Effects—A Brief Summary. *Molecules* **2023**, *28*, 6620. [CrossRef]
42. Lewis, C.; Lennon, A.M.; Eudoxie, G.; Umaharan, P. Genetic variation in bioaccumulation and partitioning of cadmium in *Theobroma cacao* L. *Sci. Total Environ.* **2018**, *640*, 696–703. [CrossRef] [PubMed]
43. FAO/WHO. Codex Commitee on Contaminants in Foods. In *Proposed Draft Maximum Levels for Cadmium in Chocolate and Cocoa-derived Products (at Step 4)*; Codex Alimentarius Commission: Utrecht, The Netherlands, 2018; Volume CF12/CRD08. Available online: https://www.fao.org/fao-who-codexalimentarius/sh-proxy/en/?lnk=1&url=https%253A%252F%252Fworkspace.fao.org%252Fsites%252Fcodex%252FMeetings%252FCX-735-12%252FWD%252Fcf12_06e.pdf (accessed on 26 April 2024).
44. Copetti, M.V.; Iamanaka, B.T.; Nester, M.A.; Efraim, P.; Taniwaki, M.H. Occurrence of ochratoxin A in cocoa by-products and determination of its reduction during chocolate manufacture. *Food Chem.* **2013**, *136*, 100–104. [CrossRef] [PubMed]
45. Coulibaly, A.; Biego, G.H.M.; Dembele, A.; Bohoussou, K.M.; Toure, A. Cocoa beans and cocoa derivatives from Cote-D'Ivoire: Investigating ochratoxin a level and assessing dietary intake adults. *Sustain. Agric. Res.* **2013**, *2*, 173–180. [CrossRef]
46. Okiyama, D.C.; Soares, I.D.; Toda, T.A.; Oliveira, A.L.; Rodrigues, C.E. Effect of the temperature on the kinetics of cocoa bean shell fat extraction using pressurized ethanol and evaluation of the lipid fraction and defatted meal. *Ind. Crops Prod.* **2019**, *130*, 96–103. [CrossRef]

47. Amezqueta, S.; Gonzalez-Penas, E.; Murillo, M.; Lopez de Cerain, A. Occurrence of ochratoxin A in cocoa beans: Effect of shelling. *Food Addit. Contam.* **2005**, *22*, 590–596. [CrossRef]
48. Serra Bonvehí, J. Occurrence of ochratoxin A in cocoa products and chocolate. *J. Agric. Food Chem.* **2004**, *52*, 6347–6352. [CrossRef]
49. Soares, I.D.; Cirilo, M.E.M.; Junqueira, I.G.; Vanin, F.M.; Rodrigues, C.E.D.C. Production of cookies enriched with bioactive compounds through the partial replacement of wheat flour by cocoa bean shells. *Foods* **2023**, *12*, 436. [CrossRef] [PubMed]
50. Copetti, M.V.; Iamanaka, B.T.; Pereira, J.L.; Lemes, D.P.; Nakano, F.; Taniwaki, M.H. Determination of aflatoxins in by-products of industrial processing of cocoa beans. *Food Addit. Contam. Part A Chem. Anal. Control Expo. Risk Assess.* **2012**, *29*, 972–978. [CrossRef] [PubMed]
51. Rapid Alert System for Food and Feed. Increased Aflatoxin Content in Organic Cocoa Powder from Ghana, via the Netherlands"—Notified on May 26th 2021 by Germany. *RASSF.* 2021, *NOTIFICATION 2021.2681*. Available online: https://webgate.ec.europa.eu/rasff-window/screen/notification/479503 (accessed on 2 May 2024).
52. Mudgil, D.; Barak, S. Composition, properties and health benefits of indigestible carbohydrate polymers as dietary fiber: A review. *Int. J. Biol. Macromol.* **2013**, *61*, 1–6. [CrossRef]
53. Rogowska, A.M. *Synaesthesia and Individual Differences*; Cambridge University Press: Cambridge, MA, USA, 2015. [CrossRef]
54. Okiyama, D.C.; Soares, I.D.; Cuevas, M.S.; Crevelin, E.J.; Moraes, L.A.; Melo, M.P.; Oliveira, A.L.; Rodrigues, C.E.C. Pressurized Liquid Extraction of Flavanols and Alkaloids from Cocoa Bean Shell Using Ethanol as Solvent. *Food Res. Int.* **2018**, *114*, 20–29. [CrossRef]
55. Barbosa-Pereira, L.; Belviso, S.; Ferrocino, I.; Rojo-Poveda, O.; Zeppa, G. Characterization and classification of cocoa bean shells from different regions of venezuela using HPLC-PDA-MS/MS and spectrophotometric techniques coupled to chemometric analysis. *Foods* **2021**, *10*, 1791. [CrossRef]
56. Nsor-Atindana, J.; Zhong, F.; Mothibe, K.J.; Bangoura, M.L.; Lagnika, C. Quantification of total polyphenolic content and antimicrobial activity of cocoa (*Theobroma cacao* L.) bean shells. *Pak. J. Nutr.* **2012**, *11*, 574–579. [CrossRef]
57. Martins, T.F.; Palomino, O.M.; Álvarez-Cilleros, D.; Martín, M.A.; Ramos, S.; Goya, L. Cocoa flavanols protect human endothelial cells from oxidative stress. *Plant Foods Hum. Nutr.* **2020**, *75*, 161–168. [CrossRef] [PubMed]
58. Kim, H.; Keeney, P.G. (-)-Epicatechin content in fermented and unfermented cocoa beans. *J. Food Sci.* **1984**, *49*, 1090–1092. [CrossRef]
59. Halliwell, B. Free radicals and antioxidants: Updating a personal view. *Nutr. Rev.* **2012**, *70*, 257–265. [CrossRef]
60. Prior, R.L. Fruits and vegetables in the prevention of cellular oxidative damage. *Am. J. Clin. Nutr.* **2003**, *78*, 570S–578S. [CrossRef]
61. Shahidi, F.; Janitha, P.K.; Wanasundara, P.D. Phenolic antioxidants. *Crit. Rev. Food Sci. Nutr.* **1992**, *32*, 67–103. [CrossRef] [PubMed]
62. Chernov, V.; Alander, J.; Bochko, V. Integer-based accurate conversion between RGB and HSV color spaces. *Comput. Electr. Eng.* **2015**, *46*, 328–337. [CrossRef]
63. Wikipedia. Hue. *Wikimedia Foundation, Inc.* Available online: https://en.wikipedia.org/wiki/Hue (accessed on 13 January 2024).
64. Jusoh, Y.M.; Chin, N.L.; Yusof, Y.A.; Rahman, R.A. Bread crust thickness measurement using digital imaging and L ab colour system. *J. Food Eng.* **2009**, *94*, 366–371. [CrossRef]
65. Dhen, N.; Rejeb, I.B.; Boukhris, H.; Damergi, C.; Gargouri, M. Physicochemical and sensory properties of wheat-Apricot kernels composite bread. *LWT-Food Sci. Technol.* **2018**, *95*, 262–267. [CrossRef]
66. de Erive, M.O.; He, F.; Wang, T.; Chen, G. Development of β-glucan enriched wheat bread using soluble oat fiber. *J. Cereal Sci.* **2020**, *95*, 103051. [CrossRef]
67. Matos, M.E.; Rosell, C.M. Relationship between instrumental parameters and sensory characteristics in gluten-free breads. *Eur. Food Res. Technol.* **2012**, *235*, 107–117. [CrossRef]
68. Germaine, K.A.; Samman, S.; Fryirs, C.G.; Griffiths, P.J.; Johnson, S.K.; Quail, K.J. Comparison of in vitro starch digestibility methods for predicting the glycaemic index of grain foods. *J. Sci. Food Agric.* **2008**, *88*, 652–658. [CrossRef]
69. Granfeldt, Y.; Björck, I.; Drews, A.; Tovar, J. An in vitro procedure based on chewing to predict metabolic response to starch in cereal and legume products. *Eur. J. Clin. Nutr.* **1992**, *46*, 649–660. [CrossRef]
70. Lai, M.H.; Liu, K.L.; Chen, P.Y.; Ke, N.J.; Chen, J.J.; Sung, J.M.; Wu, Y.L.; Lin, S.D. Predicted glycemic index and glycemic index of rice varieties grown in Taiwan. *Cereal Chem.* **2016**, *93*, 150–155. [CrossRef]
71. Jenkins, D.J.A.; Wolever, T.M.S.; Taylor, R.H.; Barker, H.; Fielden, H.; Baldwin, J.M.; Bowling, A.C.; Newman, H.C.; Jenkins, A.L.; Goff, D.V. Glycemic index of foods: A physiological basis for carbohydrate exchange. *Am. J. Clin. Nutr.* **1981**, *34*, 362–366. [CrossRef]
72. Gisslen, W. *Professional Baking*, 8th ed.; John Wiley & Sons, Inc.: New York, NY, USA, 2021.
73. Bouaziz, F.; Abdeddayem, A.B.; Koubaa, M.; Ghorbel, R.E.; Chaabouni, S.E. Date seeds as a natural source of dietary fibers to improve texture and sensory properties of wheat bread. *Foods* **2020**, *9*, 737. [CrossRef] [PubMed]
74. Jiang, X.; Wang, X.; Zhou, S. Effect of flaxseed marc flour on high-yield wheat bread production: Comparison in baking, staling, antioxidant and digestion properties. *LWT-Food Sci. Technol.* **2022**, *169*, 113979. [CrossRef]
75. Martínez, M.M.; Román, L.; Gómez, M. Implications of hydration depletion in the in vitro starch digestibility of white bread crumb and crust. *Food Chem.* **2018**, *239*, 295–303. [CrossRef] [PubMed]

*foods*

MDPI

*Article*

# Effect of Intermittent Low-Pressure Radiofrequency Helium Cold Plasma Treatments on Rice Gelatinization, Fatty Acid, and Hygroscopicity

**Ziyi Cao [1,2], Xingjun Li [2,*], Hongdong Song [1], Yu Jie [2] and Chang Liu [2]**

[1] School of Health Science and Engineering, University of Shanghai for Science and Technology, Shanghai 200093, China; caoziyiz@163.com (Z.C.); cau4080@163.com (H.S.)

[2] National Engineering Research Center for Grain Storage and Transportation, Academy of National Food and Strategic Reserves Administration, Beijing 102209, China; jyu@ags.ac.cn (Y.J.); shlglc@126.com (C.L.)

* Correspondence: lxj@ags.ac.cn or lixj3714@126.com

**Abstract:** To establish the safe and reproducible effects of cold plasma (CP) technology on food products, this study evaluated the gelatinization parameters, fatty acid profile, and hygroscopic properties of rice grains repeatedly treated with low-pressure radiofrequency (RF) helium CP (13.56 MHz, 140 Pa, 120 W-20s, 0–4 times, and 300 g sample). Compared with the untreated (zero times) sample, with an increase in CP treatment times from one to four on rice, the water contact angle and cooking time decreased, while the water absorption rate and freshness index increased, and the pH value remained unchanged. CP repeating treatments essentially had no effect on the gelatinization enthalpy, but significantly increased the peak temperature of gelatinization. From the pasting profile of rice that has been repeatedly CP treated, the peak, breakdown, and setback viscosities in flour paste decreased. CP repeating treatments on rice did not change the short-range molecular order of starch. Compared with the untreated sample, the first helium CP treatment maintained the content of C18:1n9c, C18:2n6c, and C18:3n3, but the second to fourth CP treatment significantly decreased contents of these fatty acids (FAs) as the C18:0 content increased. The first three CP treatments can increase the water and sucrose solvent retention capacity in rice flours. CP repeatedly treated rice first exhibits the similar monolayer water content and solid surface area of water sorption. Principal component analysis shows that contact angle, pasting parameters, and fatty acid profile in milled rice are quite sensitive to CP treatment. Results support that the effect of low-pressure RF 120W helium CP treatment 20 s on rice grains is perdurable, and the improvement of CP intermittent treatments on rice cooking and pasting properties is an added benefit, and the hygroscopic properties of rice was kept.

**Keywords:** cold plasma; rice; helium; gelatinization parameters; fatty acid profile

## 1. Introduction

The respect for nature, demand for healthy nutrition, and thereby green technology is increasing all over the world. Incorporating cold plasma (CP) technology into the food industry is significant because it can cut the usage of chemicals [1]. Plasma reactive species are in charge of microbial decontamination and food specific structural modifications [2,3], and they can facilitate nutrients extraction from biological systems. There is an urgent need for extensive research into the impact of CP treatment upon the quality and physicochemical indexes of cereal grains and food.

CP can modify starch by increasing surface energy, introducing functional groups, cross-linking, and depolymerization, as well as changing hydrophilicity, or molecule degradation and crystal etching [3,4]. The impact of CP varies depending on the CP resources, components, production method, treatment time, and starch properties. There are several reports on CP modification of the rice kernel surface and improvement in the taste quality of cooked brown rice and milled rice [5,6]. There are, however, few studies that address the influence of intermittent treatment and the duration of the cold plasma effect.

Some research considered that air plasma is rich in reactive oxygen-based and nitrogen-based species [7], and the resultant pH reduction was used to explain the action mechanism of CP on food. Because $HNO_x$ (x = 1, 4) can be produced in the humid air of CP when dielectric barrier discharge (DBD) CP was used to treat zein powder [8], peanut protein [9], and whey isolate protein [10], CP could decrease the pH values of these food materials. However, when the argon or oxygen cold plasma from KiNPen 09 direct current (DC) plasma jets was used to treat gellan gum, significant changes were not observed to pH and the content of titratable acid [11]. After air DBD CP treatment, the pH of the rice flours significantly increased [12]. The effect of low-pressure, RF high-purity helium CP on rice pH is worth further study.

Pal et al. [12] discovered that swelling force, light transmittance, and syneresis in short-grain and long-grain flours increased significantly with increasing air DBD CP power (60, 70 kV) and treatment time (0, 5, and 10 min), whereas blue value, final viscosity of paste, and setback viscosity decreased in short-grain rice flours but increased in long-grain rice flours. As for starch retrogradation, the intermittent treatments of helium CP on rice should be evaluated.

CP can influence the modification of starch and protein in the lipid oxidation reaction, causing lipid-containing food to deteriorate [13]. Radiofrequency (RF) CP (300 W, 13.56 MHz) treatment for 13.33 min resulted in the partial oxidation of the plant wax surface and the production of carboxylic acid and aldehydes [14]. The CP oxidation was observed in many foods; lipid oxidation should be avoided in foods with high lipid content. Removing oxygen from the feed gas and decreasing the CP treatment powder and time could reduce this oxidation [15]. The impact of high-voltage atmospheric CP (HVACP) treatment on soybean oil using hydrogen gas was mainly a reduction of polyunsaturated fatty acids and the increase in saturated fatty acids, while avoiding the formation of undesirable trans-isomers [16]. The effect of CP intermittent treatment on the fatty acid profile of cereal grain and its products needs to be studied further.

Numerous effects of CP, such as the organoleptic and biomolecular impact, plant cell structure modification, and the microbial impact, have been extensively studied [17–19]. However, there are still many gaps in our understanding of the impact mechanisms of CP on the organoleptic and biomolecular properties of food. These gaps should be thoroughly examined and the potential cumulative effects or alterations can be uncovered to ensure sustained benefits before this technology will be viable for food industry. After plasma treatment, the particles' surfaces become more hydrophilic [20], and the question is whether this phenomenon tends to decay gradually during the resting period. In addition, the sorption properties of grain and food is of interest because the relationship between moisture content and water activity is essential for a design and optimization of many processes in food industry such as drying, packaging, and storage [21], and few studies deal with whether cold plasma treatment does alter the hygroscopic properties of grain kernels. The present study performed tests just after low-pressure RF helium CP treatment on short-grain milled rice and repeated the tests several times to demonstrate the changes in gelatinization temperature, fatty acid profile, and hygroscopic properties, with the aim of establishing the effective, safe, and reproducible effects of CP technology on food products, and proving the industrial applicability of this technology.

## 2. Materials and Methods

### 2.1. Plasma Apparatus and Milled Rice Sample

A cold plasma processor HD-3 N from Changzhou Hanjie Biotechnology Co., Ltd., Jiangsu province, China was used for rice treatment. The plasma reactor comprises a quart glass tube with 10 mm thickness × 300 mm height × 300 mm internal diameter [22]. The top and base plates of the reactor were made of stainless steel material. The electrodes were connected on the plates through Wilson seals. The base plate had the ports through which a gas reservoir, pirani gauge, vacuum pumps, and air admittance valves were connected. The two parallel electrodes had area of 280 mm length × 250 mm width and the distance between the electrodes was kept 30 mm in all the experiments. The electrodes were capacitively coupled to a radiofrequency (PSG-II type) power supply with a frequency of 13.56 MHz. The working power range was 0 to 1000 W. The working vacuum level was 80 to 180 Pa.

Milled rice of a short-grain variety was retrieved from a rice processing plant in Changzhou, China in December 2022. Rice samples (300 g) were spread uniformly on the mesh, which was put on a glass stand between the two electrodes. The system was first evacuated to 20 Pa using three vacuum pumps to take away any adsorbed gases or water vapors from the surface of the rice kernels. The untreated sample was also treated in vacuum before analysis. Helium was used as feed gas for plasma generation and the working pressure was then adjusted to 140 Pa using a mass flow controller. The matching network was adjusted to get the stable glow discharge.

### 2.2. The Interval Test between Cold Plasma Processing

The rice was repeatedly treated at 120 W for 20 s in the CP apparatus according to the experimental procedure in Figure 1. The untreated (CK) sample was packed into a No.9 valve bag (280 mm length × 200 mm width × 0.04 mm thickness, Apple Brand, Shanghai China Manufacturing). The CP treated rice (0 h CP) was packaged into a No.9 valve bag and No.11 valve bag (400 mm length × 280 mm width × 0.04 mm thickness), and stored at RT. After CP treatment at the 48th hour, all the samples were stored at a 4 °C refrigerator for 30 days. During the analysis, all the samples were kept in a carton box (21 × 18 × 8 cm of length × width × height) in the 4 °C refrigerator. When CP processing was complete, a portion of rice kernels was stored at a refrigerator, and ca. Then, 200 g of untreated and CP treated rice were respectively milled into rice meal at a Gaosu universal pulverizer (Kewei Yongxing Instrument Co., Ltd., Beijing, China). The pulverizer has 26,000 rpm/min of rotation rate and 100–120 mesh degree of grinding. The rice meal was maintained at −20 °C until to analyze.

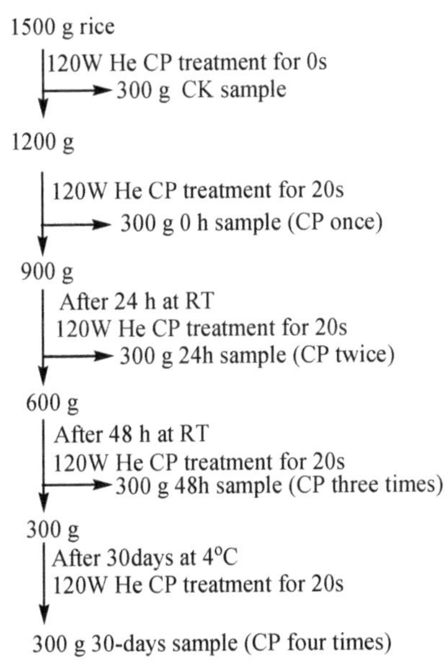


**Figure 1.** The rice sample preparations through CP intermittent treatments. Notes: He, Helium; CK, the untreated sample; CP, cold plasma; RT, room temperature.

### 2.3. The Hydrophilicity, Water Adsorption Rate, and Cooking Time of Milled Rice

The hydrophilicity of a rice kernel was assessed with a contact angle meter (V5, Yunfan Instrument Co., Ltd., Tianjin, China). A 2.5 μL droplet of deionized water was put on the kernel surface and the water contact angle was determined using sessile drop method. Analysis was started immediately after deposition of a single droplet of deionized water on the kernel surface. The dynamics of the droplet shape was further recorded at every 10 s by a video camera.

The water adsorption rate of rice grains was measured in 50 mL plastic centrifugal tubes. Four grams of rice ($m_1$) was weighed into a tube, then 40 mL deionized water was added. The tubes stood still at room temperature (RT) for 4 h. After pouring the distilled water out, the rice kernels were put on filter paper and the water on kernel surface was sucked up. The rice kernels were lastly weighed as $m_2$. The water adsorption rate of rice was computed as

$$\text{Wate radsorption rate } (\%) = \frac{m_2 - m_1}{m_1} \times 100 \tag{1}$$

Rice samples (2 g) were cooked in 20 mL of deionized water in a boiling water bath. The minimum cooking time was defined through moving out a rice kernel at different time intervals during cooking process, placing it on a glass plate, and pressing it with a little long-handle spoon. The rice was cooked until no white core was visible in any kernels when pressing, following the method of Tao et al. [23].

### 2.4. The Freshness Index and pH Value of Rice

The freshness index of rice was assessed as the difference between the absorption values at 615 nm and 690 nm, as reported by Takashi et al. [24] with some modifications. A reagent solution was prepared by dissolving about 300 mg of Bromothymol blue (BTB) in 200 mL of 75% ethanol solution, then diluted to 600 mL with deionized water, and finally adjusted to pH 7.0 with 0.2% potassium hydroxide solution. One gram of rice flour was ground in a mortar and pestle with 10 mL reagent solution. After centrifugation at 4000× $g$ rpm for 10 min, the absorption values of the supernatant were measured on a spectrophotometer (Shanghai Aoxi Instrument Co., Ltd., Shanghai, China). The greater the difference in $D_{615}$–$D_{690}$ is, the greater the freshness of the rice kernels is.

The pH value of plasma-treated rice was determined using a pH meter (S210, Mettler Seven Compact, Mettler Toledo, Zuich, Swiss). One gram (1.0000 g) of rice meal was weighed into a 50 -mL plastic centrifuge tube, then 15 mL distilled water was added and stirred with a glass rod, the pH value was measured immediately and after 4 h.

### 2.5. Gelatinization Temperature

The gelatinization temperature of the rice flours was determined with a differential scanning calorimeter (DSC; 200F3, Netzsch, Selb, Germany).The sample (5.0–5.1 mg) was weighed into an aluminum crucible and deionized water was added to produce a water/sample ratio of 2:1. The aluminum crucible was then sealed and equilibrated overnight at 4 °C. The DSC temperature was increased from 20 °C to 110 °C at the heating rate of 10 °C min$^{-1}$.

### 2.6. Pasting Property andSolvent Retention Capacity

The pasting parameters were determined with a Rapid ViscoAnalyzer (RVA) (Model RVA Tecmaster, PerkinElmer Inc., Waltham, MA, USA), using the method of Bahrami et al. [15] with some modifications. Rice flour suspension (3.5 g of flour plus 24.5 g of deionized water) were, heated from 50 to 95 °C at 6 °C min$^{-1}$ after equilibration at 50 °C for 1 min, then held at 95 °C for 5 min, and finally cooled to 50 °C at 6 °C min$^{-1}$. Peak viscosity, trough viscosity, breakdown viscosity, final viscosity, setback viscosity, and pasting temperature were obtained from a pasting profile.

The solvent retention capacity (SRC) of rice flour were determined using AACCI method 56–11.02 with some modifications [25]. Four 50-mL round-bottom centrifugal tubes with the covers were weighed as $W_0$, and then 5.000 ± 0.050 g of rice flour ($W_1$) was weighed into each tube. At the second step, to four 50 mL sharp-bottom centrifugal tubes, 25 ± 0.05 g of distilled water, 5% lactic acid, 5% sodium bicarbonate, and 50% sucrose solution were separately added, then each reagent was, respectively poured into the rice flour-containing tube and the time was recorded. The round-bottom centrifugal tubes were covered with lids and severely vortexed for 5 s in a rotary mixer (WH-866, Taicang Hualida Laboratory Equipment Co., Ltd., China), then the rice flour was allowed to swell for 20 min. Each tube was vortexed for 5 s after 5, 10, 15, and 20 min, respectively. At the third step, after the tubes were centrifuged at 4 °C and 3052× *g* rpm for 15 min, the supernatants were slowly poured out, and the tubes were placed upside-down on tissue for 10 min. Lastly, after covering with lids, the precipitate-containing tubes were weighed as $W_2$. The SRC value for each reagent was calculated using Equation (2).

$$\mathrm{SRC} = \left[\frac{W_2 - W_0}{W_1} \times \frac{86}{100(1-M)} - 1\right] \times 100\% \tag{2}$$

where $W_0$ is the weight (g) of a round-bottom centrifugal tube plus its lid; $W_1$ is the weight (g) of rice flour; $W_2$ is the weight (g) of a rice flour precipitate-containing tube and its lid; and M is the moisture content of rice flour (decimal, wet basis); 86% shows the dry matter content of the reference sample with M = 0.14.

### *2.7. Fourier Transform Infrared Spectroscopy*

Fourier transform infrared spectroscopy (FTIR) spectra of the samples were recorded, at 64 scans with a spectral resolution of 4 $cm^{-1}$, in attenuated total reflection (ATR) mode on a spectrometer (Nicolet 6700 FTIR, Thermo Electron Corporation, Waltham, MA, USA). The sample spectra were recorded against a background spectrum (without a sample in place). All measurements were analyzed using OMNIC software (Version 8.2a). The dried samples were mixed with KBr, and they were ground and pressed into pellets. FTIR spectra were collected in the wave number range of 400–4000 $cm^{-1}$.

### *2.8. Fatty Acid Profile Analysis*

The fatty acid profile was determined as described by Tao et al. [23]. For lipid extraction, rice flours (300 mg) were weighed into a 10 mL plastic centrifugal tube and mixed with 1 mL hexane and 0.25 mL of 2 mol/L KOH-methanol solution for 30 s. After being extracted for 40 min in the water bath at 30 °C using a 300 W ultrasonic probe, the sample was cooled to room temperature and mixed with 0.25 mL of 2 mol/L HCl solution for 1 min. After centrifugation at 8414× *g* rpm for 10 min, the supernatants were analyzed by gas chromatography-mass spectrometry (GC-MS).

Fatty acid methyl esters (FAME) in each sample were measured with a gas chromatography system (Agilent 8890 GC) equipped with a fused-silica column (HP-5MS UI; 30 m × 0.25 mm× 0.25 μm of length × internal diameter × film thickness) coupled with a mass selective detector (MSD; 5977B, Agilent Technologies, Santa Clara, CA, USA). Samples were injected in a split mode (0.3 μL, split ratio 10) at an initial temperature of 130 °C for 3 min, the temperature was then increased at 5 °C/min to 180 °C. After 8 min, the temperature was increased from 5 °C/min to 240 °C and stood there for 12 min. Helium was used as the carrier gas (1.0 mL/min) and the solvent delay time was 1.6 min. For MS system, the temperatures of injection port, ionization source, transfer line and quadrupole were 260 °C, 230 °C, 280 °C, and 150 °C, respectively, and electron-impact mass spectra were recorded at the 70 eV ionization voltage. The acquisitions were performed in full-scan mode (40–400 amu).

The MS information of each FAME peak was submitted to MassHunter Qualitative Analysis V10.0 software (Agilent Technologies, USA) and compared with the offline MS library of NIST08s and the online NIST Chemistry Web Book, SRD 69 (https://webbook.nist.gov/chemistry/#, accessed on 1 January 2023), considering a minimum similarity value of 80%. The measured concentration, expressed in μg/L, was quantified with the external standard method, in which 37 kinds of FAME (Sigma, Albuquerque, NM, USA) mixed with different gradients were adopted. The linear ranges of the standard curves were 0.5–400 μg/L, with $R^2 > 0.99$.

### 2.9. Moisture Sorption Isotherms and Solid Surface Area

The moisture sorption isotherms of CP treated rice was determined with dynamic water sorption analyzer (SPS11-10μ, ProUmid GmbH & Co., KG, Ulm, Germany). This instrument can automatically, gravimetrically determine the water vapor adsorption and desorption of multiple samples in a test atmosphere chamber with controlled temperature and equilibrium relative humidity (ERH). Temperature over time is ±0.1 K, ERH accuracy is ± 0.6% ERH at (23 ± 5) °C for the range of 0–100% ERH. The equilibrium moisture content (EMC) of each rice sample (*ca.* 2.0000 g) at four constant temperatures (5, 15, 25, and 35 °C) over range of 10–90% ERH were determined using deionized water producing relative humidity and high-purity nitrogen blowing dry and preventing samples from getting moldy. The interval between gravimetrical cycles was set as 10 min. The measurement cycle began at 10% ERH and at first increased with a 10% step to 20% ERH. Subsequently ERH was raised to from 30% to 80% and finally to 90% ERH. The time per cycle was set up to a minimum of 50 min and a maximum of 50 h. The default weight limit was +100%, balance band width (dm/dt) was ±0.01%/40 min. During a measurement cycle, the samples were automatically placed on an electronic balance and weighed. One sample pan remained unloaded and was always used for drift compensation of the measured values. The recorded data were shown by SPS-Toolbox Basic Rel. 1.15 software.

Experimental EMC/ERH data was used to establish isotherm curves in Kaleidagraph for Mac 4.5.2v software, with ERH and EMC data, respectively entered onto the x- and the *y*-axis. The modified Chung-pfost equation (MCPE, Equation (3)) and Guggenheim-Anderson-de Boer (GAB, Equation (4)) were employed to fit the EMC/ERH data of rice [26].

$$M = -\frac{1}{\mathrm{C}}\ln\left[-\frac{(t+\mathrm{B})\ln(ERH)}{\mathrm{A}}\right] \tag{3}$$

$$M = \frac{M_{\mathrm{m}}\cdot\mathrm{B}\cdot\mathrm{C}\cdot ERH}{(1-\mathrm{B}\cdot ERH)(1-\mathrm{B}\cdot ERH+\mathrm{B}\cdot\mathrm{C}\cdot ERH)} \tag{4}$$

where *ERH* is equilibrium RH, decimal; $M$ is equilibrium MC (% wet basis); $M_{\mathrm{m}}$ is monolayer water content (% dry basis); $t$ is temperature (°C); A, B, and Care the equation coefficients.

Fitting was investigated by non-linear regression analysis in SPSS v17.0 for Windows [27]. The criteria used for determining the equation for the EMC/ERH data were the determination coefficient ($R^2$), residue sum of squares (*RSS*), standard error (*SE*), and mean relative percentage error (*MRE*). Equations (5)–(8) were used to calculate $R^2$, *RSS*, *SE*, and *MRE*, respectively.

$$R^2 = 1 - \sum_{i=1}^{n}(m_{\mathrm{i}} - m_{\mathrm{pi}})^2 / \sum_{i=1}^{n}(m_{\mathrm{i}} - m_{\mathrm{mi}})^2 \tag{5}$$

$$RSS = \sum_{i=1}^{n}(m_{\mathrm{i}} - m_{\mathrm{pi}})^2 \tag{6}$$

$$SE = \sqrt{\sum_{i=1}^{n}(m_{\mathrm{i}} - m_{\mathrm{pi}})^2/(n-1)} \tag{7}$$

$$MRE\% = \frac{100}{n}\sum_{\mathrm{i}=1}^{\mathrm{n}}\left|\frac{m_{\mathrm{i}} - m_{\mathrm{pi}}}{m_{\mathrm{i}}}\right| \tag{8}$$

where $m_{\mathrm{i}}$ is the experimental value, $m_{\mathrm{pi}}$ is the predicted value, $m_{\mathrm{mi}}$ is the average of experimental values, and n is the number of observations. The suitability of an equation to the EMC/ERH data of rice was thought satisfactory if the MRE was below 10% [25].

The solid surface area of water sorption was determined as given by Moraes et al. [28].

$$SA = M_m \times \frac{N_a \times S_w}{MW} = 3533.3M_m \tag{9}$$

where $SA$ is the surface area of water sorption ($m^2/g$ solid), $M_m$ is the mono-molecular layer water content (% d.b.), $N_a$ is the number of Avogadro, $6 \times 10^{23}$molecules/mole, $S_w$ is the area of a water molecule ($10.6 \times 10^{-20} m^2$), and MW is the molecular weight of water (18 g/mole).

### 2.10. Data Analysis

Except for parallel samples for the determination of EMC/ERH data, at least three replicates were tested on each rice sample for physico-chemical and rice quality parameters. SPSS software (Version 17.0 [27]) was used to analyze data. One-way analysis of variance (ANOVA) and Duncan's new multiple-range test was used for comparing multiple of means. Statistical significance was stated at $p < 0.05$. Data reduction and factor method was used for principal components analysis and to make factional scatterplot.

## 3. Results

### 3.1. Effect of CP Repeating Treatments on the Water Absorption Rate and Freshness in Rice

Table 1 shows the effect of helium CP repeating treatments on the water absorption rate and freshness in milled rice. With an increase in CP treatment times, the water contact angle significantly decreased while the water absorption rate significantly increased. Compared with the untreated sample, CP treatment significantly decreased the cooking time, but the decrease in cooking time was less with an increase in CP treatment times. In addition, the rice freshness index increased with CP treatment times. The rice pH value remained unchanged with CP treatment times.

**Table 1.** Effect of CP repeating treatments on water absorption rate and freshness index in milled rice.

| CP Treatments | Water Contact Angle (°) | Water Absorption Rate (%) | Cooking Time (min) | Freshness Index (0.01) | pH Value |
|---|---|---|---|---|---|
| CK (0) | $79.7 \pm 2.90^{a}$ | $29.1 \pm 1.39^{d}$ | $20.0 \pm 0.52^{a}$ | $3.79 \pm 0.36^{d}$ | $6.59 \pm 0.03^{a}$ |
| 0 h (1) | $73.8 \pm 1.98^{b}$ | $30.9 \pm 0.19^{c}$ | $14.5 \pm 0.42^{c}$ | $4.75 \pm 0.73^{cd}$ | $6.61 \pm 0.03^{a}$ |
| 24 h (2) | $60.8 \pm 1.43^{c}$ | $36.6 \pm 1.40^{a}$ | $15.5 \pm 0.43^{b}$ | $5.59 \pm 0.05^{b}$ | $6.59 \pm 0.03^{a}$ |
| 48 h (3) | $55.2 \pm 4.42^{cd}$ | $34.8 \pm 0.20^{b}$ | $15.0 \pm 0.50^{bc}$ | $4.51 \pm 0.04^{c}$ | $6.57 \pm 0.01^{a}$ |
| 30 d (4) | $53.6 \pm 1.94^{d}$ | $37.1 \pm 0.45^{a}$ | $15.5 \pm 0.40^{b}$ | $10.25 \pm 2.06^{a}$ | $6.57 \pm 0.02^{a}$ |

Notes: CP, cold plasma; CK, the untreated sample; The number in bracket shows the CP treatment times; Data are expressed as mean± standard deviation (SD), number of repetitions—$n = 3$. Means with the different superscript letters (a, b, c, d) in a column are different significantly ($p < 0.05$) among different CP treatment times.

### 3.2. Effect of CP Repeating Treatments on Gelatinization Temperature of Rice

Table 2 shows the effect of CP repeating treatments on the gelatinization temperature of rice. CP repeating treatments significantly increased the onset ($T_o$) and peak ($T_p$) temperatures of gelatinization while maintaining the conclusion gelatinization temperature ($T_c$) and the $T_c$–$T_o$ values, indicating that the crystallinity and cross-links in the rice surface induced by CP treatment might increase with an increase in treatment times. CP repeating treatments essentially did not change gelatinization enthalpy, but significantly increased peak enthalpy. Compared with the untreated sample, the CP treatments significantly decreased the gelatinization peak width, and all repeating treatments maintained the peak width.

**Table 2.** Effect of CP repeating treatments on the gelatinization temperature of rice.

| CP Treatments | $\Delta H$ (J/g) | $T_o$ (°C) | $T_p$ (°C) | $T_c$ (°C) | $T_c$–$T_o$ (°C) | Peak Width(°C) | Peak Enthalpy (0.01 mW/mg) |
|---|---|---|---|---|---|---|---|
| CK (0) | 1.94 ± 0.00 $^{a}$ | 62.13 ± 0.15 $^{c}$ | 66.47 ± 0.32 $^{cd}$ | 70.50 ± 0.44 $^{a}$ | 8.37 ± 0.50 $^{a}$ | 5.30 ± 0.00 $^{a}$ | 5.46 ± 0.02 $^{c}$ |
| 0 h (1) | 1.82 ± 0.01 $^{b}$ | 62.37 ± 0.31 $^{bc}$ | 66.53 ± 0.06 $^{d}$ | 70.33 ± 0.86 $^{a}$ | 7.97 ± 0.76 $^{a}$ | 5.00 ± 0.10 $^{b}$ | 5.47 ± 0.07 $^{c}$ |
| 24 h (2) | 1.82 ± 0.05 $^{b}$ | 63.03 ± 0.42 $^{ab}$ | 66.83 ± 0.15 $^{bc}$ | 70.53 ± 1.19 $^{a}$ | 7.50 ± 1.05 $^{a}$ | 4.83 ± 0.21 $^{b}$ | 5.65 ± 0.18 $^{bc}$ |
| 48 h (3) | 1.93 ± 0.03 $^{a}$ | 62.93 ± 0.38 $^{ab}$ | 67.07 ± 0.12 $^{b}$ | 70.73 ± 0.81 $^{a}$ | 7.80 ± 0.52 $^{a}$ | 4.97 ± 0.06 $^{b}$ | 5.83 ± 0.10 $^{ab}$ |
| 30 d (4) | 1.93 ± 0.02 $^{a}$ | 63.77 ± 0.49 $^{a}$ | 67.53 ± 0.15 $^{a}$ | 71.07 ± 0.84 $^{a}$ | 7.30 ± 1.11 $^{a}$ | 4.77 ± 0.15 $^{b}$ | 6.04 ± 0.20 $^{a}$ |

Notes: CP, cold plasma; CK, the untreated sample; The number in bracket shows the CP treatment times; $\Delta H$, enthalpy of gelatinization; $T_o$, the onset temperature of gelatinization; $T_p$, the peak temperature of gelatinization; $T_c$, the conclusion temperature of gelatinization; $T_c$–$T_o$, gelatinization temperature range. Data are expressed as mean ± standard deviation (SD), number of repetitions—$n$ = 3. Means with the different superscript letters in a column are different significantly ($p < 0.05$) among different CP treatment times.

### 3.3. Effect of CP Repeating Treatments on Pasting Parameters

Table 3 shows the effect of CP repeating treatments on pasting parameters of milled rice. With an increase in CP treatment times, peak viscosity (PV), breakdown viscosity (BD), and setback viscosity (SB) decreased while peak time increased, whereas trough viscosity (TV), final viscosity (FV) and pasting temperature (PT) remained unchanged. The decrease in setback viscosity can be considered to reflect the less retrogradation tendency of amylose in a starch paste. Compared with the untreated sample, 120 W helium cold plasma treatments might decrease 2.6%, 0.5%, 3.3% and 7.3% amylose retrogradation in a starch paste for once (0 h), twice (24 h), three times (48 h), and four times (30 d) treatment, respectively.

**Table 3.** The pasting parameters of CP repeatedly treated rice.

| CP Treatments | PV (cp) | TV (cp) | BD (cp) | FV (cp) | SB (cp) | Peak Time (min) | PT (°C) |
|---|---|---|---|---|---|---|---|
| CK | 4597 ± 32 $^{a}$ | 2661 ± 91 $^{ab}$ | 1937 ± 72 $^{a}$ | 4106 ± 93 $^{ab}$ | 1445 ± 7 $^{a}$ | 5.89 ± 0.03 $^{c}$ | 70.93 ± 0.06 $^{a}$ |
| 0 h | 4595 ± 5 $^{a}$ | 2783 ± 45 $^{a}$ | 1812 ± 48 $^{b}$ | 4189 ± 29 $^{a}$ | 1407 ± 17 $^{b}$ | 5.98 ± 0.04 $^{b}$ | 70.65 ± 0.43 $^{a}$ |
| 24 h | 4420 ± 34 $^{b}$ | 2729 ± 27 $^{a}$ | 1691 ± 29 $^{c}$ | 4165 ± 41 $^{ab}$ | 1437 ± 62 $^{ab}$ | 6.05 ± 0.04 $^{ab}$ | 71.22 ± 0.55 $^{a}$ |
| 48 h | 4485 ± 32 $^{b}$ | 2738 ± 38 $^{a}$ | 1748 ± 57 $^{bc}$ | 4136 ± 40 $^{a}$ | 1398 ± 3 $^{b}$ | 6.02 ± 0.04 $^{ab}$ | 71.48 ± 0.51 $^{a}$ |
| 30 d | 4220 ± 54 $^{c}$ | 2625 ± 64 $^{b}$ | 1595 ± 16 $^{d}$ | 3965 ± 59 $^{b}$ | 1340 ± 7 $^{c}$ | 6.07 ± 0.01 $^{a}$ | 71.17 ± 0.46 $^{a}$ |

Notes: CP, cold plasma; CK, the untreated sample; The number in bracket shows the CP treatment times; PV, Peak viscosity; TV, trough viscosity; BD, breakdown viscosity; FV, final viscosity; SB, setback viscosity; PT, pasting temperature. These parameters were evaluated from the pasting profile. Data are expressed as mean ± standard deviation (SD), number of repetitions—$n$ = 3. Means with the different superscript letters in a column are different significantly ($p < 0.05$) among different CP treatment times.

### 3.4. Effect of CP Repeating Treatments on Solvent Retention Capacity

Table 4 shows the effect of CP repeating treatments on the SRC of rice. Compared with the untreated sample, the first three times of CP treatments significantly increased water SRC and sucrose SRC, but the fourth CP treatment no longer increased these two SRC values. The first CP treatment increased $NaHCO_3$ SRC, whereas the other 2–4 treatment times maintained the $NaHCO_3$ SRC. Compared with the untreated sample and the samples of the first, second, fourth CP treatments, only the third CP treatment increased lactic acid SRC. These results suggest that the first three helium CP treatments might induce swelling of arabinoxylan in rice flour.

### 3.5. Effect of CP Repeating Treatments on the Deconvoluted IR Spectra

The short-range molecular order of rice starch was measured by ATR-FTIR. The ATR-FTIR spectrum of rice flours from untreated and CP repeatedly treated rice exhibited similar absorption bands (The FTIR spectra were not shown). The ratios of absorbances at 1047/1022 ($R_{1047/1022}$) and 1022/995 cm$^{-1}$ ($R_{1022/995}$) acquired from deconvoluted IR spectra were given in Table 5, as the possible indicators of the degree of short-range order at the surface of starch granules. After treatments, the $R_{1047/1022}$ and $R_{1022/995}$ did not show significant ($p < 0.05$) changes compared with those of the CP untreated sample, regardless of the CP treatment times. The ratio of absorbances at 1068/1022 cm$^{-1}$ ($R_{1068/1022}$) which might show the interaction between protein and starch, did not show significant ($p < 0.05$) changes among CP untreated and treated samples.

**Table 4.** Effect of CP repeating treatments on the values of solvent retention capacity (SRC) in rice.

| CP Treatments | Water SRC (%) | $NaHCO_3$ SRC (%) | Lactic Acid SRC (%) | Sucrose SRC (%) |
|---|---|---|---|---|
| CK (0) | 105.95 ± 0.90 $^{c}$ | 113.66 ± 1.72 $^{b}$ | 110.47 ± 1.26 $^{b}$ | 131.25 ± 0.67 $^{c}$ |
| 0 h (1) | 110.09 ± 1.40 $^{b}$ | 118.90 ± 0.88 $^{a}$ | 108.01 ± 1.91 $^{b}$ | 135.33 ± 0.94 $^{a}$ |
| 24 h (2) | 110.97 ± 1.71 $^{b}$ | 112.50 ± 1.80 $^{b}$ | 110.07 ± 0.77 $^{b}$ | 132.77 ± 0.47 $^{b}$ |
| 48 h (3) | 114.77 ± 0.90 $^{a}$ | 114.98 ± 1.32 $^{b}$ | 115.43 ± 1.60 $^{a}$ | 135.71 ± 0.06 $^{a}$ |
| 30 d (4) | 106.65 ± 0.91 $^{c}$ | 114.68 ± 1.67 $^{b}$ | 109.77 ± 0.16 $^{b}$ | 131.88 ± 0.40 $^{c}$ |

Notes: CP, cold plasma; CK, the untreated sample; The number in bracket shows the CP treatment times; SRC, solvent retention capacity. Data are expressed as mean ± standard deviation (SD), number of repetitions—$n$ = 3. Means with the different superscript letters in a column are different significantly ($p < 0.05$) among different CP treatment times.

**Table 5.** IR ratios acquired from deconvoluted FTIR spectra.

| CP Treatments | IR Ratios | | |
|---|---|---|---|
| | $R_{1022/995}$ | $R_{1047/1022}$ | $R_{1068/1022}$ |
| CK (0) | 1.056 ± 0.008 $^{a}$ | 0.910 ± 0.006 $^{a}$ | 0.759 ± 0.011 $^{a}$ |
| 0 h (1) | 1.054 ± 0.021 $^{a}$ | 0.914 ± 0.008 $^{a}$ | 0.762 ± 0.012 $^{a}$ |
| 24 h (2) | 1.065 ± 0.012 $^{a}$ | 0.904 ± 0.010 $^{a}$ | 0.739 ± 0.017 $^{a}$ |
| 48 h (3) | 1.045 ± 0.013 $^{a}$ | 0.916 ± 0.007 $^{a}$ | 0.769 ± 0.013 $^{a}$ |
| 30 d (4) | 1.059 ± 0.014 $^{a}$ | 0.912 ± 0.012 $^{a}$ | 0.760 ± 0.014 $^{a}$ |

Notes: CP, cold plasma; CK, the untreated sample; The number in bracket shows the CP treatment times. Data are expressed as mean ± standard deviation (SD), number of repetitions—$n$ = 3. Means with the different superscript letters in a column are different significantly ($p < 0.05$) among different CP treatment times.

### 3.6. *Effect of CP Repeating Treatments on the Fatty Acid Profile*

Table 6 shows the effect of CP repeating treatments on the fatty acid (FA) profile of rice. Nineteen fatty acids were detected in the CP repeatedly treated rice. Five species of fatty acids with a content greater than 50 µg/g were palmitic acid (C16:0), stearic acid (C18:0),oleic acid (C18:1n9c), linoleic acid (C18:2n6c), and α-linolenic acid (C18:3n3). With an increase in CP times, the contents of C6:0, C18:1n9c, C18:2n6c, C18:3n3, C20:1, and C22:6n3 decreased, contents of C10:0, C12:0, C14:0, C14:1, C15: 0, C15:1, C18:0, C18:1n9t, C20:0, and C24:1 increased, and the contents of C16: 0, C20:5n3, and C24:0 remained unchanged. These results suggest that, compared with the untreated sample, the first helium CP treatment could maintain the content of the main monounsaturated FA (C18:1n9c) and polyunsaturated FA (C18:2n6c, C18:3n3); however, the second and more times of CP treatments significantly decreased the content of these three FAs with an increase in saturated FA (C18:0) content.

**Table 6.** Effect of CP repeatedly treatments on fatty acid profile in rice.(µg/g).

| CP Treatments | C6:0 | C10:0 | C12:0 | C14:0 | C14:1 |
|---|---|---|---|---|---|
| CK (0) | 33.0 ± 1.0 $^{a}$ | 26.8 ± 0.8 $^{b}$ | 27.7 ± 0.2 $^{c}$ | 22.1 ± 0.2 $^{b}$ | 3.8 ± 0.1 $^{c}$ |
| 0 h (1) | 32.7 ± 1.5 $^{a}$ | 27.4 ± 0.2 $^{b}$ | 29.9 ± 2.5 $^{bc}$ | 23.6 ± 3.6 $^{ab}$ | 4.3 ± 0.3 $^{ab}$ |
| 24 h (2) | 28.4 ± 0. 6 $^{c}$ | 27.7 ± 4.1 $^{b}$ | 27.5 ± 1.5 $^{bc}$ | 21.7 ± 2.1 $^{ab}$ | 4.1 ± 0.2 $^{b}$ |
| 48 h (3) | 24.4 ± 1.2 $^{d}$ | 32.1 ± 4.9 $^{ab}$ | 33.9 ± 5.8 $^{ab}$ | 20.8 ± 1.6 $^{b}$ | 4.6 ± 0.4 $^{ab}$ |
| 30 d (4) | 30.7 ± 0. 4 $^{b}$ | 34.9 ± 2.1 $^{a}$ | 39.2 ± 1.1 $^{a}$ | 23.2 ± 0.6 $^{a}$ | 4.8 ± 0.3 $^{a}$ |

Table 6. *Cont.*

| CP Treatments | C15:0 | C15:1 | C16:0 | C18:0 | C18:1n9c |
|---|---|---|---|---|---|
| CK (0) | 4.6 ± 0.2 [c] | 6.1 ± 0.4 [b] | 585.3 ± 11.2 [a] | 175.3 ± 2.5 [c] | 271.1 ± 17.3 [a] |
| 0 h (1) | 5.3 ± 0.3 [ab] | 6.4 ± 0.3 [b] | 613.6 ± 60.1 [ab] | 197.1 ± 19.1 [b] | 267.7 ± 23.2 [a] |
| 24 h (2) | 4.8 ± 0.6 [bc] | 6.3 ± 0.2 [b] | 573.1 ± 34.1 [ab] | 208.1 ± 9.7 [b] | 222.1 ± 15.9 [bc] |
| 48 h (3) | 5.7 ± 0.1 [a] | 6.7 ± 0.4 [ab] | 543.2 ± 22.2 [b] | 205.9 ± 10.2 [b] | 204.5 ± 3.1 [c] |
| 30 d (4) | 5.8 ± 0.3 [a] | 7.1 ± 0.2 [a] | 594.8 ± 5.9 [a] | 231.7 ± 4.1 [a] | 217.5 ± 3.1 [b] |
| **CP Treatments** | **C18:1n9t** | **C18:2n6c** | **C18:3n3** | **C20:0** | **C20:1** |
| CK (0) | 0.1 ± 0.1[e] | 677.2 ± 28.7 [a] | 60.2 ± 0.5 [a] | 11.4 ± 0.3 [c] | 2.7 ± 0.4 [a] |
| 0 h (1) | 9.5 ± 0.7 [d] | 657.1 ± 64.1 [a] | 59.4 ± 3.1 [ab] | 12.7 ± 0.7 [ab] | 1.6 ± 0.2 [b] |
| 24 h (2) | 14.2 ± 0.4 [c] | 551.3 ± 41.4 [b] | 55.3 ± 1.6 [bc] | 11.1 ± 0.4 [c] | 0.5 ± 0.1 [c] |
| 48 h (3) | 15.6 ± 0.6 [b] | 498.4 ± 13.1 [b] | 54.1 ± 0.6 [c] | 11.1 ± 1.2 [bc] | 1.3 ± 0.3 [b] |
| 30 d (4) | 17.7 ± 1.1 [a] | 548.3 ± 4.9 [b] | 55.1 ± 0.3 [c] | 13.5 ± 0.1 [a] | 0.1 ± 0.1 [d] |
| **CP Treatments** | **C20:5n3** | **C22:6n3** | **C24:0** | **C24:1** | **Ratio** |
| CK (0) | 2.5 ± 0.1 [ab] | 5.2 ± 0.2 [a] | 14.1 ± 0.8 [a] | 9.1 ± 0.1 [bc] | 1.15 ± 0.04 [a] |
| 0 h (1) | 2.6 ± 0.2 [ab] | 3.7 ± 0.3 [bc] | 12.9 ± 0.7 [ab] | 9.5 ± 0.4 [b] | 1.07 ± 0.01 [b] |
| 24 h (2) | 2.3 ± 0.1 [b] | 3.1 ± 0.2 [d] | 13.1 ± 1.4 [ab] | 8.7 ± 0.3 [c] | 0.95 ± 0.95 [c] |
| 48 h (3) | 2.4 ± 0.1 [ab] | 3.2 ± 0.6 [cd] | 12.1 ± 0.7 [b] | 9.7 ± 0.5 [ab] | 0.91 ± 0.03 [cd] |
| 30 d (4) | 2.7 ± 0.2 [a] | 4.1 ± 0.1 [b] | 13.1 ± 0.6 [ab] | 10.3 ± 0.1 [a] | 0.87 ± 0.02 [d] |

Notes: CP, cold plasma; CK, the untreated sample; The number in bracket shows the CP treatment times; C6:0, Methyl caproate; C10:0, Decanoic acid; C12:0, Lauric acid; C14:0, Myristic acid; C14:1, Myristoleic acid; C15:0, Pentadecanoic acid; C15:1, *cis*-10-Pentadecenoic acid; C16:0, Palmitic acid;C18:0, Stearic acid; C18:1n9c, Oleic acid; C18:1n9t, Elaidic acid; C18:2n6c, Linoleic acid; C18:3n3, α-Linolenic acid; C20:0, Arachidic acid; C20:1, *cis*-11-Eicosenoic acid; C20:5n3, *cis*-5,8,11,14,17-Eicosapentataenoic acid; C22:6n3, *cis*-4,7,10,13,16,19-Docosahexaenoic acid; C24:0, Lignoceric acid; C24:1, Nervonic acid. FA, fatty acids; Ratio is the content ratio of unsaturated to saturated fatty acids. Data are expressed as mean± standard deviation (SD), number of repetitions—$n = 3$. Means with the different superscript letters in a column are different significantly ($p < 0.05$) among different CP treatment times.

Figure 2 shows the changes in the content of fatty acid components during CP repeatedly treated rice. With an increase in treatment times from one to three, total, unsaturated, monounsaturated, polyunsaturated FAs, and the ratio of unsaturated to saturated FAs (Table 4) decreased, but the fourth CP treatment significantly increased saturated FA compared with the untreated samples.

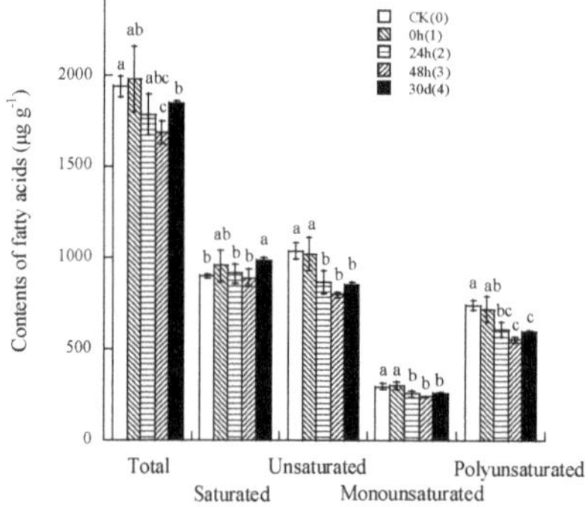


**Figure 2.** Changes in contents of fatty acid components during CP repeatedly treated rice. **Notes:** The different small letters for fatty acids species are different significantly ($p < 0.05$) among different CP treatment times.

### 3.7. Effect of CP Repeating Treatments on the Hygroscopic Property of Milled Rice

Table 7 shows the parameters of MCPE fitting to the EMC/ERH data of the CP repeatedly treated milled rice. MCPE fitted well to the MC/ERH data of CP repeatedly treated milled rice due to $R^2 > 0.97$ and MRE < 6.7%. Each coefficient of adsorption MCPE or desorption MCPE was similar among the CP repeating treatments. Figure 3 further confirms the same hygroscopic properties of CP repeatedly treated rice at the same temperature and relative humidity. For adsorption or desorption behavior of rice kernels, compared with the untreated samples, helium CP repeating treatments did not significantly increase the monolayer water content and solid surface area of water sorption in rice kernels (Table 8). At 25 °C and 70%ERH, the relative safe moisture content for the untreated sample, and the first, the second, the third, and the fourth CP treating samples were 14.36%, 14.49%, 14.57%, 14.49%, and 14.75% wet basis.

**Table 7.** Effect of CP repeating treatments on the moisture sorption equation Modified Chung-pfost (MCPE) in milled rice.

| Sorption | Cp Treatments | Mcpe Coefficients | | | Statistic Parameters | | | |
|---|---|---|---|---|---|---|---|---|
| | | A | B | C | RSS | SE | $R^2$ | MRE (%) |
| Ads. | CK (0) | 439.522 ± 49.717 [a] | 38.236 ± 6.122 [a] | 0.221 ± 0.006 [a] | 18.4802 | 0.4620 | 0.9747 | 6.0734 |
| | 0 h (1) | 449.323 ± 49.701 [a] | 39.451 ± 6.127 [a] | 0.219 ± 0.005 [a] | 17.3578 | 0.4339 | 0.9766 | 6.2175 |
| | 24 h (2) | 426.744 ± 46.921 [a] | 38.849 ± 6.041 [a] | 0.215 ± 0.005 [a] | 18.1666 | 0.4542 | 0.9763 | 6.6199 |
| | 48 h (3) | 427.376 ± 46.286 [a] | 37.201 ± 5.739 [a] | 0.218 ± 0.005 [a] | 17.9984 | 0.4499 | 0.9761 | 6.3093 |
| | 30 d (4) | 414.708 ± 41.339 [a] | 33.343 ± 4.846 [a] | 0.216 ± 0.005 [a] | 17.2339 | 0.4309 | 0.9774 | 6.1359 |
| Des. | CK (0) | 482.952 ± 41.797 [a] | 35.951 ± 4.389 [a] | 0.216 ± 0.004 [a] | 12.0133 | 0.2930 | 0.9834 | 3.6602 |
| | 0 h (1) | 502.329 ± 44.104 [a] | 36.613 ± 4.502 [a] | 0.216 ± 0.004 [a] | 12.0893 | 0.2949 | 0.9834 | 3.6179 |
| | 24 h (2) | 482.612 ± 43.177 [a] | 35.693 ± 4.507 [a] | 0.213 ± 0.004 [a] | 13.2659 | 0.3236 | 0.9822 | 3.7853 |
| | 48 h (3) | 481.107 ± 41.317 [a] | 34.809 ± 4.236 [a] | 0.215 ± 0.004 [a] | 12.3237 | 0.3006 | 0.9833 | 3.6594 |
| | 30 d (4) | 466.352 ± 36.156 [a] | 30.724 ± 3.457 [a] | 0.214 ± 0.004 [a] | 11.3338 | 0.2763 | 0.9848 | 3.549 |

Notes: CP, cold plasma; CK, the untreated sample; The number in bracket shows the CP treatment times; A, B, and C are the MCPE constants. Ads., adsorption; Des., desorption; *RSS*, residue sum of squares; *SE*, standard error; $R^2$, coefficient of determination; *MRE*, the mean relative percentage error. For moisture adsorptive or desorptive equation, A, B, and C are expressed as mean ± standard deviation (SD), number of observations—$n = 34$, the means with the different superscript letters in a column are different significantly ($p < 0.05$) among different CP treatment times.

**Table 8.** Effect of helium CP repeating treatments on the monolayer water content ($M_m$) and surface area values of the granular solids of rice kernels.

| CP Treatment (Times) | Adsorption | | Desorption | |
|---|---|---|---|---|
| | $M_m$ (% Dry Basis) | Surface Area ($m^2$/g Solid) | $M_m$ (% Dry Basis) | Surface Area ($m^2$/g Solid) |
| CK | 7.02 ± 0.51 [a] | 266.7 ± 18.0 [a] | 9.46 ± 0.82 [a] | 369.3 ± 29.4 [a] |
| 0 h (1) | 7.13 ± 0.50 [a] | 271.1 ± 17.9 [a] | 9.59 ± 0.81 [a] | 374.9 ± 28.9 [a] |
| 24 h (2) | 7.17 ± 0.53 [a] | 272.8 ± 18.8 [a] | 9.68 ± 0.86 [a] | 378.7 ± 30.6 [a] |
| 48 h (3) | 7.05 ± 0.50 [a] | 267.9 ± 17.9 [a] | 9.49 ± 0.53 [a] | 370.9 ± 29.4 [a] |
| 30 d (4) | 7.23 ± 0.52 [a] | 275.2 ± 18.6 [a] | 9.76 ± 0.88 [a] | 382.1 ± 31.3 [a] |

Notes: CK, the untreated sample; CP, cold plasma; The number in bracket shows the CP treatment times; Data are expressed as mean ± standard deviation (SD), number of observations—$n = 34$; Means with the different superscript letters in a column are different significantly ($p < 0.05$) among different CP treatment times.

### 3.8. Evaluation of Effect of CP onMilled Rice Using Principal Component Analysis

For the five samples with zero to four CP treatment times, the measured fifty-one indexes were used to principal component analysis. Figure 4 shows the loading plot of principal components after varimax rotation; the five samples were gathered together, suggesting that the effect of CP on milled rice presents difference with CP treatment times, but the difference is not huge. Further make factional scatterplot of principal components of the measured fifty-one parameters (Figure 5), contact angle, pasting parameters (PV, TV, FV, BD, SB), and fatty acid profile (C16:0; C18: 0; C18: 1n9c;

C18: 2n6c, total, saturated, unsaturated, monounsaturated, polyunsaturated FA) shows dispersed distribution, but other 36 parameters were gathered together. These results indicate that contact angle, pasting parameters and fatty acid profile in milled rice are quite sensitive to CP treatment.

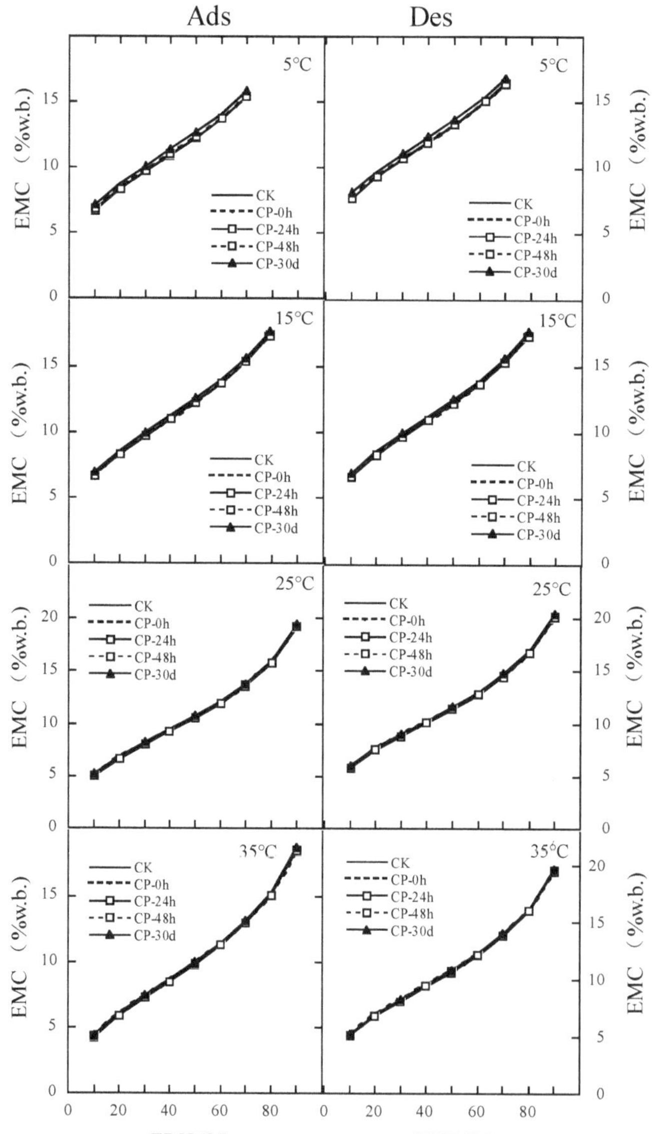


**Figure 3.** Effect of CP repeating treatments on the MCPE fitting moisture sorption isotherms of rice.

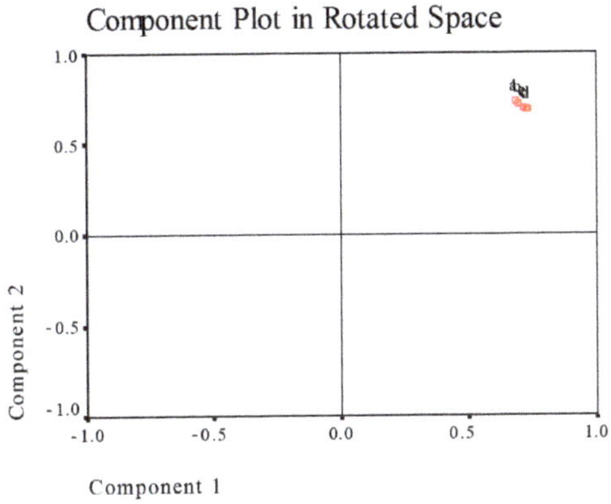


**Figure 4.** The loading plot of principal components after varimax rotation. Note: a, CK sample; b, 0 h sample; c, 24 h sample; d, 48 h sample; e, 30 d sample.

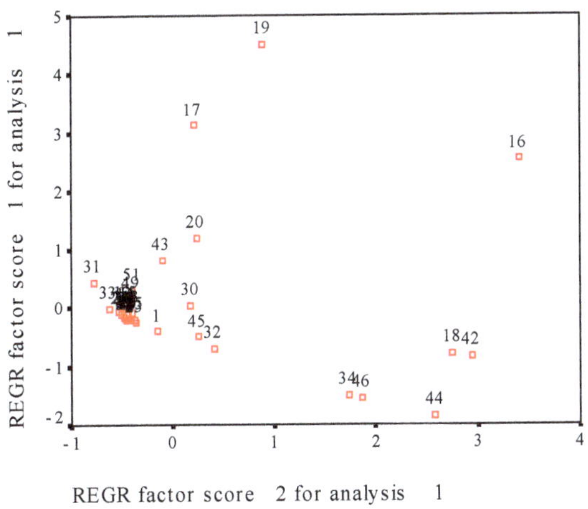


**Figure 5.** The factional scatterplot of principal components of the measured fifty-one parameters in this study. Notes: 1,water contact angle; 2, cooking time; 3, water absorption rate; 4, freshness index; 5, pH value; 6, ΔH; 7, $T_p$; 8, $T_o$; 9, $T_c$; 10, peak width; 11, peak enthalpy; 12, water SRC; 13, $NaHCO_3$ SRC; 14, Lactic acid SRC; 15, sucrose SRC; 16, PV; 17, TV; 18, BD; 19, FV; 20, SB; 21, peak time;22, PT; 23, C6:0; 24, C10:0; 25, C12:0; 26, C14:0; 27, C14:1; 28, C15:0; 29, C15:1; 30, C16:0; 31, C18:0; 32, C18: 1n9c; 33, C18:1n9t; 34, C18: 2n6c; 35, C18:3n3; 36, C20: 0; 37, C20:1; 38, C20:5n3; 39, C22:6n3; 40, C24: 0; 41, C24:1; 42, Total FA; 43, saturated FA; 44, unsaturated FA; 45, mono-unsaturated FA; 46, polyunsaturated FA; 47, ratio of unsaturated to saturated FA; 48, Mm-ads; 49, SA-des; 50, Mm-ads; 51, SA-des.

## 4. Discussion

Clarifying the underlying physical and biochemical mechanisms of plasma-cereals interaction is crucial for maximizing desired results. In combination of low-pressure radiofrequency CP power (13.56 MHz, 140 Pa) and treatment time, our previous report showed 120 W helium CP treatment 20s on 300 g rice sample had optimal results based on the cooking properties, cooked rice texture, and rice appearance quality [20], this study further confirm how long the effect of CP treatment can be kept. In the present study, the pH value of milled rice remained unchanged to 120 W helium CP treatment times; this result is similar to the argon or oxygen cold plasma from kiNPen 09 DC plasma jet on gellan gum and the air cold plasma from AC corona discharge plasma jet on milk [29,30].The effect of CP on the pH of foods depends on food resources and cold plasma species. The Chinese scientists Yan et al. [31] has measured the emission spectra of helium plasma using dielectric barrier discharge (DBD) experiment device and found that the intensity of spectral line $3^1P_1 \rightarrow 2^1S_0$ was the strongest at a wavelength of 501.6 nm, possible indication of high density of meta-stable helium atoms. When high-purity helium was adopted in dielectric discharge apparatus of the cold plasma, the helium plasma with high-energy photons (17.7 eV) could be generated [32]. The meta-stable helium atoms and high-energy photons did not change pH values in the helium CP repeatedly treated rice, then their effect on rice thermal and thermodynamic properties were analyzed in this study.

DSC is a sensitive tool for characterizing starch retrogradation, and it gives transition temperature and enthalpy change ($\triangle H$) in crystallite melting [33]. In the present study, the increase in $T_p$ induced by helium CP repeating treatments on rice kernels might indicate an increase in degree of rice starch crystallinity and a possible role of CP cross-linking [34]. The gelatinization range ($T_c$–$T_o$) insignificantly decreased with increasing helium CP treatment times, this peak-narrowing trend may be related to an easy availability of water to rice starch induced by helium CP treatments.

RVA can give the pasting viscosities during program heating and cooling of rice flour suspension [33]. Chaiwat et al. [35] used 60 W Argon (99.999% purity) CP in a semi-continuous downer reactor to treat tapioca starch for 30 min, the peak viscosity (PV) and breakdown viscosity (BD) of starch paste decreased, but gel-holding intensity and final viscosity (FV) increased, and setback viscosity (SB) and pasting temperature remained unchanged. In the present study, with an increase in helium CP treatment times on rice kernels, peak viscosity (PV), breakdown viscosity (BD), and setback viscosity (SB) in rice flour paste decreased while peak time increased, whereas trough viscosity (TV), final viscosity (FV), and pasting temperature (PT) remained unchanged. Because setback viscosity is defined as the difference between FV and TV, its magnitude could reflect the retrogradation tendency of amylose in a starch paste [36]. With an increase in CP treatment times on rice kernels, the decrease in setback viscosity might indicate CP repeating rice kernels has a less tendency for starch retrogradation. With an increase in CP treatment times on rice kernels, an insignificant increase in pasting temperature suggests that the DSC-measured $T_p$ is more sensitive than the pasting temperature from RVA curves.

Solvent retention capacity (SRC) could present swelling in four solvents that differ in their compatibility to three main polymers: 5% sodium bicarbonate to assess swelling due to starch damage, 5% lactic acid to assess gluten swelling, 50% sucrose to assess arabinoxylan-mediated swelling, and swelling in pure water influenced by all three components [37]. In the present study, the SRC values were determined to determine whether rice flour functional polymers are affected by helium plasma treatments. Compared with the untreated sample, the first three times of CP treatments significantly increased sucrose SRC and might induce arabinoxylan swelling.

The increase in $R_{1047/1022}$ acquired from deconvoluted IR spectra, probably indicating the formation of relative more ordered domains in starch granules, were observed for DBD plasma or radiofrequency plasma treated starches [38]. However, Li et al. [39] observed the similar $R_{1047/1022}$ values between DBD plasma treated chickpea starch and the native starch. The present study showed the similar $R_{1047/1022}$ values for low-pressure radiofrequency helium plasma repeatedly treated rice and the untreated samples, and the $R_{1047/1022}$ value for rice flour treated by the same cold plasma should be further measured.

Albertos et al. [40] showed a decrease in contents of oleic acid (C18:1, n-9) and eicosapentaenoic acid (20:5, n-3) in mackerel fillets after CP treatment. In the present study, compared with the untreated sample, following the increase in CP treatment times, C18:1n9c significantly decreased and C18:1n9t increased, but C20:5n3 remained unchanged. The active species during plasma discharge can initiate lipid peroxidation and produce hydroperoxide, which may be further converted into aldehydes or shorter-chain fatty acid compounds [41]. The helium CP repeating treatments increased the contents of decanoic acid (C10:0), lauric acid (C12:0), and myristic acid (C14:0) in rice while methyl caproate

(C6:0) decreased. The formation (<18 μg/g) of trans-oleic acid in rice induced by low-pressure RF helium was significantly less than the trans-fatty acid limit (3 mg/g), stipulated by China national standard GB28050–2011 [42].

Thermodynamic properties of food can have an insight into the microstructure associated with the food–water interface, and the physical structure and theoretical surface area play roles in determining the water-binding properties of a particular product [26]. We measured the solid surface area of rice water sorption to explore this relationship. In the range of 5 to 35 °C and 10–90% ERH, the monolayer water content and solid surface area of water sorption were 7.02% d.b. and 266.7 $m^2/g$ solid for adsorption, and 9.46% d.b. and 369.3$m^2/g$ solid for desorption, respectively. The helium CP repeating treatments did not significantly increase these values. It is very interesting that helium CP could increase the hydrophilic property and water absorption rate of rice kernels through its etching action, but it kept the monolayer water content and solid surface area of water sorption. In the measured fifty-one parameters of the present study, principal component analysis shows that water contact angle, pasting parameters and fatty acid profile in milled rice are quite sensitive to CP repeated treatment. The interaction of reactive plasma species with rice surface is a complex process that contains mass and energy transfer. Starch as a sugar polymer is difficult to be polarized, but acidic or basic amino acids in proteins are easily polarized. Further research will determine the protein conformational changes and amino acid composition in helium CP treated rice.

## 5. Conclusions

The intermittent one to four treatments of low-pressure RF helium CP on rice kernels increased hydrophilicity, water absorption rate, freshness index, and gelatinization peak temperature, but they maintained pH value, gelatinization enthalpy, FTIR spectrum, and hygroscopic properties, and decreased rice cooking time and peak, breakdown, and setback viscosity of rice flour paste. Contents of the main monounsaturated FA (C18:1n9c) and polyunsaturated FA (C18:2n6c, C18:3n3) in rice kernels were kept by the first helium CP treatment but were reduced by more CP treatments whilst the content ratio of unsaturated to saturated fatty acids decreased. The first three helium CP treatments can increase water and sucrose SRC values. The principal component analysis shows that water contact angle, pasting parameters and fatty acid profile in milled rice are quite sensitive to CP repeated treatment. It is concluded that the effect of low-pressure RF helium CP treatment on rice kernels is perdurable, and the improvement of helium CP repeating treatments on rice cooking and pasting properties is additive. The helium CP could increase the hydrophilic property and water absorption rate of rice kernels through its etching action, but it kept the monolayer water content and solid surface area of water sorption. This study is useful for using CP treatments to improve rice quality.

**Author Contributions:** Z.C., data collection, methodology, writing, editing; X.L., writing, reviewing and editing, methodology, investigation, supervision, results interpretation, data curation; H.S., supervision, reviewing; Y.J., methodology, data collection; C.L., methodology, data collection. All authors have read and agreed to the published version of the manuscript.

**Funding:** This research received was funded by the Operating Expenses of Basic Scientific Research Project of Central Public-interest Scientific Institution, China [JY2007].

**Data Availability Statement:** The original contributions presented in the study are included in the article, further inquiries can be directed to the corresponding author.

**Conflicts of Interest:** The authors declare that there are no conflict of interest.

## References

1. Zadeh, J.H.; Pazir, F. Invesigation of the potential application of cold plasma technology in food safety. *GIDA* **2023**, *48*, 614–626. [CrossRef]
2. Warne, G.R.; Williams, P.M.; Pho, H.Q.; Tran, N.N.; Hessel, V.; Fisk, I.D. Impact of cold plasma on the biomolecules and organoleptic properties of foods: A review. *J. Food Sci.* **2021**, *86*, 1–16. [CrossRef] [PubMed]

3. Mehr, H.M.; Koocheki, A. Effects of short-term and long-term cold plasma treatment on the color, structure, and pickering foaming properties of grass pea protein particles. *Food Hydrocoll.* **2023**, *143*, 108846. [CrossRef]
4. Ekezie, F.C.; Sun, D.W.; Cheng, J.H. A review on recent advances in cold plasma technology for the food industry: Current applications and future trends. *Trends Food Sci. Technol.* **2017**, *69*, 46–58. [CrossRef]
5. Chen, H.H.; Chang, H.C.; Chen, Y.K.; Huang, C.L.; Lin, S.Y.; Chen, Y.S. An improved process for high nutrition of germinated brown rice production: Low-pressure plasma. *Food Chem.* **2016**, *191*, 120–127. [CrossRef]
6. Thirumdas, R.; Deshmukh, R.R.; Annapure, U.S. Effect of low temperature plasma processing on physicochemical properties and cooking quality of basmati rice. *Innov. Food Sci. Emerg. Technol.* **2015**, *31*, 83–90. [CrossRef]
7. Murakam, T.; Niemi, K.; Gans, T.; O'Connell, D.; Graham, W.G. Chemical kinetics and reactive species in atmospheric pressure helium-oxygen plasma with humid air impurities. *Plasma Sources Sci. Technol.* **2013**, *22*, 1–29. [CrossRef]
8. Dong, S.; Gao, A.; Xu, H.; Chen, Y. Effects of dielectric barrier discharge (DBD) cold plasma treatment on physicochemical and structural properties of zein powders. *Food Bioprocess Technol.* **2017**, *10*, 434–444. [CrossRef]
9. Ji, H.; Dong, S.; Han, F.; Li, Y.; Chen, G.; Li, L.; Chen, Y. Effects of dielectric barrier discharge (DBD) cold plasma treatment on physicochemical and functional properties of peanut protein. *Food Bioprocess Technol.* **2018**, *11*, 344–354. [CrossRef]
10. Segat, A.; Misra, N.N.; Cullen, P.J.; Innocente, N. Atmospheric pressure cold plasma (ACP) treatment of whey protein isolate model solution. *Innov. Food Sci. Emerg. Technol.* **2015**, *29*, 247–254. [CrossRef]
11. Rowan, N.J.; Espie, S.; Harrower, J.; Anderson, J.G.; Marsili, L.; MacGregor, S.J. Pulsed-plasma gas-discharge inactivation of microbial pathogens in chilled poultry wash water. *J. Food Prot.* **2007**, *70*, 2805–2810. [CrossRef] [PubMed]
12. Pal, P.; Kaur, P.; Singh, N.; Misra, N.N.; Tiwari, B.K.; Cullen, P.J.; Virdi, A.S. Effect of nonthermal plasma on physicochemical, amino acid composition, pasting and protein characteristics of short and long grain rice flour. *Food Res. Int.* **2016**, *81*, 50–57. [CrossRef]
13. Muhammad, A.I.; Liao, X.J.; Cullen, P.J.; Liu, D.; Xiang, Q.; Wang, J.; Chen, S.; Ye, X.; Ding, T. Effects of nonthermal plasma technology on functional food components. *Compr. Rev. Food Sci. Food Saf.* **2018**, *17*, 1379–1394. [CrossRef] [PubMed]
14. Zhang, M.; Oh, J.K.; Cisneros-Zevallos, L.; Akbulut, M. Bactericidal effects of nonthermal low-pressure oxygen plasma *S. typhimurium* LT2 attached to fresh produce surfaces. *J. Food Eng.* **2013**, *119*, 425–432. [CrossRef]
15. Bahrami, N.; Bayliss, D.; Chope, G.; Penson, S.; Perehinec, T.; Fisk, I.D. Cold plasma: A new technology to modify wheat flour functionality. *Food Chem.* **2016**, *202*, 247–253. [CrossRef]
16. Yepez, X.V.; Baykara, H.; Xu, L.; Keener, K.M. Cold plasma treatment of soybean oil with hydrogen gas. *J. Am. Oil Chem. Soc.* **2021**, *98*, 103–113. [CrossRef]
17. Zhang, J.; Du, Q.; Yang, Y.; Zhang, J.; Han, R.; Wang, J. Research prgress and future trends of low temperature plasma application in food industry: A review. *Molecules* **2023**, *28*, 4714. [CrossRef]
18. Usman, I.; Afzaal, M.; Imran, A.; Saeed, F.; Afzal, A.; Ashfaq, I.; Ahah, Y.A.; Islam, F.; Azam, I.; Tariq, I.; et al. Recent updates and perspectives of plasma in food processing: A review. *Int. J. Food Prop.* **2023**, *26*, 552–566. [CrossRef]
19. Misnal, M.F.I.; Redzuan, N.; Zainal, M.N.F.; Ahmad, N.; Ibrahim, K.R.; Agun, L. Cold plasma: A potential alternative for rice grain postharvest treatment management in Malaysia. *Rice Sci.* **2022**, *29*, 1–15. [CrossRef]
20. Liu, J.J.; Wang, R.L.; Chen, Z.T.; Li, X.J. Effect of cold plasma treatment on cooking, thermomechanical and surface structural properties of Chinese milled rice. *Food Bioprocess Technol* **2021**, *14*, 866–886. [CrossRef]
21. Chen, Z.T.; Wang, R.L.; Li, X.J.; Zhu, J.; Xu, Y.N.; Liu, J.J. Sorption equilibrium moisture and isosteric heat of cold plasma treated milled rice. *Innov. Food Sci. Emerg. Technol.* **2019**, *55*, 35–47. [CrossRef]
22. Gao, G.B.; Wu, J.Z.; Wei, Z.; Li, X.J. Effects of low pressure radio-frequency cold plasma on the biochemical parameters and fatty acid profile of wheat flours. *Cereal Chem.* **2023**, *110*, 393–413. [CrossRef]
23. Tao, L.S.; Qin, W.; Wei, Z.; Li, X.J.; Zhang, H.Q. Effect of small-scale storage on the cooking property and fatty acid profile. *Appl. Food Res.* **2022**, *2*, 100175. [CrossRef]
24. Takashi, M.; Koji, K.; Tatsuhiko, O. Method of Measuring Freshness of Cereals and Beans and Apparatus Therafor. *EP 1612542A1. 1 April* 2006.
25. *AACC International Approved Methods of Analysis*, 11th ed.; Methods: 54-21.01 (Rheological Behaviour of Flour by Farinograph), 56-11.02 (Solvent Retention Capacity Profile); American Association of Cereal Chemists International: St. Paul, MN, USA, 2012.
26. Li, X.J.; Han, X.; Tao, L.S.; Jiang, P.; Qin, W. Sorption equilibrium moisture and isosteric heats of Chinese wheat bran products added to rice to increase its dietary fiber content. *Grain Oil Sci. Technol.* **2021**, *4*, 149–161. [CrossRef]
27. SPSS Inc. *SPSS for Windows*; Release 17.0.1; SPSS Inc.: Chicago, IL, USA, 2017.
28. Moraes, M.A.; Rosa, G.S.; Pinto, L.A. Moisture sorption isotherms and thermodynamic properties of apple Fuji and garlic. *Int. J. Food Sci. Technol.* **2008**, *43*, 1824–1831. [CrossRef]
29. Surowsky, B.; Fischer, A.; Schlueter, O.; Knorr, D. Cold plasma effects on enzyme activity in a model food system. *Innov. Food Sci. Emerg. Technol.* **2013**, *19*, 146–152. [CrossRef]
30. Gurol, C.; Ekinci, F.Y.; Aslan, N.; Korachi, M. Low temperature plasma for decontamination of E. *coli in milk. Int. J. Food Microbiol.* **2012**, *157*, 1–5. [CrossRef]
31. Yan, Y.H.; Feng, X.P.; Tang, X.L.; Xie, H.K. The basic character of emission spectra of He plasma from normal pressure to ultra-low pressure. *Chin. J. Light Scatt.* **2004**, *16*, 242–245. (In Chinese) [CrossRef]

32. Franchina, F.A.; Maimone, M.; Sciarrone, D.; Purcaro, G.; Tranchida, P.Q.; Mondello, L. Evaluation of a novel helium ionization detector within the context of (low-)flow modulation comprehensive two-dimensional gas chromatography. *J. Chromatogr. A.* **2015**, *1402*, 102–109. [CrossRef]
33. Wang, S.J.; Li, C.L.; Copeland, L.; Niu, Q.; Wang, S. Starch retrogradation: A comprehensive review. *Compr. Rev. Food Sci. Food Safety.* **2015**, *14*, 568–584. [CrossRef]
34. Thirumdas, R.; Kadam, D.; Annapure, U.S. Cold plasma: An alternative technology for the starch modification. *Food Biophys.* **2017**, *12*, 129–139. [CrossRef]
35. Chaiwat, W.; Wongsagonsup, R.; Tangpanichyanon, N.; Jariyaporn, T.; Deeyai, P.; Suphantharika, M.; Fuongfuchat, A.; Nisoa, M.; Dangtip, S. Argon plasma treatment of tapioca starch using a semi-continuous downer reactor. *Food Bioprocess Technol.* **2016**, *9*, 1125–1134. [CrossRef]
36. Zaidul, I.S.; Norulaini, N.A.; Omar, A.K.; Yamauchi, H.; Noda, T. RVA analysis of mixtures of wheat flour and potato, sweet potato, yam, and cassava starches. *Carbohydr. Polym.* **2007**, *69*, 784–791. [CrossRef]
37. Kweon, M.; Slade, L.; Levine, H. Solvent retention capacity (SRC) testing of wheat flour: Principles and value in predicting flour functionality in different wheat-based food processes and in wheat breeding—A review. *Cereal Chem.* **2011**, *88*, 537–552. [CrossRef]
38. Sifuentes-Nieves, I.; Mendez-Montealvo, G.; Flores-Silva, P.C.; Nieto-Pérez, M.; Neira-Velazquez, G.; Rodriguez-Fernandez, O.; Hernández-Hernández, E.; Velazquez, G. Dielectric barrier discharge and radiofrequency plasma effect on structural properties of starches with different amylose content. *Innov. Food Sci. Emerg. Technol.* **2021**, *68*, 102630. [CrossRef]
39. Li, R.H.; Li, X.Y.; Tu, A.; Jing, C.T.; Dong, S.; Chen, S.F.; Guo, P. Effects of dielectric barrier discharge plasma treatment on the structural orders and functional properties of chickpea starch. *Int. J. Food Sci. Technol.* **2022**, 1–13. [CrossRef]
40. Albertos, I.; Martin-Diana, A.; Cullen, P.; Brijesh-Kumar, T.; Ojha, S.; Bourke, P.; Álvarez, C.; Rico, D. Effects of dielectric barrier discharge (DBD) generated plasma on microbial reduction and quality parameters of fresh mackerel (*Scomber scombrus*) fillets. *Innov. Food Sci. Emerg. Technol.* **2017**, *44*, 117–122. [CrossRef]
41. Thirumdas, R. Partial hydrogenation of oils using cold plasma technology and its effect on lipid oxidation. *J. Food Sci. Technol.* **2023**, *60*, 1674–1680. [CrossRef]
42. *China Nation Standard, GB 28050-2011*; Food Safety Nation Standard–General Rules for Nutrition Labeling of Prepackaged Foods. China National Standardization Administration: Beijing, China, 2011.

*foods*

MDPI

*Article*

# Characterization of Volatile Flavor Compounds and Aroma Active Components in Button Mushroom (*Agaricus bisporus*) across Various Cooking Methods

**Limei Xie [1], Shaoli Guo [1], Hongting Rao [1], Bingying Lan [1], Baodong Zheng [1,2] and Ningning Zhang [1,2,*]**

[1] College of Food Science, Fujian Agriculture and Forestry University, Fuzhou 350002, China
[2] Fujian Provincial Key Laboratory of Quality Science and Processing Technology in Special Starch, Fuzhou 350002, China
* Correspondence: znnfst@163.com

**Abstract:** To investigate the impact of various cooking methods on the volatile aroma compounds of button mushroom, gas chromatography-mass spectrometry (GC-MS) and electronic nose (E-nose) were utilized for aroma analysis. The results indicated that the E-nose was able to effectively distinguish between the samples prepared using different cooking methods. In the raw, steamed, boiled and baked samples, 37, 23, 33 and 35 volatiles were detected, respectively. The roasting process significantly contributed to the production of flavor compounds, giving button mushroom its distinctive flavor. Sixteen differential aromas were identified based on the *p*-value and VIP value. Additionally, the cluster analysis of differential aroma substances revealed a stronger odor similarity between the steamed and raw groups, consistent with the results of the OPLS-DA analysis of overall aroma components. Seven key aromas were identified through OAV analysis and omission experiments. In addition, 1-octen-3-one was identified as the main aroma component of cooked button mushroom. The findings of the study can be valuable for enhancing the flavor of cooked button mushroom.

**Keywords:** button mushroom; cooking methods; GC-MS; E-nose; OAV; aroma recombination; aroma omission; aroma active compounds

## 1. Introduction

In recent years, button mushroom (*Agaricus bisporus*) has been widely studied for its flavor, high amino acid score, biological value and essential amino acid index, both in terms of medicinal and economic benefits [1–3]. Button mushrooms are typically not consumed raw but instead undergo various cooking methods such as frying, baking, steaming and boiling before being eaten. Although button mushroom is widely used in culinary applications, there is still a scarcity of studies focusing on the alterations in flavor compounds during its culinary preparation.

Different cooking methods can impart distinct effects on the aroma of food due to differences in temperature, heat transfer and heating medium [4]. In their research, Zhou et al. [5] explored the impact of boiling, steaming, microwaving, deep-frying and pressure cooking on the volatile constituents of shiitake mushrooms and found that pressure cooking and deep-frying produced more aroma active compounds. Similarly, Cheng et al. [6] observed that gas-frying treatments significantly contributed to the creation of flavorful substances, imparting a unique flavor to Tibetan pork. Aroma is a fundamental characteristic of food, and the cooking operation primarily affects the overall aroma of food through mechanisms such as the Maillard reaction [7], lipid oxidation, amino acid degradation [8] and their combined reactions [9]. As of now, approximately 80 volatile compounds, such as aldehydes, ketones, alcohols, acids, terpenes, esters and heterocyclic compounds, have been identified in raw button mushroom [10]. Although button mushroom does not have a strong aroma on its own, the process of cooking can lead to Strecker degradation and Maillard reactions, generating a variety of volatile and non-volatile intermediates, which contribute to the development of a delicious mushroom flavor [11,12]. However, there is little literature that systematically investigates the changes in the flavor of button mushroom during steaming, boiling and roasting.

GC-MS is the most widely used technique for detecting flavor substances, while the E-nose is an instrument comprising a series of electronic chemical sensors with partial specificity. Recently, E-nose and GC-MS methods have been widely used to analyze the volatile components of various food products to identify different species, qualities and processing and storage conditions, such as *Pleurotus eryngii* [13], skim milk [14], low-salt fermented sour fish [15], wine [16], fruits [17] and vegetables [18]. In addition, the E-nose sensor accurately responded to eight different dried commercial mushrooms, indicating its potential as a tool for identifying edible mushrooms [19]. Simultaneously, E-nose has been utilized to distinguish between various cultivated fresh mushrooms, such as *Flammulina velutipes*, *Agaricus bisporus*, *Lentinus edodes*, *Pleurotus eryngii*, *Grifola frondosa* and *Hypsizygus marmoreus* [20]. Gómez et al. utilized an E-nose, GC-MS technique and sensory panel method to analyze the odor of the freeze-dried powder of cultivated *Agaricus bisporus*. The results indicated that the E-nose has analytical utility and can be employed for assessing different types of freeze-dried *Agaricus bisporus* powder [21]. These studies showed that E-noses are highly sensitive in detecting a wide range of food products, and the aroma results obtained through both techniques are strongly correlated. Given that button mushroom is a widely consumed edible mushroom, it undergoes flavor changes during cooking. Consequently, it is essential to comprehend the impact of various cooking methods on the flavor of button mushroom.

Accordingly, the present study selected button mushroom treated with different cooking methods and analyzed their aroma characteristics using E-nose and GC-MS techniques. This study aimed to enrich the existing research on the effects of cooking methods on the flavor and aroma of edible mushrooms, providing additional insights into how cooking affects the aroma of food.

## 2. Materials and Methods

### *2.1. Chemicals*

For identification and quantitation of aroma compounds, the following authentic flavor standards were purchased from Macklin Biochemical Technology Co., Ltd., Shanghai, China: methyl valerate (99%), isovaleraldehyde (98%), furfuryl mercaptan (98%), heptanal (97%), hexanal (99%) and 2-butylfuran (98%). Benzyl alcohol (AR, 99%), 1-octen-3-one (98%), 1-octen-3-ol (98%), methyl palmitate (AR, 97%) and benzoic acid (99%) were purchased from Yuanye Bio-Technology Co., Ltd., Shanghai, China. Phenylacetaldehyde (95%) and benzaldehyde (AR, ≥98.5%) were obtained from Sinopharm Chemical Reagent Co., Ltd., Beijing, China. The internal standard, 1-decanol (GC, ≥99.5%), was purchased from Aladdin Biochemical Technology Co., Ltd., Shanghai, China.

### *2.2. Button Mushroom Samples and Cooking Methods*

Button mushrooms were purchased from the local Yonghui supermarket (Fuzhou, China). Fresh button mushrooms with non-rotting, non-moldy bodies were carefully selected and washed under running water to remove any loose growth medium from the cap. The washing process lasted for 5–10 s to avoid excessive water retention. The button mushrooms were rinsed and placed in a steel strainer metal bowl to drain well, and excess water was absorbed with a paper towel before removing their stalks and discarding them. Using a slicer, the caps were sliced longitudinally into thin slices, approximately 0.5 cm thick. The first and last slices were discarded due to uneven cooking compared to the other caps. All the sliced mushrooms were randomly divided into four groups, one of which was raw, and the rest were cooked using three methods (steaming, boiling and baking).

We chose cooking conditions according to the methods of Sun et al. [22], Lee et al. [23] and Irene et al. [24] with slight modifications, and the specific cooking treatment conditions were as follows:

1. Steamed: Room temperature water was added at a 1:20 (g/mL) ratio in a steamer. The water was heated to boiling, and the mushroom slices were placed on the steamer rack in the steamer and steamed for 7 min.
2. Boiled: Room temperature water was added at a ratio of 1:20 (g/mL) in a pan and heated on an induction cooker (Guangdong Midea Life Electric Appliance Manufacturing Co., Foshan, China) until the water boiled. Then, the sliced mushrooms were poured into the pan and were cooked for 7 min.
3. Baked: The oven (Guangdong Midea Life Electric Appliance Manufacturing Co., Foshan, China) was preheated to 163 °C, and the mushroom slices were baked in the 163 °C oven for about 4 min on each side, totaling 8 min.

The cooking time was optimized through systematic experiments based on sensory evaluation and time efficiency. Initial tests covered baking durations of 5, 7, 10 and 15 min, followed by trials at intervals of 7, 8, 9 and 10 min. An 8 min baking time was found to offer superior sensory scores and eliminate raw mushroom flavor. Similar methodologies were applied in determining steaming and boiling times.

After cooking, a portion of all 4 samples (raw, boiled, steamed and baked mushroom) was used for subsequent experiments.

### 2.3. E-Nose Measurement

The flavor characteristics of button mushroom were analyzed according to the method of Guo et al. [25] with minor modifications. Button mushroom samples (2.00 g) were placed in a 20 mL headspace bottle at room temperature for 45 min. The E-nose (PEN3 Airsense, Schwerin, Germany) contains 10 metal oxide gas sensors: W1C (aromatic), W5S (nitrogen oxides), W3C (ammonia and aromatic), W6S (hydrogen), W5C (alkanes and aromatic), W1S (short-chain alkanes), W1W (inorganic sulfur), W2S (alcohols, ethers, aldehydes and ketones), W2W (organic sulfur) and W3S (long-chain alkanes). The E-nose test conditions were as follows: sensor cleaning time of 120 s, injection preparation waiting time of 5 s; sampling time of 120 s, injection flow rate of 400 mL/min and operating ambient temperature of 26 °C. The signal from 81 s to 83 s of the smoothed data was utilized as the time point for the E-nose analysis. Each sample was measured in triplicate, and the average value was used for further analysis.

### 2.4. Headspace Solid-Phase Micro-Extraction (HS-SPME) of Aroma Compounds

By referring to previous studies with some modifications, we employed HS-SPME method to determine volatile compounds in button mushroom [26]. A total of 4 g of samples was homogenized with 10 mL of distilled water, and 5 mL of slurry was mixed with 10 μL (165.94 μg/mL) of 1-decanol internal standard solution before being sealed in a 50 mL vial using a headspace jaw cover with spacer. Nest, the samples were shaken in a 60 °C water bath for 30 min. To extract volatile compounds from button mushroom, a 75 μm divinylbenzene/carboxen/polydimethylsiloxane (DVB/CAR/PDMS, 75 μm) (Agilent Technologies Inc., Santa Clara, CA, USA) fiber was used and preconditioned at 250 °C for 30 min prior to analysis. The aged HS-SPME fiber was exposed to the sample for 40 min at 60 °C to collect the analytes. Afterward, the fiber was inserted into the injection port of the GC-MS system for thermal desorption.

### 2.5. GC-MS Analysis

GC-MS analysis was performed on the Agilent 8860/5977B GC-MS instrument (Agilent Technologies Inc., Santa Clara, CA, USA). Analytes were separated on a DB-5ms capillary column (30 m × 0.25 mm × 0.25 μm) from Agilent Technologies Inc., Santa Clara, CA, USA, in nonsplit mode. The heating-up procedure was as follows: held at 35 °C for 5 min, raised to 100 °C at 5 °C/min, held for 1 min, raised to 230 °C at 10 °C/min and held for 10 min. The total time of the procedure was 42 min. The carrier gas was helium (purity not less than 99.999%) at a constant flow rate of 1.4 mL/min. Mass spectra were acquired in an electron impact mode. MS was taken at 70 eV ionization energy in the 35–550 amu mass range, with a solvent delay of 1.00 min. The ion source temperature was 230 °C, with the quadrupole temperature set to 150 °C. All experiments were performed in triplicate.

Compounds with a match percentage greater than 85% were selected for confirmation based on the NIST standard mass spectrometry data. Through comparison with the peak area of the internal standard, the content of the aroma substance can be quantitatively calculated in μg/g using the following formula: Aroma substance content = (peak area of aroma substance × content of internal standard)/peak area of internal standard.

### 2.6. Calculation of Odor Activity Values (OAVs)

The odor activity value (OAV) was employed to evaluate the contribution of each volatile compound to the overall aroma, thereby enabling a better understanding of the significance of individual aroma substances in the overall aroma profile. The OAV was calculated as the ratio of the semiquantitative analysis result of a single volatile compound to the perception threshold of the substance in water, expressed as OAV = aroma component content/odor threshold. Thresholds and odorant descriptions for volatile compounds were obtained from http://www.odour.org.uk (accessed on 19 February 2024) and the *Compilations of Odour Threshold Values in Air, Water and Other Media*. In general, compounds with OAV greater than or equal to 1 contribute significantly to the overall aroma, while compounds with OAV less than 1 contribute less [27].

### 2.7. Aroma Recombination and Omission Experiments

To further validate the key aroma active compounds of button mushroom, the aroma was reconstituted and compared with the original button mushroom. Aroma reorganization and omission experiments were performed with appropriate modifications based on the methods of Liu et al. [28] and Fan et al. [29]. In summary, the reconstitution experimental model comprised a standard aroma substance with an OAV greater than 1 and ultrapure water, with all compounds present in natural amounts. Further omission experiments were conducted based on the results of the aroma reorganization experiments. By removing a compound from the recombination model, an omission model was obtained. We could then use the E-nose to evaluate the distinctions between each omission model and the recombination model.

### 2.8. Statistical Analysis

Every experiment was performed in triplicate, and results were expressed as the mean value ± standard deviation (SD) of three replicates. The estimation of differences between means was conducted using SPSS 24.0 (SPSS Inc., Chicago, IL, USA) for one-way ANOVA ($p < 0.05$). Radar map was drawn through Origin 2023 (OriginLab Co., Northampton, MA, USA). Clustered heat map analysis was performed using https://www.omicstudio.cn/tool (accessed on 19 February 2024). Simca 14.1 (Umetrics AB, Umea, Vasterbotten, Sweden) was used for OPLS-DA (the scaling type was Par) analyses. Additionally, we employed supervised regression modeling using OPLS-DA to obtain variable importance in projection (VIP). Differential aroma components were screened at $p < 0.05$, VIP > 1.

## 3. Results and Discussion

### 3.1. E-Nose Analysis

The E-nose is an electronic chemical sensing system that simulates the human sense of smell to rapidly analyze and detect complex volatile components in food. It offers the advantages of accuracy, sensitivity and objectivity [19]. From Figure 1a, the response values of the 10 sensors to the volatile components of button mushroom with different cooking methods were observed to be different. The samples exhibited the strongest response values at sensors W1W and W1S, followed by W2W and W2S, which suggested that sulfides, short-chain alkanes, alcohols, ethers, aldehydes and ketones were the primary volatiles of button mushroom. Meanwhile, high levels of aldehydes and alcohols were found in the roasted group in the GC-MS test. This outcome could be linked to the Maillard reaction, where polysaccharides and proteins in button mushroom undergo chemical changes at high temperatures, resulting in the generation of numerous aromatic compounds during the roasting process [30].

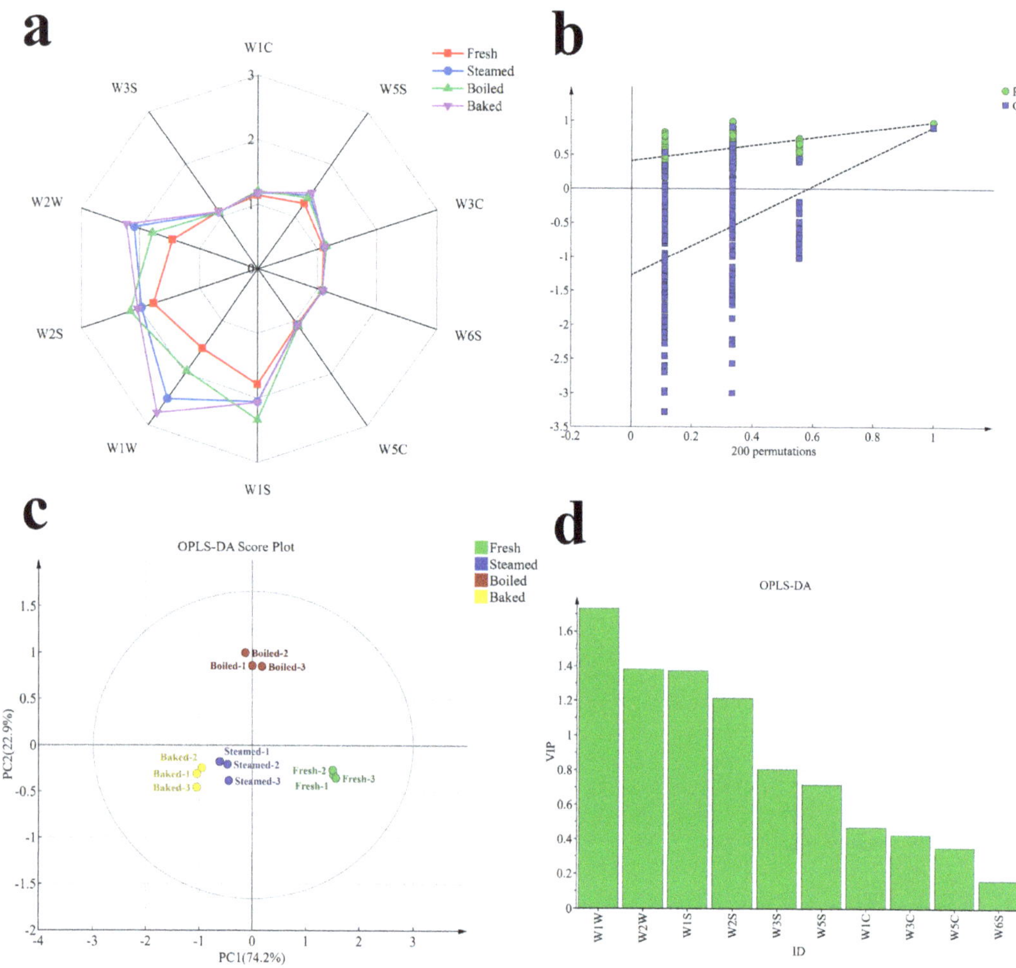


**Figure 1.** E-nose analysis of button mushroom with different cooking methods. (**a**) E-nose radar diagram. (**b**) Two hundred permutation test chart of OPLS-DA model. (**c**) OPLS-DA analysis of Enose. (**d**) VIP values of OPLS-DA model.

To delve deeper into the impact of distinct cooking methods on the odor of button mushroom, an OPLS-DA model was constructed for analysis, with the findings depicted in Figure 1b–d. According to the analysis results in Figure 1c, $PC_1$ and $PC_2$ explained 74.2% and 22.9% of the variance, respectively, contributing a total of 97.1% of the variance. This indicated that these two principal components could effectively reflect most of the characteristics of the volatile odors of the samples. The OPLS-DA analysis demonstrated an independent variable fit index ($R^2X$) of 0.999, a dependent variable fit index ($R^2Y$) of

0.882 and a model prediction index ($Q^2$) of 0.584. Both $R^2$ and $Q^2$ exceeding 0.5 indicated that the model fit was acceptable [31]. Figure 1b indicates that, after 200 permutation tests, the intersection of the $Q^2$ regression line with the vertical axis was less than zero, which suggested that there was no overfitting of the model, and the model was valid. Therefore, the results were deemed useful for the discriminant analysis of the cooking style of button mushroom. In Figure 1c, the raw group was depicted as distributed in the positive semi-axis of $PC_1$, while the cooking-treated samples were overall distributed in the negative semi-axis of $PC_1$. This suggested that the E-nose technique effectively distinguished between raw and cooked samples, highlighting a significant alteration in the overall distribution of volatile aromas of button mushroom due to the cooking treatment. Furthermore, button mushroom successfully differentiated between the three cooking styles on the score scatterplot. The roasted and steamed groups were evenly distributed on the negative half-axis of $PC_2$, with a small distance between them. This could mean that in terms of overall flavor, the roasted and steamed samples had some similarity in some components. Conversely, the boiled samples were uniformly distributed in the positive semi-axis of $PC_2$, indicating a significant difference in flavors compared to the other two groups. The reason may be that the exposure of button mushroom to drying at high temperatures during roasting resulted in the rapid evaporation of water and significant shrinkage of the mushroom surface, which may have allowed better retention and release of volatile components. On the other hand, the boiling treatment may have led to the partial dissolution of flavor substances into water, resulting in losses [5,32].

In addition, Figure 1d displays the VIP values based on the OPLS-DA model, which measured the response value of each sensor and characterized the degree of variable contribution. The higher the VIP value (VIP > 1), the greater the amount of substances in the flavor of button mushroom. All four sensors, W1W, W2W, W1S and W2S, had VIP values greater than 1, suggesting that these four types of odors were stronger in button mushroom, and it was thought that these sensors may have contributed more to the OPLS-DA model. In the meantime, the W1W and W2W sensors showed higher sensitivity to sulfide and contributed the most in this test, which indicated the presence of sulfide in button mushroom. In addition, this was verified by the GC-MS analysis results, where higher levels of benzyl mercaptan were detected in both roasted and boiled samples. Benzyl mercaptan is a common sulfide with a distinctive odor, often described as similar to that of onions or garlic [10]. Therefore these findings were crucial for understanding how the odor characteristics of button mushroom relate to its culinary treatments.

### 3.2. Characterization of Volatile Compound Species

The E-nose differs from commonly used methods for identifying flavor substances, such as GC-MS. It provides an overall flavor map rather than a map of individual components, and thus, it cannot qualify or quantify a single flavor component [33]. Table 1 displays the results of the analysis of the aroma components of button mushroom using GC-MS. A total of 73 compounds were detected, including 12 alcohols, 20 esters, 12 aldehydes, 5 acids, 9 ketones and 15 others.

In order to further grasp the effect of cooking treatments on aldehydes, alcohols, ketones, acids and esters in button mushroom, their types and contents are summarized in Table 2. The results show that 37, 23, 33 and 35 volatile components were identified in the four groups of samples, which indicates that different cooking treatments had a significant effect on the types of volatile components of button mushroom. In particular, the steaming treatment led to a significant reduction in the variety of volatiles. Meanwhile, the grilling treatment helped to retain the volatile flavor compounds in button mushroom.

**Table 1.** Qualitative and quantitative analysis of volatile compounds in raw and differently cooked button mushroom.

| No. | Volatile Compounds | CAS | RT (Min) | Odorant Description | Concentration (μg/g) Fresh | Steamed | Boiled | Baked | p-Value | VIP |
|---|---|---|---|---|---|---|---|---|---|---|
| Alcohols | | | | | | | | | | |
| 1 | **Benzyl alcohol** | 100-51-6 | 24.097 | floral | 4.55 ± 0.12 [c] | 2.32 ± 0.20 [d] | 7.26 ± 0.20 [b] | 47.56 ± 1.50 [a] | $4.01 \times 10^{-12}$ | 3.24 |
| 2 | 2,3-Butanediol | 513-85-9 | 18.286 | - | 0.30 ± 0.02 | ND | ND | ND | $2.33 \times 10^{-10}$ | 0.37 |
| 3 | 1,3-Butanediol | 107-88-0 | 18.527 | - | 1.01 ± 0.09 | ND | ND | ND | $4.38 \times 10^{-9}$ | 0.67 |
| 4 | 3-Methyl-1-butanol | 123-51-3 | 24.711 | alcoholic, fruity, fatty, banana-like | 0.31 ± 0.02 | ND | ND | ND | $5.31 \times 10^{-10}$ | 0.37 |
| 5 | Isosorbide | 652-67-5 | 22.972 | - | ND | ND | ND | 0.33 ± 0.03 | $2.26 \times 10^{-9}$ | 0.29 |
| 6 | Xylitol | 87-99-0 | 20.564 | - | ND | ND | ND | 0.54 ± 0.09 | $7.25 \times 10^{-7}$ | 0.37 |
| 7 | 1-Octen-3-ol | 3391-86-4 | 21.946 | mushroom-like | ND | ND | 1.66 ± 0.17 [a] | 0.42 ± 0.03 [b] | $5.03 \times 10^{-8}$ | 0.72 |
| 8 | **Geranylgeraniol** | 24034-73-9 | 33.572 | - | ND | ND | ND | 5.96 ± 0.13 | $8.98 \times 10^{-14}$ | 1.22 |
| 9 | **3-Octanol** | 589-98-0 | 22.481 | earthy, mushroom-like | 2.07 ± 0.12 [a] | ND | 1.45 ± 0.20 [b] | ND | $3.17 \times 10^{-8}$ | 1.02 |
| 10 | 1-Butanol | 71-36-3 | 20.236 | floral, fragrant, fruity | ND | ND | 0.79 ± 0.04 | ND | $4.75 \times 10^{-9}$ | 0.52 |
| 11 | 1-Nonanol | 143-08-8 | 22.671 | green, fatty, sweet | 0.63 ± 0.06 | ND | ND | ND | $8.74 \times 10^{-9}$ | 0.53 |
| 12 | 1-Tetradecanol | 112-72-1 | 30.917 | waxy | 0.25 ± 0.09 | ND | ND | ND | $4.19 \times 10^{-4}$ | 0.32 |
| Esters | | | | | | | | | | |
| 13 | Methyl palmitate | 112-39-0 | 29.961 | - | 0.31 ± 0.10 [b] | ND | 0.49 ± 0.12 [a] | 0.19 ± 0.07 [b] | $1.00 \times 10^{-3}$ | 0.42 |
| 14 | Dibutyl phthalate | 84-74-2 | 29.412 | - | 0.60 ± 0.06 | ND | ND | ND | $8.65 \times 10^{-9}$ | 0.52 |
| 15 | 4-Hydroxybutyl acrylate | 2478-10-6 | 28.818 | - | ND | ND | ND | 1.85 ± 0.07 | $6.62 \times 10^{-12}$ | 0.68 |
| 16 | Ethyl acetate | 141-78-6 | 19.33 | fruity, sweet | 0.93 ± 0.07 [a] | 0.46 ± 0.11 [b] | 0.44 ± 0.08 [b] | 0.31 ± 0.05 [b] | $6.10 \times 10^{-5}$ | 0.45 |
| 17 | Methyl linoleate | 112-63-0 | 37.345 | - | 0.75 ± 0.05 [a] | ND | ND | 0.40 ± 0.05 [b] | $1.26 \times 10^{-8}$ | 0.59 |
| 18 | Methyl valerate | 624-24-8 | 22.158 | - | ND | ND | ND | 0.28 ± 0.08 | $5.60 \times 10^{-5}$ | 0.26 |
| 19 | Heptadecanoic acid, heptadecyl ester | 36617-50-2 | 27.682 | - | ND | ND | 2.44 ± 0.08 [a] | 0.62 ± 0.12 [b] | $3.07 \times 10^{-10}$ | 0.88 |
| 20 | 1,3-Dipalmitin | 502-52-3 | 25.549 | - | ND | ND | ND | 0.91 ± 0.05 | $2.47 \times 10^{-10}$ | 0.48 |
| 21 | **L-ascorbyl dipalmitate** | 28474-90-0 | 26.106 | - | ND | ND | 1.47 ± 0.15 [b] | 17.82 ± 0.51 [a] | $1.21 \times 10^{-12}$ | 2.06 |
| 22 | Glycerol trioctanoate | 538-23-8 | 25.363 | - | 0.69 ± 0.08 [b] | 0.42 ± 0.09 [c] | 0.93 ± 0.03 [a] | ND | $4.77 \times 10^{-7}$ | 0.51 |
| 23 | Ethyl 2-chlorobenzoate | 7335-25-3 | 23.582 | - | ND | 0.06 ± 0.01 | ND | ND | $8.01 \times 10^{-8}$ | 0.17 |
| 24 | Ethyl isovalerate | 108-64-5 | 22.814 | fruity, sweet, grape-like | ND | 0.49 ± 0.07 | ND | ND | $3.89 \times 10^{-7}$ | 0.50 |
| 25 | bis(2-ethylhexyl) phthalate | 117-81-7 | 25.117 | - | ND | 1.98 ± 0.10 [b] | 2.82 ± 0.12 [a] | 2.74 ± 0.07 [a] | $5.58 \times 10^{-10}$ | 0.96 |
| 26 | Methyl stearate | 112-61-8 | 24.522 | - | ND | ND | 0.36 ± 0.08 | ND | $7.00 \times 10^{-6}$ | 0.35 |
| 27 | Diethylene glycol monostearate | 106-11-6 | 24.346 | - | 0.53 ± 0.09 | ND | ND | ND | $9.45 \times 10^{-7}$ | 0.49 |
| 28 | 2-Hydroxy-n-butyric acid ethyl ester | 52089-54-0 | 21.094 | - | 1.87 ± 0.05 | ND | ND | ND | $3.17 \times 10^{-13}$ | 0.91 |
| 29 | Diisobutyl phthalate | 84-69-5 | 31.856 | - | 0.24 ± 0.05 | ND | ND | ND | $9.00 \times 10^{-6}$ | 0.32 |
| 30 | Dimethyl sulfite | 616-42-2 | 14.059 | - | 0.17 ± 0.02 | ND | ND | ND | $2.67 \times 10^{-7}$ | 0.27 |
| 31 | Ethyl benzoate | 93-89-0 | 20.911 | - | 0.19 ± 0.07 | ND | ND | ND | $4.31 \times 10^{-4}$ | 0.28 |
| 32 | Methyl myristate | 124-10-7 | 28.569 | - | 0.20 ± 0.02 | ND | ND | ND | $2.32 \times 10^{-8}$ | 0.30 |
| 33 | Benzaldehyde | 100-52-7 | 17.227 | almond-like | 0.29 ± 0.09 [c] | 0.22 ± 0.02 [c] | 0.45 ± 0.09 [b] | 2.80 ± 0.05 [a] | $1.50 \times 10^{-10}$ | 0.78 |
| 34 | **Phenylacetaldehyde** | 122-78-1 | 17.785 | honey-like | ND | ND | ND | 14.53 ± 0.31 | $7.42 \times 10^{-14}$ | 1.90 |
| 35 | Hexanal | 66-25-1 | 23.865 | green | ND | ND | ND | 0.30 ± 0.04 | $5.87 \times 10^{-8}$ | 0.27 |
| 36 | (2E)-2-Octadecenal | 51534-37-3 | 33.088 | - | ND | ND | 2.71 ± 0.03 | ND | $1.11 \times 10^{-15}$ | 0.97 |

**Table 1.** *Cont.*

| No. | Volatile Compounds | CAS | RT (Min) | Odorant Description | Concentration (μg/g) Fresh | Steamed | Boiled | Baked | *p*-Value | VIP |
|---|---|---|---|---|---|---|---|---|---|---|
| 37 | 2-Carboxybenzaldehyde | 119-67-5 | 31.285 | - | 0.17 ± 0.03 | ND | ND | ND | $1.00 \times 10^{-6}$ | 0.27 |
| 38 | **Furfural** | 98-01-1 | 21.348 | bread-like, sweet, fatty, almond-like | 1.98 ± 0.12 $^{c}$ | 0.26 ± 0.07 $^{d}$ | 4.06 ± 0.20 $^{a}$ | 2.59 ± 0.11 $^{b}$ | $4.55 \times 10^{-9}$ | 1.12 |
| 39 | **Isovaleraldehyde** | 590-86-3 | 23.368 | apple-like, malty | 7.34 ± 0.23 $^{c}$ | 3.27 ± 0.12 $^{d}$ | 14.33 ± 0.26 $^{b}$ | 15.92 ± 0.14 $^{a}$ | $1.93 \times 10^{-12}$ | 1.88 |
| 40 | Decanal | 112-31-2 | 32.153 | floral, green, fruity, waxy, orange-like, | ND | ND | 0.45 ± 0.10 $^{b}$ | 1.89 ± 0.06 $^{a}$ | $3.61 \times 10^{-10}$ | 0.65 |
| 41 | oct-2-enal | 2363-89-5 | 29.133 | fruity, green, nutty | ND | ND | 2.17 ± 0.09 $^{b}$ | 2.87 ± 0.06 $^{a}$ | $5.93 \times 10^{-12}$ | 0.85 |
| 42 | Heptanal | 111-71-7 | 27.4 | green, fatty | 0.29 ± 0.06 $^{d}$ | 1.15 ± 0.13 $^{c}$ | 3.42 ± 0.07 $^{b}$ | 3.85 ± 0.07 $^{a}$ | $4.54 \times 10^{-11}$ | 0.94 |
| 43 | Vanillin | 121-33-5 | 28.166 | - | 0.41 ± 0.02 $^{a}$ | ND | 0.33 ± 0.02 $^{b}$ | ND | $3.11 \times 10^{-10}$ | 0.46 |
| 44 | Octanal | 124-13-0 | 29.744 | citrus-like, fatty | 0.25 ± 0.04 $^{b}$ | 0.22 ± 0.03 $^{b}$ | 0.30 ± 0.02 $^{a}$ | ND | $2.00 \times 10^{-6}$ | 0.27 |
| Acids | | | | | | | | | | |
| 45 | Acetic acid | 64-19-7 | 16.007 | sour | ND | 1.42 ± 0.04 $^{a}$ | 0.72 ± 0.11 $^{b}$ | 0.69 ± 0.07 $^{b}$ | $5.57 \times 10^{-8}$ | 0.82 |
| 46 | **Benzoic acid** | 65-85-0 | 20.728 | bitter gourd-like | 6.66 ± 0.21 $^{a}$ | 0.89 ± 0.07 $^{d}$ | 2.18 ± 0.14 $^{c}$ | 2.56 ± 0.13 $^{b}$ | $1.92 \times 10^{-10}$ | 1.63 |
| 47 | Lauric acid | 143-07-7 | 30.203 | tallow-like | 0.79 ± 0.06 $^{b}$ | 0.55 ± 0.08 $^{c}$ | 1.09 ± 0.16 $^{a}$ | 0.99 ± 0.16 a $^{b}$ | $3.00 \times 10^{-3}$ | 0.40 |
| 48 | **Palmitic acid** | 57-10-3 | 34.048 | - | 4.64 ± 0.12 $^{c}$ | 6.26 ± 0.09 $^{b}$ | 18.12 ± 0.16 $^{a}$ | 0.36 ± 0.11 $^{d}$ | $7.22 \times 10^{-15}$ | 2.25 |
| 49 | Heptanoic acid | 111-14-8 | 26.343 | - | ND | ND | ND | 0.25 ± 0.11 | $1.00 \times 10^{-3}$ | 0.24 |
| Ketones | | | | | | | | | | |
| 50 | Methylheptenone | 110-93-0 | 23.159 | sweet, fruity | ND | ND | 0.58 ± 0.11 | ND | $2.00 \times 10^{-6}$ | 0.45 |
| 51 | **3-Octanone** | 106-68-3 | 11.184 | earthy, mushroom-like, resinous | 0.96 ± 0.17 $^{b}$ | 1.08 ± 0.14 $^{b}$ | 5.25 ± 0.22 $^{a}$ | ND | $6.32 \times 10^{-10}$ | 1.26 |
| 52 | **3-Heptanone** | 106-35-4 | 11.013 | fruity | ND | 1.97 ± 0.17 | ND | ND | $5.26 \times 10^{-9}$ | 1.00 |
| 53 | Geranylacetone | 3796-70-1 | 12.536 | green, fruity | 0.20 ± 0.06 | ND | ND | ND | $3.70 \times 10^{-5}$ | 0.30 |
| 54 | 2-Undecanone | 112-12-9 | 12.267 | floral | 0.31 ± 0.04 | ND | ND | ND | $4.51 \times 10^{-8}$ | 0.37 |
| 55 | Carvone | 99-49-0 | 12.043 | - | 1.34 ± 0.13 $^{a}$ | 1.23 ± 0.14 $^{a}$ | ND | ND | $1.07 \times 10^{-7}$ | 0.62 |
| 56 | **1-Octen-3-one** | 4312-99-6 | 11.641 | metallic, mushroom-like | ND | ND | 4.21 ± 0.14 $^{a}$ | 2.48 ± 0.24 $^{b}$ | $5.60 \times 10^{-10}$ | 1.12 |
| 57 | 2-Nonanone | 821-55-6 | 11.835 | fruity, green, cheese-like | ND | ND | ND | 0.08 ± 0.01 | $1.30 \times 10^{-8}$ | 0.14 |
| 58 | Acetophenone | 98-86-2 | 11.329 | hawthorn-like | ND | ND | 0.53 ± 0.20 | ND | $3.26 \times 10^{-4}$ | 0.42 |
| Others | | | | | | | | | | |
| 59 | Undecane | 1120-21-4 | 15.804 | - | 0.90 ± 0.09 | ND | ND | ND | $9.88 \times 10^{-9}$ | 0.63 |
| 60 | **Squalene** | 111-02-4 | 36.81 | - | 3.29 ± 0.26 $^{c}$ | 4.32 ± 0.15 $^{b}$ | 6.49 ± 0.22 $^{a}$ | ND | $7.04 \times 10^{-10}$ | 1.27 |
| 61 | 2,6-Dimethoxyphenol | 91-10-1 | 33.76 | clove-like, smoky | ND | 0.26 ± 0.11 | ND | ND | $1.00 \times 10^{-3}$ | 0.35 |
| 62 | Isoeugenol | 97-54-1 | 34.525 | clove-like | ND | 0.43 ± 0.12 | ND | ND | $4.70 \times 10^{-5}$ | 0.46 |
| 63 | **4-Methylimidazole** | 822-36-6 | 16.783 | - | 2.37 ± 0.20 | ND | ND | ND | $4.45 \times 10^{-9}$ | 1.03 |
| 64 | **Indole** | 120-72-9 | 33.252 | musty | 11.51 ± 0.26 | ND | ND | ND | $1.00 \times 10^{-13}$ | 2.26 |
| 65 | 2-Butoxyethanol | 111-76-2 | 18.824 | - | 0.20 ± 0.08 | ND | ND | ND | $1.00 \times 10^{-3}$ | 0.29 |
| 66 | 2-Methylpyrazine | 109-08-0 | 31.076 | nutty, sweet | 0.21 ± 0.12 $^{a}$ | ND | ND | 0.27 ± 0.07 $^{a}$ | $2.00 \times 10^{-3}$ | 0.32 |
| 67 | 4-Isopropylphenol | 99-89-8 | 31.623 | - | ND | 0.24 ± 0.07 $^{b}$ | 0.31 ± 0.03 $^{a}$ | ND | $9.00 \times 10^{-6}$ | 0.35 |
| 68 | Butylated Hydroxytoluene | 128-37-0 | 34.209 | - | ND | ND | 0.55 ± 0.07 | ND | $7.17 \times 10^{-8}$ | 0.43 |
| 69 | 1-ethoxyethylbenzene | 3299-05-6 | 35.403 | - | ND | ND | ND | 0.32 ± 0.04 | $8.24 \times 10^{-8}$ | 0.28 |
| 70 | Cycloheptatriene | 544-25-2 | 13.493 | - | ND | ND | ND | 0.40 ± 0.02 | $3.56 \times 10^{-10}$ | 0.31 |
| 71 | **Benzyl mercaptan** | 100-53-8 | 37.082 | - | ND | ND | 40.51 ± 1.19 $^{b}$ | 45.38 ± 0.46 $^{a}$ | $2.88 \times 10^{-13}$ | 3.57 |
| 72 | 2-Butylfuran | 4466-24-4 | 33.938 | noncharacteristic | ND | 0.60 ± 0.06 $^{a}$ | 0.39 ± 0.02 $^{b}$ | 0.44 ± 0.02 $^{b}$ | $2.23 \times 10^{-7}$ | 0.52 |
| 73 | Furfuryl mercaptan | 98-02-2 | 25.795 | - | ND | ND | ND | 0.24 ± 0.12 | $3.00 \times 10^{-3}$ | 0.23 |

Each value is expressed as the mean ± SD ($n = 3$) of triplicate determinations. Significant differences between samples from different cooking methods were analyzed using ANOVA and Duncan's test, and specific *p*-values are presented in the table. We employed supervised regression modeling using OPLS-DA to obtain variable importance in projection (VIP). Means with different letters within a row are significantly different ($p < 0.05$) in different cooking methods. ND means that the compounds were not detected. RT: Retention time. The use of "-" indicates that the odorant description of the compounds was not found. Differential aroma components are shown in bold.

**Table 2.** Classification and content of volatile substances.

| Type of Compounds | Fresh | Contents (μg/g) | Steamed | Contents (μg/g) | Boiled | Contents (μg/g) | Baked | Contents (μg/g) |
|---|---|---|---|---|---|---|---|---|
| Alcohols | 7 | 9.11 | 1 | 2.32 | 4 | 11.16 | 5 | 54.81 |
| Esters | 11 | 6.48 | 5 | 3.41 | 7 | 8.94 | 9 | 25.13 |
| Aldehydes | 7 | 10.73 | 5 | 5.12 | 9 | 28.22 | 8 | 44.76 |
| Acids | 3 | 12.09 | 4 | 9.12 | 4 | 22.12 | 5 | 4.86 |
| Ketones | 4 | 2.82 | 3 | 4.28 | 4 | 10.58 | 2 | 2.56 |
| Others | 5 | 18.49 | 5 | 5.84 | 5 | 48.24 | 6 | 47.04 |
| Total | 37 | 59.72 | 23 | 30.09 | 33 | 129.26 | 35 | 179.16 |

The aroma content of each sample is shown as a mean value.

Based on the findings in Table 1, the primary volatile components of fresh button mushroom contained isovaleraldehyde, benzoic acid, palmitic acid, benzyl alcohol, 3-octanol and others. However, recent studies have indicated that the major volatile constituents of fresh button mushroom were eight-carbon volatiles, such as 1-octen-3-ol, 3-octanol, 2-octen-1-ol, 1-octanol and 3-octanone. These compounds have been found to vary in types and quantities [34]. For example, Maga's [35] study demonstrated that the content of 1-octen-3-ol and 1-octen-3-one increased with storage time, where 1-octen-3-ol was associated with the aroma of mushrooms, while 1-octen-3-one was not. Çağlarırmak conducted an enzymatic reaction to extract volatile compounds and subsequently measured them. The findings revealed that the volatile components in fresh button mushroom predominantly comprised C18 or C16 compounds, such as octadecanoic acid, hexadecanoic acid derivatives and other significant volatiles [36]. The study by Pei et al. [37] investigated the impact of two drying methods on the volatile constituents of button mushroom. Their findings revealed that fresh mushrooms contained high concentrations of C8 compounds (e.g., 1-octen-3-ol, 3-octanol, 2-octen-1-ol and 3-octanone), as well as benzyl alcohol, benzaldehyde and aromatic esters. Therefore, the main volatile components of fresh button mushroom found in the present study were slightly different from the previous studies, which might be due to the differences in freshness and extraction methods. Furthermore, future experiments should be conducted to investigate the specific effects of freshness and extraction methods on the flavor of button mushroom, with the aim of further elucidating its primary flavor substances.

Alcohols, commonly characterized by a sweet, fruity and vegetal aroma, are an essential component of the flavor profile of edible mushrooms [38]. According to Table 2, seven alcohols were detected in the raw group. After the cooking treatment, the number of alcohol substances decreased in all groups. Notably, the roasted group exhibited the highest concentration of alcohols at 54.81 μg/g, suggesting that the roasting process contributed to the release of more alcohols. This may be due to the fact that the grilling treatment increased the contact area of hot air with the lipids in button mushroom, which induced peroxidation reactions to produce more alcohols [39]. Benzyl alcohol is an important aromatic alcohol that imparts floral and rosy flavor to food [40]. According to Table 1, benzyl alcohol was the most abundant alcohol compound. Compared to the raw group, boiling and roasting treatments led to a significant increase in benzyl alcohol content, with the highest level of 47.56 μg/g in the roasting group. Conversely, the steamed group exhibited a notably lower benzyl alcohol content, suggesting a loss of this compound due to the steaming treatment. Furthermore, we observed that the abundance of the eight-carbon volatiles 3-octanol and 1-octen-3-ol was lower in the cooked group, which could be attributed to the instability of these volatiles under high-temperature conditions, leading to their degradation and loss [41]. These findings were consistent with previous studies by Du et al. [42], Ashmore et al. [43] and Huang et al. [44]. In summary, roasting facilitated the release of alcohols from button mushroom, thereby enhancing the overall flavor profile.

Esters, the third most abundant volatile organic compounds in edible mushrooms, are primarily formed through the esterification of acids with alcohols [45]. These compounds usually present sweet and fruity and significantly influence the overall flavor profile of food due to their low odor threshold [46]. As shown in Table 2, cooking treatments significantly reduced the number of ester compounds, causing an adverse impact on the ester compounds. The ester compounds mainly include ethyl acetate, bis(2-ethylhexyl) phthalate, L-ascorbyl dipalmitate, heptadecanoic acid, heptadecyl esterr, etc., which may be derived from the oxidation reaction of aldehydes [27]. Specifically, ethyl acetate, known for its fruity flavor profile, exhibited a significant decrease following cooking. In summary, the cooking treatments induced changes in ester compounds, potentially impacting the flavor profile of button mushroom.

Aldehydes are crucial in shaping the overall odor of food due to their low odor threshold and potent odor properties [44]. According to Table 2, only the steamed group showed a significant decrease ($p < 0.05$) in aldehydes compared to the fresh group, while both boiling and baking treatments resulted in a significant increase in aldehydes. In particular, the total aldehyde content in fresh samples increased from 10.73 μg/g to 44.76 μg/g after the roasting treatment. According to Table 1, the main aldehydes found in button mushroom were benzaldehyde, isovaleraldehyde, furfural and heptanal. Benzaldehyde, known for its sweet almond flavor, was a product of benzoic acid and had been demonstrated to contribute to the flavor of mushrooms [47]. Compared to the raw group, the boiled and roasted groups showed a significant increase in benzaldehyde content, especially in the roasted group. This could be attributed to the degradation of benzoic acid to benzaldehyde in fresh samples during roasting [48]. Furthermore, the baking treatment resulted in a notably higher isovaleraldehyde content compared to the other groups. This increase could be attributed to the high temperature and dry environment of the oven, which induced amino acid catabolism in the samples, leading to the Strecker degradation reaction of leucine, resulting in the production of isovaleraldehyde in substantial quantities [49]. Isovaleraldehyde, known for its low-threshold flavor, with apple aroma and malt odor [50], could significantly enhance the overall flavor of the food and was popular among consumers. Furthermore, Schmidberger et al. [51] identified isovaleraldehyde as an aroma active compound in truffles. Similarly, Vahdatzadeh and Splivallo [52] demonstrated that leucine, isoleucine, phenylalanine and methionine were important flavor precursors in edible mushrooms through $^{13}$C-labeling experiments. They found that benzaldehyde was a product of phenylalanine catabolism, while isovaleraldehyde was a catabolism product of leucine. In addition, furfural, which had fatty and almond flavors, was found to be the highest in the boiled samples (4.06 μg/g). All three cooking methods significantly increased the content of heptanal compared to the raw group, indicating that cooking treatments contributed to the production of heptanal. In conclusion, aldehydes played a crucial role in the odor characteristics of button mushroom. The significant production of aldehydes during cooking may contribute to the overall flavor of button mushroom, whereas grilling treatments may contribute to the release of aldehydes from button mushroom.

According to Table 2, cooking treatments significantly altered the content of acids. Boiling significantly increased the content of acids, while steaming and roasting significantly decreased the content of acids. Most of the acids had high odor thresholds and were associated with unpleasant odors, so they had relatively little effect on the aroma profile of button mushroom [53].

After boiling, the highest ketone content was observed, with 3-octanone and 1-octen-3-one being the most abundant ketones produced. 3-octanone presented an earthy and mushroomy flavor, while 1-octen-3-one offered a strong mushroom aroma, enhancing the fragrance of button mushroom [54]. Ketones, resulting from amino acid degradation and fatty acid oxidation, primarily contributed a plant fragrance smell to button mushroom and played a role in the Maillard reaction [55]. Consequently, boiling was more conducive to the release of ketones.

Other compounds such as phenols, terpenes and alkanes also influenced the flavor of button mushroom. Particularly, elevated levels of benzyl mercaptan were observed in the boiled and roasted groups, aligning with the response values of the W1W and W2W sensors in the E-nose analysis. Furthermore, 2-methylpyrazine and cycloheptatriene were detected in the baked group, suggesting that the high-temperature convective air of the oven prompted the production of additional heterocyclic molecules in button mushroom.

Based on the above discussion, it seemed that the two methods, boiling and baking, were more conducive to the release of flavor substances from button mushroom. However, in summary, the most abundant types and contents of flavor substances were released from button mushroom after the baking treatment.

### 3.3. Differential Aroma Analysis

The species characterization of volatiles clarified that cooking treatments had a significant effect on the aroma of button mushroom. To effectively differentiate the flavor of button mushroom under different cooking methods, we performed an OPLS-DA analysis with 73 aroma components as dependent variables and different cooking methods as independent variables (see Figure 2). From Figure 2a, the steamed group was positioned on the positive semi-axis of $PC_1$, closely resembling the raw group, indicating a similarity in volatile components between the steamed and raw groups. In addition, the roasted and boiled groups were uniformly distributed on the negative half-axis of $PC_1$, which implied a significant difference in flavor compared to the other samples, probably due to the sulfides detected in the roasted and boiled samples. The VIP values are shown in Figure 2b in accordance with the OPLS-DA model. Based on the criteria of VIP > 1 and $p < 0.05$, 16 differential aroma compounds (as shown in bold in Table 1) were screened in button mushroom samples, including benzyl alcohol, 3-octanol, 3-octanone, 1-octen-3-one, isovaleraldehyde and furfural.

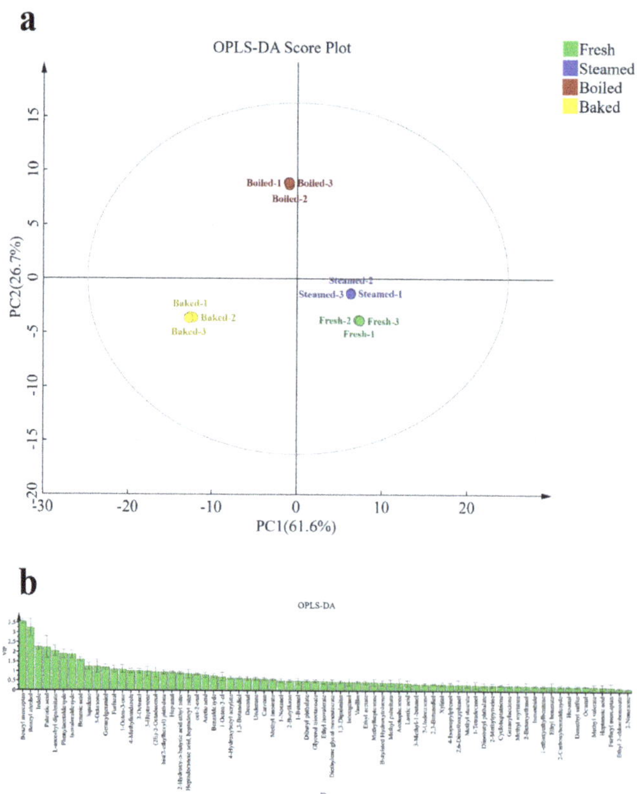


**Figure 2.** OPLS–DA analysis of volatile substances. (**a**) Plot of OPLS–DA scores of volatile flavor components. (**b**) VIP values of volatile flavor components.

To analyze the cooking-induced flavor differences more intuitively, a clustering heat map analysis was performed using the relative contents of the 16 differential aromas, and the results are shown in Figure 3. The figure reveals that clustering initially occurred between the raw and steamed samples at the minimum distance level, aligning with the OPLS-DA analysis plot of aroma components. This suggested a higher similarity in volatiles between the raw and steamed groups. As the Euclidean distance increased, the steamed and boiled groups gradually clustered together, indicating some similarities between these two clusters. Ultimately, the roasted samples were grouped with the other samples. However, the red part of the figure highlights differences in characteristic aroma substances among the four sample groups. These variations could be attributed to the increase in heterocyclic compounds during cooking, the initial loss of volatile compounds and the thermal degradation reactions of amino acids and fatty acids [37].

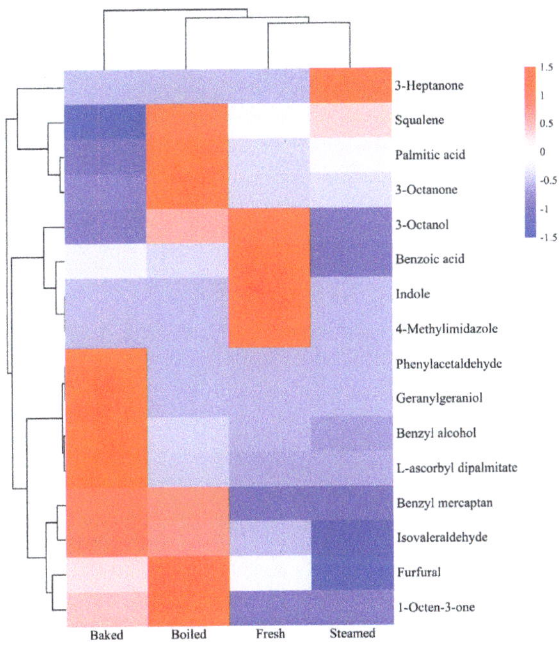


**Figure 3.** Clustering heat map of differential aroma components.

### 3.4. OAV Analysis

After OPLS-DA modeling and clustering heat map analysis, we found that there were significant differences in the aroma characteristics of button mushroom from various cooking methods. However, the level of aroma components alone cannot determine their contribution to the overall aroma because the contribution of volatiles to the overall aroma depends not only on their concentration but also on their odor threshold [56]. For this reason, the degree of volatiles' contribution can be measured by the OAV (see Table 3 for details). The investigation into thresholds revealed that 29 out of 73 aroma compounds of button mushroom lacked odor thresholds and were consequently not listed in Table 3. However, the 44 screened aroma compounds could be pivotal in evaluating the impact of various cooking methods on the flavor profile of button mushroom. Table 3 presents the published thresholds for aroma compounds in water, since water was the main medium in this study. Usually, substances with OAVs ranging from 0.1 to 1 act as flavor modifiers and have less impact on the overall aroma, while substances with OAVs greater than 1 are most likely to be the primary contributors to the overall aroma [46]. According to Table 3, we found that 15, 7, 15 and 13 volatiles had OAVs greater than 1 in raw, steamed, boiled and roasted samples, respectively. Among these volatiles, 1-octen-3-one exhibited the highest OAV, aligning with the findings of Zhang et al. [57] regarding the key aroma compounds of Boletus edulis, where 1-octen-3-one also had the highest OAV. Consequently, we concluded that 1-octen-3-one was the primary contributing substance to the overall aroma, followed by phenylacetaldehyde and methyl palmitate, which may have contributed to the aroma of button mushroom.

**Table 3.** Changes in OAV of aroma compounds in button mushroom during cooking.

| No. | Compounds | Odor Threshold μg/g | OAV | | | |
|---|---|---|---|---|---|---|
| | | | Fresh | Steamed | Boiled | Baked |
| 1 | Benzyl alcohol | 2.55 | 1.78 | 0.91 | 2.85 | 18.65 |
| 2 | 2,3-Butanediol | 0.0951 | 3.18 | ND | ND | ND |
| 3 | 1,3-Butanediol | 10 | 0.10 | ND | ND | ND |
| 4 | 3-Methyl-1-butanol | 0.25 | 1.24 | ND | ND | ND |
| 5 | 1-Octen-3-ol | 0.1 | ND | ND | 16.60 | 4.17 |
| 6 | 3-Octanol | 0.042 | 49.26 | ND | 34.55 | ND |
| 7 | 1-Butanol | 0.5 | ND | ND | 1.58 | ND |
| 8 | 1-Nonanol | 0.09 | 7.04 | ND | ND | ND |
| 9 | 1-Tetradecanol | 559 | 0.00 | ND | ND | ND |
| 10 | Methyl palmitate | 0.003 | 103.67 | ND | 163.33 | 62.67 |
| 11 | Ethyl acetate | 5 | 0.19 | 0.09 | 0.09 | 0.06 |
| 12 | Methyl valerate | 0.02 | ND | ND | ND | 14.00 |
| 13 | 2-Hydroxy-n-butyric acid ethyl ester | 0.8 | 2.34 | ND | ND | ND |
| 14 | Ethyl benzoate | 0.05 | 3.80 | ND | ND | ND |
| 15 | Benzaldehyde | 0.35 | 0.83 | 0.63 | 1.29 | 8.00 |
| 16 | Phenylacetaldehyde | 0.004 | ND | ND | ND | 3632.50 |
| 17 | Hexanal | 0.02 | ND | ND | ND | 15.00 |
| 18 | Furfural | 9.56 | 0.21 | 0.03 | 0.42 | 0.27 |
| 19 | Isovaleraldehyde | 13 | 0.56 | 0.25 | 1.10 | 1.22 |
| 20 | Decanal | 33 | ND | ND | 0.01 | 0.06 |
| 21 | oct-2-enal | 394 | ND | ND | 0.01 | 0.01 |
| 22 | Heptanal | 0.06 | 4.83 | 19.17 | 57.00 | 64.17 |
| 23 | Octanal | 0.008 | 31.25 | 27.50 | 37.50 | ND |
| 24 | Acetic acid | 22 | ND | 0.06 | 0.03 | 0.03 |
| 25 | Benzoic acid | 1 | 6.66 | 0.89 | 2.18 | 2.56 |
| 26 | Lauric acid | 59 | 0.01 | 0.01 | 0.02 | 0.02 |
| 27 | Palmitic acid | 503 | 0.01 | 0.01 | 0.04 | 0.00 |
| 28 | Heptanoic acid | 10.4 | ND | ND | ND | 0.02 |
| 29 | Methylheptenone | 0.85 | ND | ND | 0.68 | ND |
| 30 | 3-Octanone | 0.07 | 13.71 | 15.43 | 75.00 | ND |
| 31 | 3-Heptanone | 0.08 | ND | 24.63 | ND | ND |
| 32 | Geranylacetone | 0.06 | 3.33 | ND | ND | ND |
| 33 | 2-Undecanone | 0.182 | 1.70 | ND | ND | ND |
| 34 | Carvone | 0.1 | 13.40 | 12.30 | ND | ND |
| 35 | 1-Octen-3-one | 0.00005 | ND | ND | 84,200.00 | 49,600.00 |
| 36 | 2-Nonanone | 0.2 | ND | ND | ND | 0.40 |
| 37 | Acetophenone | 0.065 | ND | ND | 8.15 | ND |
| 38 | Undecane | 10 | 0.09 | ND | ND | ND |
| 39 | 2,6-Dimethoxyphenol | 1.85 | ND | 0.14 | ND | ND |
| 40 | Isoeugenol | 0.1 | ND | 4.30 | ND | ND |
| 41 | Indole | 25 | 0.46 | ND | ND | ND |
| 42 | Butylated Hydroxytoluene | 0.161 | ND | ND | 3.42 | ND |
| 43 | 2-Butylfuran | 0.005 | ND | 120.00 | 78.00 | 88.00 |
| 44 | Furfuryl mercaptan | 0.01 | ND | ND | ND | 23.80 |

Odor thresholds for volatile compounds were obtained from http://www.odour.org.uk (accessed on 21 February 2024) and the *Compilations of Odour Threshold Values in Air, Water and Other Media*. "ND" means not detected and therefore not analyzed in the OAV calculation.

In summary, the different treatments (raw, steamed, boiled and roasted) affected the OAVs of volatiles. 1-Octen-3-one may be the main contributor to the aroma of button mushroom, while other volatiles such as phenylacetaldehyde and methyl palmitate also played a role in the formation of the aroma. These findings provide valuable insights for a better understanding and utilization of the flavor characteristics of button mushroom.

### 3.5. Aroma Recombination and Omission Test

After the comprehensive analysis, the following conclusion can be drawn: the roasted group was the richest in flavor substances and presented the most distinct flavor profile in the E-nose radargram. Subsequently, aroma reorganization and omission experiments were carried out to further identify the key aroma components of the roasted samples. To assess the contribution and significance of aroma compounds with OAVs exceeding 1 to the overall flavor in the roasted group, we performed aroma reconstitution and omission tests based on standard aroma substances and ultrapure water. As depicted in Figure 4, the overall aroma of the reconstitution experiments had good similarity with the original samples, indicating that the preparation of the reconstitution model was successful. Building upon the successful aroma reconstitution, we also used omission tests to verify the contribution of 13 compounds with OAV $\geq 1$ to the overall aroma. If the omission model after the removal of a single compound showed no significant difference from the fully reconstituted samples, it suggests that the removed compound contributed less to the overall aroma; conversely, it indicates a greater contribution to the overall aroma [58]. The results are shown in Table 4. The omission of 1-octen-3-one, benzyl alcohol, methyl valerate, 1-octen-3-ol, isovaleraldehyde, furfuryl mercaptan and heptanal exhibited significant differences in the E-nose response value compared with the recombination sample. These compounds that significantly affected the overall aroma of the reconstituted samples were considered to be the most crucial aroma substances, essential for the aroma quality of roasted button mushroom.

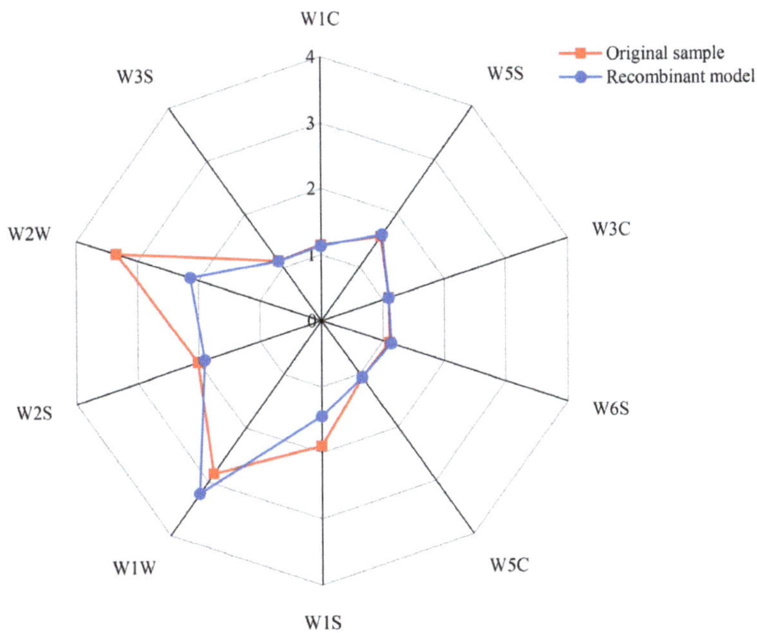


**Figure 4.** Aroma profiles of original sample and recombination model.

**Table 4.** The results of the omission tests for the recombination model.

| Compound(s) Omitted | Degree of Odor Difference |
|---|---|
| 1-Octen-3-one | W2S↓ * |
| Benzyl alcohol | W2S↓ ** |
| Methyl valerate | W2S↓ ***, W2W↓ *** |
| Phenylacetaldehyde | ns |
| 1-Octen-3-ol | W2S↓ ***, W2W↓ *** |
| Benzoic acid | ns |
| Isovaleraldehyde | W2S↓ *** |
| Hexanal | ns |
| Methyl palmitate | ns |
| Furfuryl mercaptan | W2S↓ ***, W2W↓ *** |
| 2-Butylfuran | ns |
| Heptanal | W2S↓ *** |
| Benzaldehyde | ns |

ns, not significant; *, $p < 0.05$; **, $p < 0.01$; ***, $p < 0.001$. ↓, decrease in odor response values.

## 4. Conclusions

The experiment was conducted to study the aroma components of button mushroom under different cooking methods, and a total of 73 aroma components were detected. The present study revealed significant differences in the aroma components of button mushroom under various cooking methods, exhibiting unique culinary characteristics. In particular, the most intense and attractive aroma was produced by the grilled treatment. This research identified 16 differential aroma substances related to the aroma quality of button mushroom for the first time, based on the *p*-value and VIP value. In addition, OAV analysis suggested that 1-octen-3-one, phenylacetaldehyde and methyl palmitate could be the primary contributors to the aroma of roasted button mushroom. The seven key aromas of the roasted samples identified through OAV analysis and omission experiments were 1-octen-3-one, benzyl alcohol, methyl valerate, 1-octen-3-ol, isovaleraldehyde, furfuryl mercaptan and heptanal. The study revealed several volatiles with OAVs exceeding 1, notably 1-octen-3-one with OAVs ranging from 49,600 to 84,200, indicating its significance as a key aroma active compound. These findings offer valuable insights into the aroma composition and potential reactions of cooked button mushroom, which is important for culinary and food industry applications.

Given the limitations of GC-MS technology, further combination with other analytical techniques such as GC-O (gas chromatography-olfactometry) and GC-IMS (gas chromatography-ion mobility spectrometry) is necessary to enhance the study. Additionally, the pathways of flavor substance generation in button mushroom during cooking were explored to establish an objective basis for the high-quality cooking of button mushroom.

**Author Contributions:** Conceptualization, N.Z. and B.Z.; Data curation, L.X.; Formal analysis, L.X.; Investigation, L.X. and S.G.; Supervision, N.Z. and B.Z.; Validation, H.R.; Visualization, B.L.; Writing—original draft, L.X.; Writing—Review and editing, N.Z. and B.Z. All authors have read and agreed to the published version of the manuscript.

**Funding:** This work was supported by Integrated Research Base of Edible Fungi Processing and Comprehensive Utilization Technology (PTJH22017) and Horizontal Scientific Research Project: Research and Application of Resistance to Freezing Deformation in Aquatic Frozen Foods (KH230175A). And the APC was funded by Horizontal Scientific Research Project: Research and Application of Resistance to Freezing Deformation in Aquatic Frozen Foods (KH230175A).

**Institutional Review Board Statement:** Not applicable.

**Informed Consent Statement:** Not applicable.

**Data Availability Statement:** The original contributions presented in the study are included in the article, further inquiries can be directed to the corresponding author.

**Conflicts of Interest:** The authors declare no conflicts of interest.

## References


1. Feng, T.; Yang, M.; Ma, B.; Zhao, Y.; Zhuang, H.; Zhang, J.; Chen, D. Volatile profiles of two genotype *Agaricus bisporus* species at different growth stages. *Food Res. Int.* **2021**, *140*, 109761. [CrossRef] [PubMed]
2. Krishnamoorthi, R.; Srinivash, M.; Mahalingam, P.U.; Malaikozhundan, B. Dietary nutrients in edible mushroom, *Agaricus bisporus* and their radical scavenging, antibacterial, and antifungal effects. *Process Biochem.* **2022**, *121*, 10–17. [CrossRef]
3. Rauf, A.; Joshi, P.B.; Ahmad, Z.; Hemeg, H.A.; Olatunde, A.; Naz, S.; Hafeez, N.; Simal-Gandara, J. Edible mushrooms as potential functional foods in amelioration of hypertension. *Phytother. Res.* **2023**, *37*, 2644–2660. [CrossRef] [PubMed]
4. Bassam, S.M.; Noleto-Dias, C.; Farag, M.A. Dissecting grilled red and white meat flavor: Its characteristics, production mechanisms, influencing factors and chemical hazards. *Food Chem.* **2022**, *371*, 131139. [CrossRef] [PubMed]
5. Zhou, X.; Guan, Q.; Wang, Y.; Lin, D.; Du, B. Effect of Different Cooking Methods on Nutrients, Antioxidant Activities and Flavors of Three Varieties of *Lentinus edodes*. *Foods* **2022**, *11*, 2713. [CrossRef] [PubMed]
6. Cheng, L.; Li, X.; Tian, Y.; Wang, Q.; Li, X.; An, F.; Luo, Z.; Shang, P.; Liu, Z.; Huang, Q. Mechanisms of cooking methods on flavor formation of Tibetan pork. *Food Chem. X* **2023**, *19*, 100873. [CrossRef]
7. Shakoor, A.; Zhang, C.; Xie, J.; Yang, X. Maillard reaction chemistry in formation of critical intermediates and flavour compounds and their antioxidant properties. *Food Chem.* **2022**, *393*, 133416. [CrossRef]
8. Flaskerud, K.; Bukowski, M.; Golovko, M.; Johnson, L.; Brose, S.; Ali, A.; Cleveland, B.; Picklo, M.; Raatz, S. Effects of cooking techniques on fatty acid and oxylipin content of farmed rainbow trout (*Oncorhynchus mykiss*). *Food Sci. Nutr.* **2017**, *5*, 1195–1204. [CrossRef]
9. Li, W.; Li, R.; Chen, W.; Feng, J.; Wu, D.; Zhang, Z.; Zhang, J.; Yang, Y. The anabolism of sulphur aroma volatiles responds to enzymatic and non-enzymatic reactions during the drying process of shiitake mushrooms. *Food Chem.* **2022**, *371*, 131123. [CrossRef]
10. Hou, Z.; Xia, R.; Li, Y.; Xu, H.; Wang, Y.; Feng, Y.; Pan, S.; Wang, Z.; Ren, H.; Qian, G.; et al. Key components, formation pathways, affecting factors, and emerging analytical strategies for edible mushrooms aroma: A review. *Food Chem.* **2024**, *438*, 137993. [CrossRef]
11. Davila, M.; Muniz, A.; Du, X. The impact of roasting and steaming on savory flavors contributed by amino acids, 5′-nucleotides, and volatiles in *Agaricus bisporus* mushrooms. *Int. J. Gastron. Food Sci.* **2022**, *30*, 100590. [CrossRef]
12. Selli, S.; Guclu, G.; Sevindik, O.; Kelebek, H. Variations in the key aroma and phenolic compounds of champignon (*Agaricus bisporus*) and oyster (*Pleurotus ostreatus*) mushrooms after two cooking treatments as elucidated by GC–MS-O and LC-DAD-ESI-MS/MS. *Food Chem.* **2021**, *354*, 129576. [CrossRef] [PubMed]
13. Zhu, M.; Hu, Z.; Liang, M.; Song, L.; Wu, W.; Li, R.; Li, Z.; Zhang, J. Evaluation of the flavor compounds of *Pleurotus eryngii* as affected by baking temperatures using HS-SPME-GC–MS and electronic nose. *J. Food Process. Preserv.* **2022**, *46*, e17056. [CrossRef]
14. Chi, X.; Shao, Y.; Pan, M.; Yang, Q.; Yang, Y.; Zhang, X.; Ai, N.; Sun, B. Distinction of volatile flavor profiles in various skim milk products via HS-SPME–GC–MS and E-nose. *Eur. Food Res. Technol.* **2021**, *247*, 1539–1551. [CrossRef]
15. Zhang, X.; Gao, P.; Xia, W.; Jiang, Q.; Liu, S.; Xu, Y. Characterization of key aroma compounds in low-salt fermented sour fish by gas chromatography-mass spectrometry, odor activity values, aroma recombination and omission experiments. *Food Chem.* **2022**, *397*, 133773. [CrossRef] [PubMed]
16. Capone, S.; Tufariello, M.; Francioso, L.; Montagna, G.; Casino, F.; Leone, A.; Siciliano, P. Aroma analysis by GC/MS and electronic nose dedicated to *Negroamaro* and *Primitivo* typical Italian Apulian wines. *Sens. Actuators B* **2013**, *179*, 259–269. [CrossRef]
17. Wang, Y.; Wang, D.; Lv, Z.; Zeng, Q.; Fu, X.; Chen, Q.; Luo, Z.; Luo, C.; Wang, D.; Zhang, W. Analysis of the volatile profiles of kiwifruits experiencing soft rot using E-nose and HS-SPME/GC–MS. *LWT* **2023**, *173*, 114405. [CrossRef]
18. Jin, G.; Zhu, Z.; Wu, Z.; Wang, F.; Li, J.; Raghavan, V.; Li, B.; Song, C. Characterization of volatile components of microwave dried perilla leaves using GC–MS and E-nose. *Food Biosci.* **2023**, *56*, 103083. [CrossRef]
19. Zhou, J.; Feng, T.; Ye, R. Differentiation of Eight Commercial Mushrooms by Electronic Nose and Gas Chromatography-Mass Spectrometry. *J. Sens.* **2015**, *2015*, 374013. [CrossRef]
20. Fujioka, K.; Shimizu, N.; Manome, Y.; Ikeda, K.; Yamamoto, K.; Tomizawa, Y. Discrimination Method of the Volatiles from Fresh Mushrooms by an Electronic Nose Using a Trapping System and Statistical Standardization to Reduce Sensor Value Variation. *Sensors* **2013**, *13*, 15532–15548. [CrossRef]
21. Gomez, I.; Lavega Gonzalez, R.; Tejedor-Calvo, E.; Perez Clavijo, M.; Carrasco, J. Odor Profile of Four Cultivated and Freeze-Dried Edible Mushrooms by Using Sensory Panel, Electronic Nose and GC-MS. *J. Fungi* **2022**, *8*, 953. [CrossRef]
22. Sun, Y.; Lv, F.; Tian, J.; Ye, X.Q.; Chen, J.; Sun, P. Domestic cooking methods affect nutrient, phytochemicals, and flavor content in mushroom soup. *Food Sci. Nutr.* **2019**, *7*, 1969–1975. [CrossRef]
23. Lee, K.; Lee, H.; Choi, Y.; Kim, Y.; Jeong, H.S.; Lee, J. Effect of Different Cooking Methods on the True Retention of Vitamins, Minerals, and Bioactive Compounds in Shiitake Mushrooms (*Lentinula edodes*). *Food Sci. Technol. Res.* **2019**, *25*, 115–122. [CrossRef]
24. Roncero-Ramos, I.; Mendiola-Lanao, M.; Perez-Clavijo, M.; Delgado-Andrade, C. Effect of different cooking methods on nutritional value and antioxidant activity of cultivated mushrooms. *Int. J. Food Sci. Nutr.* **2017**, *68*, 287–297. [CrossRef]
25. Guo, Q.; Adelina, N.M.; Hu, J.; Zhang, L.; Zhao, Y. Comparative analysis of volatile profiles in four pine-mushrooms using HS-SPME/GC-MS and E-nose. *Food Control* **2022**, *134*, 108711. [CrossRef]

26. Adelina, N.M.; Wang, H.; Zhang, L.; Zhao, Y. Comparative analysis of volatile profiles in two grafted pine nuts by headspace-SPME/GC–MS and electronic nose as responses to different roasting conditions. *Food Res. Int.* **2021**, *140*, 110026. [CrossRef]
27. Lu, B.; Perez-Moreno, J.; Zhang, F.; Rinaldi, A.C.; Yu, F. Aroma profile of two commercial truffle species from Yunnan and Sichuan, China: Inter- and intraspecific variability and shared key compounds. *Food Sci. Hum. Wellness* **2021**, *10*, 163–173. [CrossRef]
28. Liu, H.; Wang, Z.; Zhang, D.; Shen, Q.; Pan, T.; Hui, T.; Ma, J. Characterization of Key Aroma Compounds in Beijing Roasted Duck by Gas Chromatography–Olfactometry–Mass Spectrometry, Odor-Activity Values, and Aroma-Recombination Experiments. *J. Agric. Food Chem.* **2019**, *67*, 5847–5856. [CrossRef] [PubMed]
29. Fan, M.; Xiao, Q.; Xie, J.; Cheng, J.; Sun, B.; Du, W.; Wang, Y.; Wang, T. Aroma Compounds in Chicken Broths of Beijing Youji and Commercial Broilers. *J. Agric. Food Chem.* **2018**, *66*, 10242–10251. [CrossRef] [PubMed]
30. Deng, G.; Li, J.; Liu, H.; Wang, Y. Volatile compounds and aroma characteristics of mushrooms: A review. *Crit. Rev. Food Sci. Nutr.* **2023**, 1–18. [CrossRef] [PubMed]
31. Yun, J.; Cui, C.; Zhang, S.; Zhu, J.; Peng, C.; Cai, H.; Yang, X.; Hou, R. Use of headspace GC/MS combined with chemometric analysis to identify the geographic origins of black tea. *Food Chem.* **2021**, *360*, 130033. [CrossRef]
32. Zhang, J.; Li, M.; Zhang, Y.; Qin, Y.; Li, Y.; Su, L.; Li, L.; Bian, Z.; Lu, T. E-eye, flash GC E-nose and HS-GC-MS combined with chemometrics to identify the adulterants and geographical origins of Ziziphi Spinosae Semen. *Food Chem.* **2023**, *424*, 136270. [CrossRef] [PubMed]
33. Wei, G.; Dan, M.; Zhao, G.; Wang, D. Recent advances in chromatography-mass spectrometry and electronic nose technology in food flavor analysis and detection. *Food Chem.* **2023**, *405*, 134814. [CrossRef] [PubMed]
34. Grosshauser, S.; Schieberle, P. Characterization of the Key Odorants in Pan-Fried White Mushrooms (*Agaricus bisporus* L.) by Means of Molecular Sensory Science: Comparison with the Raw Mushroom Tissue. *J. Agric. Food Chem.* **2013**, *61*, 3804–3813. [CrossRef]
35. Maga, J.A. Influence of Maturity, Storage and Heating on the Flavor of Mushroom (*Agaricus bisporus*) Caps and Stems. *J. Food Process Preserv.* **1981**, *5*, 95–101. [CrossRef]
36. Çağlarırmak, N. Determination of nutrients and volatile constituents of *Agaricus bisporus* (brown) at different stages. *J. Sci. Food Agric.* **2009**, *89*, 634–638. [CrossRef]
37. Pei, F.; Yang, W.; Ma, N.; Fang, Y.; Zhao, L.; An, X.; Xin, Z.; Hu, Q. Effect of the two drying approaches on the volatile profiles of button mushroom (*Agaricus bisporus*) by headspace GC–MS and electronic nose. *LWT-Food Sci. Technol.* **2016**, *72*, 343–350. [CrossRef]
38. Wang, L.; Zhou, Y.; Wang, Y.; Bu, H.; Dong, T. Changes in cell wall metabolism and flavor qualities of mushrooms (*Agaricus bernardii*) under EMAP treatments during storage. *Food Packag. Shelf Life* **2021**, *29*, 100732. [CrossRef]
39. Gao, F.; Xie, W.; Zhang, H.; Li, S.; Li, T. Variations of quality and volatile components of morels (*Morchella sextelata*) during storage. *J. Plant Physiol.* **2023**, *290*, 154094. [CrossRef]
40. Feng, Y.; Xu, H.; Sun, Y.; Xia, R.; Hou, Z.; Li, Y.; Wang, Y.; Pan, S.; Li, L.; Zhao, C.; et al. Effect of light on quality of preharvest and postharvest edible mushrooms and its action mechanism: A review. *Trends Food Sci. Technol.* **2023**, *139*, 104119. [CrossRef]
41. Zhang, K.; Pu, Y.; Sun, D. Recent advances in quality preservation of postharvest mushrooms (*Agaricus bisporus*): A review. *Trends Food Sci. Technol.* **2018**, *78*, 72–82. [CrossRef]
42. Du, X.; Sissons, J.; Shanks, M.; Plotto, A. Aroma and flavor profile of raw and roasted *Agaricus bisporus* mushrooms using a panel trained with aroma chemicals. *LWT-Food Sci. Technol.* **2021**, *138*, 110596. [CrossRef]
43. Ashmore, L.; Craske, J.D.; Srzednicki, G. Volatile compounds in fresh, cooked fresh, dried and cooked dried *Agaricus bisporus* using ambient temperature vacuum distillation. *Int. Food Res. J.* **2014**, *21*, 263–268.
44. Huang, J.; Xiao, L.; Yi, Y.; Li, B.; Sun, R.; Deng, H. Preservation mechanism and flavor variation of postharvest button mushroom (*Agaricus bisporus*) coated compounds of protocatechuic acid-$CaCl_2$-NaCl-pullulan. *LWT-Food Sci. Technol.* **2022**, *169*, 114020. [CrossRef]
45. Liu, Q.; Cui, X.; Song, Z.; Kong, W.; Kang, Y.; Kong, W.; Ng, T.B. Coating shiitake mushrooms (*Lentinus edodes*) with a polysaccharide from *Oudemansiella radicata* improves product quality and flavor during postharvest storage. *Food Chem.* **2021**, *352*, 129357. [CrossRef] [PubMed]
46. Xu, X.; Xu, R.; Jia, Q.; Feng, T.; Huang, Q.; Ho, C.; Song, S. Identification of dihydro-β-ionone as a key aroma compound in addition to C8 ketones and alcohols in *Volvariella volvacea* mushroom. *Food Chem.* **2019**, *293*, 333–339. [CrossRef]
47. Li, X.; Zhang, Y.; Hengchao, E.; He, X.; Li, J.; Zhao, X.; Zhou, C. Characteristic fingerprints and comparison of volatile flavor compounds in *Morchella sextelata* under different drying methods. *Food Res. Int.* **2023**, *172*, 113103. [CrossRef]
48. Cheng, Y.; Wei, Y.; Zhang, M.; Wang, H. Effect of micro-perforated film packing on physicochemical quality and volatile profile of button mushroom (*Agaricus bisporus*) during postharvest shelf-life. *J. Food Process. Preserv.* **2022**, *46*, e16648. [CrossRef]
49. Zhu, R.; Wen, Y.; Wu, W.; Zhang, L.; Salman Farid, M.; Shan, S.; Wen, J.; Farag, M.A.; Zhang, Y.; Zhao, C. The flavors of edible mushrooms: A comprehensive review of volatile organic compounds and their analytical methods. *Crit. Rev. Food Sci. Nutr.* **2022**, 1–15. [CrossRef] [PubMed]
50. Combet, E.; Henderson, J.; Eastwood, D.C.; Burton, K.S. Influence of Sporophore Development, Damage, Storage, and Tissue Specificity on the Enzymic Formation of Volatiles in Mushrooms (*Agaricus bisporus*). *J. Agric. Food Chem.* **2009**, *57*, 3709–3717. [CrossRef]

51. Schmidberger, P.C.; Schieberle, P. Characterization of the Key Aroma Compounds in White Alba Truffle (*Tuber. magnatum pico*) and Burgundy Truffle (*Tuber uncinatum*) by Means of the Sensomics Approach. *J. Agric. Food Chem.* **2017**, *65*, 9287–9296. [CrossRef] [PubMed]
52. Vahdatzadeh, M.; Splivallo, R. Improving truffle mycelium flavour through strain selection targeting volatiles of the Ehrlich pathway. *Sci. Rep.* **2018**, *8*, 9304. [CrossRef] [PubMed]
53. Pueschel, V.A.; Schieberle, P. Changes in the key aroma compounds of matsutake mushroom (*Tricholoma matsutake* Sing.) from Canada during pan-frying elucidated by application of the sensomics approach. *Eur. Food Res. Technol.* **2020**, *247*, 51–65. [CrossRef]
54. Li, M.; Yang, R.; Zhang, H.; Wang, S.; Chen, D.; Lin, S. Development of a flavor fingerprint by HS-GC–IMS with PCA for volatile compounds of *Tricholoma matsutake* Singer. *Food Chem.* **2019**, *290*, 32–39. [CrossRef] [PubMed]
55. Schmidberger, P.C.; Schieberle, P. Changes in the Key Aroma Compounds of Raw Shiitake Mushrooms (*Lentinula edodes*) Induced by Pan-Frying as Well as by Rehydration of Dry Mushrooms. *J. Agric. Food Chem.* **2020**, *68*, 4493–4506. [CrossRef] [PubMed]
56. Murray, A.F.; Wickramasinghe, P.C.K.; Munafo, J.P. Key Odorants from the Fragrant Bolete, *Suillus punctipes*. *J. Agric. Food Chem.* **2020**, *68*, 8621–8628. [CrossRef]
57. Zhang, H.; Pu, D.; Sun, B.; Ren, F.; Zhang, Y.; Chen, H. Characterization and comparison of key aroma compounds in raw and dry porcini mushroom (*Boletus edulis*) by aroma extract dilution analysis, quantitation and aroma recombination experiments. *Food Chem.* **2018**, *258*, 260–268. [CrossRef]
58. Verma, D.K.; Srivastav, P.P. A paradigm of volatile aroma compounds in rice and their product with extraction and identification methods: A comprehensive review. *Food Res. Int.* **2020**, *130*, 108924. [CrossRef]

foods

*Article*

# Nutritional and Technological Properties of Albino Peach Palm (*Bactris gasipaes*) from the Amazon: Influence of Cooking and Drying

**Stephanie Dias Soares [1], Orquídea Vasconcelos dos Santos [1,2], Leyvison Rafael Vieira da Conceição [3], Hilton Túlio Costi [4], José Otávio Carrera Silva Júnior [5], Francisco das Chagas Alves do Nascimento [2] and Rosinelson da Silva Pena [1,6,*]**

[1] Graduate Program in Food Science and Technology, Federal University of Pará, Belém 66075-110, PA, Brazil; s203084@dac.unicamp.br (S.D.S.); orquideavs@ufpa.br (O.V.d.S.)
[2] Faculty of Nutrition, Institute of Health Sciences, Federal University of Pará, Belém 66075-110, PA, Brazil; fcan@ufpa.br
[3] Faculty of Chemistry, Institute of Exact and Natural Sciences, Federal University of Pará, Belém 66075-110, PA, Brazil; rafaelvieira@ufpa.br
[4] Emílio Goeldi Museum, Coordination of Earth Sciences and Ecology, Belém 66077-830, PA, Brazil; tulio@museu-goeldi.br
[5] Faculty of Pharmacy, Institute of Health Sciences, Federal University of Pará, Belém 66075-110, PA, Brazil; carrera@ufpa.br
[6] Faculty of Food Engineering, Institute of Technology, Federal University of Pará, Belém 66075-110, PA, Brazil
* Correspondence: rspena@ufpa.br

**Abstract:** This study aimed to subject the albino peach palm to cooking and drying processes and characterize the raw pulp (RP), cooked pulp (CP), raw pulp flour (RPF), and cooked pulp flour (CPF). The product's chemical composition, bioactive compounds, and physicochemical, color, thermal, morphological, and functional–technological properties were evaluated. The proximate composition showed that carbohydrates were the main constituents of all the products (69.59–72.08 g/100 g). The cooking process decreased the lipids (10.21 to 8.63 g/100 g), dietary fiber (13.64 to 12.81 g/100 g), and total sugar content (59.18 to 49.10 g/100 g) of the CP. The colorimetric parameters indicated a significant browning of the CP and CPF, which can be attributed to the Maillard reaction and lipid oxidation. After cooking, the total phenolic compound and ascorbic acid content decreased in the pulp. The RPF and CPF displayed different thermogravimetric behaviors. The spectral patterns in the infrared region showed the characteristic bands of organic compounds that are present in the structure of starches. The scanning electron microscopy showed amyloplast and fiber bundles with starches in the RP and gelatinized starch granules in the CP and CPF. The RPF presented small and heterogeneous starch granules with isolated amyloplast. The RPF and CPF showed different granulometric patterns and technological indices. The results suggest that the pulp and flour from the pulp of albino peach palms can be exploited by the food, pharmaceutical, and biotechnological industries.

**Keywords:** white peach palm; Amazon fruit; nutritional composition; technological properties; processing; Brazilian fruit; Arecaceae

## 1. Introduction

The Amazon region is rich in native fruits of various species, which are of great importance for the local population's culture, food, economy, and health [1]. These fruits have been a source of research related to micronutrients, bioactive compounds, antioxidant capacity, and biological activities [2–4]. More than 40% of the native Brazilian fruits are found in the Amazonian region, especially the northern region, which produces approximately 220 species of plants that produce fruit for human consumption. However, the Arecaceae family stands out with its fruits produced by palm trees, such as peach palm

(*Bactris gasipaes*), *açaí* (*Euterpe oleracea*), *bacaba* (*Oenocarpus bacaba*), and *buriti* (*Mauritia flexuosa*). These species can be used by directly consuming its fruits and obtaining products and byproducts, such as drinks, creams, flour, and oils [5,6].

The peach palm fruit is known worldwide as "pupunha" (in Brazil), "pejibaye" (in Costa Rica), and "chontaduro" (in Colombia) [2,7,8]. This fruit has ovoid and globose anatomic characteristics, but it is best known for its different colorations of green (immature fruit), yellow, orange, and red, provided by the presence of carotenoid pigments. Recently, our research group presented research on an albino peach palm variety from the Amazon not previously explored or reported [3], corroborating what was reported by Cornelius et al. [9], who found the presence of peach palm fruits with yellow peel and white pulp for the first time.

The peach palm fruit needs to be cooked in water before consumption to inhibit anti-nutritional factors (trypsin inhibitors, phytates, and tannins), eliminate oxalate crystals present in the peel [10,11], and inactivate the pulp peroxidase enzyme (105 °C for 20 min), thus avoiding the irritation of the throat mucosa [12]. On the other hand, this fruit is considered a fruit with high nutritional value because it has a significant content of oil and proteins, and a high content of carbohydrates and insoluble dietary fiber [11,13,14]. In addition, the peach palm fruit has all the essential amino acids [15]. The flour and starch obtained from the pulp of peach palms (with or without peel) have been extensively studied for their nutritional, functional, and biotechnological properties. Reports have shown the biological potential of using these products to improve metabolic properties [16,17], to increase the bioaccessibility of carotenoids in foods [18,19], and to obtain a biodegradable film from starch [20].

The cooking process required for fruit consumption and the drying to obtain flour could lead to significant changes. These processes promote changes in food's physical and chemical characteristics, such as browning and protein denaturation, nutritional composition, and cellular structure. In addition, these processes affect food products' sensory and color characteristics [21,22]. Despite that, drying is an important process for food preservation, which ensures the product's microbiological and chemical stability, promoting a longer shelf life [23–25].

Therefore, this work aimed to evaluate the chemical composition, bioactive compounds, and physicochemical, color, thermal, morphological, and functional–technological properties of albino peach palms before and after the cooking and drying processes. To the best of our knowledge, this work comprises the first report on the influence of the cooking and drying of albino peach palms.

## 2. Materials and Methods

### 2.1. Fruits

Five bunches (~20 kg each) of albino peach palm (~100 kg of fruits) were acquired at a fair located in the metropolitan region of Belém, the state of Pará, Brazil (latitude: −1.45502, longitude: −48.5024, 1°27′18″ S, 48°30′9″ W), and which were harvested during the period September to December 2020. This research was registered with access activity in the National System for the Management of Genetic Heritage and Associated Traditional Knowledge (SisGen, AA5BEE7). The voucher specimen was identified and deposited in the Herbarium "HF Professora Normélia Vasconcelos" at the Federal University of Pará (Belém, Brazil) and notified under register 4070.

### 2.2. Sample Preparation

At this stage, half of the fruits (with peel) were cooked in water in a pressure cooker for 15 min, according to the traditional procedure performed in Northern Brazil. The other half of the fruits were separated without cooking. Both fruit samples (raw and cooked) were peeled, and the pulp was separated from the seed using a stainless steel knife. The edible portion of peach palm (raw and cooked pulps) was crushed separately in a multiprocessor (Philco, 800 W, Manaus, Brazil) and then placed in plastic bags and packed in a vacuum sealer (Cetro, model DZ-280, São Paulo, Brazil). The samples were stored at −20 °C and wrapped in aluminum foil for protection against light until the analysis. To obtain raw and cooked peach palm flour, a part of the samples of their respective pulps was subjected to drying at 55 °C [14] in an oven with forced air circulation (Thoth, model 510.150, São Paulo, Brazil) until a constant weight was obtained (~26 h). The dried samples were ground in a Willey-type knife mill (Fortinox, model Start FT 50, São Paulo, Brazil) coupled to a ten-mesh sieve.

### 2.3. Physicochemical Characterization

The analyses of moisture, ash, total proteins (total nitrogen by the Kjeldahl method and conversion factor of 5.75), lipids, total titratable acidity (TTA; %citric acid), hydrogen potential, and total and reducing sugars were performed according to the Association of Official Analytical Chemists [26]. Dietary fibers (total and insoluble fractions) were determined using the enzymatic/gravimetric method [27]. The soluble dietary fiber content was determined by difference. Total carbohydrates were obtained by difference, and the caloric value was measured using the Atwater coefficients [28]. The water activity ($a_w$) was determined with a direct measure on a water activity analyzer (Aqualab 4TEV, Decagon Devices Inc., Pullman, WA, USA).

The instrumental color parameters were analyzed in a digital colorimeter (Chroma Meter CR-300, Konica Minolta, Tokyo, Japan), using the CIELAB system to assess the chromaticity coordinates ($L^*$ for luminosity, $a^*$ for red color intensity, $b^*$ for yellow color intensity). The chromaticity ($C^*$) and the Hue angle ($h°$) were calculated according to McLellan et al. [29]. To know how close a sample is to an ideal white, the whiteness index (WI) was calculated [30].

### 2.4. Mineral Composition

This analysis was performed according to Alves et al. [31]. A mass of 0.25 g was weighed for each sample in triplicate ($n$ = 3) and digested with 5.0 mL of $HNO_3$ 25% ($v/v$) and 5.0 mL of $H_2O_2$ 30% ($w/w$) in a microwave oven. The heating program consisted of the following three steps: 800 W, 180 °C for 10 min; 800 W, 180 °C for 15 min; and ventilation for 50 min. After digestion, the digested part and the residue (silicate compounds that were not digested) were quantitatively transferred to a 14 mL volumetric flask, and 10 mL of ultrapure water was added. The same procedure was used to prepare the analytical blank, without adding the sample. For the analysis of Al, As, Ba, Ca, Cd, Cr, Cu, Fe, Mg, Mn, P, Se, and Zn, an inductively coupled plasma optical emission spectrometer (iCAP 6500, Thermo Scientific, Cambridge, UK) was used. All solutions and dilutions were prepared with ultrapure water (resistivity 18.2 MΩ cm) obtained from a water purification system (Synergy UV, Millipore, Darmstadt, Germany). Nitric acid (Sigma-Aldrich, Darmstadt, Germany) was used at 65% $v/v$, previously purified by a distillation system (BSB 939-IR, Berghof, Eningen, Germany).

### 2.5. Ascorbic Acid, Phenolic Compound, Flavonoid, and Carotenoid Content

The titrimetric method quantified the ascorbic acid content with the reagent 2,6-dichlorophenolindophenol (DCFI) [26]. The total phenolic compound (TPC) content was spectrophotometrically quantified using the Folin–Ciocalteau reagent according to the colorimetry method described by Singleton and Rossi [32]. An analytical curve of gallic acid (1 to 100 mg/L) was used. The total flavonoid content was determined using the aluminum flavonoid complexation reaction method described by Pękal and Pyrzynska [33]. An analytical curve of quercetin in the 1.56 to 25 µg/mL range was used. Finally, the total carotenoid content was quantified with spectrophotometry using the method proposed by Godoy and Rodriguez-Amaya [34]; the sample was subjected to a measurement in the UV-Vis spectrophotometer at 450 nm. The total carotenoid content was calculated using the specific absorption coefficient of β-carotene in petroleum ether ($A_{1cm}^{1\%} = 2592$).

### 2.6. Granulometric Analysis

A digital electromagnetic stirrer (Bertel, model VP-01, São Paulo, Brazil) was used, with a constant stirring speed and frequency of 50/60 HZ. Each flour was sieved in a set of 9-, 20-, 28-, 60-, 80-, and 100-mesh sieves (with mesh openings of 2000, 850, 600, 250, 180, and 150 μm, respectively), and a pan, for 10 min [35]. The samples' mass retained on the sieve and the initial mass were considered for calculation.

### 2.7. Thermogravimetric and Absorption Spectroscopy Analysis

Thermogravimetric analyses of the raw and cooked pulps and flour were performed with a thermobalance (Shimadzu, model DTG-60 H, Kyoto, Japan) under the following conditions: airflow of 60 mL/min, a heating ramp of 10 °C/min, in the temperature range of 20 to 800 °C, in an alumina crucible with a mass of 5 mg (±0.5).

In the infrared region, the absorption spectroscopy was carried out using the same samples with a spectrometer (Shimadzu Corporation IR Prestige 21 Cat. No. 206-73600-36, Kyoto, Japan) with records in the spectral frequency range of absorption from 4000 to 500 $cm^{-1}$. The sample was incorporated into potassium bromide pellets, with Scan 100 and a resolution of 4 $cm^{-1}$. All bands were analyzed by the software Origin 8.0.

### 2.8. Scanning Electron Microscopy

The micrographs were obtained using an electron microscope (TESCAN, model Mira3, Brno, Czech Republic) with an electron gun-type FEG (field emission gun). The samples were metalized with Au for 2.5 min for the deposition of a film on the sample, with a 10 to 15 nm thickness. The images were generated by detecting secondary electrons, using a voltage acceleration between 5 and 10 kV and working distances between 10 and 15 mm.

### 2.9. Functional Technological Properties

The functional technological properties: water solubility index (WSI), water absorption index (WAI), and oil absorption capacity (OAC), were performed and calculated according to Anderson et al. [36], with some modifications. The foam forming ability (FFA) and foam stability (FS) were determined according to Coffmann and Garcia [37].

### 2.10. Statistical Analysis

All analyses were performed in triplicate, and the results were expressed as mean ± standard deviation. Data were submitted for analysis of variance (ANOVA) and Tukey's complementary test for means comparison ($p \leq 0.05$) using the software the version number for software Statistica 10.0 (TIBCO Software Inc., Palo Alto, CA, USA).

## 3. Results and Discussion

### 3.1. Physicochemical Properties of Pulp and Flour

The proximate composition and physicochemical properties of the pulp and flour of the raw and cooked albino peach palm are presented in Table 1. The cooked pulp (CP) presented a higher moisture content than the raw pulp (RP), which could be attributed to the incorporation of water during cooking. On the other hand, the cooked pulp flour (CPF) reached a lower moisture content than the raw pulp flour (RPF), indicating that cooking favored the elimination of the water from the pulp during drying. The pulp moistures were similar to the values observed by Pires et al. [14] for fruits of the microcarp (62.43 g/100 g) and mesocarp (63.96 g/100 g) varieties. In turn, the moisture contents of the RPF and CPF were lower than the values found by the same authors for the pulp flour of the microcarp, mesocarp, and macrocarp varieties (9.60–12.58 g/100 g) obtained by convective drying at 55 °C. The differences observed may be attributed to the different peach palm varieties evaluated, the drying system used, and the origin of the fruits.

**Table 1.** Proximate composition and physicochemical properties of raw and cooked albino peach palm pulp and the respective flour.

| Components | Pulp | | Flour | |
|---|---|---|---|---|
| | Raw | Cooked | Raw | Cooked |
| Moisture (g/100 g wet basis—wb) | 61.00 ± < 0.01 [b] | 70.50 ± 0.50 [a] | 9.67 ± 0.29 [c] | 5.83 ± 0.29 [d] |
| Ash (g/100 g dry matter—dm) | 9.68 ± 1.30 [b] | 13.93 ± 0.65 [a] | 2.21 ± < 0.01 [c] | 2.12 ± < 0.01 [c] |
| Proteins (g/100 g dm) | 7.90 ± < 0.01 [c] | 7.59 ± 0.23 [c] | 16.96 ± < 0.01 [b] | 19.99 ± 0.45 [a] |
| Lipids (g/100 g dm) | 10.21 ± 1.12 [a] | 8.63 ± 0.84 [a] | 10.81 ± 0.31 [a] | 7.86 ± 0.52 [a] |
| Total carbohydrates (g/100 g dm) | 72.08 ± 2.94 [a] | 70.58 ± 0.14 [a] | 69.92 ± 0.69 [a] | 69.59 ± 0.53 [a] |
| Total dietary fiber (g/100 g dm) | 13.64 ± 2.27 [ab] | 12.81 ± 2.18 [ab] | 16.89 ± 0.84 [a] | 8.19 ± 0.19 [b] |
| Insoluble dietary fiber (g/100 g dm) | 9.30 ± 0.72 [b] | 7.32 ± 0.01 [b] | 14.72 ± 0.70 [a] | 6.03 ± 1.38 [b] |
| Soluble dietary fiber (g/100 g dm) | 5.41 ± 1.85 [ab] | 6.74 ± 0.18 [a] | 2.66 ± 0.11 [b] | 2.16 ± 0.19 [b] |
| Total sugars (g/100 g dm) | 59.18 ± 1.11 [a] | 49.10 ± 0.57 [b] | 24.74 ± 0.45 [c] | 12.46 ± 0.12 [d] |
| Reducing sugars (g/100 g dm) | 12.58 ± 0.21 [a] | 3.86 ± < 0.01 [c] | 12.92 ± < 0.01 [a] | 6.98 ± 0.29 [b] |
| Non-reducing sugars (g/100 g dm) | 46.60 ± 1.02 [a] | 45.24 ± 0.61 [b] | 11.82 ± 0.45 [c] | 5.48 ± 0.26 [d] |
| Total energetic value (kcal/100 g wb) | 160.58 ± 0.50 [b] | 115.19 ± 1.20 [c] | 401.85 ± 0.10 [a] | 400.00 ± 1.70 [a] |
| $a_w$ | 0.980 ± < 0.01 [a] | 0.981 ± < 0.01 [a] | 0.206 ± < 0.01 [c] | 0.246 ± < 0.01 [b] |
| pH | 6.44 ± 0.08 [a] | 6.18 ± 0.03 [a] | 6.15 ± 0.05 [b] | 6.25 ± 0.13 [a] |
| Total titratable acidity (g citric acid/100 g wb) | 8.96 ± < 0.01 [c] | 4.48 ± < 0.01 [d] | 21.32 ± 0.30 [b] | 23.45 ± 0.60 [a] |

Mean values of three replicates ± standard deviation. Different letters on the same line represent differences with 95% significance for Tukey's test ($p \leq 0.05$). $a_w$: water activity.

For the ash content, a significant difference was observed ($p \leq 0.05$) between the pulps, indicating that cooking caused significant changes in this component. The ash contents found in both flour (2.12–2.21 g/100 g dm) allowed for the classification of these samples as having a low ash content (2–6 g/100 g) [38]. The results found for the RPF and CPF were close to those reported by Martínez-Girón et al. [39] for peach palm peel flour (1.95 g/100 g dm). On the other hand, the ash contents observed in the pulps (9.68–13.93 g/100 g dm) were higher than the values reported by dos Santos [3] for cooked albino peach palm (5.43 g/100 g dm) from other varieties.

Regarding proteins, the content observed in both flour was higher than in the respective pulps ($p \leq 0.05$). This behavior can be attributed to the decrease in other flour constituents, such as total sugars. The protein content found in both albino peach palm flour (RPF and CPF) was higher than that found (8.64 g/100 g dm) for a peach palm by-product (stem portion) [40]. From a nutritional point of view, 100 g of the CP provided approximately 4% of the recommended daily protein intake for a healthy adult (body weight = 70 kg) and 8% for an elderly person. The CPF (100 g), in turn, contributed approximately 32 and 64% of the recommended daily protein intake for these respective populations [41]. Thus, peach palm flour can provide excellent protein levels, whether consumed directly or in culinary preparations fortified with this flour [18,42,43].

The lipid contents were statistically equal for the four samples (RP, CP, RPF, and CPF) ($p > 0.05$), which suggests that neither cooking nor drying caused the loss of this constituent. However, a tendency towards a reduction in the lipid content was observed in the CP and CPF, compared to the values of the RP and RPF, respectively, which can be attributed to a small loss of lipids due to leaching during cooking. The lipid values in the RP and CP (10.21 and 8.63 g/100 g dm) were within the range observed by Melo et al. [11] (4.21–21.74 g/100 g dm), but were higher than the values observed by Basto et al. [42] (6.6–7.5 g/100 g dm). In our recent study [3], we reported for the first time the fatty acid profile in peach palm oil, which is mainly composed of saturated fatty acids (51.75%), especially palmitic acid (49.71%)—with 42.29% being monounsaturated fatty acids and 4.86% being polyunsaturated fatty acids.

The carbohydrates were the major macronutrient in the pulps (RP and CP) and flour (RPF and CPF), and the values were not statistically different ($p > 0.05$), indicating that neither cooking nor drying caused the loss of this macronutrient. In turn, the total dietary fiber content was not statistically different ($p > 0.05$) between pulps, but the values were statistically different ($p \leq 0.05$) between flours. For the total and insoluble fibers, a decreasing trend in the CP and CPF values was observed concerning the RP and RPF, which can be attributed to the leaching of these constituents. The insoluble dietary fiber represented 68.2%, 57.1%, 87.1%, and 73.6% of the total dietary fiber in the RP, CP, RPF, and CPF, respectively. These results corroborated those Pires et al. [14] reported for peach palm flour of the mesocarp variety. In our recent study [3] we found 10.4, 7.2, and 3.2 g/100 g dm of total, insoluble, and soluble dietary fiber, respectively, in the CP. Dietary fibers can act as functional foods that, when consumed regularly and combined with healthy lifestyle habits, are related to reducing insulin levels, managing cardiometabolic diseases, reducing cholesterol levels, and improving intestinal health [44–46]. Therefore, fiber-rich flour can provide functionality to food products [47].

A lower content of total sugars was observed in the CP and CPF compared with the respective RP and RPF ($p \leq 0.05$). Similar behavior was observed for reducing and non-reducing sugar fractions. These results indicated that during cooking, the dissolution and leaching of sugars occurred, as the same was found by Giombelli et al. [40] for the by-product of heart of palm production (stem portion). The content of non-reducing sugars in the RP and CP was higher than that of reducing sugars. On the other hand, the content of reducing sugars in both flour was higher than that of non-reducing sugars, which suggests that, during the drying of the pulp at 55 °C, the depolymerization of non-reducing sugars may have occurred through the selective hydrolysis of glycosidic bonds [48].

A higher total energetic value (TEV) was observed for the RP (160.58 kcal/100 g wb) than for the CP (115.19 kcal/100 g wb) ($p \leq 0.05$), while, for the RPF and CPF, the TEV was statistically equal (average TEV of 403.4 kcal/100 g wb) ($p > 0.05$). The TEV observed in both flour was higher than the values reported for wheat (366.6 kcal/100 g) and oat (363.7 kcal/100 g) flour [49], which indicates that the albino peach palm flour is an excellent source of calories, which can contribute to energy balance [50].

The data showed that the reduction in the moisture promoted by drying ensured the obtainment of flour with low $a_w$ values ($a_w < 0.25$), capable of guaranteeing the microbiological stability of the products ($a_w < 0.6$). On the other hand, the $a_w$ values in the RP and CP (0.98) were highly favorable to the activity of microorganisms, chemical and enzymatic reactions, and physical changes in the products [23–25]. The pH values indicated that the pulps and flour are products with low acidity (pH > 4.5). On the other hand, the total titratable acidity (TTA) of the RP (8.96 g citric acid/100 g wb) was approximately twice the observed value in the CP (4.48 g citric acid/100 g wb). This reduction can be attributed to the leaching of the acidic constituents during cooking [51], such as the tannins and phytates in peach palm fruits [10].

### 3.2. Color Parameters

The instrumental color results observed for the pulps and flour of the albino peach palm are shown in Table 2. The RP and RPF showed a different colorimetric pattern than the respective CP and CPF ($p \leq 0.05$). On the other hand, the RP and RPF, as well as the CP and CPF, in general, showed the same colorimetric pattern ($p > 0.05$). Cooking caused a decrease in the brightness ($L^*$) and whiteness index (WI) and an increase in the parameters $b^*$ (yellow color) and $C^*$ (chromaticity) of the pulp. The $L^*$ and WI values decreased significantly during cooking ($p \leq 0.05$), but, during drying, no reduction in these parameters was observed ($p > 0.05$). Therefore, there occurred a browning of the pulp during cooking (Figure S1) which can be attributed to lipid oxidation and the Maillard reaction caused by the interaction between the amino acids and reducing sugars (glucose and fructose) of the pulp [40].

**Table 2.** Color parameters for raw and cooked albino peach palm pulp and flour.

| Parameter | Pulp | | Flour | |
|---|---|---|---|---|
| | Raw | Cooked | Raw | Cooked |
| $L^*$ | 76.52 ± 4.78 [a] | 70.37 ± 1.26 [b] | 80.23 ± 2.2 [a] | 72.03 ± 1.61 [b] |
| $a^*$ | −0.43 ± 0.21 [a] | −1.59 ± 0.37 [b] | −1.39 ± 0.04 [b] | −0.47 ± 0.12 [a] |
| $b^*$ | +18.56 ± 0.86 [b] | +25.88 ± 0.67 [a] | +16.0 ± 0.72 [b] | +24.93 ± 1.66 [a] |
| C* | 18.57 ± 0.86 [b] | 25.93 ± 0.69 [a] | 15.17 ± 0.1 [c] | 24.93 ± 1.64 [a] |
| $h°$ | 91.35 ± 0.75 [b] | 93.50 ± 0.73 [a] | 94.92 ± 0.02 [a] | 90.94 ± 0.27 [b] |
| WI | 70.00 ± 3.46 [a] | 60.60 ± 0.43 [b] | 74.58 ± 0.61 [a] | 62.48 ± 0.23 [b] |

Mean values of three replicate ± standard deviation. Different letters on the same line represent a difference, with 95% significance ($p \leq 0.05$). WI: whiteness index.

The increase in the $b^*$ value corroborated the difference observed in the color of the RP and CP (Supplementary Material Figure S1) and visually indicated that cooking the fruit promoted an increase in the concentration of the yellow pigments in the pulp. Despite the whitish visual aspect of the RP (Supplementary Material Figure S1), the $b^*$ values indicated the presence of yellow pigments, both in the RP ($b^*$ = +18.56) and in the RPF ($b^*$ = +16.0), which was confirmed by the values of the C* and $h°$ that were attributed to the light yellow color of the RP (C* = 18.57 and $h°$ = 91.35°) and RPF (C* = 15.17 and $h°$ = 94.92°) [28]. In turn, the increase in the C* values indicated an increase in the intensity of the yellow color of the CP (C* = 25.93) and CPF (C* = 24.93).

### 3.3. Mineral Composition

Table 3 shows that Mg, P, and Ca were the major minerals in the pulps and flours of the albino peach palm, while Fe, Cr, Cu, and Mn were the main trace elements. Our research group [3] reported, for the first time, an elemental composition with Ca (1500 mg), P (860 mg), Mg (815.8 mg), Se (76.2 mg), Zn (45.3 mg), Fe (23.0 mg), Cu (13.9 mg), and Mn (7.2 mg) per kilogram of cooked albino peach palm (in dry matter). These values were generally lower than those observed in the CP, except for Cu, Se, and Zn. Variations in the mineral composition can be explained by differences in the soil composition, agricultural practices, and environmental conditions [52]. The contents of Cu, Fe, and Zn in the RP were higher than in fruits such as *jambo* (*Syzygium jambos*) (5.71 mg/kg for Cu), mangosteen (*Garcinia mangostana*) (5.71 mg/kg for Cu and 23.3 mg/kg for Fe), *muruci* (*Byrsonima scrispa*) (1.8 mg/kg for Cu and <0.2 mg/kg for Zn), *ajuru* (*Chrysobalanus ícaro*) (<0.6 mg/kg for Cu, <2.5 mg/kg for Fe, and <0.2 mg/kg for Zn), and *umari* (*Andira spinulosa*) (<0.6 mg/kg for Cu, 20.9 mg/kg for Fe, and <0.2 mg/kg for Zn) [31].

The minerals found in the pulps and flours of albino peach palm are essential for human body metabolism. Generally, the macro minerals are related to nerve cell functions, and the micro minerals are associated with the formation of erythrocyte cells [53]. Mg, the most abundant mineral found in the samples, can aid in glucose transport, control peripheral insulin resistance in type 2 diabetes mellitus, and keep tyrosine kinase activity functioning [54]. The Ca is deposited in the body as hydroxyapatite crystals in the matrix composed of collagen and directly assists in the mineralization of the skeleton [55,56]. Therefore, dietary Ca intake may directly aid in bone metabolism and development. The ingestion of RP, CP, RPF, and CPF can also supplement the daily requirement of P, which was shown to play a role in preventing sarcopenia [57] and was also related to the structural composition of teeth and bones [53]. These important elements present in 100 g of the CP supply almost 91% of the Mg, 33% of the Ca, and 23% of the P recommended for adults [58].

**Table 3.** Average contents of Al, As, Ba, Ca, Cr, Cu, Fe, Mg, Mn, P, Se, and Zn in raw and cooked albino peach palm pulp and flour.

| Elements (mg/kg dm) | Pulp | | Flour | | Reference Values [a] (mg/day) |
|---|---|---|---|---|---|
| | Raw | Cooked | Raw | Cooked | |
| Major minerals | | | | | |
| Ca | 1068.32 ± 137.44 [b] | 3278.34 ± 60.05 [a] | 891.70 ± 82.89 [b] | 1046.88 ± 116.06 [b] | 1000 |
| Mg | 1443.48 ± 132.16 [b] | 3632.41 ± 283.22 [a] | 1921.23 ± 192.58 [b] | 1669.45 ± 172.22 [b] | 400 |
| P | 1811.35 ± 198.27 [a] | 2237.89 ± 266.32 [a] | 1993.78 ± 19.35 [a] | 1647.83 ± 238.29 [a] | 1000 |
| Minor minerals | | | | | |
| Al | 23.59 ± 1.23 [c] | 24.50 ± 2.12 [c] | 89.31 ± 1.07 [b] | 127.91 ± 13.59 [a] | |
| As | 4.46 ± 0.20 [a] | 6.64 ± 0.67 [a] | <LOD | <LOD | |
| Ba | 1.11 ± 0.08 [b] | 2.39 ± 0.75 [ab] | 3.2 ± 0.14 [a] | 2.16 ± 0.16 [ab] | |
| Cd | <LOD | 0.09 ± <0.01 | <LOD | <LOD | |
| Cr | 20.15 ± 0.53 [b] | 21.36 ± 0.16 [b] | 1.62 ± 0.12 [c] | 32.85 ± 0.47 [a] | |
| Cu | 18.96 ± 0.38 [a] | 5.56 ± 1.38 [b] | 13.89 ± 0.74 [a] | 16.18 ± 1.77 [a] | 2 |
| Fe | 66.28 ± 2.30 [a] | 43.08 ± 3.96 [b] | 17.80 ± 4.49 [b] | 52.53 ± 4.49 [ab] | 18 |
| Mn | 19.34 ± 0.27 [b] | 49.61 ± 3.27 [a] | 15.35 ± 0.81 [b] | 22.17 ± 1.44 [b] | 2 |
| Se | 4.08 ± 0.57 [a] | 4.90 ± 0.24 [a] | 4.79 ± 1.40 [a] | 5.26 ± 1.15 [a] | 70 [b] |
| Zn | 9.04 ± 0.48 [b] | 22.40 ± 0.30 [a] | 16.33 ± 2.30 [ab] | 10.47 ± 1.15 [b] | 15 |

Mean values of three replicates ± standard deviation. Different letters on the same line represent a difference with 95% significance. Al: aluminum; As: arsenic; Ba: barium; Ca: calcium; Cr: chrome; Cu: copper; Fe: iron; Mg: magnesium; Mn: manganese; P: phosphor; Se: selenium; Zn: zinc. LOD: limit of detection. [a] FAO [58]; [b] µg/day for men and 55 µg/day for women.

### 3.4. Bioactive Compounds Contents

The contents of TPC, total flavonoids, ascorbic acid, and total carotenoids in the samples are shown in Table 4. In general, the cooking and drying processes promoted the degradation of the analyzed compounds ($p \leq 0.05$). Regarding the cooking, the reduction observed can be attributed to the leaching of these compounds by the cooking water and the cooking temperature (~100 °C). During drying, in turn, the degradation can be attributed to the process temperature as well as the exposure time of the samples.

**Table 4.** Composition of bioactive compounds in the raw and cooked albino peach palm pulp and flour.

| Components | Pulp | | Flour | |
|---|---|---|---|---|
| | Raw | Cooked | Raw | Cooked |
| TPC (mg GAE/100 g dry matter—dm) | 63.10 ± 3.10 [a] | 51.39 ± 5.47 [b] | 48.47 ± 0.98 [b] | 12.16 ± 2.27 [c] |
| Total flavonoids (mg quercetin/100 g dm) | 19.31 ± 0.13 [a] | 10.17 ± 0.42 [b] | 2.37 ± 0.18 [c] | 2.16 ± < 0.01 [c] |
| Ascorbic acid (mg/100 g dm) | 88.15 ± < 0.1 [a] | 70.63 ± 6.13 [b] | 7.90 ± < 0.1 [c] | 6.06 ± < 0.1 [c] |
| Total carotenoids (µg β-carotene/100 g dm) | 4.92 ± 0.20 [ab] | 4.62 ± 0.65 [ab] | 4.39 ± 0.02 [b] | 5.53 ± 0.21 [a] |

Mean values of three replicates ± standard deviation. Different letters on the same line represent a difference, with 95% significance ($p \leq 0.05$). TPC: total phenolic compound; GAE: gallic acid equivalent.

Rojas-Garbanzo et al. [59] were the first authors to study the effect of processing on the TPC content of peach palms. They observed 70.0 mg GAE/100 g dm in CP and 63.0 mg GAE/100 g dm in CPF. These values were higher than those observed in the CP (51.39 mg GAE/100 g) and CPF (12.16 mg GAE/100 g) of the albino peach palm. The higher TPC content found in other peach palm varieties can be attributed to the lower production of these secondary metabolites in the albino variety [60].

Flavonoids, compounds included in the group of phenolic compounds, showed similar results in the TPC during cooking, but the degradation of the TPC was 18.6%, while for total flavonoids, it was 47.3%. No significant difference ($p > 0.05$) was observed for the content of total flavonoids between the RPF and CPF, but the drying process degraded 87.7% of the flavonoids from the RP, while the CP just lost 78.8% of these compounds. These results showed that drying the peach palm pulp at 55 °C caused a greater degradation of the flavonoids in the RP than in the CP, indicating that cooking increased the thermostability of this class of compounds.

The drastic reduction in the flavonoids observed between the CP and CPF, and between the RP and RPF was different from that found for the RP and RPF concerning the TPC. This behavior can be attributed to the specificity of the Folin–Ciocalteau method for phenolics with reducing properties. Furthermore, phenolic compounds have distinct behaviors (they can be stable or associated with oxidative and enzymatic processes) when exposed to different processes such as storage, cooking, and heating [61].

Ascorbic acid (AA) showed the same behavior observed for the flavonoids during cooking. The CP showed a 19.9% reduction in the AA content compared to the RP. The AA contents in the RPF and CPF were statistically the same ($p > 0.05$). AA was shown to be susceptible to reduction at high temperatures and, as it is a water-soluble compound, it may have been leached from the pulp during cooking. However, the AA contents in the flour were approximately ten times lower than those observed in the respective pulps, which allows us to state that drying at 55 °C caused a highly significant reduction in the AA content of the raw and cooked pulps ($p \leq 0.05$). In this way, it could be seen that the temperature and duration of drying had a much greater impact on the degradation of AA than cooking.

The AA content found by Contreras-Calderón et al. [62] in the pulp of raw peach palm (33.7 mg/100 g wb) was close to the value observed in the RP (34.38 mg/100 g dm). Prado et al. [63] found a higher AA content in raw orange peach palm flour (20.33 mg/100 g dm).

The RP and CP did not show carotenoid contents statistically different ($p > 0.05$), although there was a 6% reduction in the total carotenoid content after cooking. Regarding the flour, it was observed that the CPF had a total carotenoid content 26% higher than that observed in the RPF. This result indicated a greater fixation of the carotenoids in the CP during the drying at 55 °C. On the other hand, the results found after cooking the albino peach palm were contrary to those of some authors [64–66], who found higher levels of carotenoids after cooking some food matrices. These differences may be due to different botanical characteristics.

The availability of carotenoids is directly linked to the color of the fruit, the location and planting conditions, the treatment applied, the content and type of fibers, and the interaction between different carotenoids [42,59]. Because it is an albino variety, the peach palm studied was classified as a low source of carotenoids (4.92 µg β-carotene/100 g dm in RP), according to the classification proposed by Britton and Khachik [67]. This total carotenoid content was lower than that found in medlar (15.82 µg/100 g dm) [68] with whitish pulp.

### 3.5. Thermogravimetric Analysis

The thermogravimetric behaviors observed for the RP, CP, RPF, and CPF of albino peach palm are shown in Figure 1. The thermogravimetric (TG) curves show the loss of mass suffered by the pulps (Figure 1A,B) and flours (Figure 1C,D), depending on the temperature, and the derivative thermogravimetric (DTG) curves show the first derivative of the TG curve. The first event observed for all the samples in the DTG curve was represented by a discrete endothermic peak, characteristic of water loss, at approximately 100 °C.

The second DTG event was an endothermic peak of greater intensity, observed for all the pulps and flours, with the onset of mass loss at 250 °C and the maximum mass loss at approximately 300 °C. This degradation pattern can be attributed to the loss of the glycosidic O-H bonds in glucose, the unit that forms amylose and amylopectin [69–71]. Similar behavior was observed for peach palm native starch, which showed a degradation peak, starting at 251.7 °C and a maximum weight loss at 320 °C [70].

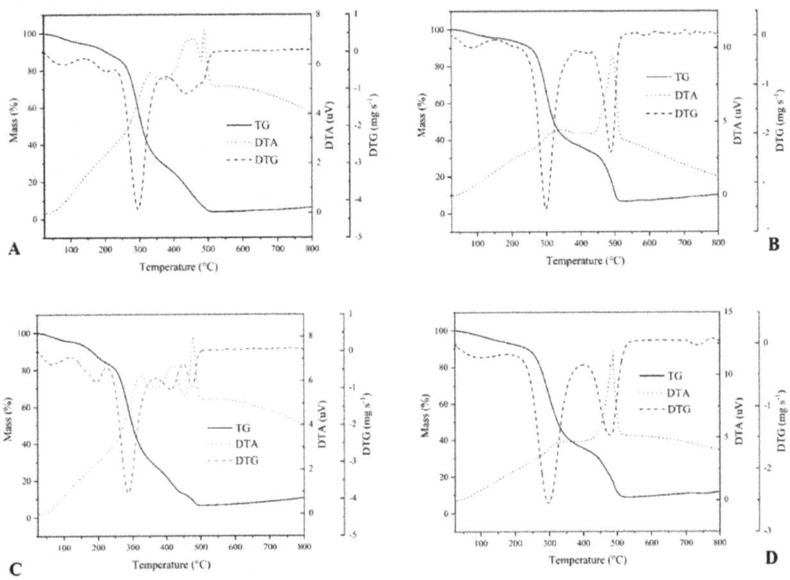


**Figure 1.** Behavior of the thermogravimetric (TG), differential thermal (DTA), and derivative thermogravimetric (DTG) curves of albino peach palm. (**A**) raw pulp; (**B**) cooked pulp; (**C**) raw pulp flour; and (**D**) cooked pulp flour.

In the differential thermal analysis (DTA), two exothermic peaks were observed for all products (pulps and flours), the first at approximately 350 °C and the second at approximately 500 °C. These peaks are related to the release of energy in the same region of the main mass decays of the TG/DTG curves. The first peak showed a low intensity for the RP (Figure 1A) and RPF (Figure 1C) and was practically suppressed for the CP (Figure 1B) and CPF (Figure 1D). On the other hand, the second peak showed a low intensity for the RP (Figure 1A), a moderate intensity for the RPF (Figure 1C), and a higher intensity for the CP (Figure 1B) and CPF (Figure 1D).

The exothermic peaks observed for the RP were related to the release of energy resulting from successive degradations and oxidations of organic-based macronutrients (proteins, carbohydrates, lipids, and fibers) and inorganic macro- and microminerals. On the other hand, for the CP, RPF, and CPF, the increase in the intensity, mainly of the second peak, was attributed to the release of energy from the carbonization process, since these products were submitted to thermal processes (cooking and/or drying), which can promote starch gelatinization, protein denaturation, and lipid oxidation [72].

The exothermic peak at approximately 500 °C paralleled a mass decay observed in the TG curve, representing the final mass loss. Finally, the thermal behavior presented by the RP and CP indicated that the hydrothermal treatment to which the fruits were submitted (cooking with water at high temperature and pressure) and drying at 55 °C, used to obtain the flour, interfered with the differential thermogravimetry profiles of the products.

### 3.6. Absorption Spectroscopy in the Infrared Region

Figure 2 shows the Fourier transform infrared spectroscopy (FTIR) results for the RP, CP, RPF, and CPF. The spectral patterns presented in the evaluated frequency range evidenced the presence of peaks between 3350 and 1055 cm$^{-1}$ for the RP (Figure 2A); 3350 and 997 cm$^{-1}$ for the CP (Figure 2B); 3375 and 1018 cm$^{-1}$ for the RPF (Figure 2C); and between 3415 and 1010 cm$^{-1}$ for the CPF (Figure 2D). In general, the peaks detected for the pulps (RP and CP) and flours (RPF and CPF) were similar to those found by Giombelli et al. [40] for the dietary fiber concentrates from peach palms obtained via different extraction procedures.

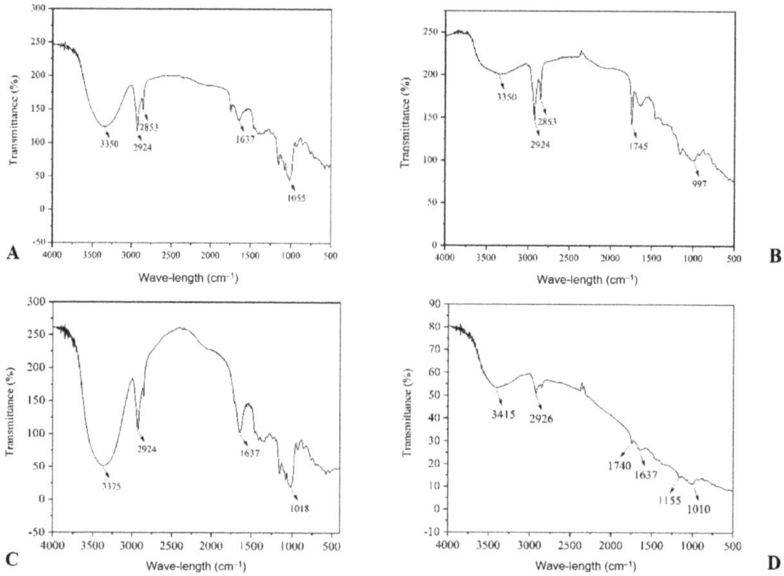


**Figure 2.** Infrared spectroscopy of albino peach palm. (**A**) raw pulp; (**B**) cooked pulp; (**C**) raw pulp flour; (**D**) cooked pulp flour.

The spectra of the RP (Figure 2A) and CP (Figure 2B) exhibited a medium width and medium absorption peak at 3350 $cm^{-1}$, while the RPF (Figure 2C) showed a broad and medium absorption band at 3375 $cm^{-1}$. This band is characteristic of hydrogen bonds, with the presence of the O–H of hydroxyl groups, present in the D-glucose units and their glycosidic bonds, units that constitute the starch structures of plant species [72], like peach palms. On the other hand, the CPF (Figure 2D) had a peak at 3415 $cm^{-1}$, with stretching characteristic of primary and secondary amines and amides (N–H).

The RP and RPF showed a strong absorption peak at 1637 $cm^{-1}$, which was practically suppressed in the CP and CPF. This peak is characteristic of the presence of primary amide bonds (primary elongation C=O). For secondary amides, the characteristic band occurred at 1155 $cm^{-1}$, which is present in proteins [73,74]. These results were consistent with the lower protein content observed in the CP compared to the RP (Table 1).

The four samples showed medium and low absorption peaks between 2926 and 2853 $cm^{-1}$, which correspond to the stretching of alkanes ($-CH_3$) and the stretching of the bonds of the C–O glycosidic groups, related to the bonds of the starch-forming units (D-glucose molecules) [74,75]. Peaks in this absorption range were detected in the flour from other peach palm varieties [14].

The observed absorptions at 1740 and 1745 $cm^{-1}$ for the CP (Figure 2B) and CPF (Figure 2D) were attributed to the C=O stretching of carbonyl esters present in functional groups such as ketones and aldehydes. The fact that these peaks were revealed only for the CP and CPF indicates that the hydrothermal process promoted the release of proteins in the albino peach palm.

The absorption bands at 1055 $cm^{-1}$ for the RP and 997 $cm^{-1}$ for the CP are characteristic of alcohols, esters, ethers, and carboxylic acids, among other functional groups. The smallest band at 997 $cm^{-1}$ can be attributed to the sequence of the aliphatic chains and aromatic rings of fatty acids in PC [5,73,74]. These peaks were also observed by Pires et al. [14] for flour from other peach palm varieties.

The interaction of the hydrogen bonds of water with the glycosidic bonds (α 1-4 and α 1-6) of the starch-forming units (D-glucose units) can cause changes in the linear (amylose) and branched (amylopectin) chains of starch. These alterations can promote structural and functional changes in starch, attributed to gelation, paste formation, and retrogradation [70,76].

### 3.7. Morphological Aspects

Scanning electron microscopy (SEM) was used to evaluate whether the hydrothermal process (cooking in water under high temperature and pressure) applied to peach palm and if drying the RP and CP at 55 °C promoted morphological structure changes in the CP, RPF, and CPF. The micrographs are shown in Figures 3–5. Figure 3A presents an overview of the RP, highlighting the plastids or amyloplasts, which are non-pigmented organelles specialized in synthesizing starch via the polymerization of glucose molecules, storing them in the form of starch granules (amyliferous) [77]. The presence of fibrous structures in bundles with attached starch granules (Figure 3B) may indicate functions other than energy, as starches that participate in this type of coupling are considered resistant and perform functions similar to fibers [78]. Thus, part of the starches attached to the fiber bundles may not be digested by the action of enzymes in the small intestine, which can promote beneficial effects on the human body, such as accelerating the digestion process, increasing satiety, and the consequent loss of weight [79,80].

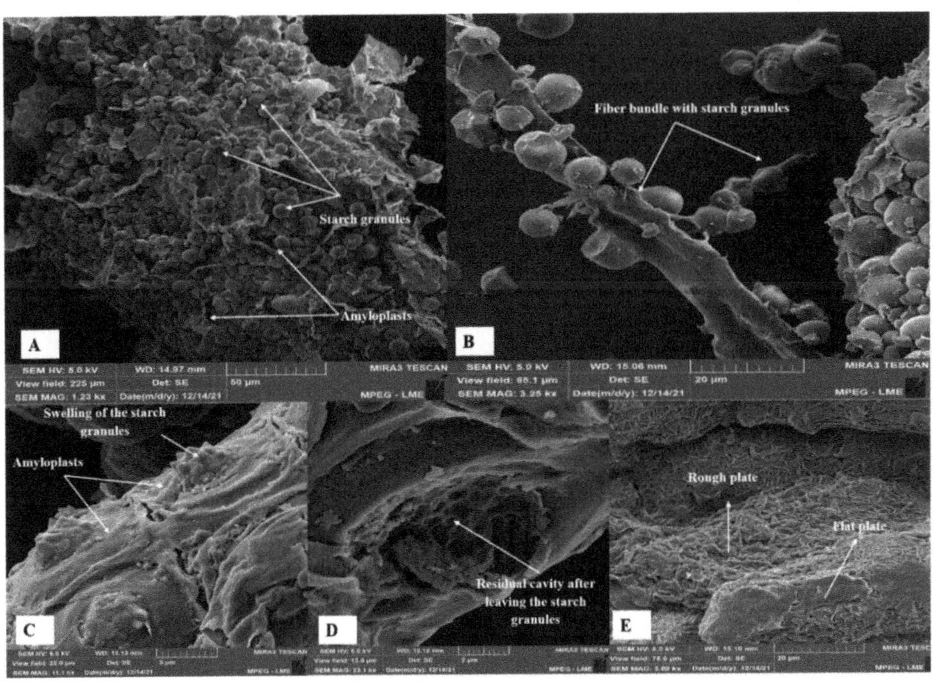


**Figure 3.** Micrographs of the raw pulp of albino peach palm: (**A**) 1.23 k× resolution and (**B**) 3.25 k× resolution; cooked pulp: (**C**) 11.1 k× resolution; (**D**) 23.1 k× resolution; (**E**) 3.69 k× resolution.

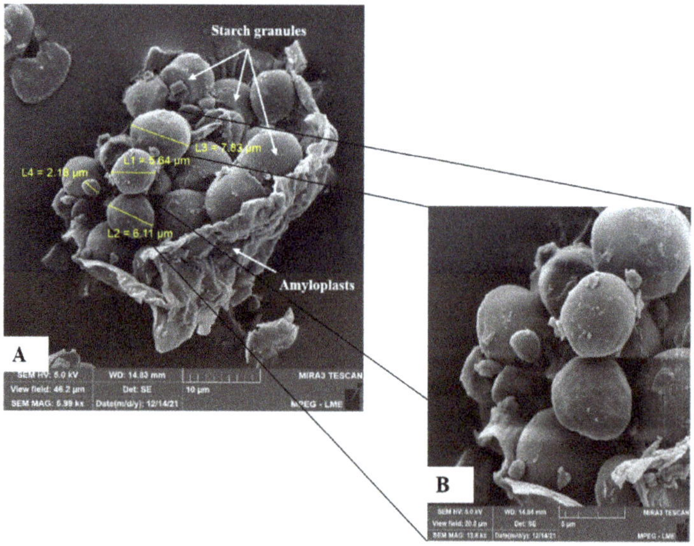


**Figure 4.** Micrographs of raw pulp flour of albino peach palm. (**A**) at 5.99 k× resolution and (**B**) 13.8 k× resolution.

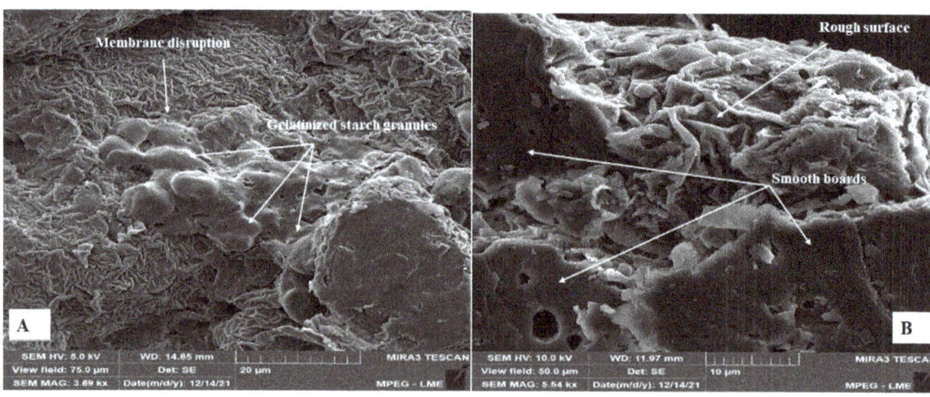


**Figure 5.** Micrographs of the cooked pulp flour of albino peach palm. (**A**) at 3.69 k× resolution and (**B**) 5.54 k× resolution.

The micrographs showed that the CP's morphological structure had undergone more modifications compared to the RP. During the hydrothermal process, heat in the presence of water promoted the rupture of amyloplasts, with the consequent release of some starch granules (Figure 3C). This destruction of the parenchymal architecture promoted the formation of residual spongy structures in the membranes (Figure 3D). Figure 3E shows an intense disruption of the parenchyma architecture due to ruptured amyloplasts, which is attributed to starch gelatinization during the hydrothermal process and to a probable retrogradation effect after cooking [72]. There was a loss of the compartmentalization of the structures, resulting in the stacking of plates with rough membranes and some smooth plates.

The micrographs of the RPF (Figure 4) showed the presence of an amyloplast containing multiple heterogeneous starch grains, with diameters between 2.18 µm and 7.83 µm and predominantly oval and spherical shapes, with a smooth surface and no cracks. The granule shapes that remained after drying at 55 °C compared to the RP (Figure 3A,B) were similar to the profiles observed by Ferrari Felisberto et al. [69] for the starch granules from another peach palm variety. Lindeboom et al. [72] proposed a classification for starch granules, according to their size, as very small (<5 µm), small (5–10 µm), medium (10–25 µm), and large (>25 µm). Based on this classification, the peach palm starch granules were classified as small (5–10 µm). Neto et al. [70] reported the diameters of the same order of magnitude (1.6–8.5 µm) for the starch granules from a peach palm species, while Valencia et al. [76] observed larger starch granules (5.2–12.5 µm) in the peach palm flour of the red variety. Knowing the structures of the starch and the size of the granules is relevant for defining the applications for starch, since these properties can interfere with digestion after cooking (contact surface), water retention, and starch efficiency as an encapsulating agent [70].

In general, the morphological aspects of the CPF (Figure 5) were similar to those observed for the CP (Figure 3C–E), confirming that the main morphological changes occurred during hydrothermal cooking. Torres-Vargas et al. [43] observed that green peach palm flour, raw and cooked under pressure (121 °C) for 1 h, showed the same botanical architecture for the starch granules, but there was also a disruption of the amyloplast membranes.

### 3.8. Particle size Analysis of Flour

The results of the particle size distribution for both the RPF and CPF (Figure 6) showed that the flour had slightly different granulometric profiles, but the highest retention occurred in the mesh with an opening of 425 µm for both flour. On the other hand, 89.4% of the CPF was retained on this sieve, while the retention was only 52.4% for the RPF. Regarding the remaining fraction, the largest amount of RPF (43%) had an average particle size greater than 425 µm, while the largest amount of CPF (~8%) had an average particle size smaller than 425 µm.

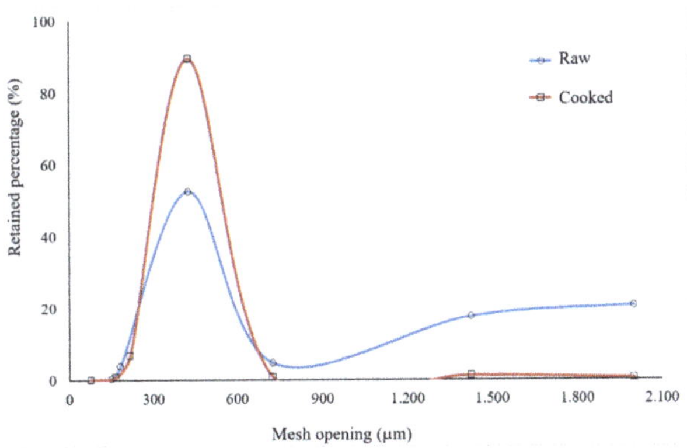


**Figure 6.** Grain size distribution for raw pulp flour and cooked pulp flour of albino peach palm.

As the conditions used in drying the RP and CP and in grinding the dry products were the same, the granulometric profiles of the flour allow us to state that the hydrothermal process used in cooking the peach palm caused the weakening of the molecular structures of the albino peach palm pulp, making the CPF less resistant to milling than the RPF. The observed behavior can be attributed to the disruption of the amyloplast membranes during the hydrothermal process, evidenced in the micrographs of the CP (Figure 3) and CPF (Figure 5). The fact that the granulometric distributions of both flour presented a single peak, with the same amplitude and great intensity indicated that the flour presented uniform granulometry. However, the peak of greater intensity observed for the CPF allows us to state that this flour had a greater granulometric uniformity, which directly interfered with its sensory quality (texture, flavor, and appearance) [81].

### 3.9. Technological Functional Properties

The values of the technological functional properties analyzed in the RPF and CPF are shown in Table 5. The CPF had a lower water solubility index (WSI) and oil absorption capacity (OAC), and a higher water absorption index (WAI) than the RPF ($p \leq 0.05$). These results indicated that the hydrothermal process (cooking) caused changes in the main technological functional properties of the albino peach palm flour. Zhou et al. [82] reported that the WSI of starchy matrices was related to the presence of starch and soluble proteins after cooking in the presence of water. According to Hatamian et al. [83], cooking promoted damage to starch granules and, consequently, an increase in the WSI. This behavior was not observed for the albino peach palm under the conditions studied and can be attributed to the leaching of water-soluble constituents from the pulp, such as sugars (Table 1), during the hydrothermal process.

**Table 5.** Functional technological properties of raw pulp and cooked pulp flour of albino peach palm.

| Property | Flour | |
|---|---|---|
| | Raw Pulp | Cooked Pulp |
| WSI (%) | $26.57 \pm < 0.01^{a}$ | $18.05 \pm 1.84^{b}$ |
| WAI (g/g dry matter) | $1.46 \pm 0.04^{b}$ | $2.07 \pm 0.08^{a}$ |
| OAC (g/g dry matter) | $1.04 \pm 0.04^{a}$ | $0.89 \pm 0.03^{b}$ |
| FC (%) | $11.11 \pm 2.24$ | nr |
| FS | nr | nr |

Mean values of three replicates ± standard deviation. Different letters on the same line represent a difference with 95% significance. WSI: water solubility index; WAI: water absorption index; OAC: oil absorption capacity; FC: foaming capacity; FS: foam stability; nr: no response for method.

Regarding the WAI, according to Silva [81], flour with a smaller granulometry adsorbs more water than flour with larger particle sizes, which is in agreement with the results observed in the present study, in which the CPF that presented a greater number of smaller particles (Figure 6), was the one with the highest WAI value ($p \leq 0.05$). Additionally, the highest WAI value for the CPF indicated that during the hydrothermal process, starch molecules might have been converted into lower-molecular-weight molecules (e.g., dextrins), which have more affinity for water molecules [83,84].

The results showed that during the hydrothermal process, there was a greater increase in the hydrophilic constituents than in the hydrophobic constituents in the pulp, which justifies a lower OAC and a higher WAI in the CPF. The OAC value is directly related to the protein content in the food; therefore, the lower the protein content, the lower the food's ability to absorb oil [85]. The OAC is an index to be considered when using flour in formulations to improve their sensory characteristics [86]. Proteins present on the surface of the product increase the hydrophobic interaction with flavor compounds [87]. However, the albino peach palm presented behavior different to what was expected (Table 1).

Finally, the RPF showed foaming capacity (FC), which indicates that the applied cooking eliminated the foaming compounds from the pulp. Likewise, proteins also influence the foaming capacity of the flour due to their amphiphilic properties and because they are surfactants [88]. The ability of flour to form foam directs its use in manufacturing products to improve the texture, consistency, and appearance, mainly of desserts and cocktails [89].

## 4. Conclusions

This study comprises the first report on the chemical composition, bioactive compounds, and physicochemical, functional–technological, color, thermal, and morphological properties of the raw pulp (RP), cooked pulp (CP), raw pulp flour (RPF), and cooked pulp flour (CPF) of albino peach palm. The results showed that the hydrothermal treatment (cooking) decreased the contents of lipids, total dietary fiber, and sugars of the albino peach palm pulp, while drying at 55 °C caused the concentration of these nutrients in the flour (except for carbohydrates and sugar). Cooking and drying, in turn, caused the browning of the CP and CPF. The RP turned out to be an excellent source of calcium, magnesium, and phosphorus, contributing to the requirements for human health, even though this latter mineral was decreased by cooking. The contents of phenolic compounds, flavonoids, ascorbic acid, and carotenoids were negatively impacted by cooking. The thermal analysis showed that the cooking and drying processes modified the thermogravimetric profile of the products, while the FTIR showed that there were little changes in the functional groups of the chemical structures of the pulp after cooking, which was more representative in the dry pulp (flour). The morphological aspects proved the existence of fibrous structures, including those linked to starch granules in the albino peach palm pulp, and evidence of the starch gelatinization caused by cooking. The CPF had smaller particle sizes than the RPF, showing that cooking caused the softening of the more resistant structures of the albino peach palm pulp. On the other hand, the RPF had a higher water solubility index, oil absorption capacity, and foam stability, and lower water absorption index than the CPF. The results of this research should encourage studies on the other components of the albino peach palm, as well as the technological and functional application of albino peach palm flour to enrich foods, especially in dietary fibers, calcium, magnesium, and phosphorus. In addition, it is also important to better explore the bioactive potential of extracts from this peach palm variety in biological assays.

**Supplementary Materials:** The following supporting information can be downloaded at: https://www.mdpi.com/article/10.3390/foods12234344/s1.

**Author Contributions:** S.D.S.: conceptualization, formal analysis, investigation, writing—original draft, writing—review and editing, and visualization; O.V.d.S.: conceptualization, resources, writing—review and editing, formal analysis, investigation, visualization, and project administration; L.R.V.d.C.: formal analysis, investigation, and visualization; H.T.C.: formal analysis and visualization; J.O.C.S.J.: formal analysis; F.d.C.A.d.N.: conceptualization, supervision, investigation, and funding acquisition; R.d.S.P.: investigation, resources, data curation, writing—review and editing, supervision, and funding acquisition. All authors provided critical feedback. All authors have read and agreed to the published version of the manuscript.

**Funding:** This research was funded by the Conselho Nacional de Desenvolvimento Científico e Tecnológico (CNPq, Brazil) for a scholarship of S. D. Soares (132850/2020-6).

**Data Availability Statement:** Data are contained within the article.

**Acknowledgments:** The authors are thankful for the financial support provided by the Federal University of Pará (UFPA, Brazil), through PROPESP. They also thank Renan Campos Chisté and Carolina Marques Mandaji for the support with the phenolics and flavonoids analysis, and Kelly das Graças Fernandes Dantas and Allan da Silva Cruz for the mineral analysis.

**Conflicts of Interest:** The authors declare no conflict of interest.

## References

1. Aguiar, J.P.L.; Yuyama, K.; Souza, F.C.A. Caracterização dos frutos de pupunheira (*Bactris gasipaes* Kunth) cultivada na Vila do Equador, RR: O que há de novo? *Sci. Amazon.* **2019**, *8*, CA1–CA5.
2. Chisté, R.C.; Costa, E.L.N.; Monteiro, S.F.; Mercadante, A.Z. Carotenoid and phenolic compound profiles of cooked pulps of orange and yellow peach palm fruits (*Bactris gasipaes*) from the Brazilian Amazonia. *J. Food Compos. Anal.* **2021**, *99*, 103873. [CrossRef]
3. dos Santos, O.V.; Soares, S.D.; Dias, P.C.S.; do Nascimento, F.D.C.A.; da Conceição, L.R.V.; da Costa, R.S.; da Silva Pena, R. White Peach Palm (*Pupunha*) a New *Bactris gasipaes* Kunt variety from the amazon: Nutritional composition, bioactive lipid profile, thermogravimetric and morphological characteristics. *J. Food Compos. Anal.* **2022**, *112*, 104684. [CrossRef]
4. Dos Santos, O.V.; Langley, A.C.D.C.P.; de Lima, A.J.M.; Moraes, V.S.V.; Soares, S.D.; Teixeira-Costa, B.E. Nutraceutical potential of amazonian oilseeds in modulating the immune system against COVID-19—A narrative review. *J. Funct. Foods* **2022**, *94*, 105123. [CrossRef] [PubMed]
5. dos Santos, O.V.; Soares, S.D.; Dias, P.C.S.; de Paula de Almeida Duarte, S.; dos Santos, M.P.L.; das Chagas Alves do Nascimento, F. Chromatographic profile and bioactive compounds found in the composition of pupunha oil (*Bactris gasipaes* Kunth): Implications for human health. *Rev. Nutr.* **2020**, *33*, e190146. [CrossRef]

6. Muñoz, A.M.; Casimiro-Gonzales, S.; Gómez-Coca, R.B.; Moreda, W.; Best, I.; Cajo-Pinche, M.I.; Loja, J.F.; Ibañez, E.; Cifuentes, A.; Ramos-Escudero, F. Comparison of four oil extraction methods for sinami fruit (*Oenocarpus mapora* H. Karst): Evaluating quality, polyphenol content and antioxidant activity. *Foods* **2022**, *11*, 1518. [CrossRef]
7. Rojas-Garbanzo, C.; Pérez, A.M.; Vaillant, F.; Pineda-Castro, M.L. Physicochemical and antioxidant composition of fresh peach palm (*Bactris gasipaes* Kunth) fruits in Costa Rica. *Braz. J. Food Technol.* **2016**, *19*, e2015097. [CrossRef]
8. Pinzón-Zárate, L.X.; Hleap-Zapata, J.I.; Ordóñez-Santos, L.E. Análisis de los parámetros de color en salchichas Frankfurt adicionadas con extracto oleoso de residuos de Chontaduro (*Bactris gasipaes*). *Inf. Tecnol.* **2015**, *26*, 45–54. [CrossRef]
9. Cornelius, J.P.; Weber, J.C.; Sotelo-Montes, C.; Ugarte-Guerra, L.J. Phenotypic correlations and site effects in a peruvian landrace of peach palm (*Bactris gasipaes* Kunth). *Euphytica* **2010**, *173*, 173–183. [CrossRef]
10. Pinheiro, R.C.; Ballesteros, L.F.; Cerqueira, M.A.; Rodrigues, A.M.C.; Teixeira, J.A.; Silva, L.H.M. Peach palm (*Bactris gasipaes* Kunth) and mammee apple (*Mammea americana* L.) seeds: Properties and potential of application in industry. *Lwt* **2022**, *170*, 114089. [CrossRef]
11. Melo, M.C.T.; Costa, L.L.; Pereira, F.C.; Castro, L.M.; Neponuceno, S. Physical and chemical analysis of fruit "in Natura" of pupunha. *Ciência E Tecnol. Aliment.* **2017**, *3*, 13–17.
12. dos Santos, M.A.S.; Protázio, D.C.; da Costa, G.P.; Rebello, F.K.; Martins, C.M.; Bezerra, A.S.; da Silva Nogueira, A. Profile of peach palm fruit consumers in the metropolitan region of Belém, Pará, Brazilian Amazon. *Int. J. Innov. Educ. Res.* **2021**, *9*, 550–560. [CrossRef]
13. Santos, I.L.; Steel, C.J.; Aguiar, J.P.L.; Schmiele, M.; de Sales Ferreira, C.J.; do Amaral Souza, F.D.C. Sensory analysis of extruded corn-based breakfast cereals with whole peach palm fruit (*Bactris gasipaes*, Kunth) powder. *Afr. J. Food Sci.* **2017**, *11*, 310–317. [CrossRef]
14. Pires, M.B.; Amante, E.R.; Lopes, A.S.; da Cruz Rodrigues, A.M.; da Silva, L.H.M. Peach palm flour (*Bactris gasipae* Kunth): Potential application in the food industry. *Food Sci. Technol.* **2019**, *39*, 613–619. [CrossRef]
15. Leterme, P.; García, M.F.; Londoño, A.M.; Rojas, M.G.; Buldgen, A.; Souffrant, W.B. Chemical composition and nutritive value of peach palm (*Bactris gasipaes* Kunth) in rats. *J. Sci. Food Agric.* **2005**, *85*, 1505–1512. [CrossRef]
16. dos Santos, A.B.; Pereira, M.L.A.; de Oliveira Silva, H.G.; de Carvalho, G.G.P.; de Jesus Pereira, T.C.; Ribeiro, L.S.O.; Azevêdo, J.A.G.; das Graças Conceição Parada Costa Silva, M.; Sousa, L.B.; Sousa, L.B.; et al. Intake, digestibility and performance of lambs fed diets containing peach palm meal. *Trop. Anim. Health Prod.* **2016**, *48*, 509–515. [CrossRef] [PubMed]
17. Faria, J.V.; Valido, I.H.; Paz, W.H.P.; da Silva, F.M.A.; de Souza, A.D.L.; Acho, L.R.D.; Lima, E.S.; Boleti, A.P.A.; Marinho, J.V.N.; Salvador, M.J.; et al. Comparative evaluation of chemical composition and biological activities of tropical fruits consumed in Manaus, Central Amazonia, Brazil. *Food Res. Int.* **2021**, *139*, 109836. [CrossRef]
18. Martínez-Girón, J.; Figueroa-Molano, A.M.; Ordóñez-Santos, L.E. Effect of the addition of peach palm (*Bactris gasipaes*) peel flour on the color and sensory properties of cakes. *Food Sci. Technol.* **2017**, *37*, 418–424. [CrossRef]
19. de Souza Mesquita, L.M.; Neves, B.V.; Pisani, L.P.; de Rosso, V.V. Mayonnaise as a model food for improving the bioaccessibility of carotenoids from *Bactris gasipaes* fruits. *LWT* **2020**, *122*, 109022. [CrossRef]
20. de Melo Neto, B.A.; Junior, C.C.M.F.; da Silva, E.G.P.; Franco, M.; dos Santos Reis, N.; Ferreira Bonomo, R.C.; de Almeida, P.F.; Pontes, K.V. Biodegradable thermoplastic starch of peach palm (*Bactris gasipaes* Kunth) fruit: Production and characterisation. *Int. J. Food Prop.* **2018**, *20*, S2429–S2440. [CrossRef]
21. Singo, T.M.; Beswa, D. Effect of roselle extracts on the selected quality characteristics of ice cream. *Int. J. Food Prop.* **2019**, *22*, 42–53. [CrossRef]
22. Salazar, D.; Arancibia, M.; Raza, K.; López-Caballero, M.E.; Montero, M.P. Influence of underutilized unripe banana (*Cavendish*) flour in the formulation of healthier chorizo. *Foods* **2021**, *10*, 1486. [CrossRef] [PubMed]
23. Lang, G.H.; Lindemann, I.D.S.; Ferreira, C.D.; Pohndorf, R.S.; Vanier, N.L.; de Oliveira, M. Influence of drying temperature on the structural and cooking quality properties of black rice. *Cereal Chem.* **2018**, *95*, 564–574. [CrossRef]
24. Lang, G.H.; Lindemann, I.D.S.; Ferreira, C.D.; Pohndorf, R.S.; Vanier, N.L.; de Oliveira, M. Fluidized-bed drying of black rice grains: Impact on cooking properties, in vitro starch digestibility, and bioaccessibility of phenolic compounds. *J. Food Sci.* **2020**, *85*, 1717–1724. [CrossRef] [PubMed]
25. Vimercati, W.C.; da Silva Araújo, C.; Macedo, L.L.; Maradini Filho, A.M.; Saraiva, S.H.; Teixeira, L.J.Q. Influence of drying temperature on drying kinetics, energy consumption, bioactive compounds and cooking quality of pasta enriched with spinach. *J. Food Process Eng.* **2020**, *43*, e13571. [CrossRef]
26. AOAC. *Official Methods of Analysis of the Association of Official Analytical Chemists International*; AOAC: Washington, DC, USA, 2016.
27. AOAC. *Official Methods of Analysis of the Association of Official Analytical Chemists International*; AOAC: Gaithersburg, MD, USA, 2000.
28. FAO/WHO. *Food Energy: Methods of Analysis and Conversion Factors*; FAO: Rome, Italy, 2002.
29. Mclellan, M.R.; Lind, L.R.; Kime, R.W. Hue angle determinations and statistical analysis for multiquadrant hunter L,a,b data. *J. Food Qual.* **1995**, *18*, 235–240. [CrossRef]
30. Palavecino, P.M.; Penci, M.C.; Calderón-Domínguez, G.; Ribotta, P.D. Chemical composition and physical properties of sorghum flour prepared from different sorghum hybrids grown in Argentina. *Starch-Stärke* **2016**, *68*, 1055–1064. [CrossRef]
31. Alves, B.S.F.; Pereira Junior, J.B.; Carvalho, F.I.M.; Dantas Filho, H.A.; Fernandes Dantas, K.G. Mineral composition of Amazonian fruits by flame atomic absorption spectrometry using multivariate analysis. *Biol. Trace Elem. Res.* **2019**, *189*, 259–266. [CrossRef]

32. Singleton, V.L.; Rossi, J.A. Colorimetry of total phenolics with phosphomolybidic-phosphotungstic acid reagent. *Am. J. Enol. Vitic.* **1965**, *16*, 144–158. [CrossRef]
33. Pękal, A.; Pyrzynska, K. Evaluation of aluminium complexation reaction for flavonoid content assay. *Food Anal. Methods* **2014**, *7*, 1776–1782. [CrossRef]
34. Godoy, H.T.; Rodriguez-Amaya, D.B. Occurrence of cis-isomers of provitamin A in Brazilian fruits. *J. Agric. Food Chem.* **1994**, *42*, 1306–1313. [CrossRef]
35. Zanotto, D.L.; Bellaver, C. Método de Determinação da Granulometria de Ingredientes Para Uso em Rações de Suínos e Aves. Embrapa/CNPSA. 1996. Available online: https://www.infoteca.cnptia.embrapa.br/handle/doc/433741 (accessed on 20 October 2022).
36. Anderson, R.A.; Conway, H.F.; Pfeifer, V.; Griffin, E.L. Gelatinization of corn grits by roll and extrusion cooking. *Cereal Sci. Today* **1969**, *14*, 4–12.
37. Coffmann, C.N.; Garcia, V. V Functional properties and amino acid content of a protein isolate from mung bean flour. *Int. J. Food Sci. Technol.* **1977**, *12*, 473–484. [CrossRef]
38. Silva, R.B.; Silva-Júnior, E.V.; Rodrigues, L.C.; Andrade, L.H.C.; da Silva, S.I.; Harand, W.; Oliveira, A.F.M. A comparative study of nutritional composition and potential use of some underutilized tropical fruits of Arecaceae. *An. Acad. Bras. Cienc.* **2015**, *87*, 1701–1709. [CrossRef]
39. Martínez-Girón, J.; Rodríguez-Rodríguez, X.; Pinzón-Zárate, L.X.; Ordóñez-Santos, L.E. Caracterización fisicoquímica de harina de residuos del fruto de chontaduro (*Bactris gasipaes* Kunth, Arecaceae) obtenida por secado convectivo. *Corpoica Cienc. Y Tecnol. Agropecu.* **2017**, *18*, 599–613. [CrossRef]
40. Giombelli, C.; Raspe, D.; Donadone, D.; da Silva, C.; Barros, B. Chemical composition and functional properties of dietary fiber concentrates obtained from peach palm by-product. *J. Braz. Chem. Soc.* **2023**, *34*, 927–936. [CrossRef]
41. FAO/WHO. *Dietary Protein Quality Evaluation in Human Nutrition: Report of an FAO Expert Consultation*; FAO: Rome, Italy, 2013.
42. Basto, G.J.; Carvalho, C.W.P.; Soares, A.G.; Costa, H.T.G.B.; Chávez, D.W.H.; de Oliveira Godoy, R.L.; Pacheco, S. Physicochemical properties and carotenoid content of extruded and non-extruded corn and peach palm (*Bactris gasipaes*, Kunth). *LWT-Food Sci. Technol.* **2016**, *69*, 312–318. [CrossRef]
43. Torres-Vargas, O.L.; Luzardo-Ocampo, I.; Hernandez-Becerra, E.; Rodríguez-García, M.E. Physicochemical characterization of unripe and ripe chontaduro (*Bactris gasipaes* Kunth) fruit flours and starches. *Starch-Stärke* **2021**, *73*, 2000242. [CrossRef]
44. Dong, Y.; Chen, L.; Gutin, B.; Zhu, H. Total, insoluble, and soluble dietary fiber intake and insulin resistance and blood pressure in adolescents. *Eur. J. Clin. Nutr.* **2019**, *73*, 1172–1178. [CrossRef]
45. Aune, D.; Chan, D.S.M.; Greenwood, D.C.; Vieira, A.R.; Navarro Rosenblatt, D.A.; Vieira, R.; Norat, T. Dietary fiber and breast cancer risk: A systematic review and meta-analysis of prospective studies. *Ann. Oncol.* **2012**, *23*, 1394–1402. [CrossRef]
46. Sumiyoshi, M.; Kimura, Y. Low molecular weight chitosan inhibits obesity induced by feeding a high-fat diet long-term in mice. *J. Pharm. Pharmacol.* **2010**, *58*, 201–207. [CrossRef] [PubMed]
47. Negu, A.; Zegeye, A.; Astatkie, T. Development and quality evaluation of wheat based cookies enriched with fenugreek and oat flours. *J. Food Sci. Technol.* **2020**, *57*, 3573–3580. [CrossRef] [PubMed]
48. Giombelli, C.; Iwassa, I.J.; da Silva, C.; Bolanho Barros, B.C. Valorization of peach palm by-product through subcritical water extraction of soluble sugars and phenolic compounds. *J. Supercrit. Fluids* **2020**, *165*, 104985. [CrossRef]
49. Suzauddula, M.; Hossain, M.B.; Farzana, T.; Orchy, T.N.; Islam, M.N.; Hasan, M.M. Incorporation of oat flour into wheat flour noodle and evaluation of its physical, chemical and sensory attributes. *Braz. J. Food Technol.* **2021**, *24*, e2020252. [CrossRef]
50. USDA (United State Department of Agriculture). *Nutrient Database for Standard—Reference—Release 15: Nutrient*; USDA: Washington, DC, USA, 2002.
51. Mamiro, P.S.; Mwanri, A.W.; Mongi, R.J.; Chivaghula, T.J.; Nyagaya, M.; Ntwenya, J. Effect of cooking on tannin and phytate content in different bean (*Phaseolus vulgaris*) varieties grown in Tanzania. *Afr. J. Biotechnol.* **2017**, *16*, 1186–1191. [CrossRef]
52. Akin, M.; Eyduran, S.P.; Eyduran, E.; Reed, B.M. Analysis of macro nutrient related growth responses using multivariate adaptive regression splines. *Plant Cell Tissue Organ Cult.* **2020**, *140*, 661–670. [CrossRef]
53. Gharibzahedi, S.M.T.; Jafari, S.M. The importance of minerals in human nutrition: Bioavailability, food fortification, processing effects and nanoencapsulation. *Trends Food Sci. Technol.* **2017**, *62*, 119–132. [CrossRef]
54. Kostov, K. Effects of magnesium deficiency on mechanisms of insulin resistance in type 2 diabetes: Focusing on the processes of insulin secretion and signaling. *Int. J. Mol. Sci.* **2019**, *20*, 1351. [CrossRef]
55. Vannucci, L.; Fossi, C.; Quattrini, S.; Guasti, L.; Pampaloni, B.; Gronchi, G.; Giusti, F.; Romagnoli, C.; Cianferotti, L.; Marcucci, G.; et al. Calcium Intake in Bone Health: A focus on calcium-rich mineral waters. *Nutrients* **2018**, *10*, 1930. [CrossRef]
56. Bhattarai, H.K.; Shrestha, S.; Rokka, K.; Shakya, R. Vitamin D, calcium, parathyroid hormone, and sex steroids in bone health and effects of aging. *J. Osteoporos.* **2020**, *2020*, 9324505. [CrossRef]
57. van Dronkelaar, C.; van Velzen, A.; Abdelrazek, M.; van der Steen, A.; Weijs, P.J.M.; Tieland, M. Minerals and sarcopenia; the role of calcium, iron, magnesium, phosphorus, potassium, selenium, sodium, and zinc on muscle mass, muscle strength, and physical performance in older adults: A systematic review. *J. Am. Med. Dir. Assoc.* **2018**, *19*, 6–11.e3. [CrossRef] [PubMed]
58. FAO. Food and Agriculture Organization of the United Nations. Available online: http://faostat.fao.org/ (accessed on 10 October 2022).

59. Rojas-Garbanzo, C.; Pérez, A.M.; Pineda Castro, M.L.; Vaillant, F. Major physicochemical and antioxidant changes during peach-palm (*Bactris gasipaes* H.B.K.) flour processing. *Fruits* **2012**, *67*, 415–427. [CrossRef]
60. Haminiuk, C.W.I.; Maciel, G.M.; Plata-Oviedo, M.S.V.; Peralta, R.M. Phenolic compounds in fruits—An overview. *Int. J. Food Sci. Technol.* **2012**, *47*, 2023–2044. [CrossRef]
61. Lang, G.H.; da Silva Lindemann, I.; Ferreira, C.D.; Hoffmann, J.F.; Vanier, N.L.; de Oliveira, M. Effects of drying temperature and long-term storage conditions on black rice phenolic compounds. *Food Chem.* **2019**, *287*, 197–204. [CrossRef] [PubMed]
62. Contreras-Calderón, J.; Calderón-Jaimes, L.; Guerra-Hernández, E.; García-Villanova, B. Antioxidant capacity, phenolic content and vitamin C in pulp, peel and seed from 24 exotic fruits from colombia. *Food Res. Int.* **2011**, *44*, 2047–2053. [CrossRef]
63. de Prado, J.M.; de Passos, G.R.; dos Santos, T.B.; da Silva, C.N.S.; Lemes, A.C.; de Aleluia Batista, K.; de Souza Sora, G.T.; Polesi, L.F.; de Paula, L.C. Production and characterization of a carotenoid-rich peach palm flour. *J. Eng. Exact Sci.* **2022**, *8*, 14866-01i. [CrossRef]
64. de Castro, N.T.; de Alencar, E.R.; Zandonadi, R.P.; Han, H.; Raposo, A.; Ariza-Montes, A.; Araya-Castillo, L.; Botelho, R.B.A. Influence of cooking method on the nutritional quality of organic and conventional brazilian vegetables: A study on sodium, potassium, and carotenoids. *Foods* **2021**, *10*, 1782. [CrossRef]
65. Kourouma, V.; Mu, T.H.; Zhang, M.; Sun, H.N. Effects of cooking process on carotenoids and antioxidant activity of orange-fleshed sweet potato. *Lwt* **2019**, *104*, 134–141. [CrossRef]
66. Zhang, S.; Ji, J.; Zhang, S.; Guan, C.; Wang, G. Effects of three cooking methods on content changes and absorption efficiencies of carotenoids in maize. *Food Funct.* **2020**, *11*, 944–954. [CrossRef]
67. Britton, G.; Khachik, F. Carotenoids: Nutrition and Health. In *Carotenoids in Food*; Birkhäuser Basel: Basel, Switzerland, 2009; pp. 45–66.
68. Farina, V.; Cinquanta, L.; Vella, F.; Niro, S.; Panfili, G.; Metallo, A.; Cuccurullo, G.; Corona, O. Evolution of carotenoids, sensory profiles and volatile compounds in microwave-dried fruits of three different loquat cultivars (*Eriobotrya japonica* Lindl.). *Plant Foods Hum. Nutr.* **2020**, *75*, 200–207. [CrossRef]
69. Ferrari Felisberto, M.H.; Souza Costa, M.; Villas Boas, F.; Lopes Leivas, C.; Maria Landi Franco, C.; Michielon de Souza, S.; Pedrosa Silva Clerici, M.T.; Mach Côrtes Cordeiro, L. Characterization and technological properties of peach palm (*Bactris gasipaes* Var. *Gasipaes*) fruit starch. *Food Res. Int.* **2020**, *136*, 109569. [CrossRef] [PubMed]
70. de Melo Neto, B.A.; Fernandes, B.S.; Junior, C.C.M.F.; Franco, M.; Bonomo, R.C.F.; de Almeida, P.F.; Pontes, K.V. Thermal-morphological characterisation of starch from peach-palm (*Bactris gasipaes* Kunth) fruit (*Pejibaye*). *Int. J. Food Prop.* **2017**, *20*, 1007–1015. [CrossRef]
71. de Melo Neto, B.A.; Bonomo, R.C.F.; Franco, M.; de Almeida, P.F.; Pontes, K.V. Starch extraction from the peach palm (*Bactris gasepaes* Kunth.) fruit: A model approach for yield increase. *Eng. Agric.* **2017**, *37*, 148–159. [CrossRef]
72. Lindeboom, N.; Chang, P.R.; Tyler, R.T. Analytical, biochemical and physicochemical aspects of starch granule size, with emphasis on small granule starches: A review. *Starch-Stärke* **2004**, *56*, 89–99. [CrossRef]
73. Elzey, B.; Pollard, D.; Fakayode, S.O. Determination of adulterated neem and flaxseed oil compositions by FTIR spectroscopy and multivariate regression analysis. *Food Control* **2016**, *68*, 303–309. [CrossRef]
74. Rai, A.; Mohanty, B.; Bhargava, R. Supercritical extraction of sunflower oil: A central composite design for extraction variables. *Food Chem.* **2016**, *192*, 647–659. [CrossRef] [PubMed]
75. Lei, M.; Jiang, F.-C.; Cai, J.; Hu, S.; Zhou, R.; Liu, G.; Wang, Y.-H.; Wang, H.-B.; He, J.-R.; Xiong, X.-G. Facile microencapsulation of olive oil in porous starch granules: Fabrication, characterization, and oxidative stability. *Int. J. Biol. Macromol.* **2018**, *111*, 755–761. [CrossRef] [PubMed]
76. Valencia, G.A.; Moraes, I.C.F.; Lourenço, R.V.; Bittante, A.M.Q.B.; Sobral, P.J.D.A. Physicochemical, morphological, and functional properties of flour and starch from peach palm (*Bactris gasipaes* K.) fruit. *Starch-Stärke* **2015**, *67*, 163–173. [CrossRef]
77. Richardson, S.; Gorton, L. Characterisation of the Substituent distribution in starch and cellulose derivatives. *Anal. Chim. Acta* **2003**, *497*, 27–65. [CrossRef]
78. Koksel, H.; Masatcioglu, T.; Kahraman, K.; Ozturk, S.; Basman, A. Improving effect of lyophilization on functional properties of resistant starch preparations formed by acid hydrolysis and heat treatment. *J. Cereal Sci.* **2008**, *47*, 275–282. [CrossRef]
79. Jefferson, A.; Adolphus, K. The effects of intact cereal grain fibers, including wheat bran on the gut microbiota composition of healthy adults: A systematic review. *Front. Nutr.* **2019**, *6*, 33. [CrossRef] [PubMed]
80. Tosh, S.M.; Bordenave, N. Emerging science on benefits of whole grain oat and barley and their soluble dietary fibers for heart health, glycemic response, and gut microbiota. *Nutr. Rev.* **2021**, *78*, 13–20. [CrossRef] [PubMed]
81. SILVA, L.H. Desenvolvimento de pão de fôrma com a adição de farinha de "Okara". *Braz. J. Food Technol.* **2010**, *12*, 315–322. [CrossRef]
82. Zhou, S.; Reddy, C.K.; Du, B.; Xu, B. Pasting, thermal, and functional properties of wheat flour and rice flour formulated with chestnut flour. *Bioact. Carbohydr. Diet. Fibre* **2021**, *26*, 100290. [CrossRef]
83. Hatamian, M.; Noshad, M.; Abdanan-Mehdizadeh, S.; Barzegar, H. Effect of roasting treatment on functional and antioxidant properties of chia seed flours. *NFS J.* **2020**, *21*, 1–8. [CrossRef]
84. Lin, Y.; An, F.; He, H.; Geng, F.; Song, H.; Huang, Q. Structural and rheological characterization of pectin from passion fruit (*Passiflora edulis f. flavicarpa*) peel extracted by high-speed shearing. *Food Hydrocoll.* **2021**, *114*, 106555. [CrossRef]
85. Awuchi, C.G.; Igwe, V.S.; Echeta, C.K. The functional properties of foods and flours. *Int. J. Adv. Acad. Res. Sci.* **2019**, *5*, 2488–9849.

86. Das, P.C.; Khan, M.J.; Rahman, M.S.; Majumder, S.; Islam, M.N. Comparison of the physico-chemical and functional properties of mango kernel flour with wheat flour and development of mango kernel flour based composite cakes. *NFS J.* **2019**, *17*, 1–7. [CrossRef]
87. Ofori, J.; Tortoe, C.; Agbenorhevi, J.K. Physicochemical and functional properties of dried okra (*Abelmoschus esculentus* L.) seed flour. *Food Sci. Nutr.* **2020**, *8*, 4291–4296. [CrossRef]
88. Kanetro, B.; Riyanto, M.; Pujimulyani, D.; Huda, N. Improvement of functional properties of jack bean (*Canavalia ensiformis*) flour by germination and its relation to amino acids profile. *Curr. Res. Nutr. Food Sci.* **2021**, *9*, 812–822. [CrossRef]
89. Zielinska, E. Evaluating the Functional characteristics of certain insect flours (non-defatted/defatted flour) and their protein preparations. *Molecules* **2022**, *27*, 6339. [CrossRef]

foods

MDPI

Article

# Correlation Analysis between Volatile Compounds and Quality Attributes in Pork Tenderloin in Response to Different Stir-Frying Processes

**Ziqiang Wang [1], Tianjie Nie [1], Huiying Zhang [1,2,*], Wenqian Wang [1], Haitao Chen [1,2], Shuqi Wang [1,2] and Baoguo Sun [1,2]**

[1] Beijing Key Laboratory of Flavor Chemistry, Beijing Technology and Business University, Beijing 100048, China; qhysdeqj@126.com (Z.W.); taotaodeleyuan@126.com (T.N.); wwwenqian1998@163.com (W.W.); chenht@th.btbu.edu.cn (H.C.); wsq930312@163.com (S.W.); sunbg@btbu.edu.cn (B.S.)

[2] China Food Flavor and Nutrition Health Innovation Center, Beijing Technology and Business University, Beijing 100048, China

* Correspondence: zhanghuiying@btbu.edu.cn; Tel./Fax: +86-10-010-68984685

**Abstract:** Volatile compounds and physicochemical properties of meat are significantly changed by cooking processes. This study explored the influence of different stir-frying temperatures and times on the dynamic changes of the physicochemical characteristics and volatiles of pork tenderloin and determined the correlation between them. Results showed that time played more of a role than temperature. At the same temperature, the water content decreased ($p < 0.05$) and the cooking loss increased ($p < 0.05$) with stir-frying time extending. The L* value and the b* value showed first an increasing and then decreasing trend ($p < 0.05$), while the a* value significantly increased ($p < 0.05$). The higher the cooking temperature of sample, the faster the indexes changed. In stir-fried samples, 50 volatiles were identified. Correlation analysis showed that among the quality attributes, b* value and water content had the strongest impact on volatiles. The water content was negatively correlated with most of the compounds attributed to the desired aroma of stir-fried samples, while the correlation between the b* value and these volatiles was positive. Hence, changes in the types and contents of volatiles in stir-fried pork tenderloin could be predicted by detection of b* value and water content.

**Keywords:** pork tenderloin; stir frying; quality attributes; volatile compounds; correlation

## 1. Introduction

Over the last ten years, the need for the industrialization of Chinese cuisine has been growing steadily [1]. Pork is one of the most consumed meats worldwide [2]. Pork tenderloin, due to its tenderness and taste, has always been the basic ingredient of traditional stir-frying dishes, such as fish-flavored pork, in China. Aroma is one of the most crucial quality attributes which greatly affects consumer acceptance and preference [3]. The monitoring of aroma is helpful in determining the end point of the cooking process and optimizing the cooking parameters. It is of significance for improvement in the quality of pork tenderloin dishes and the development of intelligent manufacturing equipment.

The aroma qualities of meat are mainly attributed to volatile substances [4]. Aroma quality assessment based on volatile analysis methods such as gas chromatography-mass spectrometry (GC-MS) has been widely used. Wang et al. [5] investigated the influence of different heating modes on the volatiles and aroma of pork belly and found that balanced distribution and rich contents of volatiles contributed to a better aroma quality. Yang et al. [6] explored the effect of different cooking methods on the aroma compounds in pork loin. The results showed that aldehyde and pyrazine were the key contributors of cooked pork aroma. Chang et al. [7] analyzed the key aroma components in pork broth stewed for different times. Finally, the aroma profile under the optimum stewing time was obtained. Wang et al. [8] identified 12 aroma-active compounds in pork broth by gas chromatography–olfactometry–mass spectrometry and found the broth cooked at 96 °C possessed a strong meaty aroma. However, the analysis of volatile substances under different cooking processes is cumbersome, which is inapplicable in online aroma monitoring of automatic cooking in industrial production.

The changes in volatile compounds are normally accompanied by variations of other quality attributes such as color and water content in meat processing. Wang et al. [9] investigated the changes of texture, color, and volatiles of steamed pork loins. The results showed that hardness, a*, and b* showed an increasing trend with cooking time extending, whereas the L* value and the content of volatiles first increased and then decreased. Zhao et al. [10] examined the dynamic changes in the volatile chemicals and physicochemical properties (gel strength, whiteness, and hardness) of tilapia sausage in the fermentation process and found that the main physical indexes and typical volatile compounds were both significantly enhanced after fermentation. Lee et al. [11] showed that the effect of mild heat treatment of beef tenderloin reduced the formation of meat aroma but caused a remarkable increase in appearance and texture. Compared with the determination of volatiles, the analysis of quality indicators is relatively convenient. Obtaining the correlation data between them could be useful for realizing aroma prediction and quality control in the industrialized production of Chinese cuisine. Related studies have not been reported yet.

Therefore, the aims of this study were to (1) investigate the main physicochemical features that directly impact cooked pork tenderloin quality, such as the water content, cooking loss, color, and volatile compounds and (2) analyze the correlation between volatile compounds and other quality attributes in pork tenderloin in response to different stir-frying processes, which would provide data support for aroma prediction based on physicochemical indexes.

## 2. Materials and Methods

### 2.1. Materials and Chemicals

The fresh pork tenderloin, "Luhua" soybean oil, and food-grade sodium chloride were all purchased from Wu Mart supermarket in Beijing, China. The pork tenderloin was obtained from Delisi Food Co., Ltd. (Beijing, China), freshly cut and packaged in a modified-atmosphere box. To avoid the impact of storage, the samples were purchased on the same day the experiment was conducted and stored at 4 °C for a maximum of 4 h. The n-alkanes (C6–C30) used for retention indices calculation and the internal standard chemical, 2-methyl-3-heptanone, were provided by Sigma-Aldrich (Shanghai, China).

### 2.2. Sample Preparation

The pork tenderloin was washed, cleaned, and then cut into long strips of about $4.00 \times 1.00 \times 1.00$ cm; the average weight was $3.50 \pm 0.50$ g.

To stir fry the tenderloin, fresh pork samples ($20.00 \pm 1.00$ g) were put into a frying pan with soybean oil ($5.00 \pm 0.50$ g). In this study, pork tenderloin was fried at temperatures of 140 °C, 160 °C, 180 °C, 200 °C, and 220 °C. The frying times at each temperature were 50 s, 100 s, 150 s, 200 s, and 250 s. A total of 25 groups of pork tenderloin samples with different time and temperature conditions were obtained. The cooking treatment was repeated three times. Subsequently, the physicochemical qualities, sensory scores, and volatiles of the samples were obtained. Raw meat of the same size was served as the control group.

### 2.3. Cooking Loss and Water Content

After stir frying, the temperature of samples was decreased to 25 °C. Then, the samples were chopped to measure cooking loss value and water content. The weight difference between the samples before and after stir frying was used to calculate cooking loss [12]. Water content was measured by direct drying method according to the PRC National Standard 5009.3-2016 [13].

### 2.4. Color

The surface color of samples was measured with a CR-1000 Plus chromaticity meter (Konica Minolta, Tokyo, Japan). Fresh-cut pork slices were taken as the control group. Each group contained at least three slices. For each slice, at least three different positions were measured. The color parameters were determined by averaging the values of the three test samples [14]. Color of the pork tenderloin sample was expressed as the values of lightness ($L^*$), redness ($a^*$), and yellowness ($b^*$).

### 2.5. Sensory Evaluation

Twelve panelists (five males and seven females, 20–24 years old) were recruited from Beijing Key Laboratory of Flavor Chemistry of Beijing Technology and Business University. All experimental protocols were approved by the Ethics Committee of Beijing Business and Technology University, and informed consent was obtained from all subjects involved in the study. A sensory evaluation standard for stir-fried pork tenderloin was developed based on the panelists' description and discussion of the samples' sensory characteristics before the evaluation [15]. Four attributes were evaluated in the sensory part: color (golden, yellow or light brown, white or dark brown), texture (moderately tender, medium tough or tender, extremely tough or tender), aroma (rich aroma, common aroma, or light aroma) and overall flavor (good flavor, general flavor, or poor flavor). The selected attributes were scored by the panelists. The score scale ranged from 1 to 9. The scoring criteria is shown in Table S1. In this study, the stir-fried samples were placed in a thermostatic container at 60 °C. All panelists completed sensory evaluation within 30 min. To prevent the sequence effect, three random numbers were used to encrypt each sample [16]. The whole evaluation process was conducted in a specialized sensory evaluation laboratory. The team members did not contact and communicate with each other, and they rinsed their mouths with clear water between different samples [17].

### 2.6. Volatile Compounds Analysis

Volatile compounds were extracted by solid-phase micro-extraction (SPME) methods. Each of the 6.00 g stir-fried pork tenderloin samples was cut into about 2 mm granules, 10 μL internal standard (2-methyl-3-heptanone, 16.32 mg/μL) was added to them, and they were placed into a 40 mL headspace vial. Then, the headspace vial was sealed with a polytetrafluoroethylene/silicone septum and equilibrated in a water bath at 45 °C for 20 min to reach the equilibrium state. A SPME fiber coated with divinylbenzene/carboxen/polydimethylsiloxane (DVB/CAR/PDMS, 50/30 μm, Supelco, Bellefonte, PA, USA) was used to adsorb volatiles for 30 min. Eventually, volatiles were thermally desorbed at GC-MS injection port at 250 °C for 5 min.

The volatile compounds were analyzed by Thermo Fisher Trace 1310 gas chromatography (Thermo Fisher Scientific, Waltham, MA, USA) combined with a mass selection detector (Thermo Fisher Scientific, Waltham, MA, USA). DB-WAX column (30 m × 0.25 mm × 0.25 μm; Thermo Fisher Scientific, Waltham, MA, USA) was used to separate different volatiles. Helium (purity of 99.99%) was used as the carrier gas, and the flow rate was 1.0 mL/min. The programmed temperature conditions were as follows: the initial temperature of oven was 40 °C kept for 2 min, and then the oven temperature was ramped to 230 °C at 5 °C/min and maintained for 4 min. The mass spectrometry detector conditions were as follows. The ionization energy was 70 eV. The ion source and transfer line temperatures were set as 250 °C and 240 °C, respectively. The mass scanning ranges were 45–450 $m/z$.

Volatile compounds were tentatively identified by comparison with the mass spectrum of existing data in NIST 14.0 databases. Retention indexes (RIs) methodologies were also applied using n-alkane (C6–C30) as standards under the same GC conditions. Compounds with a matching probability higher than 80% and an RI difference value less than 20 were considered [18]. The concentration of each volatile compound was determined by comparing the peak area of the target component with that of the internal standard [19]. The average of the three replicates represented the final results.

*2.7. Statistical Analysis*

All of the measurements were repeated in triplicate. One-way analysis of variance (ANOVA) with a significance level of $p < 0.05$ was performed using SPSS Statistics 19.0 (IBM Corp., Chicago, IL, USA). Tukey's test was used to assess significant differences ($p < 0.05$) between separate groups. Shapiro–Wilk test was applied for testing normality, and Levene's test ($p < 0.05$) was used to test variance homogeneity. When the variance did not meet the hypothesis of homogeneity, Welch's test was applied for post hoc pairwise comparisons. Partial least squares regression analysis between physicochemical characteristics and volatile compounds was analyzed using XLSTAT 2018 (Addinsoft, New York, NY, USA). The correlation heatmap was carried out by using tools in Hiplot Pro 2.0 (Shanghai Tengyun Biotechnology Co., Ltd., Shanghai, China). Other graphics were drawn by Origin 2021 (OriginLab Corp., Northampton, MA, USA).

## 3. Result and Discussion

### *3.1. The Changes of Quality Attributes in Stir-Fried Pork Tenderloin*

#### 3.1.1. Cooking Loss and Water Content

The water content and cooking loss of stir-fried pork tenderloin are shown in Figure 1. It can be observed that the longer stir-frying time resulted in a higher cooking loss and a lower water content. This was consistent with previous research showing that cooking loss in rabbit meat increased with longer heating times [20]. The water content of the raw pork tenderloin was the highest, about 75%, and decreased to 33% at most. The higher the cooking temperature of sample, the faster the water content decreased. Due to heat processing, protein denaturation and contraction reduced the water-holding capacity of muscle fibers, resulting in a gradual loss of water with the increase in stir-frying time. The moisture content in pork has an important influence on the texture, appearance, and flavor of pork meat. Meat product with slightly higher moisture content has a better chewiness and juiciness, which were considered to be important factors in determining the quality of meat [21]. A previous study showed that during steak frying, the sensory score of juiciness would significantly decrease with the increase in moisture loss. When the moisture content dropped too low, it would lead to a decrease in palatability [22]. Considering the results, only when the water content of cooked pork is not lower than 40% could a relatively better edible quality be found. It was shown in Figure 1b that when the stir-frying time reached 250 s, the water content of most samples decreased to about 40%, so the stir frying time should not exceed 250 s.

#### 3.1.2. Color

Cooking processes in both industry and the household have an influence on meat color, which directly influences purchasing desire [23,24]. The color change in meat products during cooking mainly depends on the thermal denaturation of myoglobin. Globin denaturation in myoglobin leads to the formation of denatured globin hemichrome, which is responsible for the dull-brown color in cooked meats. In addition, Maillard reaction products formed on the meat surface also contributed to the color variation [25,26]. L* value is one of the most vital indicators of evaluating meat product quality. The average L* value of uncooked pork tenderloin was 45.00, as indicated in Figure 2a. Compared to the control group, the L* value increased significantly within 50 s in the stir-frying process. After that, with the extension of stir-frying time, the L* value began to decline gradually. When the L* value decreased to 40, the surface color of pork tenderloin turned almost black. The mean a* and b* values of the control group were, respectively, 6.4 and 10.67. The a* value of the samples exhibited a remarkable increasing trend with time, while showed a decreasing trend from 200 s to 250 s at 220 °C. The change trend of the L* value and the a* value was in agreement with that reported for fried shrimp at different temperatures [27]. The b* value exhibited a trend of initially rising and then dropping, and the b* value of each treatment was higher than that of the control group. During the stir-frying process, due to the denaturation of myoglobin and the Maillard reaction, the pork tenderloin changed from white to golden yellow and, finally, to dark brown. Under the same cooking conditions, all b* values were greater than a* values, which showed that the yellow color of stir-fried samples occupied a dominant position. The data in Figure 2 showed that when the stir-frying time reached 250 s, the L* value of the pork tenderloin surface would drop below 40. In high-temperature, short-time cooking, time played more of a role than temperature in the color change in cooked meat. Therefore, too much stir-frying time is not recommended.

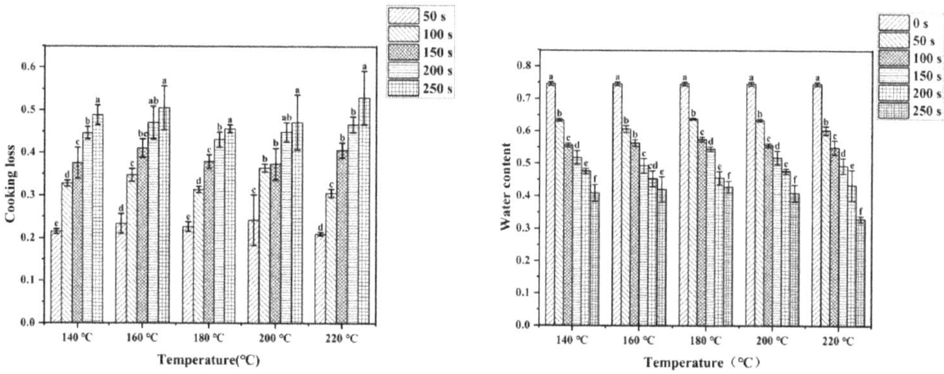


(a) (b)

**Figure 1.** Cooking loss (**a**) and water content (**b**) of pork tenderloin under various cooking conditions. Values bearing different lowercase letters within the same temperature were significantly different ($p < 0.05$).

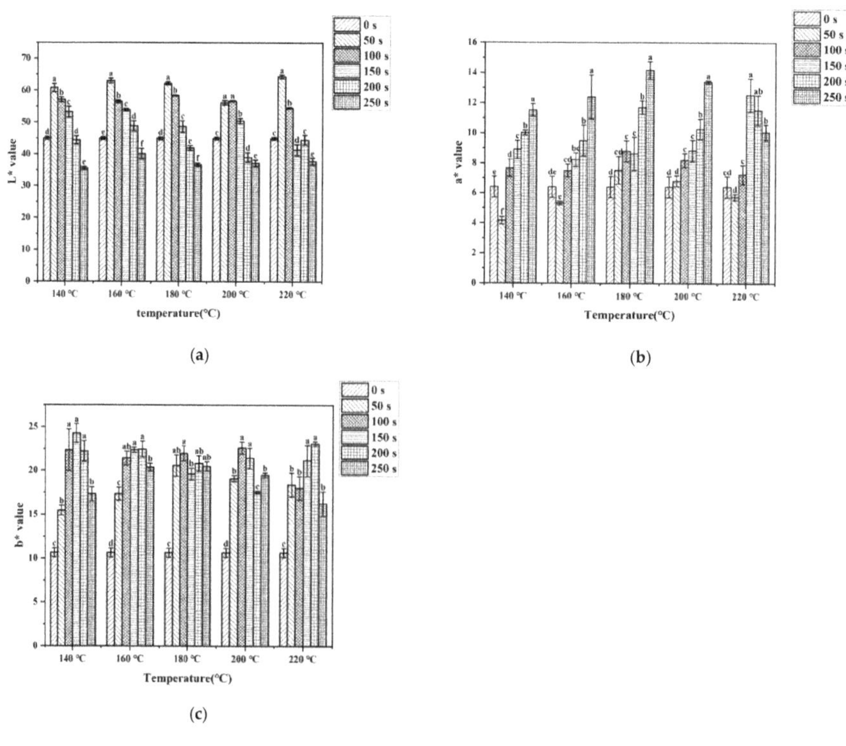


(a) (b) (c)

**Figure 2.** L* value (**a**), a* value (**b**), and b* value (**c**) of pork tenderloin under various cooking conditions. Values bearing different lowercase letters within the same temperature were significantly different ($p < 0.05$).

### 3.2. Sensory Evaluation

The four attributes (color, tenderness, aroma, and overall flavor) of each pork tenderloin sample were evaluated by a team of 12 well-trained individuals. The sensory qualities of samples were shown to change with different temperatures and stir-frying times (Table 1). As shown in Table 1, the color score of pork tenderloin samples presented an earlier increasing and later decreasing trend ($p < 0.05$) and became relatively higher from 150 to 200 s. During the stir-frying process, with the denaturation of myoglobin and the Maillard reaction with the loss of water, the meat color changed from white to golden yellow, then turned to dark brown. Tenderness has been determined to be one of the vital factors affecting acceptability for consumers [28]. In terms of the tenderness of the samples, they showed a decreased trend with the increase in stir-frying time. With regard to odor, the scores of stir-fried pork tenderloin showed a remarkable increasing trend. With the prolongation of cooking time, the Maillard reaction occurs, and lipid oxidation produces more flavor substances, which makes the aroma more pleasant [29]. For the overall flavor, the scores elevated gradually with increasing stir-frying time. However, deep stir-fried samples showed a decreasing trend in overall flavor score. In conclusion, when the stir-frying time was 150–200 s, the cooked pork tenderloin had the golden color, tender texture, and pleasant flavor of stir-fried dishes.

**Table 1.** Sensory evaluation of stir-fried pork tenderloin under different cooking processes.

| Cooking Conditions | | Sensory Scores [a] | | | |
|---|---|---|---|---|---|
| Temperature (°C) | Time (s) | Color | Tenderness | Aroma | Overall Flavor |
| 140 | 50 | 4.00 [b] | 7.00 [ab] | 4.25 [c] | 5.13 [a] |
| | 100 | 4.38 [b] | 7.50 [a] | 5.38 [abc] | 6.38 [a] |
| | 150 | 5.00 [b] | 6.38 [abc] | 5.25 [bc] | 6.25 [a] |
| | 200 | 6.88 [a] | 6.13 [bc] | 6.63 [a] | 6.63 [a] |
| | 250 | 6.25 [a] | 5.38 [c] | 6.50 [ab] | 6.50 [a] |
| 160 | 50 | 3.50 [b] | 6.63 [a] | 5.25 [a] | 5.38 [ab] |
| | 100 | 6.00 [a] | 6.88 [a] | 5.25 [a] | 5.38 [ab] |
| | 150 | 6.63 [a] | 5.25 [b] | 4.50 [a] | 6.75 [a] |
| | 200 | 3.38 [b] | 3.13 [c] | 5.75 [a] | 4.75 [b] |
| | 250 | 4.25 [b] | 3.50 [c] | 5.63 [a] | 5.50 [ab] |
| 180 | 50 | 3.25 [c] | 6.25 [ab] | 5.50 [a] | 5.38 [a] |
| | 100 | 4.00 [bc] | 6.88 [a] | 5.88 [a] | 5.75 [a] |
| | 150 | 6.38 [a] | 4.88 [b] | 6.25 [a] | 6.25 [a] |
| | 200 | 5.75 [ab] | 5.13 [b] | 6.00 [a] | 6.75 [a] |
| | 250 | 5.13 [abc] | 2.75 [c] | 6.75 [a] | 6.25 [a] |
| 200 | 50 | 5.13 [a] | 7.88 [a] | 6.13 [a] | 6.13 [a] |
| | 100 | 5.00 [a] | 6.63 [ab] | 5.50 [a] | 5.88 [a] |
| | 150 | 6.13 [a] | 5.50 [bc] | 6.63 [a] | 6.38 [a] |
| | 200 | 5.13 [a] | 4.25 [c] | 7.00 [a] | 6.63 [a] |
| | 250 | 5.00 [a] | 4.63 [c] | 6.13 [a] | 6.50 [a] |
| 220 | 50 | 4.00 [c] | 7.25 [a] | 6.00 [ab] | 6.13 [a] |
| | 100 | 4.75 [bc] | 6.75 [a] | 5.75 [b] | 5.88 [a] |
| | 150 | 6.38 [ab] | 6.50 [a] | 5.38 [b] | 6.13 [a] |
| | 200 | 6.88 [a] | 5.75 [ab] | 6.63 [ab] | 7.00 [a] |
| | 250 | 5.88 [ab] | 4.25 [b] | 7.88 [a] | 7.25 [a] |

[abc] Values bearing different lowercase letters within the same temperature are significantly different ($p < 0.05$).

### 3.3. Volatile Compounds

Odor is a crucial sensory attribute for meat products since it influences consumers' initial perception and acceptance of the product. The relationship between volatile components and odor characteristics has been well documented. To identify the difference of odor in pork tenderloin during stir frying, the SPME-GC-MS method was used to analyze the volatiles in samples stir fried at 140 °C and 220 °C for 100 s, 150 s, 200 s, and 250 s. As shown in Table 2, 51 volatile compounds in pork tenderloin were tentatively identified, including fifteen aldehydes, thirteen alcohols, five ketones, four esters, one terpenoid, two furans, three phenols, five pyrazines, one ether, and two sulfur-containing compounds. There were 24 volatiles identified in raw pork and 50 volatiles identified in stir-fried samples, which indicated the notable effect of stir frying on the generation of volatile compounds.

Aldehydes are primarily produced by the oxidation of lipids and are characterized by pleasant odors like malt, grass, cheese, and fruit [30]. It has been demonstrated that these compounds significantly improved the flavor of pork [31]. In this investigation, aldehydes were the most prevalent class of volatile chemicals found in stir-fried pork tenderloin. Among them, relatively high contents of aliphatic aldehydes, including hexanal, heptanal, octanal, and nonanal with grass and fat odors, were found. In general, over half of the volatiles in cooked meat are formed by lipid oxidation and degradation. Different cooking conditions lead to different lipid oxidation pathways, influencing the volatiles of meat products. Oleic acid is a common class of monounsaturated fatty acids found in meat, whereas arachidonic acid and linoleic acid are common polyunsaturated fatty acids. Compounds including octanal, nonanal, and decanal are produced by oleic acid oxidation. Hexanal and (*E*)-2-nonanol are oxidized from linoleic acid [32]. In fresh pork tenderloin samples, aldehydes were undetected or detected with a very low content. Under less intense cooking conditions, aldehydes in cooked pork tenderloins were present in the highest proportion. However, the relative content decreased as time went on, especially in samples cooked at a lower temperature. The changes (increase followed by a decrease) in the levels of aldehydes could be attributed to lipid oxidation and heat-induced evaporation.

Alcohol formation is also attributed to polyunsaturated fatty acid oxidization and degradation [33]. Among the detected 13 alcohol compounds, 2-ethyl-1-hexanol, 1-octen-3-ol, pentanol, hexanol, heptanol, and 1-octanol were the main alcohol compounds. These thirteen volatiles might be grouped into two clusters, as Table 1 illustrates, according to the pattern of fluctuations in their concentration in the samples. Volatiles in the first group, including 2-ethyl-1-hexanol, linalool, octanol, 2,3-butanediol, hexanol, nonanol, and dodecanol, were found in a high content in the control sample. However, the content of these compounds decreased in the fried samples. One possible explanation for the elimination of these volatiles might be evaporation during high-temperature cooking. On the other hand, volatile substances in the second cluster, including 1-octen-3-ol, pentanol, heptanol, (*E*)-2-octen-1-ol, and 4-isopropoxybutanol had a higher content in stir-fried samples compared to those in raw meat, which could be attributed to lipid oxidation during the cooking process. 1-Octen-3-ol is an odorant that is typically found in meat products. Its odor is similar to that of mushroom. Throughout the cooking process, the concentration of 1-octen-3-ol remained high in our investigation. Because of its large quantity and relatively low threshold value, 1-octen-3-ol might be regarded as a key distinctive flavor component in cooked meat [34].

Ketones are usually generated by the oxidation of unsaturated fatty acids or through the degradation of amino acids, and they have a distinctive fruity and floral aroma [35]. A total of 5 ketones were discovered throughout the cooking stage, including acetoin (cheese aroma), 6-methyl-5-hepten-2-one (fruit and grass aroma), 2-undecanone (fruit and fatty aroma), 6,10-dimethyl-5,9-undecadien-2-one (spicy aroma), and 1-(2,4-dimethylphenyl)-ethanone (floral aroma), which might contribute to the aroma formation of stir-fried pork tenderloin because of their lower thresholds. Among ketones, except for acetoin, which was more abundant in raw pork, the content of the others increased and then decreased after cooking. This might be because ketones could serve as substrates to react with amino acids, peptides, proteins, and other components to form heterocyclic chemicals, which are indirectly attributed to the development of pork flavor [36,37].

**Table 2.** Relative concentration (μg/Kg) of volatile compounds in stir-fried pork tenderloin under different cooking processes.

| No. | Compounds | Identification Methods $^{a}$ | RI $^{b}$ | | Concentration (μg/Kg) $^{c}$ | | | | | | | | | |
|---|---|---|---|---|---|---|---|---|---|---|---|---|---|---|
| | | | Calculation | Literature | Raw | 140 °C 100 s | 140 °C 150 s | 140 °C 200 s | 140 °C 250 s | Raw | 220 °C 100 s | 220 °C 150 s | 220 °C 200 s | 220 °C 250 s |
| | aldehydes | | | | | | | | | | | | | |
| 1 | hexanal | MS RI | 1067 | 1078 | 73.97 $^{b}$ | 854.68 $^{ab}$ | 790.97 $^{ab}$ | 1730.12 $^{a}$ | 1362.79 $^{a}$ | 73.97 $^{b}$ | 828.14 $^{ab}$ | 732.98 $^{ab}$ | 968.51 $^{ab}$ | 1717.63 $^{a}$ |
| 2 | heptanal | MS RI | 1182 | 1181 | 0 $^{b}$ | 70.72 $^{a}$ | 65.59 $^{a}$ | 64.67 $^{a}$ | 59.51 $^{a}$ | 0 $^{c}$ | 45.05 $^{ab}$ | 31.17 $^{bc}$ | 57.90 $^{ab}$ | 77.75 $^{a}$ |
| 3 | octanal | MS RI | 1286 | 1286 | 0 $^{c}$ | 128.28 $^{a}$ | 115.75 $^{ab}$ | 84.10 $^{ab}$ | 63.14 $^{b}$ | 0 $^{c}$ | 64.70 $^{ab}$ | 45.63 $^{b}$ | 76.42 $^{a}$ | 67.37 $^{ab}$ |
| 4 | nonanal | MS RI | 1390 | 1390 | 31.40 $^{b}$ | 518.49 $^{a}$ | 519.87 $^{a}$ | 423.28 $^{a}$ | 358.78 $^{a}$ | 31.40 $^{c}$ | 304.78 $^{ab}$ | 257.84 $^{b}$ | 369.30 $^{ab}$ | 505.43 $^{a}$ |
| 5 | (*E*)-2-octenal | MS RI | 1424 | 1434 | 0 $^{b}$ | 15.00 $^{a}$ | 18.38 $^{a}$ | 21.25 $^{a}$ | 13.54 $^{a}$ | 0 $^{c}$ | 11.47 $^{b}$ | 8.84 $^{b}$ | 10.94 $^{b}$ | 17.27 $^{a}$ |
| 6 | decanal | MS RI | 1492 | 1485 | 0 $^{c}$ | 23.53 $^{ab}$ | 25.57 $^{a}$ | 13.58 $^{b}$ | 19.09 $^{ab}$ | 0 $^{c}$ | 13.35 $^{ab}$ | 9.33 $^{b}$ | 12.09 $^{ab}$ | 17.13 $^{a}$ |
| 7 | benzaldehyde | MS RI | 1512 | 1508 | 6.41 $^{b}$ | 32.69 $^{a}$ | 32.73 $^{a}$ | 29.46 $^{ab}$ | 25.50 $^{ab}$ | 6.41 $^{c}$ | 10.56 $^{c}$ | 15.38 $^{bc}$ | 22.88 $^{ab}$ | 34.06 $^{a}$ |
| 8 | (*E*)-2-nonenal | MS RI | 1528 | 1530 | 0 $^{b}$ | 6.71 $^{a}$ | 3.66 $^{a}$ | 6.83 $^{a}$ | 5.11 $^{a}$ | 0 $^{b}$ | 8.10 $^{a}$ | 3.66 $^{ab}$ | 4.64 $^{a}$ | 7.19 $^{a}$ |
| 9 | undecanal | MS RI | 1599 | 1587 | 0 $^{c}$ | 4.66 $^{ab}$ | 7.05 $^{a}$ | 3.23 $^{bc}$ | 2.43 $^{bc}$ | 0 $^{c}$ | 2.66 $^{ab}$ | 1.75 $^{b}$ | 3.29 $^{a}$ | 3.94 $^{a}$ |
| 10 | (*E*)-2-decenal | MS RI | 1638 | 1630 | 0 $^{c}$ | 0 $^{c}$ | 0 $^{c}$ | 15.38 $^{a}$ | 10.83 $^{b}$ | 0 $^{c}$ | 10.22 $^{b}$ | 7.05 $^{b}$ | 12.41 $^{ab}$ | 18.29 $^{a}$ |
| 11 | 4-ethyl-benzaldehyde | MS RI | 1699 | 1719 | 0 $^{d}$ | 2.29 $^{c}$ | 2.69 $^{bc}$ | 3.50 $^{a}$ | 3.11 $^{ab}$ | 0 $^{b}$ | 2.48 $^{a}$ | 1.82 $^{ab}$ | 2.21 $^{a}$ | 2.43 $^{a}$ |
| 12 | dodecanal | MS RI | 1705 | 1710 | 0 $^{b}$ | 4.54 $^{a}$ | 4.41 $^{a}$ | 2.34 $^{ab}$ | 2.03 $^{ab}$ | 0 $^{c}$ | 2.04 $^{ab}$ | 1.24 $^{bc}$ | 1.85 $^{ab}$ | 2.98 $^{a}$ |
| 13 | 2-undecenal | MS | 1746 | - | 0 $^{b}$ | 6.95 $^{a}$ | 0 $^{b}$ | 0 $^{b}$ | 0 $^{b}$ | 0 $^{d}$ | 5.40 $^{bc}$ | 3.49 $^{cd}$ | 9.01 $^{ab}$ | 13.84 $^{a}$ |
| 14 | (*E*,*E*)-2,4-decadienal | MS RI | 1802 | 1813 | 0 $^{d}$ | 16.29 $^{c}$ | 39.03 $^{a}$ | 21.06 $^{b}$ | 16.88 $^{bc}$ | 0 $^{c}$ | 5.07 $^{c}$ | 8.96 $^{c}$ | 33.69 $^{b}$ | 53.78 $^{a}$ |
| 15 | 4-methoxy-benzaldehyde | MS RI | 2012 | 2011 | 0 $^{b}$ | 0.55 $^{b}$ | 0.75 $^{b}$ | 1.91 $^{a}$ | 0.51 $^{b}$ | 0 $^{b}$ | 0.49 $^{a}$ | 0.30 $^{ab}$ | 0 $^{b}$ | 0 $^{b}$ |
| | alcohols | | | | | | | | | | | | | |
| 16 | 1-pentanol | MS RI | 1252 | 1252 | 29.62 $^{b}$ | 60.22 $^{ab}$ | 85.01 $^{a}$ | 88.68 $^{a}$ | 68.40 $^{ab}$ | 29.62 $^{b}$ | 52.42 $^{ab}$ | 38.00 $^{b}$ | 60.92 $^{ab}$ | 102.87 $^{a}$ |
| 17 | 3-methyl-2-buten-1-ol | MS RI | 1325 | 1320 | 0.01 a | 0 $^{a}$ | 0 $^{a}$ | 0 $^{a}$ | 0 $^{a}$ | 0.01 $^{a}$ | 0 $^{a}$ | 0 $^{a}$ | 0 $^{a}$ | 0 $^{a}$ |
| 18 | 1-hexanol | MS RI | 1354 | 1369 | 47.06 $^{a}$ | 14.49 $^{b}$ | 18.99 $^{b}$ | 20.22 $^{b}$ | 16.40 $^{b}$ | 47.06 $^{a}$ | 18.92 $^{b}$ | 12.94 $^{b}$ | 12.04 $^{b}$ | 20.55 $^{b}$ |
| 19 | 1-octen-3-ol | MS RI | 1449 | 1451 | 42.97 $^{a}$ | 136.01 $^{b}$ | 191.93 $^{b}$ | 185.01 $^{b}$ | 145.14 $^{b}$ | 42.97 $^{c}$ | 114.99 $^{b}$ | 88.15 $^{b}$ | 94.33 $^{b}$ | 168.01 $^{a}$ |
| 20 | 1-heptanol | MS RI | 1455 | 1447 | 16.16 $^{b}$ | 24.57 $^{ab}$ | 31.57 $^{a}$ | 19.92 $^{ab}$ | 15.06 $^{b}$ | 16.16 $^{ab}$ | 18.67 $^{ab}$ | 11.11 $^{b}$ | 20.43 $^{ab}$ | 32.57 $^{a}$ |
| 21 | 2-ethyl-1-hexanol | MS RI | 1487 | 1484 | 2263.71 $^{a}$ | 20.14 $^{c}$ | 31.29 $^{b}$ | 21.99 $^{c}$ | 28.27 $^{b}$ | 2263.71 $^{a}$ | 37.34 $^{b}$ | 18.43 $^{b}$ | 20.18 $^{b}$ | 17.59 $^{b}$ |
| 22 | 1-octanol | MS RI | 1559 | 1561 | 55.27 $^{a}$ | 48.70 $^{a}$ | 0 $^{b}$ | 0 $^{b}$ | 0 $^{b}$ | 55.27 $^{b}$ | 0 $^{a}$ | 0 $^{a}$ | 0 $^{a}$ | 0 $^{a}$ |
| 23 | 2,3-butanediol | MS RI | 1574 | 1570 | 133.87 $^{a}$ | 0 $^{b}$ | 1.09 $^{b}$ | 4.60 $^{b}$ | 15.86 $^{b}$ | 133.87 $^{a}$ | 8.43 $^{b}$ | 21.45 $^{b}$ | 0 $^{b}$ | 50.35 $^{b}$ |
| 24 | (*E*)-2-octen-1-ol | MS RI | 1614 | 1618 | 7.20 $^{b}$ | 25.32 $^{a}$ | 29.13 $^{a}$ | 24.15 $^{ab}$ | 22.20 $^{ab}$ | 7.20 $^{c}$ | 13.76 $^{b}$ | 12.32 $^{bc}$ | 10.21 $^{bc}$ | 22.18 $^{a}$ |
| 25 | 1-nonanol | MS RI | 1659 | 1673 | 7.53 $^{a}$ | 3.07 $^{b}$ | 2.85 $^{b}$ | 2.19 $^{b}$ | 1.95 $^{b}$ | 7.53 $^{a}$ | 2.15 $^{b}$ | 1.35 $^{b}$ | 1.41 $^{b}$ | 3.51 $^{ab}$ |
| 26 | α-terpineol | MS | 1694 | - | 18.50 $^{a}$ | 5.08 $^{b}$ | 7.59 $^{b}$ | 6.67 $^{b}$ | 5.47 $^{b}$ | 18.50 $^{a}$ | 5.24 $^{a}$ | 3.15 $^{a}$ | 3.95 $^{a}$ | 20.09 $^{a}$ |
| 27 | 1-dodecanol | MS RI | 1965 | 1953 | 7.10 $^{a}$ | 4.04 $^{a}$ | 5.21 $^{a}$ | 4.50 $^{a}$ | 3.91 $^{a}$ | 7.10 $^{a}$ | 3.35 $^{ab}$ | 1.93 $^{b}$ | 2.52 $^{b}$ | 5.18 $^{ab}$ |
| 28 | 4-(1-methylethoxy)-1-butanol | MS | 2484 | - | 0 $^{c}$ | 0.75 $^{bc}$ | 2.10 $^{a}$ | 0.87 $^{b}$ | 0.99 $^{b}$ | 0 $^{c}$ | 1.37 $^{a}$ | 0.71 $^{b}$ | 0.70 $^{b}$ | 0.82 $^{b}$ |
| | ketones | | | | | | | | | | | | | |
| 29 | acetoin | MS RI | 1279 | 1273 | 1486.72 $^{a}$ | 1.60 $^{b}$ | 5.21 $^{b}$ | 23.31 $^{b}$ | 47.57 $^{b}$ | 1486.72 $^{a}$ | 71.56 $^{b}$ | 84.53 $^{b}$ | 6.24 $^{b}$ | 184.13 $^{b}$ |
| 30 | 6-methyl-5-hepten-2-one | MS RI | 1334 | 1341 | 19.53 $^{a}$ | 4.65 $^{b}$ | 2.21 $^{b}$ | 2.25 $^{b}$ | 3.24 $^{b}$ | 19.53 $^{a}$ | 5.09 $^{b}$ | 3.70 $^{b}$ | 1.49 $^{b}$ | 5.10 $^{b}$ |
| 31 | 2-undecanone | MS RI | 1594 | 1599 | 0 $^{b}$ | 0 $^{b}$ | 0 $^{b}$ | 3.42 $^{a}$ | 2.72 $^{a}$ | 0 $^{b}$ | 2.62 $^{a}$ | 1.84 $^{a}$ | 0 $^{b}$ | 0 $^{b}$ |
| 32 | 6,10-dimethyl-5,9-undecadien-2-one | MS | 1849 | - | 0 $^{b}$ | 1.13 $^{a}$ | 0.84 $^{a}$ | 0.71 $^{a}$ | 1.01 $^{a}$ | 0 $^{c}$ | 0.86 $^{a}$ | 0.79 $^{ab}$ | 0.39 $^{bc}$ | 1.15 $^{a}$ |
| 33 | 1-(2,4-dimethylphenyl)-ethanone | MS | 2223 | - | 2.41 $^{a}$ | 0 $^{b}$ | 2.98 $^{a}$ | 1.55 $^{ab}$ | 1.26 $^{ab}$ | 2.41 $^{a}$ | 1.16 $^{ab}$ | 0.66 $^{b}$ | 0.91 $^{b}$ | 1.20 $^{ab}$ |
| | esters | | | | | | | | | | | | | |
| 34 | γ-hexalactone | MS RI | 1694 | 1678 | 0 $^{c}$ | 0.85 $^{bc}$ | 2.96 $^{a}$ | 2.28 $^{ab}$ | 2.06 $^{ab}$ | 0 $^{d}$ | 1.97 $^{ab}$ | 0.89 $^{cd}$ | 1.03 $^{bc}$ | 2.72 $^{a}$ |
| 35 | γ-octanoic lactone | MS RI | 1905 | 1898 | 0 $^{d}$ | 2.35 $^{a}$ | 1.96 $^{a}$ | 1.31 $^{b}$ | 0.78 $^{c}$ | 0 $^{a}$ | 0.57 $^{a}$ | 0.77 $^{a}$ | 1.58 $^{a}$ | 5.18 $^{a}$ |
| 36 | γ-nonanolactone | MS RI | 2018 | 2007 | 3.70 $^{a}$ | 1.06 $^{b}$ | 2.11 $^{ab}$ | 1.45 $^{b}$ | 1.89 $^{ab}$ | 3.70 $^{a}$ | 1.22 $^{b}$ | 0.92 $^{b}$ | 1.00 $^{b}$ | 1.87 $^{ab}$ |

**Table 2.** *Cont.*

| No. | Compounds | Identification Methods $^{a}$ | RI $^{b}$ | | Concentration (µg/Kg) $^{c}$ | | | | | | | | | |
|---|---|---|---|---|---|---|---|---|---|---|---|---|---|---|
| | | | Calculation | Literature | Raw | 140 °C 100 s | 140 °C 150 s | 140 °C 200 s | 140 °C 250 s | Raw | 220 °C 100 s | 220 °C 150 s | 220 °C 200 s | 220 °C 250 s |
| 37 | hexadecanoic acid ethyl ester | MS RI | 2239 | 2220 | 0 $^{b}$ | 3.67 $^{a}$ | 2.63 $^{a}$ | 1.96 $^{ab}$ | 1.68 $^{ab}$ | 0 $^{c}$ | 1.58 $^{ab}$ | 1.22 $^{b}$ | 1.07 $^{b}$ | 1.86 $^{a}$ |
| | terpenoid | | | | | | | | | | | | | |
| 38 | linalool | MS RI | 1544 | 1540 | 23.70 $^{a}$ | 1.10 $^{b}$ | 0.96 $^{b}$ | 1.76 $^{b}$ | 2.60 $^{b}$ | 23.70 $^{a}$ | 1.70 $^{b}$ | 1.22 $^{b}$ | 1.45 $^{b}$ | 3.93 $^{b}$ |
| | furans | | | | | | | | | | | | | |
| 39 | 2-pentyl-furan | MS RI | 1230 | 1230 | 0 $^{c}$ | 22.89 $^{b}$ | 43.84 $^{a}$ | 30.12 $^{ab}$ | 24.64 $^{b}$ | 0 $^{c}$ | 8.26 $^{c}$ | 11.91 $^{c}$ | 46.90 $^{b}$ | 78.27 $^{a}$ |
| 40 | 2(5*H*)-furanone | MS RI | 1741 | 1767 | 0 $^{c}$ | 1.33 $^{a}$ | 1.09 $^{ab}$ | 0.45 $^{abc}$ | 0.30 $^{bc}$ | 0 $^{c}$ | 0.26 $^{b}$ | 0.32 $^{b}$ | 0.34 $^{b}$ | 0.48 $^{a}$ |
| | phenols | | | | | | | | | | | | | |
| 41 | phenol | MS RI | 1995 | 1953 | 7.38 $^{a}$ | 2.23 $^{b}$ | 3.26 $^{b}$ | 2.18 $^{b}$ | 2.02 $^{b}$ | 7.38 $^{a}$ | 1.98 $^{b}$ | 1.22 $^{b}$ | 1.63 $^{b}$ | 3.37 $^{ab}$ |
| 42 | p-cresol | MS RI | 2073 | 2078 | 2.07 $^{a}$ | 0.93 $^{ab}$ | 1.04 $^{ab}$ | 0.62 $^{b}$ | 0.51 $^{b}$ | 2.07 $^{a}$ | 0.70 $^{b}$ | 0.41 $^{b}$ | 0.41 $^{b}$ | 0.58 $^{b}$ |
| 43 | 2,4-di-tert-butylphenol | MS RI | 2305 | 2312 | 18.73 $^{a}$ | 0 $^{b}$ | 0.42 $^{b}$ | 0.82 $^{b}$ | 0.70 $^{b}$ | 18.73 $^{a}$ | 0.30 $^{b}$ | 0.33 $^{b}$ | 0.24 $^{b}$ | 0.63 $^{b}$ |
| | pyrazines | | | | | | | | | | | | | |
| 44 | methyl-pyrazine | MS RI | 1267 | 1265 | 0 $^{b}$ | 3.15 $^{a}$ | 1.10 $^{a}$ | 0 $^{b}$ | 0 $^{b}$ | 0 $^{a}$ | 0 $^{a}$ | 0 $^{a}$ | 0.22 $^{a}$ | 0 $^{a}$ |
| 45 | 2,3-dimethyl-pyrazine | MS RI | 1343 | 1357 | 0 $^{d}$ | 2.62 $^{a}$ | 2.13 $^{b}$ | 0.72 $^{c}$ | 0.75 $^{c}$ | 0 $^{b}$ | 0.87 $^{b}$ | 0.36 $^{b}$ | 0.80 $^{b}$ | 2.40 $^{a}$ |
| 46 | 2-ethyl-6-methyl-pyrazine | MS RI | 1383 | 1382 | 0 $^{d}$ | 3.81 $^{a}$ | 2.07 $^{b}$ | 0 $^{d}$ | 1.02 $^{c}$ | 0 $^{b}$ | 0.35 $^{b}$ | 0 $^{b}$ | 0.85 $^{b}$ | 5.63 $^{a}$ |
| 47 | trimethyl-pyrazine | MS RI | 1402 | 1406 | 0 $^{c}$ | 12.50 $^{a}$ | 12.33 $^{a}$ | 0 $^{c}$ | 7.02 $^{b}$ | 0 $^{c}$ | 4.01 $^{bc}$ | 2.46 $^{bc}$ | 5.74 $^{b}$ | 23.04 $^{a}$ |
| 48 | 3-ethyl-2,5-dimethyl-pyrazine | MS RI | 1443 | 1430 | 0 $^{c}$ | 8.90 $^{a}$ | 13.76 $^{b}$ | 7.08 $^{b}$ | 6.94 $^{b}$ | 0 $^{c}$ | 7.49 $^{ab}$ | 2.15 $^{bc}$ | 4.26 $^{abc}$ | 9.97 $^{a}$ |
| | sulfur-containing compounds | | | | | | | | | | | | | |
| 49 | dimethyl trisulfide | MS RI | 1373 | 1378 | 0 $^{b}$ | 0 $^{b}$ | 0 $^{b}$ | 0.18 $^{a}$ | 0.18 $^{a}$ | 0 $^{c}$ | 0 $^{c}$ | 0.12 $^{bc}$ | 0.22 $^{ab}$ | 0.30 $^{a}$ |
| 50 | 4-methyl-5-thiazoleethanol | MS | 2299 | - | 0 $^{c}$ | 0 $^{c}$ | 1.09 $^{a}$ | 0.39 $^{b}$ | 0.32 $^{b}$ | 0 $^{b}$ | 0.44 $^{a}$ | 0.22 $^{ab}$ | 0.22 $^{ab}$ | 0.16 $^{b}$ |
| | ether | | | | | | | | | | | | | |
| 51 | 2-(2-ethoxyethoxy)-ethanol | MS RI | 1617 | 1622 | 17.11 $^{a}$ | 4.76 $^{b}$ | 12.67 $^{ab}$ | 8.42 $^{ab}$ | 7.82 $^{ab}$ | 17.11 $^{a}$ | 7.65 $^{b}$ | 3.94 $^{b}$ | 4.17 $^{b}$ | 6.59 $^{b}$ |

[a] Mode of identification: RI, retention index published in the literature data; MS, mass spectrum verified by comparison with mass spectra database (NIST 14.0). [b] Retention index. [c] Values bearing different lowercase letters within the same temperature are significantly different ($p < 0.05$).

Esters are formed through the esterification of organic acids and alcohols. Only four esters with relatively low contents were found in this investigation. γ-Nonanolactone was found in raw pork samples, whereas the other three esters were generated throughout the successive stir-frying phases.

Furan derivatives are characteristic compounds generated in the Maillard reaction and fat degradation processes [27]. In all samples, only two furans were identified. 2(5*H*)-Furanone is a Maillard reaction product with a typical caramel-like and roasted aroma. 2-Pentylfuran, which has a beany and fruity smell, contributed significantly to the overall flavor of pork and might be caused by the oxidation of linoleic acid [8,38]. Throughout the stir-frying stage of pork tenderloin, the content of 2-pentyl-furan in samples was significantly increased after cooking.

The heterocyclic compounds containing sulfur are the most significant substances created by the Maillard reaction and contribute to the meaty, savory, and roasted flavor characteristics of cooked meat [29]. Sulfur-containing compounds have a low threshold and a vital influence on the flavor formation of pork at low content. In the study, one sulfur-containing heterocyclic compound, namely, 4-methyl-5-thiazoleethanol, could be detected with the prolongation of cooking time. The reason why the other sulfur compounds were not detected might be that the content of sulfur amino acids in pork tenderloin is too low, which leads to a low content of sulfur-containing compounds generated during the reaction. It was also possible that the method applied in this investigation was insufficient to detect these sulfur components [39].

Pyrazines have a distinctive roasted smell. They are typical Maillard reaction products which are formed by the aldehyde condensation reaction and Strecker degradation during cooking [27,40]. Pyrazines were not identified in the control group. During stir frying, the content of pyrazines could judge the progress of the Maillard reaction [31]. The findings demonstrated that heating at a higher temperature and, to a lesser extent, prolonged cooking time, promoted the formation of volatile compounds from the Maillard reaction, which stimulated desirable flavors.

The most likely source of phenolic compounds in pork tenderloin is animal feed [41]. The phenolic compounds were found in raw pork samples and decreased after cooking.

### *3.4. The Correlations between Physical Index and Volatile Compounds*

Meat's physicochemical characteristics and volatile components can be drastically altered by heat treatment. The aroma attributes' changes are accompanied by variations of other quality attributes. Normally, the detection of volatile substances in meat processing is cumbersome, and the determination of quality indicators is relatively convenient. Obtaining the correlation data between them could better realize the aroma prediction in the industrial production process. In this study, the correlations among quality attributes and volatile components were studied using partial least squares regression (PLSR) analysis, and the results are shown in Figure 3. As shown in Figure 3, the physical index and most volatile compounds were located between the inner and outer ellipses, demonstrating that they were remarkably explained by the PLSR model [42]. The x variables ($R^2$ X = 0.894) explained the variation in the y variables ($R^2$ Y = 0.505) according to the first two factors in the correlation loading plot for the volatile chemicals and the physicochemical indices. Factor 1 corresponded to increasing sample stir-frying times. Factor 1 explained a majority of the variance in both data sets, indicating that the effects of time in stir-frying processes were responsible for the majority of the covariance between the two data sets.

Cooking loss, a*, and b* descriptors were positively correlated with factor 1 and were associated with samples stir fried ≥150 s under 220 °C and samples stir fried ≥200 s under 140 °C, while water content and L* value were associated with the control and stir-fried <150 s samples and negatively correlated with factor 1 (Figure 3). Regarding factor 1, volatile chemicals in the headspace were divided into two primary groups that correlated with early and late stir-frying periods. The majority of volatile substances negatively correlated with factor 1 and positively correlated with water content attribute were mainly alcohols and phenols. These volatiles showed high correlation with raw samples. The other groups of volatiles most positively associated with factor 1, cooking loss, a*, and b* values mainly included volatiles particularly those formed from Maillard reactions and lipid oxidation, such as pyrazines, sulfur-containing components, aldehydes, five to eight-carbon alcohols, six to eight-carbon esters, and furans. As shown in Figure 3, the volatiles in the second group were plotted in the vicinity of samples 140–200, 220–150, and 220–200, indicating these volatiles and samples were correlated with each other.

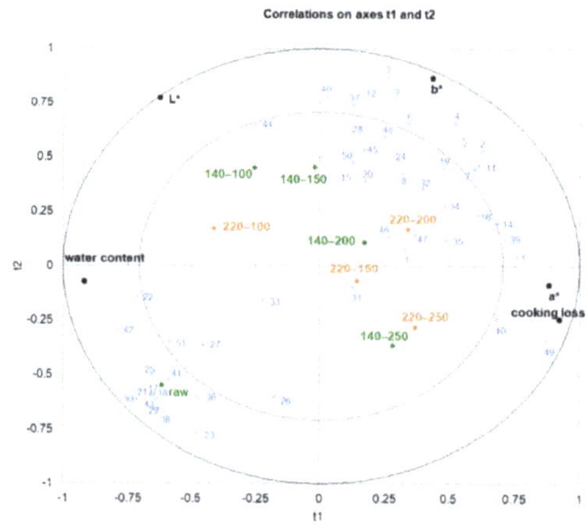


**Figure 3.** Partial least squares regression (PLSR) analysis between physical indexes and volatile compounds in pork tenderloin stir fried at 140 °C (indicated by green) and 220 °C (indicated by orange). The No. (indicated by blue) in the figure is in accordance with the No. of compounds in Table 2.

The correlation coefficients between the 50 detected volatiles and quality attributes, including L*, a*, b*, water content, and cooking loss, are shown in Figure 4. Among the quality attributes, b* value and water content had the strongest impact on the volatiles, while cooking loss, L* value, and a* value only showed strong correlations with a few volatile compounds, including hexanal, (*E*)-2-decenal, 1-nonanol, linalool, p-cresol, and dimethyl trisulfide. As shown in Figure 4, aldehydes showed a negative correlation with water content, while being positively correlated with the b* value. During the stir-frying process, the heat transferred rapidly to the pork samples. With an increased temperature, the water decreased, and the surface color changed. This process accelerated the oxidation reaction of fatty acids and resulted in the accumulation of aldehydes. The correlation coefficients for indexes of water content and b* value with alcohols were also higher than the others. Among alcohols, 1-pentanol, 1-octen-3-ol, 1-heptanol, (*E*)-2-octen-1-ol, and 4-(1-methylethoy)-1-butanol were negatively related to water content and showed a positive correlation with the b* value. As the products of lipid oxidation and heat degradation, these alcohols accumulated and helped to form the desired aroma of stir-fried samples. Among the ketones with high correlation with the physicochemical indexes, acetoin and 6-methyl-5-hepten-2-one were positively correlated with water content and negatively related to b* value. This could be linked to the abundance of these compounds in raw pork samples. The compound of 6,10-dimethyl-5,9-undecadien-2-one was negatively correlated to water content, but there was no obvious correlation between 6,10-dimethyl-5,9-undecadien-2-one and b* value. For esters, γ-nonanolactone was positively correlated with water content and had a negative correlation with b* value. γ-Nonanolactone, with milk aroma, was usually detected in the raw pork. However, other esters presented negative correlations with water content, which might be due to their formation through the esterification of organic acids and alcohols throughout the successive stir-frying phases. Phenolic compounds also showed positive correlations with water content and negatively correlated with b* value.

The most likely source of phenolic compounds in pork tenderloin is animal feed, and they were found in raw pork samples and decreased after cooking. Pyrazines were negatively related to water content, but the correlation between pyrazines and physicochemical attributes was relatively small. This might be due to the coexistence of multiple reactions in the food processing system and the complexity and interaction between different reactions, so the compounds derived from the same reaction type or the same reaction pathway cannot present an ideal correlation. To sum up, the water content and b* value could be used as potential indicators of flavor in pork tenderloin during stir frying.

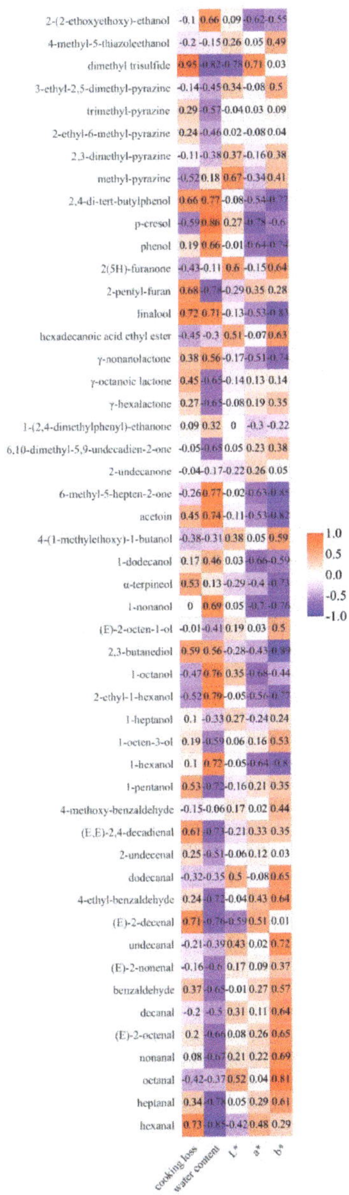


**Figure 4.** The correlation heatmap between physical index and volatile compound.

## 4. Conclusions

This paper showed the influence of different stir-frying temperatures and times on the dynamic changes of the physicochemical characteristics and volatile compounds of pork tenderloin and determined the correlation between them. The main physical indexes (water content, cooking loss, brightness, yellowness, and redness) and characteristic volatiles of pork tenderloin changed significantly with different stir-frying temperatures and times. Volatiles, particularly those formed from Maillard reactions and lipid oxidation, such as aldehydes, five to eight-carbon alcohols, six to eight-carbon esters, and pyrazines, accumulated and helped to form the desired aroma of the stir-fried samples. Correlation analysis showed that among the quality attributes, b* value and water content had the strongest impact on these volatiles. The water content was negatively correlated with most of the compounds in stir-fried samples, while the correlation between the b* value and these volatiles was positive. The study provides data support for volatile prediction based on physicochemical indexes. This is an important basis for future attempts to realize flavor prediction and quality control in the industrial production of Chinese dishes and to provide ideas for developing an online flavor control system.

**Supplementary Materials:** The following supporting information can be downloaded at: https://www.mdpi.com/article/10.3390/foods12234299/s1, Table S1: Sensory evaluation standard for stir-fried pork tenderloin.

**Author Contributions:** Z.W., investigation, methodology, writing—original draft preparation; T.N., investigation, methodology; H.Z., conceptualization, writing—review and editing, funding acquisition; W.W., writing—review and editing; H.C., investigation, resources; S.W., methodology, investigation; B.S., conceptualization, resources. All authors have read and agreed to the published version of the manuscript.

**Funding:** This work was funded by the National Key R&D Program of China [No. 2022YFD2100303]; the National Natural Science Foundation of China [No. 32001823]; the National Natural Science Foundation of China [No. 32072347].

**Institutional Review Board Statement:** The study was conducted in accordance with the Declaration of Helsinki and approved by the Ethics Committee of Ethics Committee of Beijing Business and Technology University (protocol code 2023-68; date of approval 22 May 2023).

**Informed Consent Statement:** Informed consent was obtained from all subjects involved in the study.

**Data Availability Statement:** Data are contained within the article.

**Conflicts of Interest:** The authors declare no conflict of interest.

## References

1. Li, W.; Zheng, L.; Xiao, Y.; Li, L.; Wang, N.; Che, Z.; Wu, T. Insight into the aroma dynamics of Dongpo pork dish throughout the production process using electronic nose and GC×GC-MS. *LWT-Food Sci. Technol.* **2022**, *169*, 113970. [CrossRef]
2. Miller, R. Drivers of consumer liking for beef, pork, and lamb: A review. *Foods* **2020**, *9*, 428. [CrossRef] [PubMed]
3. dos Santos Rocha, C.; Magnani, M.; de Paiva Anciens Ramos, G.L.; Bezerril, F.F.; Freitas, M.Q.; Cruz, A.G.; Pimentel, T.C. Emerging technologies in food processing: Impacts on sensory characteristics and consumer perception. *Curr. Opin. Food Sci.* **2022**, *47*, 100892. [CrossRef]
4. Sohail, A.; Al-Dalali, S.; Wang, J.; Xie, J.; Shakoor, A.; Asimi, S.; Shah, H.; Patil, P. Aroma compounds identified in cooked meat: A review. *Food Res. Int.* **2022**, *157*, 1111385. [CrossRef] [PubMed]
5. Wang, X.; Fan, C.; Wang, X.; Feng, T.; Zhang, X.; Yu, J.; Cui, H.; Xia, S. Microwave heating and conduction heating pork belly: Influence of heat transfer modes on volatile compounds and aroma attributes. *Food Biosci.* **2023**, *52*, 102438. [CrossRef]
6. Yang, Z.; Lu, R.; Song, H.; Zhang, Y.; Tang, J.; Zhou, N. Effect of different cooking methods on the formation of aroma components and heterocyclic amines in pork loin. *J. Food Process. Pres.* **2017**, *41*, e12981. [CrossRef]
7. Chang, Y.; Wang, S.; Chen, H.; Zhang, N.; Sun, J. Characterization of the key aroma compounds in pork broth by sensory-directed flavor analysis. *J. Food Sci.* **2021**, *86*, 4932–4945. [CrossRef]
8. Wang, Y.; Song, H.; Zhang, Y.; Tang, J.; Yu, D. Determination of aroma compounds in pork broth produced by different processing methods. *Flavour Fragr. J.* **2016**, *31*, 319–328. [CrossRef]
9. Wang, R.; Huang, F.; Zhang, L.; Liu, Q.; Zhang, C.; Zhang, H. Changes in the texture, microstructures, colour and volatile compounds of pork meat loins during superheated steam cooking. *Int. J. Food Sci. Technol.* **2019**, *54*, 2821–2830. [CrossRef]

10. Zhao, Y.; Wang, Y.; Li, C.; Li, L.; Yang, X.; Wu, Y.; Chen, S.; Zhao, Y. Novel insight into physicochemical and flavor formation in naturally fermented tilapia sausage based on microbial metabolic network. *Food Res. Int.* **2021**, *141*, 110122. [CrossRef]
11. Lee, I.Y.; Kim, B.; Joo, N. Comparison between the volatile compounds and physicochemical and sensory characteristics of reverse-seared and conventionally seared beef steaks. *Foods* **2022**, *11*, 2135. [CrossRef] [PubMed]
12. Roobab, U.; Zeng, X.-A.; Ahmed, W.; Madni, G.M.; Manzoor, M.F.; Aadil, R.M. Effect of pulsed electric field on the chicken meat quality and taste-related amino acid stability: Flavor simulation. *Foods* **2023**, *12*, 710. [CrossRef]
13. *GB 5009.3-2016*; National Food Safety Standard—Determination of Moisture in Foods. National Health and Family Planning Commission of the People's Republic of China: Beijing, China, 2016.
14. Parvin, R.; Zahid, M.A.; Seo, J.-K.; Park, J.; Ko, J.; Yang, H.-S. Influence of reheating methods and frozen storage on physicochemical characteristics and warmed-over flavor of nutmeg extract-enriched precooked beef meatballs. *Antioxidants* **2020**, *9*, 9080670. [CrossRef] [PubMed]
15. Yu, S.; Huang, X.; Wang, L.; Chang, X.; Ren, Y.; Zhang, X.; Wang, Y. Qualitative and quantitative assessment of flavor quality of Chinese soybean paste using multiple sensor technologies combined with chemometrics and a data fusion strategy. *Food Chem.* **2023**, *405*, 134859. [CrossRef] [PubMed]
16. Wang, J.; Huang, X.-H.; Zhang, Y.-Y.; Nie, C.; Zhou, D.; Qin, L. Mechanism of salt effect on flavor formation in lightly-salted large yellow croaker by integrated multiple intelligent sensory and untargeted lipidomics analyses. *Food Chem.* **2024**, *435*, 137542. [CrossRef] [PubMed]
17. Xiao, X.; Hou, C.; Zhang, D.; Li, X.; Ren, C.; Ijaz, M.; Hussain, Z.; Liu, D. Effect of pre- and post-rigor on texture, flavor, heterocyclic aromatic amines and sensory evaluation of roasted lamb. *Meat Sci.* **2020**, *169*, 108220. [CrossRef] [PubMed]
18. Zhang, T.; Ni, H.; Qiu, X.-J.; Li, T.; Zhang, L.-Z.; Li, L.-J.; Jiang, Z.-D.; Li, Q.-B.; Chen, F.; Zheng, F.-P. Suppressive interaction approach for masking stale note of instant ripened Pu-Erh tea products. *Molecules* **2019**, *24*, 4473. [CrossRef]
19. Liu, S.; Tu, Y.; Sun, J.; Cai, P.; Zhou, Y.; Huang, Y.; Zhang, S.; Chen, W.; Wang, L.; Du, M.; et al. Fermented mixed feed regulates intestinal microbial community and metabolism and alters pork flavor and umami. *Meat Sci.* **2023**, *201*, 109177. [CrossRef]
20. Rao, J.W.; Meng, F.B.; Li, Y.C.; Chen, W.J.; Liu, D.Y.; Zhang, J.M. Effect of cooking methods on the edible, nutritive qualities and volatile flavor compounds of rabbit meat. *J. Sci. Food Agric.* **2022**, *102*, 4218–4228. [CrossRef]
21. Suleman, R.; Wang, Z.; Aadil, R.M.; Hui, T.; Hopkins, D.L.; Zhang, D. Effect of cooking on the nutritive quality, sensory properties and safety of lamb meat: Current challenges and future prospects. *Meat Sci.* **2020**, *167*, 108172. [CrossRef]
22. Costa, R.; Albergamo, A.; Bua, G.D.; Saija, E.; Dugo, G. Determination of flavor constituents in particular types of flour and derived pasta by heart-cutting multidimensional gas chromatography coupled with mass spectrometry and multiple headspace solid-phase microextraction. *LWT-Food Sci. Technol.* **2017**, *86*, 99–107. [CrossRef]
23. Ceylan, Z.; Atıcı, C.; Unal, K.; Meral, R.; Kutlu, N.; Babaoğlu, A.S.; Dilek, N.M. A novel material for the microbiological, oxidative, and color stability of salmon and chicken meat samples: Nanofibers obtained from sesame oil. *Food Res. Int.* **2023**, *170*, 112952. [CrossRef] [PubMed]
24. Gagaoua, M.; Suman, S.P.; Purslow, P.P.; Lebret, B. The color of fresh pork: Consumers expectations, underlying farm-to-fork factors, myoglobin chemistry and contribution of proteomics to decipher the biochemical mechanisms. *Meat Sci.* **2023**, *206*, 109340. [CrossRef] [PubMed]
25. Bıyıklı, M.; Akoğlu, A.; Kurhan, Ş.; Akoğlu, İ.T. Effect of different Sous Vide cooking temperature-time combinations on the physicochemical, microbiological, and sensory properties of turkey cutlet. *Int. J. Gastron Food Sci.* **2020**, *20*, 100204. [CrossRef]
26. Roldan, M.; Loebner, J.; Degen, J.; Henle, T.; Antequera, T.; Ruiz-Carrascal, J. Advanced glycation end products, physico-chemical and sensory characteristics of cooked lamb loins affected by cooking method and addition of flavour precursors. *Food Chem.* **2015**, *168*, 487–495. [CrossRef]
27. Zhou, M.; Shi, G.; Deng, Y.; Wang, C.; Qiao, Y.; Xiong, G.; Wang, L.; Wu, W.; Shi, L.; Ding, A. Study on the physicochemical and flavor characteristics of air frying and deep frying shrimp (crayfish) meat. *Front. Nutr.* **2022**, *9*, 1022590. [CrossRef]
28. Bolumar, T.; Toepfl, S. Application of shockwaves for meat tenderization. In *Innovative Food Processing Technologies*; Elsevier: Amsterdam, The Netherlands, 2016; pp. 231–258. [CrossRef]
29. Wang, H.; He, F.; He, H.; Du, P.; Liu, Y.; Wang, W.; Wang, S.; Ma, Y.; Chu, X.; Wang, Y.; et al. Predicting the formation of 2-amino-3-methyl-imidazole[4,5-f]quinoline (IQ) in the Maillard reaction model system under various reaction conditions. *LWT-Food Sci. Technol.* **2023**, *176*, 114551. [CrossRef]
30. Feng, Y.; Su, G.; Zhao, H.; Cai, Y.; Cui, C.; Sun-Waterhouse, D.; Zhao, M. Characterisation of aroma profiles of commercial soy sauce by odour activity value and omission test. *Food Chem.* **2015**, *167*, 220–228. [CrossRef]
31. Wang, Z.; Cai, R.; Yang, X.; Gao, Z.; Yuan, Y.; Yue, T. Changes in aroma components and potential Maillard reaction products during the stir-frying of pork slices. *Food Control* **2021**, *123*, 107855. [CrossRef]
32. Yu, Y.; Wang, G.; Yin, X.; Ge, C.; Liao, G. Effects of different cooking methods on free fatty acid profile, water-soluble compounds and flavor compounds in Chinese Piao chicken meat. *Food Res. Int.* **2021**, *149*, 110696. [CrossRef]
33. Zhu, W.; Luan, H.; Bu, Y.; Li, X.; Li, J.; Ji, G. Flavor characteristics of shrimp sauces with different fermentation and storage time. *LWT-Food Sci. Technol.* **2019**, *110*, 142–151. [CrossRef]
34. Liang, F.; Lin, L.; Zhu, Y.; Jiang, S.; Lu, J. Comparative study between surimi gel and surimi/crabmeat mixed gel on nutritional properties, flavor characteristics, color, and texture. *J. Aquat. Food Prod. Technol.* **2020**, *29*, 681–692. [CrossRef]

35. Zhang, J.j.; Wang, X.c.; Shi, W.z. Odor characteristics of white croaker and small yellow croaker fish during refrigerated storage. *J. Food Biochem.* **2019**, *43*, e12852. [CrossRef] [PubMed]
36. Zhou, Y.; Wang, X.; Chen, Y.; Yuan, B. Effects of different paprikas on the quality characteristics and volatile flavor components of spiced beef. *J. Food Process. Preserv.* **2021**, *45*, e15353. [CrossRef]
37. Chen, J.; Tao, L.; Zhang, T.; Zhang, J.; Wu, T.; Luan, D.; Ni, L.; Wang, X.; Zhong, J. Effect of four types of thermal processing methods on the aroma profiles of acidity regulator-treated tilapia muscles using E-nose, HS-SPME-GC-MS, and HS-GC-IMS. *LWT-Food Sci. Technol.* **2021**, *147*, 111585. [CrossRef]
38. Flores, M. Understanding the implications of current health trends on the aroma of wet and dry cured meat products. *Meat Sci.* **2018**, *144*, 53–61. [CrossRef]
39. Li, X.; Liu, S.-Q. Effect of pH, xylose content and heating temperature on colour and flavour compound formation of enzymatically hydrolysed pork trimmings. *LWT-Food Sci. Technol.* **2021**, *150*, 112017. [CrossRef]
40. Sun, A.; Wu, W.; Soladoye, O.P.; Aluko, R.E.; Bak, K.H.; Fu, Y.; Zhang, Y. Maillard reaction of food-derived peptides as a potential route to generate meat flavor compounds: A review. *Food Res. Int.* **2022**, *151*, 110823. [CrossRef]
41. Zhao, J.; Wang, M.; Xie, J.; Zhao, M.; Hou, L.; Liang, J.; Wang, S.; Cheng, J. Volatile flavor constituents in the pork broth of black-pig. *Food Chem.* **2017**, *226*, 51–60. [CrossRef]
42. Wu, W.; Zhan, J.; Tang, X.; Li, T.; Duan, S. Characterization and identification of pork flavor compounds and their precursors in Chinese indigenous pig breeds by volatile profiling and multivariate analysis. *Food Chem.* **2022**, *385*, 132543. [CrossRef]

*foods*

MDPI

*Article*

# Evaluation of the Quality and Lipid Content of Artisan Sausages Produced in Tungurahua, Ecuador

**Lander Pérez [1,2,*], Rosa Pincay [2], Diego Salazar [2], Nelly Flores [3] and Consuelo Escolastico [1]**

[1] International School of Doctorate, Sciences Doctorate, Universidad Nacional de Educación a Distancia (UNED), E-28040 Madrid, Spain; cescolastico@ccia.uned.es

[2] G+ Biofood and Engineering Research Group, Food and Biotechnology Faculty, Technical University of Ambato (UTA), Av. Los Chasquis y Río Payamino, Ambato 180206, Ecuador; rpincay5588@uta.edu.ec (R.P.); dm.salazar@uta.edu.ec (D.S.)

[3] Research and Development Directorate, Food and Biotechnology Faculty, Technical University of Ambato (UTA), Av. Los Chasquis y Río Payamino, Ambato 180206, Ecuador; ne.flores@uta.edu.ec

* Correspondence: lv.perez@uta.edu.ec

**Abstract:** The consumption of sausage worldwide increases every year; because of this increase, artisanal products have appeared and are intended to be perceived as natural and healthy. Obesity and cardiovascular diseases associated with consuming meat and meat derivatives have been estimated to be the leading cause of death in several countries. This study aimed to evaluate the nutritional quality, lipid content, and presence of saturated and unsaturated fatty acids, contributing to demonstrating the real nutritional value of artisanal sausages produced in Ecuador. Sausages from 10 factories in Ambato, Pelileo, and Píllaro, located in Tungurahua, Ecuador, were evaluated. The pH and acidity, color, proximal, sensory, microbiological, and lipid content were assessed. The pH and acidity showed a slight variation in all of the samples. Proximal analysis (moisture, protein, fat, and ash) established that the artisan sausages did not differ from the type of sausages reported in the literature. Microbiological analyses showed a good microbial quality, and there was no presence of *Staphylococcus aureus*, *Enterobacteria*, molds, or yeasts. The sensory attributes were similar for all of the sausages; the panelists did not notice any strange taste or odor. The lipid content showed that the artisanal sausages contained the highest percentage of palmitic, stearic, elaidic, and linolelaidic fatty acids. Unsaturated fatty acids were the most prevalent in all of the sausages collected from different locations. The results showed that the nutritional, microbiological, and sensory quality of the artisanal sausages did not show any parameter that would allow them to be classified as different or as having a better nutritional value.

**Keywords:** artisan sausages; fatty acids; nutritional properties; sensorial properties

## 1. Introduction

Currently, there is a growing emphasis on improving the nutritional balance of food. In this context, artisanal or organic products have gained popularity as they are perceived by the population as being more natural. However, it is important to note that not all these products are nutritious and healthy. Furthermore, these trends align with the understanding that attitudes related to environmental concerns and personal health significantly influence people's willingness to pay for organic food [1]. Historically, some of the most consumed foods have included meat products such as Frankfurter, Bratwurst, and Vienna, among other sausages, which are losing market value due to deficient health components such as fiber, or as a result of high salt, nitrite, and fat contents [2–4]. Consumers perceive this type of food as unhealthy and related to diseases such as diabetes, heart disease, stroke, and some forms of cancer [5,6]. In this sense, researchers are studying new formulations to improve the nutritional quality of products and provide new formulations, ingredients, and nutritional additives to these producers.

Recent research has revealed that adjusting sausage formulations, such as through the inclusion of bamboo fibers to reduce the fat content in bologna sausages, is a viable alternative [7], as well as the enhancement of Frankfurter-type sausages enriched with buckwheat as a source of bioactive compounds [8] and a reduction or substitution in sodium additives in cooked sausages [9]. It is widely acknowledged that the development of novel meat formulations is prohibitively expensive, leading to hesitation among some manufacturers. In emerging markets, certain producers opt for a more straightforward approach by altering the product label to create the impression of a "new" offering, thereby attracting a fresh consumer base [10].

The quality of sausages depends on critical factors, notably their fat, nitrite, and salt contents. Salt, a standard component of the human diet, has been associated with health concerns, particularly hypertension and other related issues [2]. Quality sausages are determined by the content of fat, nitrites, and salt content. Salt is a food preservative and flavoring component; however, the excessive consumption of salt in the human diet is associated with hypertension and other health problems [2,11]. However, consumers are reluctant to compromise on the taste and flavor profiles of meat products; they are not ready to change the taste and flavor of meat products [12]. Additionally, consuming meat and meat products has been linked to increased risk of colorectal cancer, primarily attributed to nitrites [13]. Notably, the fat content poses a significant challenge in the reformulation of meat products because of its influence on flavor. Moreover, it is the most extensively studied component concerning its impact on public health [14]. In sausage production, fat plays an essential role in enhancing attributes such as juiciness, tenderness, and overall palatability [15]. Traditionally, the fat in sausages is sourced from pork back fat and intramuscular fat. However, the nutritional quality of these fat sources, characterized by their high fatty acid composition, has recently become a primary concern, categorizing them as detrimental to human health [16,17]. The abundance of free fatty acids, including oleate, palmitate, stearate, myristate, palmitoleate, and linoleate, is associated with hepatic lipotoxicity, predominantly due to the excessive consumption of saturated and trans-fatty acids [18].

Consumers, based on messages seeking to improve their health, have focused on searching for foods similar to those they have usually consumed, but that somehow have a better nutritional value and that can be consumed with a low risk of developing diseases. This consumer requirement has generated fraudulent practices by labeling a false idea of nutritional foods, which, of course, are not true. In recent years, the concept of food fraud scandals has been associated with mislabeling practices [19]. Although this dishonest practice is difficult to prove, many consumers are guided by the perception that a food labeled as nutritional, organic, or artisanal has a different nutritional value that will not affect their health or generate possible illnesses [20,21]. Thus, the objective of this work is to evaluate the quality of sausages labeled as artisanal produced in 10 factories in Tungurahua, Ecuador.

## 2. Materials and Methods

### 2.1. Materials

Sausages were procured from 10 distinct factories located across various cities in Ecuador. Specifically, Ambato was the source of sausages from six factories, Pelileo contributed sausages from three factories, and one factory in Píllaro provided the remaining samples. Each factory supplied 200 g of sausages, which were subsequently refrigerated and stored at temperatures ranging from 4 to 6 °C.

To gain insight into sausage production processes, comprehensive observations were conducted at these diverse facilities. A uniform procedure for sausage production was employed across all of the evaluated factories. The common manufacturing approach involved the following steps: First, meat consisting of beef and pork, as well as pork back fat, was mechanically minced. Subsequently, the minced meat and fat, along with the requisite seasonings and crushed ice, were homogenized within a cutter. The resulting mixture was carefully stuffed into an artificial casing. The sausages were subsequently cooked in a water bath, maintaining temperatures in the range of 80–85 °C until the core of the product reached 73 °C, a process that typically lasted approximately 30 min. Following the heating phase, the sausages were subjected to rapid cooling in a cold bath, with the objective of reaching an internal temperature of 30 °C. Subsequently, the sausages were packaged in high-density polyethylene plastic food bags and stored under refrigeration, while maintaining the temperature within the 4–6 °C range.

### 2.2. Quality Determination in Sausages

#### 2.2.1. Proximal Analysis

The proximate composition (moisture, ash, protein, and fat) was evaluated following the official methods AOAC 19 927.05, AOAC 923.03, AOAC 2001.11, and AOAC 2033.06, respectively (AOAC, 2005) [22]. The carbohydrate content was estimated using difference and expressed as g/100 g. The moisture content (%g/100 g) was determined by drying in a stove at 105 °C ± 2 °C, the nitrogen was established using the Kjeldahl method, and the protein content (g/100 g) was estimated with the nitrogen portion using a 6.25 factor. The fat content (g/100 g) was determined using the Soxhlet method and the ash percentage was determined through incineration in a muffle at 550 °C. The water activity was determined using the aw equip (AQUA LAB Dew Point Water Activity Meter 4TE); the circular container of the equipment was filled with ultra-pure Milli-Q (Thermo Fisher Scientific, Waltham, MA, USA) water for calibration, and it was leveled with the sample, previously ground with a pestle, and placed into the equipment for measuring. The titratable acidity was evaluated in a titrator (Mettler Toledo G20 Compact Titrator); 10 g of sample was mixed with 60 mL of ultrapure Milli-Q (Thermo Scientific) to obtain the sample to be evaluated, and the acidity was reported as the lactic acid content (g/100 g of lactic acid). Furthermore, a potentiometer measured the pH (Mettler Toledo Seven Compact). All of the determinations were performed in triplicate using three samples for each treatment.

#### 2.2.2. Energy Value

The calorie content was estimated using 100 g of sample, where the overall sum of calories of the individual components was the energy value for each component: fat (X 9 kcal/g), protein (X 4 kcal/g), and carbohydrate (X 4 kcal/g) [23,24].

#### 2.2.3. Color

Color parameters were measured using a Hunter Lab Colorimeter (mini Scan 4500 L EZ, Hunter Associates Laboratory INC, Reston, VA, USA) calibrated with an illuminator D65 (natural light) and standard observer D10. The colorimeter was calibrated before taking the measurements, and the white tile standard was used. The chroma polar coordinate or saturation C* was calculated using the expression $C^* = \sqrt{(a^{*2} + b^{*2})}$ and hue tone using $(h^*) = \arctan(b^*/a^*)$ to a* and b* positives. The chroma and hue tone were calculated based on Salazar, Arancibia, Calderón, López-Caballero, and Montero [1]. At least 15 measurements were performed in different areas of the sample, and the averages were recorded as the reported values.

#### 2.2.4. Microbiological Analysis

Here, 10 g of artisan sausages and 90 mL of buffered 0.1% peptone water (Difco, Le Pont de Claix, France) were placed into a sterile bag and homogenized in a stomacher homogenizer (Model 400 C, Seward, London, UK) for 1 min at room temperature. Serial decimal dilutions were prepared in peptone water for each sample. *Enterobacteria* were determined on a double layer of Violet Red Bile Glucose Agar (VRBG) (Acumedia, MI, USA) incubated at 30 °C/72 h. Mesophilic aerobic bacteria were determined in PCA agar (Difco) incubated at 30 °C for 72 h. *Staphylococcus aureus* were determined on Baird Parker agar (Difco) supplemented with egg yolk tellurite incubated at 30 °C for 48 h. Mold and yeast were spread plated on Rose Bengal Agar (RBC) (Difco) and incubated at 25 °C. All of the assays were performed according to the Bacteriological Analytical Manual of the U.S. Food and Drug Administration [25]. All of the analyses were performed in triplicate.

#### 2.2.5. Sensory Quality

Sensorial parameters were evaluated according to the procedure described by Salazar, Arancibia, Calderón, López-Caballero, and Montero [1]. Thirty trained panelists assessed attributes such as the overall acceptability, taste, and odor of the sausages. The panel received training in previous sessions. The sensory evaluation was developed in a sensorial cabin illuminated with white light, the temperature in the room was 20 °C, and the panelists used a test based on a 5-point hedonic scale (1—disliked very much; 2—disliked moderately; 3—neither liked nor disliked; 4—liked moderately; 5—liked very much). For the evaluation, two cylinders (1.6 cm in diameter _ and 3 cm in length) of grilled samples without casing were provided to the panelist for assessment. A glass of water and salted crackers for mouth cleaning were also provided.

#### 2.2.6. Chlorides Determination

The determination of chlorides was carried out according to the Mohr method [26]. Approximately 1 g of each sample was weighed into an Erlenmeyer flask, and an ultrapure Milli-Q water mixture (Thermo Scientific) was added and boiled to extract the sodium chloride. The chlorides were titrated with 0.1 N of silver nitrate according to the Mohr method, using two drops of 5% potassium chromate as an indicator. The analysis of each sample was developed in triplicate.

#### 2.2.7. Nitrite Determination

Nitrite determination was developed according to the AOAC 973.331 method. The measurement was based on spectrophotometric detection of a chromophore formed by the reaction between nitrites and 1-naphthyl-ethylene diamine (NED) sulfanilamide. Approximately 1 g of previously ground sample was weighed and diluted with ultra-pure Milli-Q (Thermo Scientific) water. The mixture was transferred to 50 mL volumetric flasks, and two washes were completed with 2 mL of hot water. The volumetric flasks with the samples were kept in a water bath for 2 h, then the aqueous extract was obtained at room temperature; finally, it was bottled and filtered. Then 2 mL of the filtered fluid was transferred into a 10 mL flask containing 0.5 mL sulfanilamide; after that, the mixture was shaken. After 5 min, 0.5 mL of Eriochrome Black T (EBT) chemical indicator was added, and the absorbance was measured at 540 nm in quartz cells in a UV−VIS (Hach, DR 5000) spectrometer. Three analyses of each sample were developed.

### 2.3. *Lipid Content Quantification*

#### 2.3.1. Fat Extraction

The method proposed by Bligh and Dyer [27] was used: 6.7 mL of chloroform, 1.7 mL of extra pure methanol (99.5%), 1.4 mL of ultra-pure Milli-Q water were added; all of the tubes were prepared into the fume hood; and the tubes were covered with aluminum foil and placed in a horizontal agitator for 1 h at 280 rpm and then centrifuged for 10 min at 5000× *g* rpm. The tubes were placed to rest for 72 h to facilitate the separation of the two phases. The organic phase was placed in the rotary evaporator (Eyela N-400 H, Shanghai, China) at 33 °C with 26% vortex velocity and 400 millibars of pressure to recover the solvent.

#### 2.3.2. Methyl Esterification of Fatty Acids

Methyl esterification was developed based on the method proposed by Ichihara and Fukubayashi [28]. A sample of the previously extracted fat was taken, and 2 mL of potassium hydroxide with methanol 0.5 M was added. The mixture was placed in a boiling water bath for 10 min. It was then allowed to cool at room temperature, 1 mL of methanolic hydrochloric acid (1: 4 *v*/*v*) was placed in it, and it was subsequently maintained in a water bath of 50 °C for 25 min. Finally, 3 mL of ultra-pure Milli-Q water and 10 mL of chromatographic grade hexane were added and homogenized.

2.3.3. Chromatographic Determination

For the chromatography reading, the extract obtained previously was left to rest for a minimum of 2 h intervals and 1.5 mL of the hexane extract was obtained. Finally, the samples were filtered using microfilters (Econofilter) and placed into Agilent Technology vials for analysis [29,30]. The chromatographic conditions for quantifying the fatty acids were as follows: volume of injection of 0.3 μL and the temperature of the injector was 250 °C. Split injection at 15 mL/min. An Agilent HP-88 Column (Length: 60 m; diameter; 0.25 mm; thickness: 0.25μm). Flux of 1.4 mL/min. Carrier Gas of 99.999% pure helium. Initial temperature of 80 °C. Temperature ramp of (1) 10 °C/min until 120 °C, and then held for 4 min. Temperature ramp (2) of 20 °C/min until 140 °C, and then held 5 min. Temperature ramp (3) of 2 °C/min until 200 °C, and then held for 45 min. The temperature of the detector was set to 240 °C. The total time of the analysis was 53 min. The detection system was a mass detector.

2.3.4. Identification and Quantification of Fatty Acids

The identification and quantification of fatty acids were developed using a gaseous chromatography technique with Agilent Technologies 7890B G.C. coupled with 5977[a] GC/MSD equipment. The proportion was determined with the reference pattern (FAME MIX C4-C24, Supelco, Bellefonte, PA, USA) through a comparison with the NIST14. L library and using a calibration factor.

### *2.4. Statistical Analysis*

The data were processed and analyzed with the GraphPad Prism 5.0 program (GraphPad Software, San Diego, CA, USA). Analysis of variance was performed using the one-way or two-way ANOVA test, and, when this was significant, the means were compared using Tukey's test.

## 3. Results

The visual appearance of the different sausages evaluated in this study are shown in Figure 1. The sausages showed differences in appearance, color, and shape characteristic of the areas where they were collected. Visually, the artisanal sausages were no different from the commercial products found on the market.

### *3.1. Proximal Composition and Nutritional Estimations*

The proximal composition of the different sausages is shown in Table 1. As a result of the composition and formulation, sausages are considered a high-intermediate moisture food, and the moisture values for this type of product could be between 50–75% [1,31]; all of the sausages evaluated were in this range, oscillating between 62.50 to 70 g/100 g of sample. Also, it is essential to note that all of them had a high water content, likely attributable to the inclusion of starch, fibers, or some flour [32]. The difference in moisture percentage in the samples was possibly due to the amount of water added to each sample according to the formulation of each factory.

The ash content showed differences in all of the sausages, but in a very narrow range ($p < 0.05$) (Table 1); values ranged between 2.54 and 4.31 g/100 g of sample; the ash content in this study was similar to that reported by dos Santos Alves et al. [33] (~3–4%) in a Bologna low-fat sausage formulated with a gel pork skin emulsion, banana flour, and water (ratio of 1:2:2). The differences in ash content could be attributed to the amount of flour usually used in the production of this type of sausages and the mineral of the meat that was used to produce the artisan sausages. Ecuadorian artisan sausages were proven to be a richer source of proteins, and the results showed significant differences ($p < 0.05$). It is noticeable that in these artisan sausages, the protein content was similar to what was presented by Choe and Kim [34] (12.60%) in sausages in which a wheat fiber mixture as a fat replacer was added, and Alvarado-Ramírez et al. [35] who used carrot powder as an ingredient.

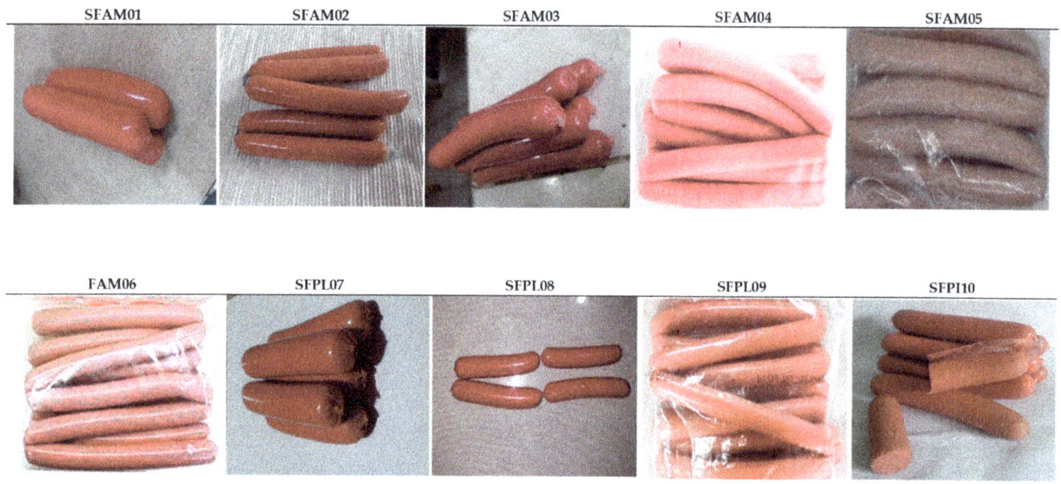


**Figure 1.** Visual appearance of artisan sausages produced in Tungurahua, Ecuador.

The fat content showed differences ($p < 0.05$) ranging from 6.61 to 12.33 g/100 g of sample. In this type of product, the variability in fat content is attributed to the meat and pork back fat used in the formulations; however, the producer decides what kind of sausages it needs to produce, and the amount of total fat content, which should be at most 28% [36]; in this sense, all of the samples were within the range. It was not expected that the sausages in this study would have a lower fat content than the sausages reported by Salazar, Arancibia, Calderón, López-Caballero, and Montero [1] the control sausage of their study; also, our values were lower than those in the studies in which fat was reduced [4]. The content of carbohydrates showed variability ($p < 0.05$); the values in samples varied from 6.65 to 14.16 g/100 g of sample; the values observed could be attributable to the use of flours, starches, and other meat extenders to improve the texture and sensory quality of products [37]. Concerning the caloric content of sausages (Table 1), it was observed that according to the components evaluated, the caloric content was comparable to the values reported by Choi et al. [38] in sausages with makgeolli lees fiber, or by Salcedo-Sandoval et al. [39] in sausages with olive oil, flax, or konjac gels. The World Health Organization [40] recommends that the energy content be composed of a variable contribution of carbohydrates, fat, and protein. In products such as sausages, the proportion is very far from the desirable energy balance; however, it is essential to evaluate which fat is detrimental to human health and to establish a lipid profile to know the natural effect of fat on the nutrition of customers.

**Table 1.** Proximate composition (g/100 g), energy values (Kcal/100 g), and pH and acidity (g/100 g of lactic acid) of artisan sausages produced in Tungurahua, Ecuador.

| Parameters | Samples | | | | | | | | | |
|---|---|---|---|---|---|---|---|---|---|---|
| | SFAM01 | SFAM02 | SFAM03 | SFAM04 | SFAM05 | SFAM06 | SFPL07 | SFPL08 | SFPL09 | SFPI10 |
| Moisture | $66.72 \pm 0.01^{c}$ | $66.28 \pm 0.01^{cd}$ | $66.70 \pm 0.32^{c}$ | $70.00 \pm 0.17^{a}$ | $63.21 \pm 0.09^{f}$ | $67.27 \pm 0.01^{b}$ | $62.50 \pm 0.21^{g}$ | $64.27 \pm 0.01^{e}$ | $65.84 \pm 0.14^{d}$ | $64.13 \pm 0.21^{e}$ |
| Protein | $9.54 \pm 0.05^{h}$ | $12.97 \pm 0.05^{c}$ | $10.85 \pm 0.01^{e}$ | $10.67 \pm 0.02^{f}$ | $15.70 \pm 0.02^{a}$ | $13.64 \pm 0.01^{b}$ | $10.95 \pm 0.09^{e}$ | $9.66 \pm 0.05^{g}$ | $12.17 \pm 0.02^{d}$ | $12.19 \pm 0.02^{d}$ |
| Fat | $8.69 \pm 0.30^{ef}$ | $8.01 \pm 0.30^{f}$ | $8.31 \pm 0.32^{f}$ | $6.61 \pm 0.28^{g}$ | $10.18 \pm 0.56^{c}$ | $9.91 \pm 0.21^{cd}$ | $12.33 \pm 0.01^{a}$ | $9.27 \pm 0.03^{de}$ | $9.41 \pm 0.16^{cde}$ | $11.09 \pm 0.06^{b}$ |
| Ash | $2.68 \pm 0.01^{de}$ | $3.25 \pm 0.01^{c}$ | $3.50 \pm 0.10^{b}$ | $3.45 \pm 0.01^{b}$ | $4.31 \pm 0.01^{a}$ | $2.71 \pm 0.05^{d}$ | $2.73 \pm 0.09^{d}$ | $2.80 \pm 0.05^{d}$ | $2.54 \pm 0.04^{e}$ | $3.37 \pm 0.06^{bc}$ |
| Carbohydrates | $12.39 \pm 0.03^{b}$ | $9.51 \pm 0.03^{f}$ | $11.10 \pm 0.03^{d}$ | $9.56 \pm 0.02^{f}$ | $6.99 \pm 0.02^{g}$ | $6.65 \pm 0.02^{h}$ | $11.76 \pm 0.29^{c}$ | $14.16 \pm 0.13^{a}$ | $10.26 \pm 0.04^{e}$ | $9.34 \pm 0.05^{f}$ |
| Calories | $165.93 \pm 1.20^{g}$ | $162.01 \pm 0.82^{h}$ | $162.59 \pm 0.35^{h}$ | $140.42 \pm 0.65^{i}$ | $182.42 \pm 0.21^{c}$ | $170.37 \pm 0.38^{f}$ | $201.84 \pm 0.57^{a}$ | $178.67 \pm 0.99^{d}$ | $174.47 \pm 0.71^{e}$ | $185.89 \pm 0.82^{b}$ |
| pH | $7.16 \pm 0.01^{a}$ | $7.14 \pm 0.01^{a}$ | $7.10 \pm 0.07^{ab}$ | $6.64 \pm 0.47^{cde}$ | $6.24 \pm 0.01^{e}$ | $6.48 \pm 0.01^{de}$ | $6.65 \pm 0.03^{cde}$ | $6.47 \pm 0.01^{de}$ | $6.97 \pm 0.01^{abc}$ | $6.70 \pm 0.01^{bcd}$ |
| Acidity | $0.39 \pm 0.02^{e}$ | $0.48 \pm 0.01^{b}$ | $0.42 \pm 0.01^{c}$ | $0.36 \pm 0.01^{e}$ | $0.70 \pm 0.01^{a}$ | $0.42 \pm 0.01^{c}$ | $0.31 \pm 0.01^{f}$ | $0.31 \pm 0.01^{f}$ | $0.47 \pm 0.01^{b}$ | $0.42 \pm 0.01^{c}$ |

The results are the mean ± standard deviation. One-way ANOVA: Different letters ($^{a.b.}$...) in the same row indicate significant differences among samples ($p \leq 0.05$). Values of proximal composition are presented as g/100 g of sausage.

### 3.2. Acidity and pH

Acidity variations between sausages were minimal ($p < 0.01$) (Table 1). According to the percentage of lactic acid, the values were between 0.31% to 0.70%. The difference in acidity could be due to the glycogen content in the meat before elaborating the sausages, as lactic acid is produced from glycogen, according to Wang et al. [41]. Lactic acid produced by the fermentation of carbohydrates generates a decrease in pH, inhibiting the growth of harmful microorganisms, precipitating dehydration, slowing water retention, and affecting the color of the finished products. Similar results were reported by Salazar, Arancibia, Calderón, López-Caballero, and Montero [1] in sausages enriched with green banana flour. However, the relationship between lactic acid and pH has not been determined because pH does not only rely on lactic acid, as in cured and refrigerated products. It has been observed that the velocity of the reduction in residual nitrite is related exponentially to the pH and temperature. The reduction rate doubles with each 12 °C temperature increment and each reduction of 0.86 pH units [42,43]. SFAM06 showed a low concentration of nitrites and low pH value; according to the pH in the samples, it was possible to observe that these values were close to neutral. The values obtained in this study were lower than those of the sausages in rural markets in the center of Mexico [44].

### 3.3. Water Activity, Chlorides, and Nitrites

The sausage water activity did not show significant differences ($p > 0.05$) (Table 2). However, although there were no differences, it is essential to note that samples with a high-water activity had a shorter shelf life. The water activity values were typical for these types of gel-emulsion cooked meat products [45]. The growth of *Salmonella* spp was inhibited in water activity values below 0.94, *L. monocytogenes* $\leq$ 0.94, and *Staphylococcus aureus* < 0.83. The reduction in water activity in meat derivatives depended on the temperature, relative moisture, and, mainly, the formulation [46–48]. On the other hand, chlorides ranged between 1.81% to 3.41% ($p < 0.05$). Sample SFAM06 showed the highest value; also, this value was outside of the average allowed by the Ecuadorian Institute for Standardization [36], for which the values ranged between 1.5% to 3.0%; sample SFAM06 was unique and did not comply with the rules; this result was probably attributable to the different quantities of salt added by the sausage producers.

**Table 2.** Results of the physicochemical parameters of the sausage samples.

| Sample | Water Activity ($a_w$) | Nitrites (ppm) | Chlorides (%) |
|---|---|---|---|
| SFAM01 | $0.98 \pm 0.01^{a}$ | $100 \pm 5.00^{d}$ | $1.81 \pm 0.09^{d}$ |
| SFAM02 | $0.98 \pm 0.01^{a}$ | $100 \pm 8.00^{d}$ | $2.60 \pm 0.12^{b}$ |
| SFAM03 | $0.98 \pm 0.01^{a}$ | $140 \pm 6.00^{a}$ | $2.62 \pm 0.17^{b}$ |
| SFAM04 | $0.97 \pm 0.01^{a}$ | $130 \pm 4.00^{ab}$ | $2.87 \pm 0.20^{b}$ |
| SFAM05 | $0.96 \pm 0.01^{a}$ | $140 \pm 7.00^{a}$ | $3.41 \pm 0.01^{a}$ |
| SFAM06 | $0.98 \pm 0.01^{a}$ | $110 \pm 5.00^{cd}$ | $1.98 \pm 0.09^{cd}$ |
| SFPL07 | $0.98 \pm 0.01^{a}$ | $122 \pm 7.00^{bc}$ | $2.16 \pm 0.02^{d}$ |
| SFPL08 | $0.98 \pm 0.01^{a}$ | $122 \pm 6.00^{bc}$ | $2.10 \pm 0.02^{cd}$ |
| SFPL09 | $0.98 \pm 0.01^{a}$ | $116 \pm 5.00^{bcd}$ | $1.81 \pm 0.06^{d}$ |
| SFPI10 | $0.98 \pm 0.01^{a}$ | $121 \pm 4.00^{bc}$ | $2.14 \pm 0.01^{c}$ |

The results are the mean ± standard deviation. One-way ANOVA: Different letters ($^{a.b}$....) in the same column indicate significant differences among samples ($p \leq 0.05$).

The concentration of nitrites in artisan sausages showed different values ($p < 0.05$); the results showed that the addition of nitrites was likely a problem because the producers were not dosing correctly, even though the values were below those allowed by the Food And Drug Administration [49] and the Codex Alimentarius. Samples SFAM06 and SFPL10 showed the lowest concentration and SFAM05 showed the highest, given the different amounts of nitrites added to the sausage according to each formula. Adding nitrites also produced nitrosomyoglobin, a characteristic red pigment generated when this salt decomposes [50]. The evaluated sausages presented as characteristic red sausages; however, the characteristic red color could not be evaluated because industries use artificial coloring to generate homogeneity in their products.

*3.4. Color Properties*

The visual appearance of the sausages obtained from different factories showed a homogeneous aspect, with differences in color found in some samples (Table 3). The color of meat derivatives was due to biochemical reactions between myoglobin, hemoglobin, and oxygen, as well as the action of nitrites and nitrates [51]. Lightness (L*) values showed significant differences ($p < 0.05$) between the analyzed samples. The sausages showed values of ~50, which means they could not be classified as light or dark. The lightness, as well as other color properties, were difficult to compare with other sausages because of the influence of the formulation, the use of different colorants, and the process. In this sense, the data obtained showed specific information about artisan sausages produced in Tungurahua, Ecuador.

**Table 3.** Color parameters of artisan sausages from Tungurahua, Ecuador.

| Sample | L* | a* | b* | C* | h* |
|---|---|---|---|---|---|
| SFAM01 | 50.09 ± 0.52 $^{de}$ | 21.45 ± 0.16 $^{c}$ | 16.97 ± 0.16 $^{bc}$ | 27.35 ± 0.53 $^{bc}$ | 38.35 ± 0.17 $^{de}$ |
| SFAM02 | 46.45 ± 0.33 $^{g}$ | 23.08 ± 0.08 $^{b}$ | 16.52 ± 0.18 $^{c}$ | 28.38 ± 0.34 $^{b}$ | 35.59 ± 0.09 $^{f}$ |
| SFAM03 | 52.20 ± 0.48 $^{bc}$ | 16.15 ± 0.27 $^{e}$ | 14.82 ± 0.27 $^{d}$ | 21.92 ± 0.49 $^{e}$ | 42.54 ± 0.28 $^{b}$ |
| SFAM04 | 53.98 ± 0.84 $^{a}$ | 14.65 ± 0.79 $^{f}$ | 13.23 ± 0.77 $^{f}$ | 19.74 ± 0.85 $^{f}$ | 42.08 ± 0.80 $^{b}$ |
| SFAM05 | 49.45 ± 0.51 $^{e}$ | 25.91 ± 0.38 $^{a}$ | 17.98 ± 0.17 $^{b}$ | 31.54 ± 0.52 $^{a}$ | 34.76 ± 0.39 $^{fg}$ |
| SFAM06 | 47.56 ± 0.44 $^{fg}$ | 20.48 ± 0.13 $^{cd}$ | 16.27 ± 0.33 $^{c}$ | 26.16 ± 0.45 $^{cd}$ | 38.46 ± 0.14 $^{cd}$ |
| SFPL07 | 51.64 ± 0.14 $^{bc}$ | 12.01 ± 0.17 $^{g}$ | 22.92 ± 0.48 $^{a}$ | 25.88 ± 0.15 $^{d}$ | 62.35 ± 0.18 $^{a}$ |
| SFPL08 | 51.28 ± 0.47 $^{cd}$ | 20.32 ± 0.48 $^{d}$ | 16.73 ± 0.37 $^{bc}$ | 26.32 ± 0.48 $^{cd}$ | 39.47 ± 0.49 $^{c}$ |
| SFPL09 | 48.85 ± 0.56 $^{ef}$ | 25.54 ± 0.21 $^{a}$ | 17.05 ± 0.97 $^{bc}$ | 30.71 ± 0.57 $^{a}$ | 33.73 ± 0.22 $^{g}$ |
| SFPI10 | 52.70 ± 0.29 $^{ab}$ | 21.51 ± 0.46 $^{c}$ | 16.41 ± 0.42 $^{c}$ | 27.05 ± 0.30 $^{bcd}$ | 37.34 ± 0.47 $^{e}$ |

The results are the mean ± standard deviation. One-way ANOVA: Different letters ($^{a,b,\ldots}$) in the same column indicate significant differences among samples ($p \leq 0.05$). L* (lightness); a* (red/green chromaticity); b* (yellow/blue chromaticity); C* (chroma); h* (Hue tone).

The samples were grouped in the same quadrant of CIE*L*a*b* space based on their positive a* and b* values. The a* values ($p < 0.05$) showed a tendency towards the red color, probably attributable to the application of natural or artificial colorants [49], which are used to obtain homogeneous products and satisfy customers [14,33]. Regarding the b* values, the SFPL07 sample showed the most pronounced values ($p < 0.05$).

Samples SFAM05 and SFPL09 showed higher chroma (C*) values ($p < 0.05$). That is, these sausages had more saturated or intense colors compared with the rest of the samples. On the other hand, samples SFAM03 and SFPM04 showed lower chroma (C*) values ($p < 0.05$). The presence of the meat proteins (beef and pork) used for emulsification and the addition of permitted additives (such as salt replacers) considerably influenced this parameter [52]. Typically, this characteristic is adjusted with the use of coloring according to consumer preferences. For example, sausages with more intense colors in some developing countries are considered better quality [53,54], while in developed countries, the trend is completely the opposite [55].

The hue tone (*h) showed a similar tendency to the chroma (*C). Samples SFAM05 and SFPL09 presented lower h* values ($p < 0.05$) because they were closer to the intense reds in the CIE*L*a*b* space. The rest of the samples show higher values ($p < 0.05$), which indicates they were in the pink zone (around h* = 45°). Also, the different binders used in the formulation played an important role in high C* and h* values, according to a study by Jin et al. [56].

The different coloration in the samples could be due to the variation in the color of the mass formed by the emulsified muscle protein [57]. As for the results obtained by other authors, variable effects have been found in chroma when fat is replaced with another ingredient in meat products [58].

### 3.5. Microbiological Analysis

The counts of mesophilic aerobic microorganisms were 2.5–3.8 log CFU/g for all of the lots. Similar results were found by Salazar, Arancibia, Calderón, López-Caballero, and Montero [1] in sausages with green banana flours, which reported 3 log CFU/g, and Ranucci et al. [59] in sausages with pork enriched with emmer wheat, hazelnut, and almond, which reported 3.05 log CFU/g. These were studies on cooked sausages; however, the formulations were so different that it took effort to compare the results. Despite this, similar results were observed in other meat products; Vienna-homogenized sausages manufactured with different commercial functional additives reached 3.35 and 3.61 log CFU/g for aerobic bacteria when AFX and FPRX commercial additives at minimum concentrations were used [60], and bologna-type fat-reduced turkey sausage with potato starch and pea fiber did not exceed 4 log CFU/g (depending of the formulation) [61]. On the other hand, Cerón-Guevara et al. [62] reported counts that ranged from 4.52 to 6.12 log CFU/g in low-fat Frankfurter-type sausages with mushroom flour (2.5 and 5%); the authors attributed the results to the spore-forming bacteria present in agricultural products.

Mold and yeast counts were evaluated due to the possible inclusion of flour in the formulations producers used for meat extenders, which is common practice in these types of artisan sausages. No microorganisms were found the artisan sausages; however, it is essential to note that the product quality could be affected by the inclusion of flour, as well as its starch content, due to the presence and increment of mold and yeasts, given that these microorganisms use starch as a growth substrate [63]. Macedo et al. [64] reported counts of molds and yeast between 2.78 to 4.41 log CFU/g in hot dog sausages; these results were attributable to the pH, high water activity, and nutrient composition [65]. Likewise, Nkekesi et al. [66] reported fungal counts ranging from 0.0 CFU/g to $9.83 \times 10^3$ CFU/g for street-vended grilled beef sausages. The presence of molds and yeasts was usually related to food contamination and the production of mycotoxins, which, in most cases, were not easy to eliminate during processing.

In *Enterobacteria* and *Staphylococcus aureus*, no presence was detected, probably as a result of the heat treatment (85 °C during 30 min) and the addition of nitrites into the formulation, because they inhibited the growth of these microorganisms due to a halt in active transport, oxygen absorption, and oxidative phosphorylation in bacteria such as *S. aureus*, *Escherichia coli*, and *Clostridium botulinum* [67].

### 3.6. Sensory Analysis

Sensory properties are an important parameter for all types of foods; however, when products receive the qualification of being artisan, these parameters become dependent on the point of view of the consumers who purchase the product. Many consumers and producers are not willing to make concessions regarding the aroma, flavor, texture, and juiciness of meat products, especially sausages. The sensory attributes of sausages of 10 factories are shown in Figure 2. The sausages registered an acceptability average score of ~5, indicating that consumers liked the sausages; in this sense, the acceptability results could be attributed to the formulation used by each factory in order to be accepted in the market. It was assumed that each factory had developed a sensory evaluation that allowed them to create the most appropriate formulation that was accepted by consumers. It is important to remember that the objective of this work was to evaluate the quality of these types of artisanal sausages and that they were already for sale on the market. The sensory evaluation allowed us to establish that these products were accepted in the market. No strange odors or flavors were detected in the sausages during the evaluation. The scores tended to be the same as acceptability; that is to say, the judges showed their perception that they liked the aroma of the sausages. These results were attributed to the fact that these types of sausages used similar condiments in their preparation. A literature revision was developed to compare the sausages produced using nutritional components such as fibers, vegetable flour, fruit flour, and other types of flour. The results in different studies showed that when a non-traditional ingredient was added, the sensorial properties tended to decrease. Muchekeza et al. [68] reported that sausages made with quinoa as a binder obtained acceptability values that ranged from 6.71 to 7.14 on a nine-point hedonic scale (extremely like); however, Amaranth had a range of 5.4–5.94, which reflected a neither like nor dislike by consumers. Another study of Frankfurter-type sausages with the addition of carrot paste showed that the aroma and flavor decreased the acceptability when compared with the control [69]. Zaini et al. [70] used banana peel powders to study the effect on the technological functionality, sensory quality, and nutritional quality of chicken sausage; with respect to sensorial properties, the results showed that that the addition of 2–6% of banana peel powder significantly reduced the aroma and appearance of the treated sausage. Choi et al. [71] reported that the inclusion of 3–6% brown rice fiber reduced the overall acceptability of the sausage. The artisan sausages evaluated in this study showed a different behavior compared with the sausages with a differentiated nutritional value and were not only called artisanal to influence consumer perception.

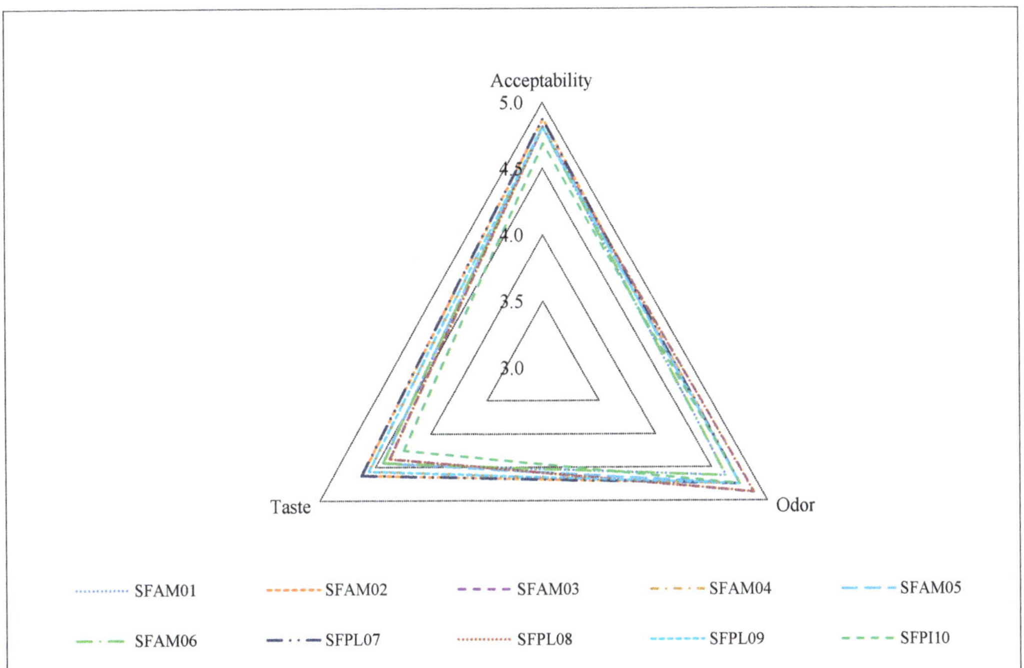


**Figure 2.** Sensory attributes of commercial sausages obtained from 10 different factories located in Ecuador.

### 3.7. *Lipidic Content of Sausages*

The variation in the composition of the fatty acids in meat products depends fundamentally on the diet of the animals or the overall dietary patterns [72]. In the fatty acids detected in the artisan sausages of this study (Table 4), the highest percentages corresponded to elaidic, palmitic, linolelaidic, and stearic fatty acids. Palmitic and stearic acids correspond to the saturated fatty acid group, while elaidic and linolelaidic acids are trans-fatty acids. Concerning the different places the sausages were collected from, it was observed that sausages from Pelileo (SFPL07, SFPL08, and SFPL09) and one area of Ambato (SFAM03) showed the highest content of elaidic fatty acid; this fatty acid has attracted much attention because it is the principal trans-fatty acid found in hydrogenated vegetable oils, and it is linked to heart disease [73]. It is important to note that in the other sausages from other places, the elaidic content was also high; likewise, the content of palmitic, stearic, and linolelaidic acids were high in contrast with the other fatty acids, which could probably be attributed to the diets animals received, usually consisting of a particular type of hydrogenated vegetable oil. These types of fatty acid are often created by the partial hydrogenation or elaidinization of vegetable oils [74]. The tendency in composition observed in the sausages could probably show that, independent of the place where sausages were produced, the meat and other components used for sausage production were similar throughout Tungurahua province. The results obtained in this study were close to those reported by Franco et al. [75] in traditional sausages from Galicia; the authors reported the highest total level of fatty acids and included oleic (45% of the total fatty acids), palmitic (21%), linoleic (14%), stearic (13%), and palmitoleic acids (2%). The presence of fatty acids such as oleic acid allowed us to infer that, despite assuming that the fat content was detrimental to health, components were also observed, such as the case of oleic acid, which counteracted the intensity of specific inflammatory processes, as it reduced the production of chemotactic mediators of inflammation [76].

Saturated fatty acids, monounsaturated acids, polyunsaturated acids, and trans-fatty acids showed differences ($p < 0.05$) (Table 5). The sum of monounsaturated and polyunsaturated fatty acids determined the total amount of unsaturated fatty acids (UFA). The SFA/UFA index was calculated, as was the relationship between omega-6 and omega-3 ($\omega6/\omega3$). The atherogenicity index showed values between 3.34 to 6.36, which could be attributed to the meat used to produce the sausages. The results of this study showed that, independent of the place where the sausages were obtained, the significant components were saturated and trans-fatty acids; this indicated that the artisan sausages probably had the same harmful fat components as other commercial meat derivates, as saturated and trans-fatty acids have been linked to several adverse health effects [77]. Epidemiological and clinical evidence has shown that trans fats are essential for developing cardiovascular diseases and are linked to inflammation, diabetes, and cancer.

On the other hand, unsaturated fatty acids help preserve consumer health [78]. The results obtained in this study were different from the sausages collected in butcher shops in the center of Mexico, which contained 37.95% saturated fatty acids, 48.45% monounsaturated acids, and 13.61% of polyunsaturated acids [44]. Meat products supplied 23% of the total fat, 29% of dietary fatty acids, 21% of monounsaturated fatty acids, and 16% of polyunsaturated fatty acids [79].

The relationship between saturated and unsaturated fatty acids is recommended to be above 2.5 because it is an indicator of high nutritional value [80], Slama et al. [81] reported, for example, that pearl millet had a ratio of unsaturated fatty acid to saturated fatty acid equal to 3.39. Lower levels should be increased to optimize the relationship between polyunsaturated and saturated acids for consumption. In this sense, the relationship between saturated and unsaturated fatty acids found in the artisan sausages made in Tungurahua, Ecuador, indicated that the sausages would not harm consumer health.

The studies showed that the level of $\omega6$ fatty acids and the relationship between $\omega6/\omega3$ pointed to a higher risk of obesity related to these parameters, and it played a fundamental role in the increased risk of obesity through eicosanoid metabolites [82]. The optimal relationship between $\omega$-6/$\omega$-3 should be between 5:1 and 10:1 in weight; the samples in this study were below these referred values. The atherogenicity index by Ulbricht and Southgate defined the relationship between the fatty acid content and the capacity to elevate the levels of serum cholesterol, formed by lauric, myristic, and palmitic acids, as well as monounsaturated and polyunsaturated fatty acids, which fulfill a protective role by reducing the risk of developing coronary diseases [83].

**Table 4.** Fatty acid contents (g/100 g) in the fat of the artisan sausages produced in Ecuador.

| Fatty Acids | Artisan Sausages | | | | | | | | | |
|---|---|---|---|---|---|---|---|---|---|---|
| (g/100 g) | SFAM01 | SFAM02 | SFAM03 | SFAM04 | SFAM05 | SFAM06 | SFPL07 | SFPL08 | SFPL09 | SFPI10 |
| Lauric | $0.21 \pm 0.04^{b}$ | $0.15 \pm 0.04^{b}$ | $0.22 \pm 0.09^{b}$ | $0.12 \pm 0.02^{b}$ | $0.12 \pm 0.06^{b}$ | $0.14 \pm 0.00^{b}$ | $0.11 \pm 0.02^{b}$ | $0.30 \pm 0.15^{a}$ | $0.16 \pm 0.05^{b}$ | $0.10 \pm 0.05^{b}$ |
| Myristic | $1.34 \pm 0.07^{bcd}$ | $1.69 \pm 0.07^{b}$ | $1.05 \pm 0.32$ | $1.25 \pm 0.16^{cd}$ | $2.13 \pm 0.01^{a}$ | $1.68 \pm 0.05^{b}$ | $0.74 \pm 0.05^{e}$ | $1.37 \pm 0.24^{bc}$ | $1.54 \pm 0.04^{bc}$ | $1.30 \pm 0.18^{bc}$ |
| Myristoleic | $0.31 \pm 0.03^{b}$ | $0.42 \pm 0.01^{a}$ | - | $0.43 \pm 0.00^{a}$ | $0.41 \pm 0.03^{a}$ | $0.21 \pm 0.02^{cd}$ | $0.16 \pm 0.01^{d}$ | $0.21 \pm 0.10^{bc}$ | $0.40 \pm 0.08^{a}$ | $0.15 \pm 0.04^{d}$ |
| Pentadecyl | $0.36 \pm 0.04^{a}$ | $0.34 \pm 0.01^{abc}$ | - | $0.37 \pm 0.05^{ab}$ | $0.31 \pm 0.01^{abcd}$ | $0.21 \pm 0.04^{d}$ | $0.12 \pm 0.02^{e}$ | $0.19 \pm 0.11^{cd}$ | $0.25 \pm 0.05^{bcd}$ | $0.12 \pm 0.00^{e}$ |
| Palmitic | $23.21 \pm 0.95^{a}$ | $25.08 \pm 0.06^{a}$ | $22.04 \pm 2.40^{a}$ | $22.74 \pm 2.55^{a}$ | $23.65 \pm 0.08^{a}$ | $24.28 \pm 0.75^{a}$ | $25.22 \pm 0.60^{a}$ | $22.85 \pm 2.98^{a}$ | $18.60 \pm 5.59^{a}$ | $24.46 \pm 1.79^{a}$ |
| Palmitoleic | $3.65 \pm 0.01^{cd}$ | $4.65 \pm 0.21^{ab}$ | $4.26 \pm 0.55^{bc}$ | $4.93 \pm 0.07^{a}$ | $3.16 \pm 0.03^{de}$ | $3.04 \pm 0.16^{de}$ | $4.91 \pm 0.09^{a}$ | $2.91 \pm 0.54^{de}$ | $2.90 \pm 0.28^{de}$ | $2.36 \pm 0.39^{e}$ |
| Margaric | $0.72 \pm 0.01^{a}$ | $0.73 \pm 0.01^{a}$ | - | $0.76 \pm 0.11^{a}$ | $0.76 \pm 0.00^{a}$ | $0.58 \pm 0.11^{ab}$ | $0.25 \pm 0.02^{c}$ | $0.32 \pm 0.00^{c}$ | $0.65 \pm 0.20^{a}$ | $0.45 \pm 0.06^{b}$ |
| Cis-10-heptadecenoic | $0.28 \pm 0.03^{de}$ | $0.35 \pm 0.01^{cd}$ | - | $0.28 \pm 0.07^{de}$ | $0.38 \pm 0.03^{ab}$ | $0.28 \pm 0.03^{de}$ | - | $0.18 \pm 0.04^{e}$ | $0.43 \pm 0.04^{a}$ | $0.32 \pm 0.08^{abc}$ |
| Stearic | $12.83 \pm 0.39^{bc}$ | $13.17 \pm 0.10^{b}$ | $6.06 \pm 1.19^{d}$ | $13.92 \pm 1.35^{b}$ | $16.07 \pm 0.13^{b}$ | $13.87 \pm 0.14^{b}$ | $7.13 \pm 0.86^{d}$ | $7.84 \pm 2.47^{cd}$ | $12.43 \pm 3.33^{b}$ | $17.60 \pm 3.94^{a}$ |
| Oleic | $2.22 \pm 0.06^{a}$ | $1.30 \pm 0.10^{c}$ | $0.11 \pm 0.16^{f}$ | $1.44 \pm 0.03^{c}$ | $1.66 \pm 0.08^{b}$ | $0.87 \pm 0.16^{d}$ | $0.30 \pm 0.14^{e}$ | $0.55 \pm 0.02^{e}$ | $2.37 \pm 0.01^{a}$ | $0.32 \pm 0.07^{e}$ |
| Elaidic | $35.81 \pm 1.32^{c}$ | $38.96 \pm 0.061^{bc}$ | $44.26 \pm 3.84^{ab}$ | $39.18 \pm 2.46^{bc}$ | $39.52 \pm 0.36^{bc}$ | $40.13 \pm 1.60^{bc}$ | $43.87 \pm 0.78^{ab}$ | $43.06 \pm 1.11^{b}$ | $45.15 \pm 6.95^{a}$ | $38.54 \pm 3.13^{bc}$ |
| Linolelaidic | $17.50 \pm 2.00^{bc}$ | $11.27 \pm 0.39^{fg}$ | $21.06 \pm 0.22^{a}$ | $13.37 \pm 1.52^{ef}$ | $9.97 \pm 0.26^{g}$ | $12.36 \pm 0.03^{efg}$ | $16.37 \pm 1.13^{cd}$ | $18.70 \pm 3.00^{ab}$ | $13.74 \pm 1.66^{de}$ | $12.47 \pm 2.02^{ef}$ |
| Arachidic | $0.30 \pm 0.10^{b}$ | - | - | $0.14 \pm 0.01^{b}$ | $0.22 \pm 0.08^{b}$ | $0.39 \pm 0.36^{a}$ | $0.10 \pm 0.03^{b}$ | $0.13 \pm 0.00^{b}$ | - | $0.27 \pm 0.05^{b}$ |
| Linolenic | $0.88 \pm 0.28^{b}$ | $1.56 \pm 0.13^{a}$ | $0.92 \pm 0.17^{bc}$ | $0.67 \pm 0.04^{cde}$ | $0.63 \pm 0.17^{bcde}$ | $0.63 \pm 0.30^{bcd}$ | $0.46 \pm 0.01^{e}$ | $0.80 \pm 0.04^{bcd}$ | $0.67 \pm 0.06^{bcde}$ | $0.52 \pm 0.07^{de}$ |
| Cis-11-eicosenoic | $0.32 \pm 0.05^{cd}$ | $0.32 \pm 0.06^{cd}$ | - | $0.37 \pm 0.07^{bcd}$ | $0.49 \pm 0.10^{b}$ | $0.68 \pm 0.29^{a}$ | $0.24 \pm 0.02^{d}$ | $0.31 \pm 0.02^{d}$ | $0.39 \pm 0.020^{bcd}$ | $0.50 \pm 0.03^{bc}$ |
| Eicosadiene | - | - | - | - | $0.49 \pm 0.24^{ab}$ | $0.65 \pm 0.64^{a}$ | - | $0.27 \pm 0.07^{b}$ | $0.36 \pm 0.05^{b}$ | $0.50 \pm 0.02^{b}$ |

The results are the mean ± standard deviation. One-way ANOVA: Different letters ($^{a.b.}$...) in the same column indicate significant differences among fatty acids ($p \leq 0.05$).

**Table 5.** Results of the percentage of fatty acids, omega relationship, and atherogenicity index.

| Samples | Saturated Fatty Acids (%) | Unsaturated Fatty Acids (%) | Trans-fatty Acids (%) | Saturated/ Unsaturated | ω-6/ω-3 | Atherogenicity Index (I.A.) |
|---|---|---|---|---|---|---|
| SFAM01 | 39.01 ± 0.04 $^{e}$ | 7.67 ± 0.04 $^{c}$ | 53.32 ± 0.05 $^{e}$ | 5.08 ± 0.01 $^{de}$ | 0.00 | 3.76 ± 0.03 $^{f}$ |
| SFAM02 | 41.17 ± 0.04 $^{c}$ | 8.61 ± 0.01 $^{a}$ | 50.23 ± 0.06 $^{fg}$ | 4.78 ± 0.21 $^{de}$ | 0.00 | 3.72 ± 0.01 $^{f}$ |
| SFAM03 | 33.64 ± 0.09 $^{f}$ | 7.47 ± 0.05 $^{d}$ | 58.89 ± 0.02 $^{d}$ | 4.50 ± 0.55 $^{e}$ | 0.55 ± 0.02 $^{c}$ | 3.34 ± 0.02 $^{g}$ |
| SFAM04 | 29.38 ± 0.02 $^{h}$ | 5.30 ± 0.05 $^{h}$ | 65.32 ± 0.04 $^{a}$ | 5.55 ± 0.07 $^{cd}$ | 0.00 | 5.00 ± 0.07 $^{c}$ |
| SFAM05 | 39.30 ± 0.06 $^{d}$ | 8.14 ± 0.01 $^{b}$ | 52.56 ± 0.08 $^{e}$ | 4.83 ± 0.03 $^{de}$ | 0.00 | 3.42 ± 0.03 $^{g}$ |
| SFAM06 | 43.29 ± 0.05 $^{b}$ | 7.23 ± 0.04 $^{e}$ | 49.49 ± 0.05 $^{g}$ | 5.99 ± 0.16 $^{bc}$ | 0.78 ± 0.11 $^{b}$ | 4.47 ± 0.03 $^{e}$ |
| SFPL07 | 41.15 ± 0.02 $^{c}$ | 6.36 ± 0.02 $^{g}$ | 52.49 ± 0.06 $^{e}$ | 6.47 ± 0.09 $^{b}$ | 1.02 ± 0.02 $^{a}$ | 4.89 ± 0.02 $^{cd}$ |
| SFPL08 | 44.30 ± 0.01 $^{a}$ | 4.68 ± 0.11 $^{i}$ | 51.02 ± 0.03 $^{f}$ | 9.47 ± 0.54 $^{a}$ | 0.97 ± 0.02 $^{a}$ | 6.36 ± 0.04 $^{a}$ |
| SFPL09 | 33.68 ± 0.05 $^{f}$ | 6.70 ± 0.05 $^{f}$ | 60.25 ± 0.02 $^{c}$ | 5.55 ± 0.28 $^{cd}$ | 0.00 | 4.67 ± 0.04 $^{de}$ |
| SFPI10 | 33.01 ± 0.05 $^{g}$ | 5.22 ± 0.01 $^{h}$ | 61.77 ± 0.01 $^{b}$ | 6.32 ± 0.39 $^{bc}$ | 0.33 ± 0.06 $^{d}$ | 5.49 ± 0.08 $^{b}$ |

The results are the mean ± standard deviation. One-way ANOVA: Different letters ($^{a,b}$...) in the same column indicate significant differences among the samples ($p \leq 0.05$).

## 4. Limitations

The findings of this study are limited to the use of physicochemical parameters, microbiological characteristics, and sensory properties that are considered most suitable for establishing the quality of commercial sausages sold. Likewise, the determination of the fat content was supposed to show whether there were elements that allowed for the term artisanal to be associated with a quality somehow implicitly implied in the name given to these sausages. However, the depth of the study did not allow for a comparison with considerations regarding the development of products with better nutritional and technological quality. However, one of the concerns in developing countries is that consumers should not be deceived by terms that can generate an expectation different from the actual product attributes. The determination of the quality and lipid profile of this study aimed to contribute to the nutritional information of sausages that are marketed in Tungurahua, Ecuador.

## 5. Conclusions

This study evaluated the quality and lipid content of artisan sausages from 10 factories. The results of the nutritional quality indicated that the artisan sausages were in the ranges recommended for commercial sausages concerning moisture, fat, protein, and ash. There was no evidence in the results that indicated that the artisan sausages could have any valuable parameters to establish a better nutritional value; on the contrary, the results could indicate that "artisan" is only the name of the product. As indicators of the quality of production, the pH and acidity did not show significant changes, related to microbiological quality, because none of the microbiological indicators evaluated showed the presence of harmful microorganisms. There were no negative observations for taste or odor in the sensorial parameters, while the acceptability was assessed well. The results of the lipid content showed significant differences, and the results showed a high presence of saturated fatty acids (palmitic, stearic, elaidic, and linolelaidic). The results indicated that the same unsaturated fatty acids were present and were the highest in composition, independent of the place they were produced and collected. Also, the relationship showed that unsaturated fatty acids were primarily found in the artisan sausages. In conclusion, the results of the artisanal sausages evaluated here showed that there was no remarkable characteristic denoting that the premise of the artisanal sausages being healthier was reflected in their composition.

**Author Contributions:** Conceptualization, L.P. and C.E.; methodology, L.P. and R.P.; formal analysis, L.P., D.S. and N.F.; investigation, L.P. and R.P.; resources, L.P.; data curation, L.P., C.E., D.S. and N.F.; writing—original draft preparation, L.P., R.P., D.S. and N.F.; writing—review and editing, C.E.;

supervision, C.E.; project administration, L.P. All authors have read and agreed to the published version of the manuscript.

**Funding:** This research was funded by Dirección de Investigación y Desarrollo (DIDE) and the Technical University of Ambato-Ecuador Project 1596-CU-P-2016, "Determination of the relationship between saturated and unsaturated fatty acids in mass consumption foods through the development and implementation of the analytical method as a contribution to nutrition information".

**Data Availability Statement:** Data are contained within the article.

**Acknowledgments:** The authors thank Dirección de Investigación y Desarrollo (DIDE) and the Technical University of Ambato-Ecuador for support through the project's financial approval by resolution 1596-CU-P-2016.

**Conflicts of Interest:** The authors declare no conflict of interest.

## References

1. Salazar, D.; Arancibia, M.; Calderón, L.; López-Caballero, M.E.; Montero, M.P. Underutilized Green Banana (*Musa acuminata AAA*) flours to develop fiber enriched frankfurter-type sausages. *Foods* **2021**, *10*, 1142. [CrossRef]
2. Elias, M.; Laranjo, M.; Agulheiro-Santos, A.C.; Potes, M.E. The role of salt on food and human health. In *Salt in the Earth*; IntechOpen: London, UK, 2020; Volume 19.
3. FAO. *Global Emergence of Infectious Diseases: Links with Wild Meat Consumption, Ecosystem Disruption, Habitat Degradation and Biodiversity Loss*; FAO: Rome, Italy, 2020; p. 6.
4. Méndez-Zamora, G.; García-Macías, J.A.; Santellano-Estrada, E.; Chávez-Martínez, A.; Durán-Meléndez, L.A.; Silva-Vázquez, R.; Quintero-Ramos, A. Fat reduction in the formulation of frankfurter sausages using inulin and pectin. *Food Sci. Technol.* **2015**, *35*, 25–31. [CrossRef]
5. San-Cristobal, R.; Navas-Carretero, S.; Martínez-González, M.Á.; Ordovas, J.M.; Martínez, J.A. Contribution of macronutrients to obesity: Implications for precision nutrition. *Nat. Rev. Endocrinol.* **2020**, *16*, 305–320. [CrossRef] [PubMed]
6. Malik, V.S.; Hu, F.B. The role of sugar-sweetened beverages in the global epidemics of obesity and chronic diseases. *Nat. Rev. Endocrinol.* **2022**, *18*, 205–218. [CrossRef] [PubMed]
7. Magalhães, I.M.C.; Paglarini, C.d.S.; Vidal, V.A.S.; Pollonio, M.A.R. Bamboo fiber improves the functional properties of reduced salt and phosphate-free Bologna sausage. *J. Offood Process. Preserv.* **2020**, *44*, e14929.
8. Salejda, A.M.; Olender, K.; Zielińska-Dawidziak, M.; Mazur, M.; Szperlik, J.; Miedzianka, J.; Zawiślak, I.; Kolniak-Ostek, J.; Szmaja, A. Frankfurter-type sausage enriched with buckwheat by-product as a source of bioactive compounds. *Foods* **2022**, *11*, 674. [CrossRef]
9. Araya-Quesada, Y.; Araya-Morice, A.; Araya-Vargas, S.; Redondo-Solano, M.; Madrigal-Arias, E.; Cubero-Castillo, E. Reduction of sodium additives in cooked sausages: Effect on physicochemical, sensory and microbiological characteristics. *J. Food Sci. Technol.* **2020**, *57*, 3051–3059. [CrossRef]
10. Rees, J. *Food Adulteration and Food Fraud*; Reaktion Books: London, UK, 2020.
11. Kim, T.-K.; Yong, H.-I.; Jung, S.; Kim, H.-W.; Choi, Y.-S. Technologies for the production of meat products with a low sodium chloride content and improved quality characteristics—A review. *Foods* **2021**, *10*, 957. [CrossRef]
12. Rodrigues, I.; Gonçalves, L.A.; Carvalho, F.A.; Pires, M.; JP Rocha, Y.; Barros, J.C.; Carvalho, L.T.; Trindade, M.A. Understanding salt reduction in fat-reduced hot dog sausages: Network structure, emulsion stability and consumer acceptance. *Food Sci. Technol. Int.* **2020**, *26*, 123–131. [CrossRef]
13. Morrow, L.; Greenwald, B. Healthy Food choices, physical activity, and Screening Reduce the Risk of colorectal cancer. *Gastroenterol. Nurs.* **2022**, *45*, 113–119. [CrossRef]
14. Nasonova, V.V.; Tunieva, E.K. A comparative study of fat replacers in cooked sausages. *IOP Conf. Ser. Earth. Environ. Sci.* **2019**, *333*, 012085. [CrossRef]
15. Patel, A.R.; Nicholson, R.A.; Marangoni, A.G. Applications of fat mimetics for the replacement of saturated and hydrogenated fat in food products. *Curr. Opin. Food Sci.* **2020**, *33*, 61–68. [CrossRef]
16. Karwowska, M.; Kononiuk, A.D.; Borrajo, P.; Lorenzo, J.M. Comparative studies on the fatty acid profile and volatile compounds of fallow deer and beef fermented sausages without nitrite produced with the addition of acid whey. *Appl. Sci.* **2021**, *11*, 1320. [CrossRef]
17. de Carvalho, F.A.L.; Munekata, P.E.S.; Pateiro, M.; Campagnol, P.C.B.; Domínguez, R.; Trindade, M.A.; Lorenzo, J.M. Effect of replacing backfat with vegetable oils during the shelf-life of cooked lamb sausages. *LWT* **2020**, *122*, 109052. [CrossRef]
18. Islam, M.A.; Amin, M.N.; Siddiqui, S.A.; Hossain, M.P.; Sultana, F.; Kabir, M.R. Trans fatty acids and lipid profile: A serious risk factor to cardiovascular disease, cancer and diabetes. *Diabetes Metab. Syndr.* **2019**, *13*, 1643–1647. [CrossRef]
19. Moreira, M.J.; García-Díez, J.; de Almeida, J.M.M.M.; Saraiva, C. Consumer knowledge about food labeling and fraud. *Foods* **2021**, *10*, 1095. [CrossRef]
20. Pleadin, J.; Lešić, T.; Vujačić, V.; Milićević, D.; Buneta, A.; Šušnić, S.; Lukanić, I.; Krešić, G. Comparison of Chemical Composition and Fatty Acid Profile of Traditional Meat Products from Croatia and Montenegro. *J. Food Qual.* **2021**, *2021*, 5586436. [CrossRef]

21. López-Pedrouso, M.; Lorenzo, J.M.; Gullón, B.; Campagnol, P.C.B.; Franco, D. Novel strategy for developing healthy meat products replacing saturated fat with oleogels. *Curr. Opin. Food Sci.* **2021**, *40*, 40–45. [CrossRef]
22. AOAC. *Official Methods of Analysis*, 18th ed.; AOAC: Rockville, MD, USA, 2005.
23. Canadian Food Inspection Agency (CFIA). Food Labelling for Industry. Available online: https://www.inspection.gc.ca/food-label-requirements/labelling/industry/nutrition-labelling/elements-within-the-nutrition-facts-table/eng/1389206763218/1389206811747?chap=1 (accessed on 10 September 2023).
24. Food Drug Administration (FDA). Nutrition Labeling of Food. 21 CFR 101.9. Available online: https://www.ecfr.gov/current/title-21/chapter-I/subchapter-B/part-101/subpart-A/section-101.9 (accessed on 10 September 2023).
25. Bennett, R.W.; Monday, S.R. Staphylococcus aureus. In *International Handbook of Foodborne Pathogens*; CRC Press: Boca Raton, FL, USA, 2003; pp. 61–80.
26. Korkmaz, D. Precipitation titration: "Determination of chloride by the Mohr method". *Methods* **2001**, *2*, 1–6.
27. Bligh, E.G.; Dyer, W.J. A rapid method of total lipid extraction and purification. *Can. J. Biochem. Physiol.* **1959**, *37*, 911–917. [CrossRef]
28. Ichihara, K.i.; Fukubayashi, Y. Preparation of fatty acid methyl esters for gas-liquid chromatography. *J. Lipid Res.* **2010**, *51*, 635–640. [CrossRef]
29. Orata, F. *Derivatization Reactions and Reagents for Gas Chromatography Analysis*; Intech: Penang, Malaysia, 2012; Volume 91.
30. Antolín, E.M.; Delange, D.M.; Canavaciolo, V.G. Evaluation of five methods for derivatization and GC determination of a mixture of very long chain fatty acids (C24:0–C36:0). *J. Pharm. Biomed. Anal.* **2008**, *46*, 194–199. [CrossRef]
31. Chin, K.B.; Lee, H.L.; Chun, S.S. Product characteristics of comminuted sausages as affected by various fat and moisture combinations. *Asian-Australas. J. Anim. Sci.* **2004**, *17*, 538–542. [CrossRef]
32. Rosero-Chasoy, G.; Serna-Cock, L. Effect of plantain (*Musa paradisiaca* L. cv. Dominico Harton) peel flour as binder in frankfurter-type sausage. *Acta Agron.* **2017**, *66*, 305–310. [CrossRef]
33. dos Santos Alves, L.A.A.; Lorenzo, J.M.; Gonçalves, C.A.A.; dos Santos, B.A.; Heck, R.T.; Cichoski, A.J.; Campagnol, P.C.B. Production of healthier bologna type sausages using pork skin and green banana flour as a fat replacers. *Meat Sci.* **2016**, *121*, 73–78. [CrossRef] [PubMed]
34. Choe, J.; Kim, H.-Y. Quality characteristics of reduced fat emulsion-type chicken sausages using chicken skin and wheat fiber mixture as fat replacer. *Poult. Sci.* **2019**, *98*, 2662–2669. [CrossRef] [PubMed]
35. Alvarado-Ramírez, M.; Santana-Gálvez, J.; Santacruz, A.; Carranza-Montealvo, L.D.; Ortega-Hernández, E.; Tirado-Escobosa, J.; Cisneros-Zevallos, L.; Jacobo-Velázquez, D.A. Using a functional carrot powder ingredient to produce sausages with high levels of nutraceuticals. *J. Food Sci.* **2018**, *83*, 2351–2361. [CrossRef] [PubMed]
36. Ecuadorian Institue of Normalization. Carne y Productos Cárnicos. In *Productos Cárnicos Crudos, Productos Cárnicos Curados-Madurados y Productos Cárnicos Precocidos-Cocidos*; Ecuadorian Institue of Normalization: Quito, Ecuador, 2012.
37. de los Angeles Vivar-Vera, M.; Pérez-Silva, A.; Ruiz-López, I.I.; Hernández-Cázares, A.S.; Solano-Barrera, S.; Ruiz-Espinosa, H.; Bernardino-Nicanor, A.; González-Cruz, L. Chemical, physical and sensory properties of Vienna sausages formulated with a starfruit dietary fiber concentrate. *J. Food Sci. Technol.* **2018**, *55*, 3303–3313. [CrossRef] [PubMed]
38. Choi, Y.-S.; Kim, H.-W.; Hwang, K.-E.; Song, D.-H.; Choi, J.-H.; Lee, M.-A.; Chung, H.-J.; Kim, C.-J. Physicochemical properties and sensory characteristics of reduced-fat frankfurters with pork back fat replaced by dietary fiber extracted from makgeolli lees. *Meat Sci.* **2014**, *96*, 892–900. [CrossRef]
39. Salcedo-Sandoval, L.; Cofrades, S.; Ruiz-Capillas Pérez, C.; Solas, M.T.; Jiménez-Colmenero, F. Healthier oils stabilized in konjac matrix as fat replacers in n−3 PUFA enriched frankfurters. *Meat Sci.* **2013**, *93*, 757–766. [CrossRef]
40. WHO. *Diet, Nutrition and the Prevention of Chronic Diseases*; Technical Report Series 916; WHO: Geneva, Switzerland, 2003.
41. Wang, Y.; Han, J.; Wang, D.; Gao, F.; Zhang, K.; Tian, J.; Jin, Y. Research update on the impact of lactic acid bacteria on the substance metabolism, flavor, and quality characteristics of fermented meat products. *Foods* **2022**, *11*, 2090. [CrossRef]
42. Sallan, S.; Yılmaz Oral, Z.F.; Kaya, M. A review on the role of lactic acid bacteria in the formation and reduction of volatile nitrosamines in fermented sausages. *Foods* **2023**, *12*, 702. [CrossRef]
43. Paudel, N.; Subedi, D.; Khanal, S.; Acharya, D.R.; Bhattarai, S. Estimation of nitrite level and effect of processing on residual nitrite level in sausages marketed in Dharan, Nepal. *Afr. J. Food Sci.* **2021**, *15*, 67–71.
44. González-Tenorio, R.; Totosaus, A.; Caro, I.; Mateo, J. Caracterización de Propiedades Químicas y Fisicoquímicas de Chorizos Comercializados en la Zona Centro de México. *Inf. Tecnol.* **2013**, *24*, 3–14. [CrossRef]
45. Perez-Baez, A.J.; Camou, J.P.; Valenzuela-Melendres, M.; Lucas-Gonzalez, R.; Viuda-Martos, M. Assessment of chemical, physico-chemical and sensorial properties of frankfurter-type sausages added with roselle (*Hibiscus sabdariffa* L.), extracts. *Proceedings* **2021**, *70*, 73.
46. Castro Bolanos, M.; Herrera Ramirez, C.H.; Lutz Cruz, G. Composition, characterization and atherogenic potential of oils, fats and other by products produced or marketed in Costa Rica. *Acta Med. Costarric* **2005**, *47*, 36–42.
47. International Commission on Microbiological Specifications for Foods. *Microorganisms in Foods 5: Characteristics of Microbial Pathogens*; Springer Science & Business Media: Berlin/Heidelberg, Germany, 1996; Volume 5.
48. Kamboj, S.; Gupta, N.; Bandral, J.D.; Gandotra, G.; Anjum, N. Food safety and hygiene: A review. *Int. J. Chem. Stud.* **2020**, *8*, 358–368. [CrossRef]
49. FDA. *Food Additives Permitted for Direct Addition to Food for Human Consumption*; FDA: Silver Spring, MD, USA, 2023; Volume 3.

50. Stoica, M.; Antohi, V.M.; Alexe, P.; Ivan, A.S.; Stanciu, S.; Stoica, D.; Zlati, M.L.; Stuparu-Cretu, M. New strategies for the total/partial replacement of conventional sodium nitrite in meat products: A review. *Food. Bioproc. Tech.* **2022**, *15*, 514–538. [CrossRef]
51. Yang, X.; Sebranek, J.G.; Luo, X.; Zhang, W.; Zhang, M.; Xu, B.; Zhang, Y.; Liang, R. Effects of calcium salts on the physicochemical quality of cured beef sausages during manufacturing and storage: A potential calcium application for sausages with alginate casings. *Foods* **2021**, *10*, 2783. [CrossRef]
52. Pérez-Alvarez, J.A.; Fernández-López, J. *Color Characteristics of Meat and Poultry Processing*; CRC Press: Boca Raton, FL, USA, 2009; pp. 355–373.
53. Üren, A.; Babayiğit, D. Colour parameters of Turkish-type fermented sausage during fermentation and ripening. *Meat Sci.* **1997**, *45*, 539–549. [CrossRef]
54. Fadda, S.; Vignolo, G. *Central and South American Products*; Wiley: Medford, MA, USA, 2014; pp. 355–360.
55. Aykın-Dinçer, E.; Güngör, K.K.; Çağlar, E.; Erbaş, M. The use of beetroot extract and extract powder in sausages as natural food colorant. *Int. J. Food Eng.* **2020**, *17*, 75–82. [CrossRef]
56. Jin, S.-k.; Kim, S.-H.; Choi, J.-S.; Yim, D.-G. Effect of diverse binder materials and their addition levels on physico-chemical characteristics of sausages. *J. Food Meas. Charact.* **2019**, *13*, 1558–1565. [CrossRef]
57. Khatkar, A.B.; Kaur, A. Effect of protein incorporation on functional, thermal, textural and overall quality characteristics of instant noodles. *J. Food Meas. Charact.* **2018**, *12*, 2218–2229. [CrossRef]
58. Noguerol, A.T.; Larrea, V.; Pagán, M.J. The effect of psyllium (*Plantago ovata* Forsk) fibres on the mechanical and physicochemical characteristics of plant-based sausages. *Eur. Food Res. Technol.* **2022**, *248*, 2483–2496. [CrossRef]
59. Ranucci, D.; Roila, R.; Andoni, E.; Braconi, P.; Branciari, R. Punica granatum and Citrus spp. extract mix affects spoilage microorganisms growth rate in vacuum-packaged cooked sausages made from pork meat, emmer wheat (*Triticum dicoccum* Schübler), almond (*Prunus dulcis* Mill.) and hazelnut (*Corylus avellana* L.). *Foods* **2019**, *8*, 664. [CrossRef]
60. Fudali, A.; Chełmecka, I.; Salejda, A.M.; Krasnowska, G. Microbiological safety and organoleptic quality of homogenized sausages manufactured with commercial functional additives. *Appl. Sci.* **2021**, *11*, 11662. [CrossRef]
61. Varga-Visi, É.; Toxanbayeva, B.; Andrássyné Baka, G.; Romváari, R. Shelf life and quality of Bologna-type fat reduced turkey sausage. *Acta Aliment.* **2020**, *49*, 60–68. [CrossRef]
62. Cerón-Guevara, M.I.; Rangel-Vargas, E.; Lorenzo, J.M.; Bermúdez, R.; Pateiro, M.; Rodríguez, J.A.; Sánchez-Ortega, I.; Santos, E.M. Reduction of salt and fat in Frankfurter sausages by addition of *Agaricus bisporus* and *Pleurotus ostreatus* flour. *Foods* **2020**, *9*, 760. [CrossRef]
63. Sachindra, N.; Sakhare, P.; Yashoda, K.; Rao, D.N. Microbial profile of buffalo sausage during processing and storage. *Food Control* **2005**, *16*, 31–35. [CrossRef]
64. Macedo, R.C.B.D.S.; Rebouças, C.K.D.O.; Melo, E.C.C.D.; Freire, B.C.F.; Mendonça, L.P.D.; Santana, F.E.D.O.; Bezerra, A.C.D.S.; Soares, K.M.D.P. Evaluation of the physicochemical and microbiological quality of hot dog sausages in the city of Mossoró, Rio Grande do Norte, Brazil. *Res. Soc. Dev.* **2021**, *10*, e118101119195. [CrossRef]
65. de Marco, I.; Vogel, C.; Morandin, G.C.; Beilke, L.; Valduga, N.Z.; Zambiazi, E.T.; Martinhago, J.V.; Schittler, L. Salsichas tipo hot dog: Perfil microbiológico, isolamento e caracterização de bactérias ácido láticas (BAL) com potencial antagonista. In Proceedings of the Revista do Congresso Sul Brasileiro de Engenharia de Alimentos, Guarapuava, Brazil, 5–9 November 2018.
66. Nkekesi, B.; Amenya, P.; Aboagye, G.; Kortei, N.K. Street-vended grilled beef sausages as potential vehicles of bacterial and fungal pathogens: An exploratory survey in Ho, the capital city of the Volta Region of Ghana. *Food Sci. Nutr.* **2023**, *11*, 7013–7025. [CrossRef] [PubMed]
67. Lin, L.; Hu, J.Y.; Wu, Y.; Chen, M.; Ou, J.; Yan, W.L. Assessment of the inhibitory effects of sodium nitrite, nisin, potassium sorbate, and sodium lactate on *Staphylococcus aureus* growth and staphylococcal enterotoxin A production in cooked pork sausage using a predictive growth model. *Food Sci. Hum. Wellness.* **2018**, *7*, 83–90. [CrossRef]
68. Muchekeza, J.T.; Jombo, T.Z.; Magogo, C.; Mugari, A.; Manjeru, P.; Manhokwe, S. Proximate, physico-chemical, functional and sensory properties of quinoa and amaranth flour AS potential binders in beef sausages. *Food. Chem.* **2021**, *365*, 130619. [CrossRef]
69. Sam, F.E.; Ma, T.-Z.; Atuna, R.A.; Salifu, R.; Nubalanaan, B.-A.; Amagloh, F.K.; Han, S.-Y. Physicochemical, oxidative stability and sensory properties of frankfurter-type sausage as influenced by the addition of carrot (*Daucus carota*) paste. *Foods* **2021**, *10*, 3032. [CrossRef] [PubMed]
70. Zaini, H.B.M.; Sintang, M.D.B.; Pindi, W. The roles of banana peel powders to alter technological functionality, sensory and nutritional quality of chicken sausage. *Food Sci. Nutr.* **2020**, *8*, 5497–5507. [CrossRef] [PubMed]
71. Choi, Y.S.; Choi, J.H.; Han, D.J.; Kim, H.Y.; Kim, H.W.; Lee, M.A.; Chung, H.J.; Kim, C.J. Effects of Laminaria japonica on the physico-chemical and sensory characteristics of reduced-fat pork patties. *Meat Sci* **2012**, *91*, 1–7. [CrossRef]
72. Van Elswyk, M.E.; McNeill, S.H. Impact of grass/forage feeding versus grain finishing on beef nutrients and sensory quality: The U.S. experience. *Meat Sci.* **2014**, *96*, 535–540. [CrossRef]
73. Valenzuela, C.A.; Baker, E.J.; De Souza, C.O.; Miles, E.A.; Calder, P.C. Differential effects of ruminant and industrial 18-carbon trans-monounsaturated fatty acids (trans vaccenic and elaidic) on the inflammatory responses of an endothelial cell line. *Molecules* **2021**, *26*, 5834. [CrossRef] [PubMed]

74. Belichovska, D.; Pejkovski, Z.; Nikolova, A.S.; Belichovska, K.; Nastova, R.; Gjorgovska, N. Chemical and fatty acid composition of poultry meat and pork fatback as a raw material for the production of frankfurters. *Maced. J. Anim. Sci.* **2020**, *10*, 23–28. [CrossRef]
75. Franco, I.; Iglesias, S.; Prieto, B.; Carballo, J.; de los Alimentos, Á.d.T. Perfiles de ácidos grasos totales y libres en embutidos crudo-curados tradicionales de Galicia. *Grasas Aceites* **2004**, *55*, 273–281.
76. García, M.; García, A.; Gil Hernández, A. Importancia de los lípidos en el tratamiento nutricional de las patologías de base inflamatoria. *Nutr. Hosp.* **2006**, *21*, 30–43.
77. Pipoyan, D.; Stepanyan, S.; Stepanyan, S.; Beglaryan, M.; Costantini, L.; Molinari, R.; Merendino, N. The effect of trans fatty acids on human health: Regulation and consumption patterns. *Foods* **2021**, *10*, 2452. [CrossRef] [PubMed]
78. Tvrzicka, E.; Kremmyda, L.S.; Stankova, B.; Zak, A. *Fatty Acids as Biocompounds: Their Role in Human Metabolism, Health and Disease—A Review. Part 1: Classification, Dietary Sources and Biological Functions*; Biomedical Papers of the Medical Faculty of Palacky University in Olomouc; Palacky University in Olomouc: Olomouc, Czech Republic, 2011; Volume 155, pp. 117–130. [CrossRef]
79. Chan, D.S.; Lau, R.; Aune, D.; Vieira, R.; Greenwood, D.C.; Kampman, E.; Norat, T. Red and processed meat and colorectal cancer incidence: Meta-analysis of prospective studies. *PLoS ONE* **2011**, *6*, e20456. [CrossRef]
80. Aronson, W.J.; Glaspy, J.A.; Reddy, S.T.; Reese, D.; Heber, D.; Bagga, D. Modulation of omega-3/omega-6 polyunsaturated ratios with dietary fish oils in men with prostate cancer. *Urology* **2001**, *58*, 283–288. [CrossRef] [PubMed]
81. Slama, A.; Cherif, A.; Sakouhi, F.; Boukhchina, S.; Radhouane, L. Fatty acids, phytochemical composition and antioxidant potential of pearl millet oil. *J. Cons. Protect. Food Saf.* **2020**, *15*, 145–151. [CrossRef]
82. Simopoulos, A.P. Omega-6 and omega-3 fatty acids: Endocannabinoids, genetics and obesity. *Ocl* **2020**, *27*, 7. [CrossRef]
83. Baylin, A.; Siles, X.; Donovan-Palmer, A.; Fernandez, X.; Campos, H. Fatty acid composition of Costa Rican foods including trans fatty acid content. *J. Food Compos. Anal.* **2007**, *20*, 182–192. [CrossRef]

foods

MDPI

Article

# Variations in the Major Nutrient Composition of Dominant High-Yield Varieties (HYVs) in Parboiled and Polished Rice of Bangladesh

**Kazi Turjaun Akhter [1], Habibul Bari Shozib [2], Md. Hafizul Islam [1], Sneha Sarwar [1], Md. Mariful Islam [2], Md. Rubel Akanda [2], Muhammad Ali Siddiquee [2], Md. Mohiduzzaman [1], Abu Torab M. A. Rahim [1] and Nazma Shaheen [1,*]**

1 Institute of Nutrition and Food Science, University of Dhaka, Dhaka 1000, Bangladesh; kazi.turjaun@du.ac.bd (K.T.A.); hafizinfs19@gmail.com (M.H.I.); snehasarwar@du.ac.bd (S.S.); mohiduzzaman.md@gmail.com (M.M.); torabrahim@du.ac.bd (A.T.M.A.R.)

2 Bangladesh Rice Research Institute (BRRI), Gazipur 1701, Bangladesh; shozib11@gmail.com (H.B.S.); rupak115@gmail.com (M.M.I.); rubelakanda494@gmail.com (M.R.A.); mastpgpbd@yahoo.com (M.A.S.)

* Correspondence: nazmashaheen@du.ac.bd

**Abstract:** Rice (*Oryza sativa* L.) is the principal staple food, a fundamental component of food security, a significant source of energy and major nutrients, and a key player in the overall nutritional status in Bangladesh. Parboiling is a common rice-processing treatment in Bangladesh. Recently, polishing has also become a common practice among millers seeking to attract consumers. Polishing may influence the nutrient composition of rice. The present study aimed to investigate the impact of parboiling and polishing on the nutritional content of the five High Yield Varieties (HYVs) of rice (BR11, BRRI dhan28, BRRI dhan29, BRRI dhan49, and BRRI dhan84) and their percent contributions to the Recommended Dietary Allowances (RDA) of vitamins and minerals. All of the rice samples were analyzed for proximate parameters, vitamins (B1, B2, B3, B6, and folate), and minerals (Ca, Mg, Fe, Zn, Na, K, P). Moisture, ash, fat, and total dietary fiber (TDF) were determined gravimetrically, according to the AOAC Official Methods; protein was measured by the Kjeldahl method; B-group vitamins were measured using Ultra Pressure Liquid Chromatography; and mineral content was determined by ICP-OES. The energy, protein, fat, and total dietary fiber (TDF) content of the samples ranged between 342–357 kcal/100 g, 6.79–10.74 g/100 g, 0.31–1.69 g/100 g, and 2.59–3.92 g/100 g respectively. Thiamin, riboflavin, niacin, pyridoxin, and folate content ranged from 0.11–0.25 mg/100 g, 0.01–0.05 mg/100 g, 2.82–6.42 mg/100 g, 0.12–0.30 g/100 g, and 5.40–23.95 g/100 g respectively. In a comparison of parboiling and polishing, macronutrients and vitamin retention were higher in parboiled unpolished rice than in polished unparboiled rice. The minerals (mg/100 g) Ca, Mg, Fe, Zn, Na, K, and P were in the ranges 32.82–44.72, 30.69–58.34, 0.51–0.70,1.83–2.79, 5.00–5.36, 106.49–112.73, and 162.23–298.03. Minerals of BRRI dhan84 were unaffected by polishing and parboiling. BRRI dhan84 contributed a higher percentage of RDA of all B vitamins and minerals. Therefore, to reduce nutrient loss in rice, industries and consumers should be encouraged to avoid polishing or limit polishing to 10% DOM and to consume unpolished rice, either parboiled or unparboiled.

**Keywords:** rice varieties; parboiling; polishing; nutrient composition; Bangladesh

## 1. Introduction

Rice (*Oryza sativa* L.) is the principal staple food and a fundamental component of food security in Bangladesh. According to the recent Household Income and Expenditure Survey (HIES-2022), the per capita intake of rice is around 329 g daily in Bangladesh [1]. Thus, rice is a significant source of energy and vital nutrients and a key player in the nation's overall nutritional status. As a result of favorable geographic features, environmental adaptability, climate conditions, and lower risk of regional variation, farmers favor the production of rice over other crops [2,3]. Bangladesh, with its small land area, is placed third worldwide in rice production and consumption, behind China and India [2]. However, the indigenous rice varieties have lower yields and are less resilient to climate change compared to the high-yielding varieties (HYVs) developed by the Bangladesh Rice Research Institute (BRRI) [2,4,5]. The HYVs offer shorter growth periods, have improved heat tolerance and salt tolerance, and possess multiple pest-resistance capacities [5]. In response to the growing consumption needs of the population, the government is actively promoting the adoption of HYVs. Therefore, the majority of farmers are now replacing their traditional rice varieties with HYVs. Each rice variety has its own unique characteristics and nutritional composition [6]. Additionally, value addition to the rice crop by processing methods such as parboiling and polishing, which are frequently used in rice production and preparation, might affect the nutritional value of rice [7,8].

Polished rice is mostly preferred over unpolished rice in Bangladesh. Milling is a crucial post-production step for obtaining polished rice. When rice grains are polished, the outer layer, germ, or bran is removed from the underlying starchy endosperm, resulting in a white appearance [9,10]. By lowering the likelihood of rancidity, this technique not only improves the aesthetic appeal of rice but also increases its shelf life [10]. However, as nutrients are concentrated in the bran layers of the grain, removal of the bran results in nutrient losses from the grain [3]. Therefore, although polishing improves the physical, sensory, and storage properties of rice, it also diminishes nutritional value [11].

Parboiling, another important industrial processing technique, includes three additional steps: soaking, pressure steaming, and drying before dehusking [9,12]. The transfer of nutrients from the bran to the inner layers of the rice during the soaking process is one of the main effects of parboiling, along with the inactivation of lipase enzyme from heat exposure and starch gelatinization [9,12]. Parboiling causes high nutrient retention and improves the bioavailability of vitamins and minerals; therefore, parboiled rice is considered to have superior nutritional value [11,13]. Moreover, due to enzyme inactivation, parboiled rice has an extended shelf life. The physical structure of the grain is changed during this process, which makes it resistant to breakage during milling and increases its desirability [9].

Taken together, parboiling and polishing are believed to have a considerable impact on the nutritional composition of the rice. To ascertain the nutritional composition of food, commonly referenced databases such as the USDA National Nutrient Database and national equivalents like the Food Composition Table for Bangladesh (FCTB) are frequently employed [13]. Nevertheless, these databases often lack consideration of varietal disparities. It is anticipated that different HYVs may exhibit distinct responses to analogous processing treatments, such as parboiling and polishing. Consequently, for a more comprehensive understanding, it is imperative to account for varietal differences in order to achieve precision in nutritional assessments.

The present study aimed to investigate the impact of parboiling and polishing on the nutrient content of the five HYVs of rice in Bangladesh. This study provides extensive nutrient profiling, encompassing the proximate composition, vitamins, and mineral content of the selected dominant HYVs. It also examined the percent contribution to daily Recommended Dietary Allowances (RDA) of vitamins and minerals, especially those nutrients that have public-health importance in Bangladesh. It is anticipated that the study's findings, which highlight potential variations in important macronutrients and micronutrients among rice varieties processed in various ways, will help with the establishment of appropriate policies to reduce the loss of nutrients due to excessive polishing of rice.

## 2. Materials and Methods

### 2.1. Sample Collection and Processing

Five HYVs, namely BR11, BRRI dhan28, BRRI dhan29, BRRI dhan49, and BRRI dhan84, were collected from the Genetic Resources and Seed (GRS) Division Bank of Bangladesh Rice Research Institute (BRRI). Among these varieties, three (BRRI dhan28, BRRI dhan29,

BRRI dhan84) were Boro rice (planted in November–December and harvested in March–April) and two (BR11, BRRI dhan49) were Transplanted Aman rice (planted in March–April and harvested in November–December). Four types of sample (unparboiled unpolished, unparboiled polished, parboiled unpolished, and parboiled polished) were prepared for each of the five rice varieties. Unpolished rice is also referred to as brown rice. To obtain parboiled grains, 300 g of the rough rice (with the hulls) was placed into a beaker and one liter of distilled water was added to the beaker. The second step of the parboiling procedure involved autoclaving the hydrated rough rice for 10 min at 108 °C. Rough rice that had been autoclaved and watered was left out overnight at room temperature. Finally, a 38 °C oven dryer was used to dry the rough rice until it had a 13% moisture content. All the rice samples were dehusked using a Satake Rice Mill (THU35B, Hiroshima, Japan). To obtain polished rice, the dehusked rice grains were placed in a Grainman rice polisher (Miami, FL, USA) and polished to a 10% degree of milling (DOM) in a procedure based on a previous study [6,9] at GQN Division, BRRI. The DOM was determined using the following equation:

$$\text{Degree of milling (\%)} = (\text{Weight of total milled rice}/\text{Weight of brown rice}) \times 100$$

The same procedure was employed to dehusk and polish the dried unparboiled and parboiled rough rice grains. The samples were then crushed with liquid nitrogen to convert them into powder form. Powdered samples were kept in a −80 °C freezer at Rice Analytical Laboratory (RAL), BRRI for further analysis, including analysis of water-soluble vitamins and minerals.

### 2.2. Proximate Analysis

According to the 'Official Methods of Analysis of AOAC International' (AOAC, 2012), the content of moisture (AOAC 934.01), ash (AOAC 942.05), and fat (AOAC 963.15) were determined gravimetrically [14]. The Kjeldahl method was used to calculate total nitrogen, and the result was multiplied by 5.95 (a conversion factor specifically used for rice) for protein estimation [14]. The total dietary fiber (TDF), which is the sum of the soluble (SDF) and insoluble (IDF) fiber, was calculated using the enzymatic gravimetric method (AOAC 991.43). The content of available carbohydrates in the rice samples was calculated by subtracting the summed percentage of moisture, protein, fat, ash, and total dietary fiber. To calculate total energy content, the energy-yielding components such as crude protein, fat, fiber, and available carbohydrates were multiplied by the appropriate Atwater conversion factors [15].

### 2.3. Water-Soluble B-Vitamin Analysis

The analysis of water-soluble B vitamin (thiamine, riboflavin, niacin, pyridoxine, and folate) levels in the rice samples was carried out using Ultra-Pressure Liquid Chromatography (UPLC). A Shimadzu Prominence-i and LC-2030C 3D Plus Liquid Chromatograph were used for vitamin analysis. These water-soluble vitamins were extracted from the rice powder by acid hydrolysis followed by enzymatic hydrolysis. Briefly, 2 g of the powdered rice samples were weighed into a 250 mL Erlenmeyer flask (in triplicate). Acid hydrolysis was carried out by adding 40 mL of 0.1 M HCl and heating the mixture at 90 °C for 30 min. Thereafter, enzymatic hydrolysis was conducted by adding taka-diastase and heating at 50 °C for 2 h. After cooling, the samples were placed in tubes and centrifuged at 4000 rpm for 5 min. The supernatant was filtered through a 0.45-μm PVDF syringe filter and collected into HPLC auto-sampler vials. The aqueous extract was injected into a reverse-phase $C_{18}$ HPLC column. The fluorescence of riboflavin was measured, and thiamin content was determined after post-column derivatization with alkaline potassium ferricyanide, which converts the thiamin to thiochrome. A modified method was used to measure thiamin and riboflavin in the rice by HPLC, as described in the ASEAN Manual of Food Analysis [16].

All extracts were analyzed using the same chromatographic technique for the B3 vitamin (as nicotinamide) and the B6 vitamin [17,18], and the triplicate extraction of each sample was prepared for analysis. All vitamin standards were purchased from Sigma Aldrich (thiamine hydrochloride, vit B1; riboflavin, vit B2; niacin, vit B3; pyridoxine, vit B6; folic acid, vit B9) (Figure 1).

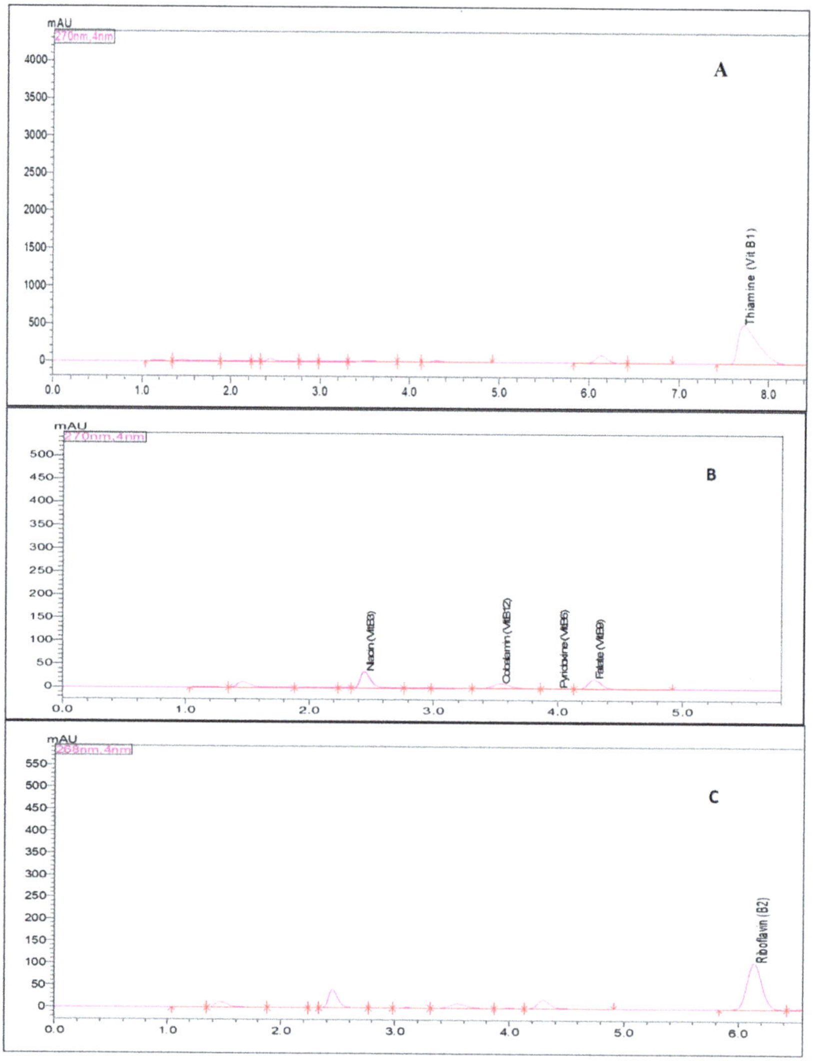


**Figure 1.** Water-soluble vitamin standards (**A**) thiamin, (**B**) niacin, pyridoxine, and folate, and (**C**) riboflavin are visualized at the common wavelength of 268–270 nm.

## 2.4. Estimation of Minerals

Calcium (Ca), magnesium (Mg), iron (Fe), zinc (Zn), sodium (Na), potassium (K), and phosphorus (P) were measured in mg/Kg (ppm) using a Shimadzu ICP-OES 9820 and later were converted to mg/100 g of edible portion. Rice samples were digested using a Milestone microwave digestor at RAL, BRRI. The dried rice-powder sample (1.0 gm) was taken from a plastic zip-lock package and weighed using an analytical balance. The rice sample was placed into a microwave vessel with a mixture of nitric acid and hydrogen peroxide in a 6:1 ratio (6 mL of $HNO_3$ and 1 mL of $H_2O_2$). The microwave-digestion program was carried out with parameters shown in Supplementary Table S1. Next, the digested contents were transferred into polypropylene vials and topped up so total digestion could be achieved. The analysis was carried out using a Shimadzu ICP-OES 9820 coupled with a mini plasma torch, concentric nebulizer, and cyclonic chamber. The detailed instrument configuration and operation parameters for analysis are summarized in Supplementary Table S2.

To create calibration curves, concentrated nitric acid (65%), and hydrogen peroxide (30%) were used in trace-metal analysis. Ultrapure water was obtained from the Milli-Q water purification system. External calibration standards were prepared by mixing single-element stocks from Merck (Rahway, NJ, USA) and Sigma-Aldrich (St. Louis, MO, USA). Calibration standards were prepared in 1% nitric acid (Table 1). Periodic table mix 1 for ICP (TraceCERT) CRM was used as a standard solution.

**Table 1.** Calibration standard concentration (mg/Kg).

| Elements | Standard 1 | Standard 2 | Standard 3 | Standard 4 | r | Wavelength (nm) |
|---|---|---|---|---|---|---|
| Ca | 20 | 100 | 200 | 500 | 0.999 | 315.8 |
| Mg | 4 | 20 | 40 | 100 | 0.999 | 285.2 |
| Fe | 0.4 | 2 | 4 | 10 | 0.999 | 238.2 |
| Zn | 0.4 | 2 | 4 | 10 | 0.999 | 213.8 |
| Na | 20 | 100 | 200 | 500 | 0.999 | 589.5 |
| K | 20 | 100 | 200 | 500 | 0.999 | 766.4 |
| P | 20 | 100 | 200 | 500 | 0.999 | 213.6 |

### 2.5. Statistical Analysis

Analytical values of proximate composition, vitamins, and mineral composition for the samples were derived in triplicate and presented as mean, and standard deviation (SD). One-way analysis of variance (ANOVA) was used to identify differences across treatments, and Tukey's means comparison test was applied with a significance threshold of 5%. Statistical Package for the Social Sciences (SPSS) version 25 was used for data analysis.

## 3. Results and Discussion

### 3.1. Rice Varieties and Their Production Yield

In this study, BRRI dhan28 and BRRI dhan29 are reported to have the highest yields and therefore to be consumed by the majority of the population (Supplementary Table S3).

### 3.2. Effects of Parboiling and Polishing on the Proximate Composition of Different Rice Varieties Analyzed

Variations in the proximate composition of different dominant rice varieties due to parboiling and polishing (10% DOM) are presented in Figure 2. The energy, protein, fat, and TDF content of the analyzed rice samples ranged between 342–357 kcal/100 g, 6.79–10.74 g/100 g, 0.31–1.69 g/100 g, and 2.59–3.92 g/100 g of edible portion, respectively.

Regardless of parboiling and polishing status, all of the rice varieties had a moisture content of 10–12% (Supplementary Table S4). According to the literature, the moisture content of rice should be below 14% in order to prevent insect infestation and microbiological decomposition, particularly fungal decomposition, and to permit safe storage [13]. The moisture content of the rice varieties that were analyzed was found to be within the recommended range.

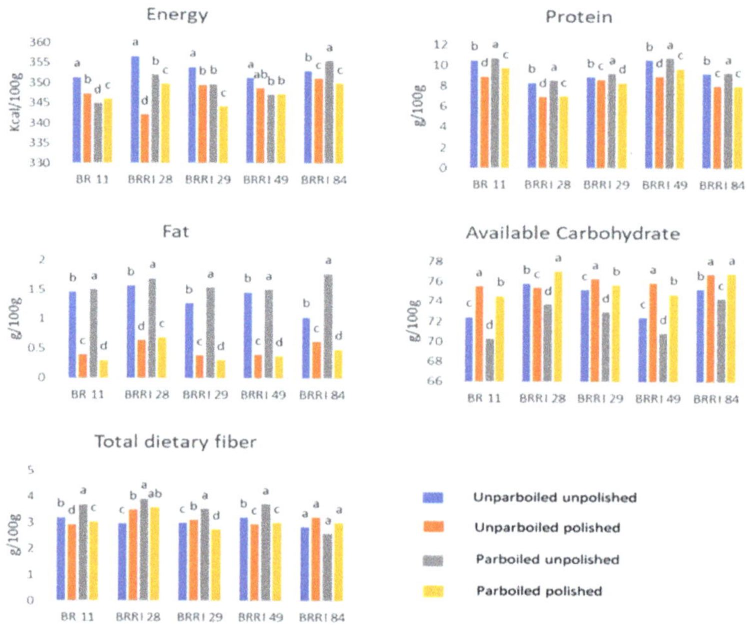


**Figure 2.** Effects of parboiling and polishing on proximate composition of different rice varieties in Bangladesh. Values with the same superscript letter within the same rice variety do not differ significantly at $p < 0.05$ by one-way ANOVA.

According to the findings, compared to brown rice (unpolished), polished rice had low energy, protein, and available carbohydrate content. For example, the polishing process significantly reduced the percent protein content in BR11 (9–15% reduction), BRRI dhan29 (3–10% reduction), BRRI dhan49 (10–15% reduction), and BRRI dhan84 (10–14% reduction) (Figure 2 and Supplementary Table S4). Previous research yielded comparable results regarding the effects of milling on Brazilian rice samples. Specifically, brown rice had a protein content of 6.85%, which decreased to 6.66% after milling [13]. Another study showed that the protein content of brown rice (9.2%) decreased with increasing DOM, indicating a decline in protein content from the rice surface to the endosperm; furthermore, the continued decrease after bran removal implies non-uniform distribution, with lower concentrations in the core endosperm compared to the outer layers [10]. In terms of fat content after milling, the fat content decreased drastically; the greatest decrease was seen in BRRI dhan29, in which fat content decreased up to 69–80%, followed by BR11 (72–78%), BRRI dhan49 (72–75%), BRRI dhan84 (40–72%), and BRRI dhan28 (59%). This finding is in line with earlier findings, in which the fat content decreased from 2.65% in brown rice to 0.50% in milled rice [13]. A reduction of about 62% in the fat content of rice (post-milling) was observed in another study [11]. Rice milling aims to eliminate the oil-rich aleurone layer, preventing rancidity during storage and increasing consumer appeal [11]. Most of the fat is distributed in the bran and outer endosperm, and these parts are removed during polishing to decrease the roughness of rice [13,19]. Therefore, a decrease in fat content can be observed in polished rice compared to unpolished rice. As a consequence of this fat reduction, milling reduces the fatty-acid content of the rice [11].

Based on this study, parboiling (without polishing) leads to high nutrient (energy, protein, fat, ash, available carbohydrate, and total dietary fiber) retention (Figure 2 and Supplementary Table S4). Parboiling causes protein to mobilize from the bran to the inner layer by disrupting the fibers of the grain, resulting in a greater percentage of protein in the final processed grain [9]. This result indicates that parboiled rice varieties contain more protein compared to unparboiled rice. In terms of fat, during parboiling, detached oil bodies leach out due to increased temperature and pressure, increasing the fat content of the rice [7,9].

However, parboiling caused high nutrient retention compared to polishing of unparboiled rice (Figure 2). The several factors that affect the nutrient content of rice varieties are the soaking time, the stage of germination, and the synthesis of enzymes, which is related to the production of amino acids during protein synthesis [7,8]. Therefore, any difference in any of these factors among the samples could account for the increased nutrient content of parboiled rice. However, when polishing and parboiling were undertaken together, overall nutrient retention was somewhat similar to that of non-parboiled polished rice.

The ash content of the rice varieties ranged between 0.4–1.48 g/100 g. It seems that polishing severely reduced the ash content in all rice varieties, viz., BR11 (61–70%), BRRI dhan49 (52–66%), BRRI dhan29 (47–53%), BRRI dhan84 (48–51%), and BRRI dhan28 (42–52%). An approximately 42% decrease in ash content upon milling has been reported [11]. As the ash is predominantly concentrated in the germ portion of rice, milling leads to a reduction in total ash content [11]. In contrast, parboiling increased the ash content in rice (Supplementary Tables S5 and S6). Parboiling causes the migration of nutrients from the bran to the inner layer of the grain. Therefore, parboiling causes high nutrient retention. The findings of the present study strongly substantiate the findings of earlier studies [13]. When polishing and parboiling are applied concurrently, the collective mineral retention surpasses that of exclusively polished rice but falls short of the levels observed in solely parboiled rice. Thus, polishing, a practice undertaken to enhance consumer appeal, augments nutrient availability when it is combined with parboiling.

### 3.3. Parboiling and Polishing Effect on the Mineral Contents of Different Rice Varieties

Figure 3 presents the effects of different processing conditions on the mineral content of different rice varieties in Bangladesh. The levels of minerals, viz. Ca, Mg, Fe, Zn, Na, K, and P ranged within 32.82–44.72 mg/100 g, 30.69–58.34 mg/100 g, 0.51–0.70 mg/100 g, 1.83–2.79 mg/100 g, 5.00–5.36 mg/100 g, 106.49–112.73 mg/100 g, and 162.23–298.033 mg/100 g, respectively. A recent study conducted in Bangladesh highlighted varietal distinctions in the overall mineral content across 35 HYVs of rice [6]. The present study showed lower levels of thiamin, riboflavin, and iron content but a higher level of zinc content compared to earlier studies [6].

The present study examined the effect of parboiling and polishing. Brown rice, in comparison to unparboiled polished rice varieties, had higher mineral content. The levels of Ca, Mg, K, and Fe sharply decreased in polished compared to unpolished unparboiled rice (Supplementary Table S5). Overall zinc content was not drastically affected by polishing. The decrease was not uniform across varieties. For instance, polishing reduced the Mg content of BR11 (reduction by 21–22.5%), BRRI dhan29 (reduction by 4.8–5.7%), and BRRI dhan84 (reduction by 14–15%). Polishing significantly reduced Fe content in unparboiled BR11 (by 7.3%), unparboiled BRRI dhan28 (by 7.2%), and parboiled BRRI dhan29 (by 3.4%). Moreover, polishing significantly reduced P content in BR11 (by 62% in unparboiled and 48% in parboiled), and BRRI dhan84 (by about 4% in unparboiled and parboiled) varieties. Polishing increased the Na content, but not at a significant level (Supplementary Table S7).

Between the rice bran and endosperm, the majority of the minerals are found in the bran (61.0%), followed by the outer endosperm (23.7%), core endosperm (11.6%), and middle endosperm (3.7%), respectively [10]. As these elements were found to be present at high levels in the external layers of the rice samples, milling, which leads to partial removal of the external layers, causes a drop in their levels [13,20]. Earlier reports revealed that major macro elements, for instance, Ca, Mg, K, and P, undergo significant loss upon milling [13]. Other evidence showed, interestingly, that mineral loss due to milling follows an order, i.e., iron > manganese > potassium > calcium > magnesium > zinc [3]. However, none of these patterns was seen exactly in any of the rice varieties examined in the present study. These differences might be due to non-uniform mineral distribution within the grain. The mineral content of rice varieties is significantly influenced by factors such as the availability and uptake of nutrients from the soil, agricultural practices, varietal distinctions, and processing conditions. These factors collectively contribute to the observed variations in mineral content between native and non-native rice varieties [13]. However, in line with the earlier studies, overall low mineral content in polished rice was observed. A low degree of milling is generally associated with elevated mineral content. Therefore, it is advisable to prioritize the consumption of rice with a lower degree of polishing (<10% Degree of Milling or DOM) over excessively polished varieties (≥10% DOM) [6].

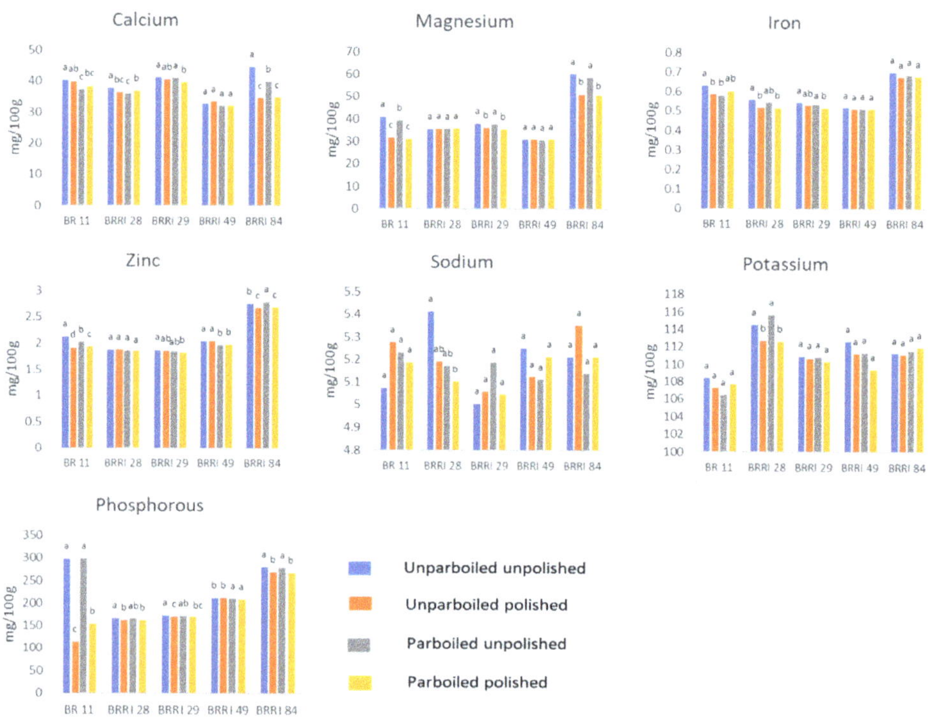


**Figure 3.** Parboiling and polishing effect on the mineral composition of different rice varieties in Bangladesh. Values with the same superscript letter within the same rice variety do not differ significantly at $p < 0.05$ by one-way ANOVA.

### *3.4. Parboiling and Polishing Effect on the Vitamin Content of Different Rice Varieties of Bangladesh*

Figure 4 presents the effect of parboiling and polishing on the vitamin content of five different dominant rice varieties in Bangladesh. Thiamin, riboflavin, niacin, pyridoxine, and folate content in the analyzed samples ranged from 0.11–0.25 mg/100 g, 0.01–0.05 mg/100 g, 2.82–6.42 mg/100 g, 0.12–0.30 g/100 g, and 5.40–23.95 g/100 g respectively. Thiamin content was notably high in all samples of BRRI dhan49, ranging from 0.21–0.26 g/100 g; riboflavin (0.042–0.046 mg/100 g) and niacin (5.573–6.42 mg/100 g) levels were also high in the same variety. Pyridoxine (0.141–0.181 mg/100 g) and folate (19.277–23.953 mcg/100 g) levels were high in BRRI 84. Most of the varieties examined in the present study showed vitamin content comparable to earlier findings [21] (Supplementary Table S8).

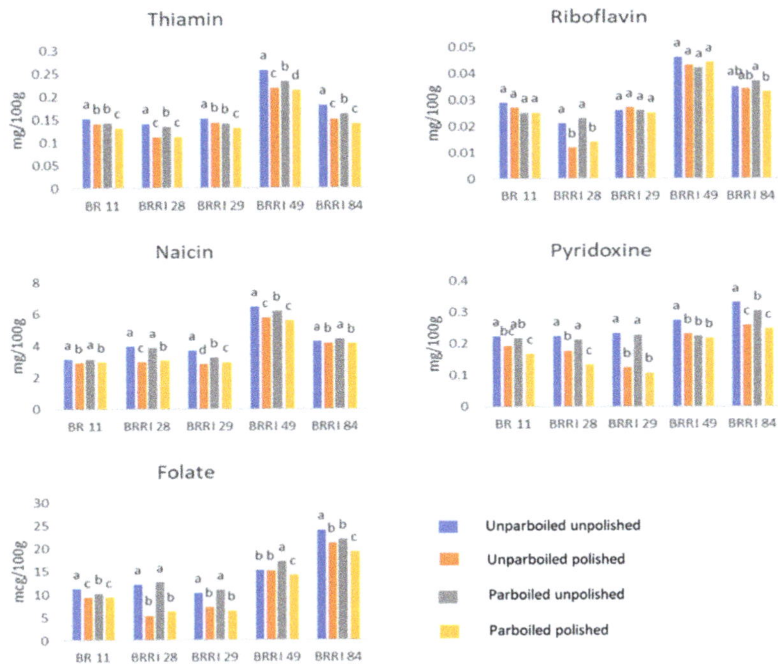


**Figure 4.** Effect of polishing and parboiling on the B-vitamins composition of different rice varieties in Bangladesh. Values with the same superscript letter within the same rice variety do not differ significantly at $p < 0.05$ by one-way ANOVA.

This study also examined the effect of parboiling and polishing on vitamin content. When brown rice (unparboiled unpolished rice) was compared with unparboiled polished rice, it was observed that thiamin, riboflavin, niacin, pyridoxine, and folate content were decreased significantly upon polishing, compared to unparboiled unpolished rice varieties (Figure 4 and Supplementary Table S8). For instance, polishing reduced the content of pyridoxine in all rice varieties, notably in BRRI dhan29 (47–53% reduction) and BRRI dhan28 (21–38% reduction) (Supplementary Table S5). Earlier literature suggests that milling causes high losses of B vitamins [3]. Folate in rice was found to decrease by 72% upon milling [11]. When rice was only parboiled without polishing, vitamin retention was generally high compared to polished unparboiled rice. However, when the same rice underwent both polishing and parboiling, the vitamin content showed a pattern similar to that observed in unparboiled polished rice. Thus, the vitamin content, in descending order, is as follows: unparboiled unpolished > parboiled unpolished > unparboiled polished ≥ parboiled polished.

In this study, parboiling (without polishing) resulted in substantially higher nutrient retention than polishing. However, when parboiling and polishing were undertaken in combination, the effect was somewhat similar to that of polishing only. The findings differed across varieties (Figure 2). In parboiling, the amount of nutrient transfer from the surface layers to the center of the grain determines how effective parboiling is at reducing milling losses [22]. Even in the case of parboiling, the effects of the process were found to be sporadic in the analyzed samples. The extent to which a nutrient migrates from the surface layers to the center of the grain determines the extent of the beneficial effects of parboiling in minimizing milling losses [22]. Several factors, including the levels of minerals in the grain, their solubility during soaking, the ratios of migration, variations in the hydrothermal process, and the parboiled grain's milling resistance also determine the parboiling effect on grain [13].

### 3.5. Percent Contribution to the RDA of B Vitamins and Minerals from the Daily Intake of Different Varieties of Parboiled Polished Rice

The people of Bangladesh usually prefer to consume parboiled polished rice, and the price of rice rises when it is parboiled and polished. According to the Bangladesh Nutrition Survey 2017–2018, an adult consumed 390 g of rice per day [23]. Table 2 shows the percentage contribution to RDA of B vitamins and minerals from the daily intake of different varieties of parboiled polished rice. Daily intake of different varieties of rice covered 64–121% of the RDA of niacin, 24–46% of the RDA of thiamin, 17–40% of the RDA of pyridoxine, 8–25% of the RDA of folate, and 2–7% of the RDA of riboflavin. In the case of minerals, daily intake of different varieties of parboiled polished rice supplied 27–45% of the RDA of Mg, 25–62% of the RDA of Zn, 13–16% of the RDA of Ca, and only 10–14% of the RDA of Fe. BRRI dhan84 contributed a higher percentage of the RDA of all B vitamins and minerals. In contrast to a preceding Indian study, the probable daily dietary intake of calcium and magnesium derived from equal amounts of a local parboiled milled rice variant exhibited notable differences compared with the Bangladeshi BRRI dhan84 variety (calcium: Indian variety 9.8% vs. Bangladeshi BRRI dhan84 13.56%; magnesium: Indian variety 27% vs. Bangladeshi BRRI dhan84 44.54%). However, the contribution of Fe and Zn from the Indian variety was observed to be higher than that from the local variety [22]. As previously elucidated, variations in processing methods and environmental factors contribute to the differences in mineral content among rice samples, consequently impacting their potential contribution to nutrient content. HYVs like BRRI dhan84, which are rich in vitamins and minerals, can play a vital role in combating particular micronutrient deficiencies in the country and attaining better nutrition for the population [6]. The contributions of other processing conditions of different rice varieties to the RDA of B vitamins and minerals are given in Supplementary Table S9.

**Table 2.** Percent contribution to RDA of B vitamins and minerals by the daily intake of different varieties of parboiled polished rice.

| Nutrients (% Contribution of RDA) | BR11 | BRRI dhan28 | BRRI dhan29 | BRRI dhan49 | BRRI dhan84 |
|---|---|---|---|---|---|
| Thiamin | 28.5 | 24.1 | 28.5 | 30.6 | 46.1 |
| Riboflavin | 4.0 | 2.2 | 3.9 | 5.1 | 6.9 |
| Niacin | 63.8 | 66.4 | 63.6 | 89.2 | 120.8 |
| Pyridoxine | 27.1 | 21.7 | 17.3 | 35.2 | 40.1 |
| Folate | 12.2 | 8.2 | 8.4 | 18.4 | 25.1 |
| Ca | 15.0 | 14.4 | 15.5 | 12.5 | 13.6 |
| Mg | 27.5 | 31.8 | 31.5 | 27.4 | 44.5 |
| Fe | 12.4 | 10.6 | 10.6 | 10.5 | 13.9 |
| Zn | 44.5 | 24.5 | 42.0 | 45.3 | 61.8 |

## 4. Conclusions

The present study, with a comprehensive varietal-level nutrient profiling of rice and changes in these nutrient levels due to processing, reveals that the levels of most nutrients were reduced by polishing. The unpolished rice of all varieties contained different nutrients in higher amounts, both in parboiled and unparboiled conditions, compared to polished rice. When parboiling and polishing were both used, macronutrient and vitamin retention were higher in polished parboiled rice than in polished unparboiled rice. The percent contribution to the RDA of vitamins and minerals showed that rice is a key source of vitamins and minerals as a staple food. The findings of the present study will help to assess the fraction of dietary requirements fulfilled according to age, sex, physical activity level, and physiological state. Thus, to combat nutritional problems in Bangladesh, rice could be considered a significant source of various nutrients. Moreover, the findings suggest that the levels of different nutrients in rice varieties are reduced by polishing the rice, while parboiling has a positive effect on nutrient composition. Therefore, to reduce nutrient loss from rice, the relevant industries and consumers should be encouraged to avoid rice that has undergone excessive polishing and recommend consuming unpolished rice in either parboiled or unparboiled conditions. The findings of the present study can help the government in the formulation of policy on the degree of milling required to prevent the loss of nutrients, especially micronutrients, from rice.

**Supplementary Materials:** The following supporting information can be downloaded at: https://www.mdpi.com/article/10.3390/foods12213997/s1, Table S1: Microwave-digestion program; Table S2: Instrument configuration and ICP parameters; Table S3: Rice varieties with their production yield; Table S4: Effects of parboiling and polishing on the proximate composition of different rice varieties in Bangladesh; Table S5: Percentage nutrient loss due to 10%-DOM treatment; Table S6: Percentage nutrient loss due to parboiling; Table S7: Effects of parboiling and polishing on the mineral content of different rice varieties in Bangladesh; Table S8: Effects of parboiling and polishing on the B-vitamin content of different rice varieties in Bangladesh; Table S9: Percent contribution to RDA of B vitamins and minerals from the daily intake of different varieties of unparboiled unpolished, unparboiled polished, parboiled unpolished, and parboiled polished rice.

**Author Contributions:** Conceptualization, K.T.A., H.B.S., M.A.S. and N.S.; Methodology, H.B.S., M.M.I., M.R.A., M.M. and N.S.; Software, M.H.I. and S.S.; Validation, H.B.S., M.M.I. and A.T.M.A.R.; Formal analysis, M.H.I. and A.T.M.A.R.; Resources, M.A.S.; Data curation, K.T.A., H.B.S., M.H.I., S.S., M.M.I., M.R.A. and M.M.; Writing—original draft, M.H.I. and S.S.; Writing—review & editing, K.T.A., A.T.M.A.R. and N.S.; Supervision, M.M.; Project administration, M.A.S. and N.S. All authors have read and agreed to the published version of the manuscript.

**Funding:** The Asian Food and Agriculture Cooperation Initiative (AFACI) funded the project "Establishment of Asian Food Composition Database (AFCD)".

**Data Availability Statement:** The datasets generated for this study are available on request to the corresponding author.

**Conflicts of Interest:** The authors declare no conflict of interest.

## References

1. BBS. *Household Income and Expenditure Survey*; HIES: Leyland, UK, 2022.
2. Rahman, M.C.; Islam, M.A.; Rahman, M.S.; Sarkar, M.A.R.; Ahmed, R.; Kabir, M. Identifying the Threshold Level of Flooding for Rice Production in Bangladesh: An Empirical Analysis. *J. Bangladesh Agric. Univ.* **2021**, *19*, 243–250.
3. Longvah, T.; Mangthya, K.; Subhash, K.; Sen, S.; Rathi, S. Comprehensive nutritional evaluation of popular rice varieties of Assam, Northeast India. *J. Food Compos. Anal.* **2021**, *101*, 103952. [CrossRef]
4. Mackill, D.J.; Khush, G.S. IR64: A high-quality and high-yielding mega variety. *Rice* **2018**, *11*, 1–11. [CrossRef] [PubMed]
5. Nayak, S.; Habib, M.A.; Das, K.; Islam, S.; Hossain, S.M.; Karmakar, B.; Neto, R.F.; Bhosale, S.; Bhardwaj, H.; Singh, S.; et al. Adoption Trend of Climate-Resilient Rice Varieties in Bangladesh. *Sustainability* **2022**, *14*, 5156. [CrossRef]
6. Shozib, H.B.; Jahan, S.; Sultan, M.Z.; Alam, S.; Das, S.C.; Amin, R.B.; Hasan, M.; Siddiquee, M.A. Nutritional Properties of Some BRRI HYV Rice in Bangladesh. *Vitam. Miner.* **2018**, *7*, 1–6. [CrossRef]
7. Alexandre, N.; Paul, H.J.; Gerardine, N. Effect of Parboiling Technique on the Nutritional Quality of Rice. *Glob. J. Nutr. Food Sci.* **2020**, *2*, 1–13. [CrossRef]
8. Pal, P.; Singh, N. Effect of Parboiling on Phenolic, Protein, and Pasting Properties of Rice from Different Paddy Varieties. *J. Food Sci.* **2018**, *83*, 2761–2771. [CrossRef]
9. Paiva, F.F.; Vanier, N.L.; Berrios, J.D.J.; Pinto, V.Z.; Wood, D.; Williams, T.; Pan, J.; Elias, M.C. Polishing and parboiling effect on the nutritional and technological properties of pigmented rice. *Food Chem.* **2016**, *191*, 105–112. [CrossRef]
10. Lamberts, L.; De Bie, E.; Vandeputte, G.E.; Veraverbeke, W.S.; Derycke, V.; De Man, W.; Delcour, J.A. Food Chemistry Effect of milling on colour and nutritional properties of rice. *Food Chem.* **2007**, *100*, 1496–1503. [CrossRef]
11. Monks, J.L.F.; Vanier, N.L.; Casaril, J.; Berto, R.M.; De Oliveira, M.; Gomes, C.B.; de Carvalho, M.P.; Guerra Dias, A.R.; Elias, M.C. Effects of milling on proximate composition, folic acid, fatty acids and technological properties of rice. *J. Food Compos. Anal.* **2013**, *30*, 73–79. [CrossRef]
12. Demont, M.; Zossou, E.; Rutsaert, P.; Ndour, M.; Van Mele, P.; Verbeke, W. Consumer valuation of improved rice parboiling technologies in Benin. *Food Qual Prefer* **2012**, *23*, 63–70. [CrossRef]

13. Heinemann, R.J.B.; Fagundes, P.L.; Pinto, E.A.; Penteado, M.D.V.; Lanfer-Marquez, U.M. Comparative study of nutrient composition of commercial brown, parboiled and milled rice from Brazil. *J. Food Compos. Anal.* **2005**, *18*, 287–296. [CrossRef]
14. AOAC Official. *Methods of Analysis of AOAC International*, 18th ed.; Association of Official Analysis Chemists International: Gaithersburg, MD, USA; Available online: https://www.aoac.org/official-methods-of-analysis/ (accessed on 25 June 2023).
15. Food and Agriculture Organization (FAO). *Calculation of the Energy Content of Foods—Energy Conversion Factors*; FAO: Rome, Italy, 2002.
16. ASEAN. *ASEAN Manual of Food Analysis 2011*; ASEAN, Institute of Nutrition, Mahidol University: Nakhon Pathom, Thailand, 2011; pp. 22–26.
17. Rybicka, I.; Gliszczyńska-Świgło, A. Gluten-Free Flours from Different Raw Materials as the Source of Vitamin B1, B2, B3 and B6. *J. Nutr. Sci. Vitaminol.* **2017**, *63*, 125–132. [CrossRef]
18. Viñas, P.; López-Erroz, C.; Balsalobre, N.; Hernández-Córdoba, M. Reversed-phase liquid chromatography on an amide stationary phase for the determination of the B group vitamins in baby foods. *J. Chromatogr. A* **2003**, *1007*, 77–84. [CrossRef] [PubMed]
19. Itani, T.; Tamaki, M.; Arai, E.; Horino, T. Distribution of Amylose, Nitrogen, and Minerals in Rice Kernels with Various Characters. *J. Agric. Food Chem.* **2002**, *50*, 5326–5332. [CrossRef] [PubMed]
20. Sellappan, K.; Datta, K.; Parkhi, V.; Datta, S.K. Rice caryopsis structure in relation to distribution of micronutrients (iron, zinc, β-carotene) of rice cultivars including transgenic indica rice. *Plant Sci.* **2009**, *177*, 557–562. [CrossRef]
21. Longvah, T.; Chauhan, A.; Mudavath, S. Nutrient variability of rice landraces (Oryza sativa L.) from Manipur, Northeast India and its nutrients supply potential in rice-based diets. *Nutr. Food Sci.* **2022**, *52*, 1100–1115. [CrossRef]
22. Doesthale, Y.G.; Devara, S.; Rao, S.; Belavady, B. Effect of Milling on Mineral and Trace Element Composition of Raw and Parboiled Rice. *J. Sci. Food Agric.* **1979**, *30*, 40–46. [CrossRef] [PubMed]
23. Shaheen, N.; Amin, R.; Islam, S.; Shimul, S.N.; Hamid, S.A. *Nutrient Density and Affordability of Habitual and Desirable Diets in Bangladesh by Life Cycle Stage, Region, and Vulnerable Groups*; University of Dhaka: Dhaka, Bangladesh, 2021. [CrossRef]

foods

MDPI

*Article*

# Effects of Casein Hydrolysate Prepared with Savinase on the Quality of Bread Made by Frozen Dough

**Hatice Bekiroglu [1,2], Gorkem Ozulku [1,*] and Osman Sagdic [1]**

[1] Department of Food Engineering, Faculty of Chemical and Metallurgical Engineering, Yildiz Technical University, Istanbul 34220, Turkey; h.bkroglu@gmail.com (H.B.); osagdic@yildiz.edu.tr (O.S.)

[2] Food Engineering Department, Agricultural Faculty, Şırnak University, Şırnak 73300, Turkey

* Correspondence: ozulkug@yildiz.edu.tr

**Abstract:** The effect of casein savinase hydrolysate (CSH) usage on frozen dough (1%, 1.5% and 2%, g/100 g flour) was investigated in terms of rheological, thermal and structural characteristics of wheat doughs and the textural and color properties of corresponding breads. Rheological measurements showed that CSH addition into dough led to a reduction in $G'$ and $G''$ values, but a similar trend was not observed in frozen dough samples. The increase in protein band intensity was observed for control dough (CD) after frozen storage (−30 °C, 28 days), while there were no increases in the band intensities of the doughs with CSH. The freezable water content of unfrozen doughs decreased gradually with the addition of CSH, dependent on concentration level. Frozen storage caused a notable reduction in the α-helices structure of the CD sample ($p < 0.05$) while no significant variation was observed for the doughs containing CSH ($p > 0.05$). The lowest specific volume reduction and hardness increment were observed for the breads containing 1.5% and 2% CSH. Frozen storage caused a significant reduction in the b* value of bread crust ($p < 0.05$), while no significant effect was observed for L* and a* value during frozen storage ($p > 0.05$). Overall, CSH incorporation into frozen dough can be an alternative that could reduce the quality deterioration of frozen bread.

**Keywords:** casein hydrolysate; frozen dough; bread quality

## 1. Introduction

Milk proteins are frequently preferred in the development of biological characteristics of foods because of their high digestibility features and abundant essential amino acid content [1]. They are also considered to be a fundamental dietary component due to their balanced and adequate nutrient content. In addition to the superior bioactive properties they provide, milk proteins also enable technological characteristics such as emulsion, oil–water retention capacity and texture [2,3].

In recent years, the importance of healthy and functional ingredients has become more widely accepted, leading to an increased demand for improving the functional characteristics of frequently consumed nutrients. This demand has generated food products with peptides containing much more biologically qualified and active components. Enriching food items with low protein content, such as bakery products, and especially bread, with milk protein and its hydrolysates can significantly increase their nutritional value [4]. The enzymatic hydrolysis-based modification of milk proteins has significant potential as a tool in food protein processing to optimize the technofunctional, biological and nutritional properties of proteins [5]. Casein hydrolyzed formulations have been sold for many years due to their exceptional nutritional content, amino acid composition, commercial availability in huge numbers, and affordable price [6]. Casein hydrolysates can be used to enhance the nutritional value, stability, shelf life, and many other characteristics of various food products, as well as protein supplements [7].

The use of frozen dough in bread and bakery products has been increasing. Basic factors like shelf life and bakery quality may be regulated with the use of frozen dough, and at the same time it enables the production of huge quantities of dough, ease of storage and transportation, and the production of fresh products in small and medium-sized enterprises [8]. It is known that the freezing and thawing process in the production of frozen bread damages the gluten network [9] and prevents the bread from rising, thus causing low volume and harder crumb formation [10,11]. Some proteins and their hydrolysates have been used in several studies to eliminate these quality deteriorations. The effects of milk protein [12], pigskin gelatin [13] and sweet potato protein hydrolysates [14] on the quality of breads made from frozen dough have been investigated. In a study of Zhou et al. [15], desalted egg white and gelatin mixture systems provided a positive effect on the quality of frozen dough by slowing down the increase in the freezable water content and the loss tangent (tanδ) value. Also, a cryo-protective effect of ice-binding peptides derived from collagen hydrolysates was reported by Cao et al. [16].

Additionally, casein hydrolysate could be an alternative for frozen dough quality improvements thanks to its inexpensive and efficient production. Therefore, we aimed to investigate the contributions of casein hydrolysates produced from savinase to frozen dough and corresponding bread quality in this study. To realize this aim, the rheological, thermal and protein analyses (SDS-PAGE and secondary structure) were studied alongside the quality of breads made from frozen dough.

## 2. Materials and Methods

### 2.1. Materials

Wheat flour (13.5% moisture, 0.73% ash, 10.8% protein and 58% water absorption) was purchased from Eksun Food Industry Co., Ltd. (Tekirdağ, Turkey). Dry yeast (Pakmaya) and salt (Billur) were purchased from a local market. The bovine casein (NaCas) was obtained from a local brand (Milkon, Çallı Food Food Industry and Trade Inc., İstanbul, Turkey). The savinase enzyme (protease from *Bacillus* sp., 16 U/g), sodium dodecyl sulfate (SDS), acrylamide, and all other chemicals and solvents were purchased from Sigma-Aldrich (St. Louis, MO, USA).

### 2.2. Preparation of Casein Hydrolysate

Casein was first dissolved with distilled water and kept at +4 °C for 10 h for complete hydration. The pH of the casein solutions was adjusted to pH 9, which is optimum for the savinase enzyme using 0.1 N NaOH. Then, the volume of casein solution was completed with distilled water to 5 g protein/100 mL. Savinase was added to the casein solutions (enzyme:substrate, 1:100) and hydrolyzed for 180 min in a water bath set at 50 °C. During the hydrolysis period, the pH of the solution was adjusted to 9 by using 0.1 N NaOH every 15 min, and the hydrolysis degree (HD) of casein was determined according to the total amount of base consumed and the principle of the pH-stat method [17–19]. The hydrolysis degree of casein hydrolysates was obtained as 24.7%. At the end of the hydrolysis period, the enzyme inactivation of the casein solutions was achieved by keeping them in boiling water for 15 min. They were centrifuged at 6500 rpm for 15 min and the supernatant fraction was freeze-dried.

### 2.3. Preparation of Bread Dough and Baking

The bread dough formulation (control) consisted of 100 g wheat flour, 2 g dry yeast, 1.5 g salt and 58 g water. The wheat flour was replaced by casein savinase hydrolysate (CSH) at the concentrations of 1%, 1.5% and 2% wheat flour dry basis (*w*/*w*) to produce hydrolysate-added dough (HD). All ingredients were kneaded in a dough mixer (KitchenAid, Benton Harbor, MI, USA) at speed 4 for 5 min. After complete mixing, the dough was sheeted, molded and panned. The unfrozen dough samples were fermented at 30 °C and 85% relative humidity (RH) for 2 h and baked at 235 °C for 25 min in an electrical oven (Maksan, Nevşehir, Turkey). The frozen dough samples were directly stored at −35 °C for 28 days after shaping. The thawing process was performed at 4 °C for 3 h.

After fermentation (30 °C and 85% RH for 2 h) of the thawed dough, the same baking procedure was applied.

*2.4. Rheological Measurements*

Unfrozen and thawed dough samples were subjected to dynamic rheological tests by using stress and temperature controlled rotational rheometer (Antonpaar MCR 302, Graz, Austria). Frequency sweep tests were conducted between 0.1 rad/s and 100 rad/s with a fixed strain of 0.1% in the linear viscoelastic region at 25 °C with a 2 mm gap. Two replications of each sample were carried out.

*2.5. Freezable Water (FW) Content*

The FW content of dough samples was determined with the differential scanning calorimeter (DSC; TA instrument Q20, USA) by using the method of Chen et al. [14]. Briefly, 10 mg of the dough sample was sealed hermetically in an aluminum tray and an empty aluminum tray was used as the reference. The following heating program was applied: cooling from 20 °C to −40 °C at a rate of 5 °C/min and keeping constant for 5 min, and then heating from −40 °C to 10 °C at a rate of 5 °C/min. The enthalpy (ΔH) of the melting peak was determined during the heating process and FW content (%) was calculated according to Equation (1):

$$\mathrm{FW}(\%) = \frac{\Delta \mathrm{H}}{\Delta \mathrm{Ho} \times \mathrm{Wc}} \times 100 \tag{1}$$

where ΔH is the enthalpy (J/g) of the melting peak of the endothermic curve; ΔHo is the enthalpy (334 J/g) of melting peak of pure water; and Wc represents the total water content (%) of the dough, which was measured using a moisture analyzer (Radwag MR50, Poznań, Poland). Two replications of each sample were performed.

*2.6. Sodium Dodecyl Sulfate–Polyacrylamide Gel Electrophoresis (SDS-PAGE) Analysis*

SDS-PAGE analyses of all dough samples were carried out according to the method described by Laemmli [20]. First, the freeze-dried dough samples were powdered via a grinder. Powder dough samples were mixed with sample buffer at 10 mg/mL and kept in a boiling water bath for 5 min. The samples were added to each loading well as 20 µL and sorted according to their molecular weights by running under a 20 mA/gel electric current in the separation gel (7.5% and 20% acrylamide) using a vertical electrophoresis system (Mini-PROTEAN®System, Bio-Rad, Hercules, CA, USA). After the electrophoresis process, the bands stained with staining solution (comassie brillant blue:methanol:acetic acid:water; 0.25%:50%:10%:39.75%) were removed from excess dye by keeping them in the destaining solution (methanol:acetic acid:water; 50%:10%:40%) and then visualized with the Biorad Gel Doc EZ imaging system.

*2.7. Fourier Transform Infrared Spectroscopy (FT-IR)*

The secondary structure of dough samples was determined by using a Bruker Tensor 27 spectrometer equipped with a DLa TGS detector (Bremen, Germany). FT-IR spectra of all samples were obtained with wave numbers from 400 to 4000 $cm^{-1}$ during 16 scans per spectra, with 2 $cm^{-1}$ resolution. The measurements were carried out 3 times. Deconvolution was applied to all spectra by the interpretation of changes in the overlapping amide I band (1600–1700 $cm^{-1}$) using Origin 2020b software [21].

*2.8. Determination of Bread Quality*

2.8.1. Specific Volume (SV)

The volume (mL) of bread was measured in duplicate by using the rapeseed displacement method (Method 10-05, AACC 2000). The SV of breads was calculated by dividing the volume by the weight (g) of bread.

2.8.2. Texture

Each bread sample was cut into slices that were 1.25 mm thick after cooling for 2 h at room temperature. Texture Profile Analysis (TPA) was performed by using a texture analyzer (SMS TA.XT2 Plus, Glasgow, UK) equipped with a 5 kg load cell and 36 mm diameter cylindrical compression probe. The TPA test was conducted with 50% compression, 5.0 mm/s test speed and 5 s delay time between the two compression cycles. Duplicate measurements were performed.

2.8.3. Color

The crust and crumb color of bread samples were measured by using a Chromameter (CR-100 Konica Minolta, Tokyo, Japan), recording the lightness (L*), redness (a*) and yellowness (b*) values. All determinations were carried out in triplicate. Three measurements were taken from each sample. The color differences (ΔE) were determined by using the following Equation (2):

$$\Delta E = \sqrt{\left(L_0^* - L^*\right) + \left(a^* - a_0^*\right) + \left(b^* - b_0^*\right)} \quad (2)$$

where $L_0^*$, $a_0^*$ and $b_0^*$ belong to control bread, which is a reference sample, while L*, a* and b* belong to bread containing CSH.

### *2.9. Statistical Analysis*

An analysis of variance (ANOVA) followed by Duncan's multiple range tests was used to determine any significant differences ($p < 0.05$) between the mean values. SPSS version 19.0 software (SSPS Inc., Chicago, IL, USA) was used.

## 3. Results and Discussion

### *3.1. Dynamic Rheological Properties of Dough*

Figure 1 exhibits the storage (G′) and loss (G″) modulus of dough samples, indicating the elastic and viscous properties, respectively. The loss factor, tan δ, was also shown to reflect the viscoelasticity of samples. For all samples, the G′ was greater than G″, showing that the dough was more elastic and exhibited solid-like behavior [22]. CSH addition into the dough led to a reduction in G′ and G″ values, but a similar trend was not observed in frozen dough samples. Frozen storage decreased the G′ and G″ values of all dough samples, and control dough (CD) had the lowest value. The same behavior was shown in G″ values. Also, the decreasing effect of frozen storage on dough containing 2% was lower than that in other doughs with CSH (1% HD and 1.5% HD) (Figure 1B). These results indicated that CSH addition caused a detrimental effect due to lowering G′ and G″ and increasing tan δ values in a fresh dough system. According to the study of Khatkar et al. [23], the flour from wheat, showing good bread quality, had a higher G′ value and lower tan δ. On the other hand, the addition of CSH improved the gluten strength of frozen dough when compared to control-frozen dough (CFD), as a result of elevated G′ and G″ values [24]. A similar observation was reported by Zhou et al. [15] for a frozen dough system containing a desalted egg white and gelatin mixture. In the dynamic rheological measurement of the frozen dough system, tan δ value (G″/G′) is considered an important parameter to monitor the viscoelasticity of dough [22]. The tan δ values obtained in this study were lower than 1, indicating that the elasticity of the samples was greater than the viscosity [15]. Frozen storage resulted in a remarkable increase in tan δ for all samples (Figure 1C). The increment of the tan δ value during freezing has previously been reported by many studies [14,15]. This was attributed to the gradual increase in ice crystals during freezing, causing damage to the gluten network and subsequently resulting in an increased viscosity of the dough. The rise of tan δ in dough with CSH was lower than CD after frozen storage (Figure 1C). This can be explained by slowing the ice crystal growth due to the addition of CSH.

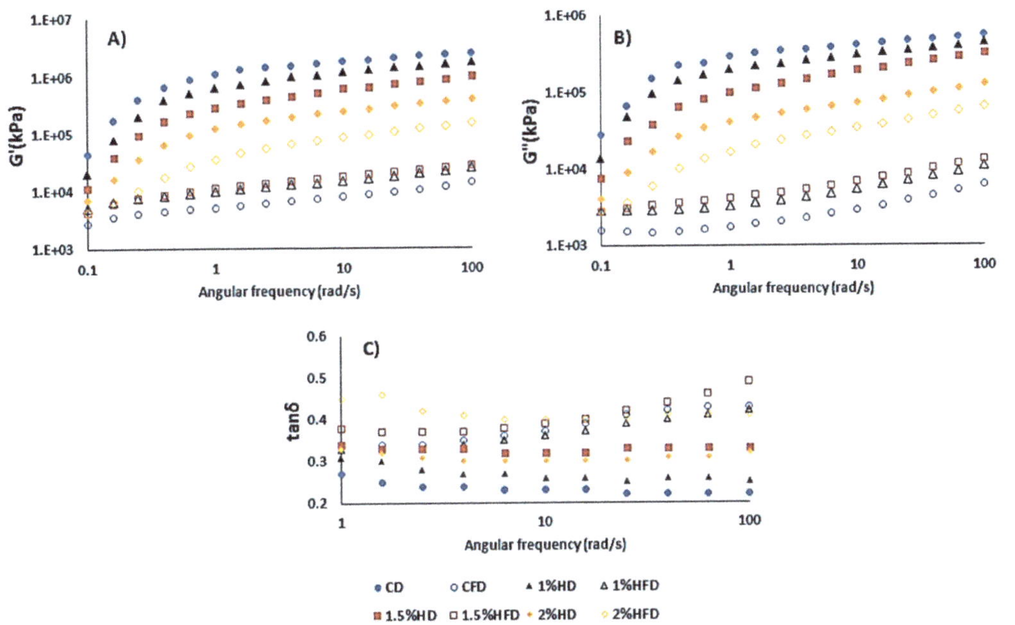


**Figure 1.** Effect of casein savinase hydrolysate (CSH) addition on dynamic rheological properties of dough samples. (**A**) Storage modulus, G′, (**B**) Loss modulus, G″, (**C**) Loss factor, tanδ. CD: control dough, CFD: control-frozen dough, HD: dough containing CSH, HFD: frozen dough containing CSH.

### 3.2. SDS-PAGE Analysis

The changes in SDS-PAGE patterns of proteins with the incorporation of casein hydrolyzate during frozen storage are presented in Figure 2. The formation of new bands and/or changes in band intensities showed polymerization/depolymerization behavior in the gluten network [25–27]. Depolymerization exists in the protein matrix of wheat dough when exposed to frozen conditions [26]. The increase in protein band intensity was observed for control dough (CD) after frozen storage (CFD), indicating that a depolymerization of the high molecular weight glutenin subunits (HMW-GS) region occurred. Similar results were previously reported by numerous studies for wheat dough under frozen conditions [13,28,29]. As for the samples including casein hydrolyzate, no depolymerization was observed due to frozen storage as there was no increases in band intensities, except for the 2% HD sample (Figure 2B). The 2% casein hydrolyzate addition increased the band intensity after frozen storage, showing that this level of addition may inhibit the crosslinking of SDS-insoluble proteins and interact with SDS-soluble proteins [13]. The incorporation of unhydrolyzed pigskin gelatin at 1% level also resulted in an increase in band intensity corresponding to the ω-gliadins region in a study by Yu et al. [13], who suggested that pigskin gelatin contributed to the formation of aggregates through non-covalent interactions. The inclusion of casein hydrolyzate in the wheat dough matrix led to a decrease in band intensities in the HMW-GS region when compared to CD. This can show that new interactions between glutenin, casein hydrolyzate and water were formed during dough processing since band intensity increment was observed in the low molecular weight glutenin subunit (LMW-GS), γ-gliadin and α/β gliadin regions. Another possible explanation of this can be the formation of protein polymers with a larger size, which were unable to migrate into the resolving gel [25]. Also, new band formation due to the addition of casein hydrolyzate into wheat dough was not observed in the present study, indicating the absence of new protein interactions. Similar results were reported in a study by Yu et al. [13], who incorporated the pigskin gelatin into frozen dough, and a study which investigated the effect of natural inulin on frozen dough [27].

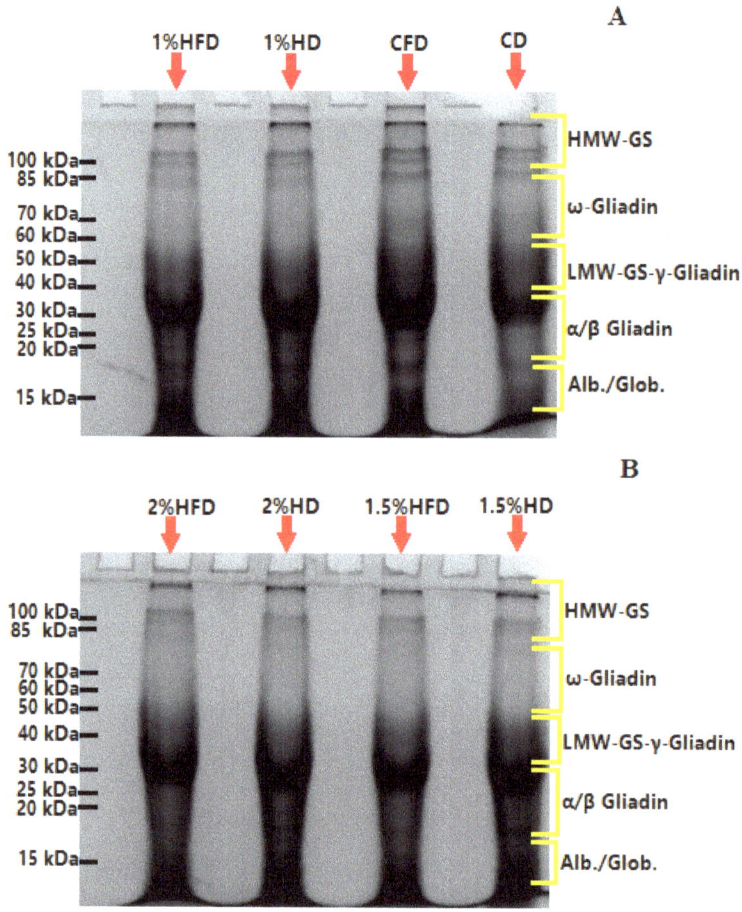


**Figure 2.** (**A**) SDS-PAGE patterns of control dough and the doughs containing casein savinase hydrolysate (CSH) at the level 1%. (**B**) SDS-PAGE patterns of the doughs containing CSH at the level 1.5% and 2% CD: control dough, CFD: control-frozen dough, HD: dough containing CSH, HFD: frozen dough containing CSH. The molecular weights of the dough samples were arranged in the bands.

### *3.3. Freezable Water Content*

Freezable water (FW) in dough produces ice crystals when subjected to freezing. The crystallization amount of ice is directly attributed to the FW content in dough and causes the weakening of the gluten structure. Therefore, it is important to measure FW content to monitor the quality of dough during frozen storage [15,24]. As shown in Figure 3, the FW (%) content of dough samples increased after frozen storage (−30 °C, 28 days) due to the formation of ice crystals. The rheological properties of dough are also influenced by FW content since ice crystals disrupted the gluten network. This was supported by the tan δ value of dough samples after frozen storage (Figure 1C) due to an increasing trend being shown. CSH incorporation gradually decreased the FW (%) content of fresh dough. This could be due to the binding of water molecules to CSH, as stated in a study by Bekiroglu et al. [7]. Also, the increasing effect of frozen storage on FW (%) was suppressed by the addition of CSH, dependent on concentration (Figure 3). The 2% HD sample exhibited the lowest FW (%) value, suggesting that a 2% CSH substitution level could change the moisture distribution within the dough [29]. These results are in line with the study carried out by Chen et al. [14], who investigated the effect of sweet potato protein hydrolysates on the quality of frozen dough. The conversion of bound water to freezable water was also inhibited when a desalted egg white and gelatin mixture was added into frozen dough in a study by Zhou et al. [15].

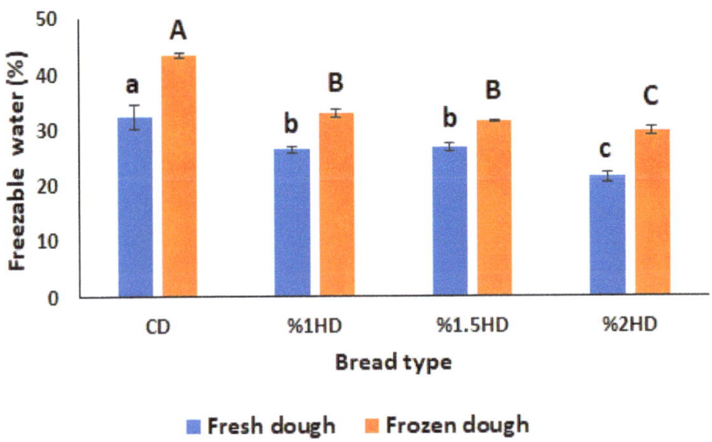


**Figure 3.** Effect of casein savinase hydrolysate (CSH) addition on freezable water content in dough samples. CD: control dough, HD: dough containing CSH. Different lowercase letters show statistically significant differences among fresh dough samples ($p < 0.05$). Different uppercase letters show statistically significant differences among frozen dough samples ($p < 0.05$).

### *3.4. Analysis of Secondary Structures of Protein*

The secondary structures of gluten proteins corresponding to the amide I region (1700–1600 cm$^{-1}$) are shown in Table 1. The FTIR spectra of the samples after baseline correction and Gaussian smoothing are shown in Figure 4. Absorption peaks were ascribed as β-sheets (1682–1696 cm$^{-1}$), β-turns (1662–1681 cm$^{-1}$), α-helices (1650–1660 cm$^{-1}$) and random coils (1640–1650 cm$^{-1}$) based on the reports of Yang et al. [30] and Zhou et al. [15]. It was reported that α-helices and β-sheets were relatively orderly and stable, while β-turns and random coils were disordered [24]. The incorporation of CSH into wheat dough formulation significantly affected the secondary structure of proteins, except for the β-sheets structure at Day 0. Frozen storage (−30 °C, 28 days) caused a notable reduction in the α-helices structure of the CD sample ($p < 0.05$), while no significant variation was observed for the samples containing CSH ($p > 0.05$). This could be due to the inhibition of recrystallization by CSH addition in the frozen storage state. The structure of β-turns significantly increased after frozen storage, indicating a transformation of α-helices into β-turns structure during frozen storage [15]. The reduction in α-helices and increment of β-turns were higher in CD than the doughs with CSH. The random coils structure was not changed during frozen storage ($p > 0.05$), similar to the study of Yu et al. [13], who investigated the effect of pigskin gelatin in frozen dough.

**Table 1.** Effects of casein savinase hydrolysate on the secondary structure of gluten protein during frozen storage.

| Storage Time | Samples | β-Sheets | β-Turns | α-Helices | Random Coils |
|---|---|---|---|---|---|
| Day 0 | CD | 12.42 ± 1.92 $^{aA}$ | 28.72 ± 1.82 $^{aB}$ | 49.46 ± 3.13 $^{aA}$ | 9.71 ± 1.34 $^{bA}$ |
| | 1% HD | 7.75 ± 1.05 $^{bA}$ | 23.12 ± 1.51 $^{bB}$ | 33.08 ± 2.87 $^{bA}$ | 36.05 ± 2.80 $^{aA}$ |
| | 1.5% HD | 15.04 ± 1.12 $^{aB}$ | 17.38 ± 1.17 $^{cA}$ | 35.10 ± 2.63 $^{bA}$ | 32.48 ± 3.35 $^{aA}$ |
| | 2% HD | 14.14 ± 2.11 $^{aA}$ | 19.75 ± 1.43 $^{bcA}$ | 32.61 ± 2.64 $^{bA}$ | 34.00 ± 2.83 $^{aA}$ |
| Day 28 | CFD | 16.30 ± 1.29 $^{bA}$ | 41.22 ± 1.94 $^{aA}$ | 35.53 ± 1.47 $^{aB}$ | 6.45 ± 1.85 $^{bA}$ |
| | 1% HFD | 10.24 ± 1.61 $^{cA}$ | 33.74 ± 1.82 $^{bA}$ | 22.14 ± 2.81 $^{bA}$ | 33.88 ± 1.58 $^{aA}$ |
| | 1.5% HFD | 23.06 ± 1.44 $^{aA}$ | 21.39 ± 2.19 $^{cA}$ | 28.05 ± 2.72 $^{bA}$ | 27.5 ± 3.54 $^{aA}$ |
| | 2% HFD | 19.73 ± 1.53 $^{abA}$ | 24.8 ± 2.33 $^{cA}$ | 24.24 ± 1.30 $^{bA}$ | 31.23 ± 4.40 $^{aA}$ |

Data are expressed as mean ± standard deviations of duplicate determinations. Means with different lowercase letters in the same column indicate significant differences ($p < 0.05$) for the samples at different casein savinase hydrolysate levels (CSH). Different uppercase letters show statistically significant effects of frozen storage among the same dough types ($p < 0.05$). CD: control dough; CFD: control-frozen dough, HD: dough containing CSH, HFD: frozen dough containing CSH. The addition levels of CSH are 1%, 1.5% and 2%.

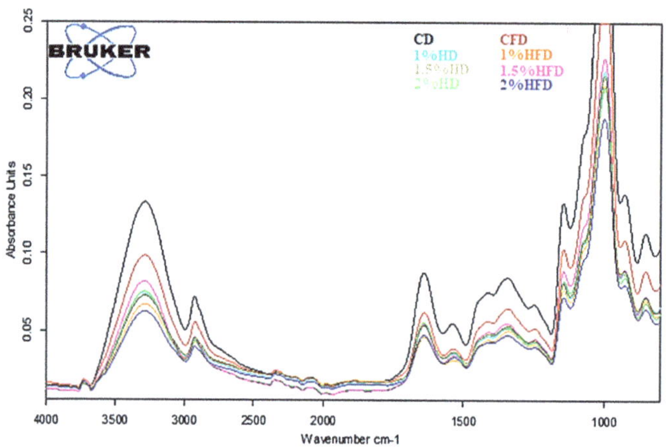


**Figure 4.** FTIR spectra of the samples containing casein savinase hydrolysate (CSH) under frozen storage. CD: control dough, CFD: control-frozen dough, HD: dough containing CSH, HFD: frozen dough containing CSH.

### 3.5. Specific Volume, Hardness and Color of Bread

The quality of bread is mostly evaluated in terms of specific volume (SV), hardness (N) and color characteristics. These criteria are also important indicators that are used for monitoring the frozen dough quality. Figure 5 presents the SV (mL/g) and hardness (N) of bread samples. CSH addition into bread formulation had no remarkable effect on bread quality ($p > 0.05$). The primary impact of frozen storage at −30 °C for 28 days on the bread was a reduction in SV values and an increase in hardness. These phenomena, which have previously been reported in many studies, suggest that during frozen storage ice crystals lead to a decrease in the gas-holding capacity of the dough by destroying the structure of the gluten [14,15]. Also, Cao et al. [16] noted that the fermentation capacity of the dough was influenced by the growth of ice crystals, which decreased the survival of yeast cells. However, the significant effect was shown only in CD and 1%HD after frozen storage ($p < 0.05$). The lowest SV reduction and hardness increment were observed for the breads containing 1.5% and 2% CSH. This can be attributed to the FW (%) content of 1.5%HD and 2%HD samples also being lower than in the CD sample (Figure 3). Mao et al. [31] reported that a reduction in the FW content in frozen dough protected the gluten network by preventing the formation of ice crystals.

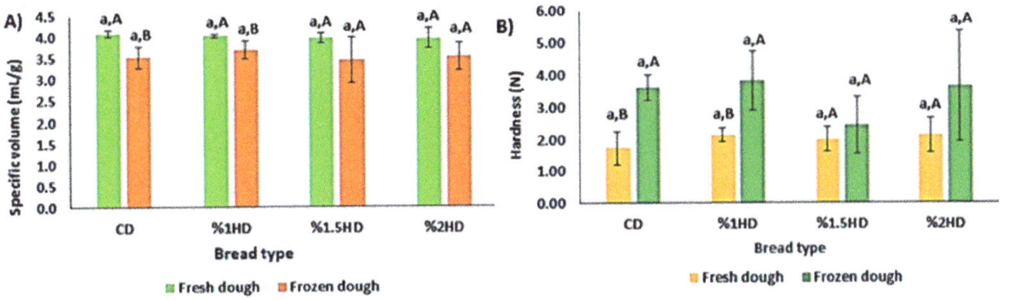


**Figure 5.** Effect of casein savinase hydrolysate (CSH) addition on specific volume (**A**) and hardness (**B**) of breads. CD: control dough, HD: dough containing CSH. Different lowercase/uppercase letters show statistically significant differences between fresh dough and frozen dough, respectively.

The effect of CSH addition and frozen storage (−30 °C, 28 days) on the crust and crumb color of bread samples is shown in Table 2. CSH addition decreased the L* value of the bread crust and a significant reduction was observed for the breads containing 1.5% and 2% CSH when compared to control ($p < 0.05$). The browning of the bread crust (decrease in L* value) with the addition of CSH as a protein source showed an intensive Maillard reaction compared to the control. In a study which investigated the effect of pigskin gelatin in frozen bread, it was indicated that the browning of the bread crust was due to the addition of pigskin gelatin [13]. Frozen storage caused a significant reduction in the b* value of bread crust ($p < 0.05$), while no significant effect was observed for L* and a* value during frozen storage ($p > 0.05$). Similar results for the b* value of the bread crust were also reported by Sharadanant and Khan [32]. In terms of color differences (ΔE), frozen storage had no significant effect on the ΔE value of the crust color, but the ΔE values of 1.5% HFD and 2% HFD were higher than those of other samples (e.g., CFD and 1% HFD). As for the crumb color, frozen storage resulted in the reduction in L* value in the crumb ($p < 0.05$). A similar trend was also noted by Yu et al. [13] at the end of 60 days of frozen storage (−18 °C). On the other hand, Shon et al. [12] reported that no significant effect was observed for crumb L* value during frozen storage (−20 °C, 60 days). The incorporation of CSH into fresh bread increased the b* value of the bread crumb ($p < 0.05$). While frozen storage increased the crumb b* value of the control fresh bread ($p < 0.05$), the breads with CSH were not affected by frozen storage ($p > 0.05$) in terms of crumb b* value. The bread made of control-frozen dough (CFD) exhibited the highest color differences (ΔE) in bread crumb when compared to the frozen dough breads containing CSH. This shows that frozen storage caused no significant effect on the breads with CSH in terms of ΔE in the bread crumb ($p < 0.05$).

**Table 2.** Color properties of breads.

| Storage Time | Samples | Crust | | | | Crumb | | | |
|---|---|---|---|---|---|---|---|---|---|
| | | L* | a* | b* | ΔE | L* | a* | b* | ΔE |
| Day 0 | CD | $58.58 \pm 3.59^{aA}$ | $11.83 \pm 0.84^{aA}$ | $29.34 \pm 1.12^{aA}$ | - | $70.44 \pm 2.01^{aA}$ | $-0.01 \pm 0.27^{aA}$ | $14.81 \pm 0.67^{cB}$ | - |
| | 1% HD | $55.78 \pm 2.71^{abA}$ | $12.87 \pm 0.70^{aA}$ | $31.58 \pm 1.53^{aA}$ | $5.60 \pm 3.35^{aA}$ | $71.45 \pm 1.10^{aA}$ | $0.06 \pm 0.05^{aA}$ | $16.46 \pm 0.26^{abA}$ | $2.99 \pm 1.15^{aA}$ |
| | 1.5% HD | $52.37 \pm 2.94^{bcA}$ | $13.50 \pm 1.23^{aA}$ | $31.99 \pm 0.95^{aA}$ | $7.09 \pm 2.14^{aA}$ | $72.83 \pm 2.16^{aA}$ | $-0.04 \pm 0.13^{aA}$ | $16.27 \pm 0.48^{bA}$ | $4.07 \pm 2.21^{aA}$ |
| | 2% HD | $49.93 \pm 3.67^{cA}$ | $13.62 \pm 1.24^{aA}$ | $28.40 \pm 1.67^{aA}$ | $9.31 \pm 6.17^{aA}$ | $72.27 \pm 0.78^{aA}$ | $0.05 \pm 0.09^{aA}$ | $17.13 \pm 0.51^{aA}$ | $3.45 \pm 1.44^{aA}$ |
| Day 28 | CFD | $58.32 \pm 0.77^{aA}$ | $11.13 \pm 0.61^{aA}$ | $24.73 \pm 1.92^{aB}$ | $6.09 \pm 1.91^{b}$ | $63.76 \pm 2.50^{bB}$ | $0.02 \pm 0.14^{aA}$ | $16.25 \pm 0.83^{aA}$ | $6.92 \pm 1.53^{aA}$ |
| | 1% HFD | $57.06 \pm 2.55^{abA}$ | $12.31 \pm 1.15^{aA}$ | $24.47 \pm 1.81^{aB}$ | $5.70 \pm 2.11^{bA}$ | $67.90 \pm 1.69^{aB}$ | $0.12 \pm 0.23^{aA}$ | $16.79 \pm 1.07^{aA}$ | $3.46 \pm 1.29^{bA}$ |
| | 1.5% HFD | $55.30 \pm 0.71^{bA}$ | $11.90 \pm 2.08^{aA}$ | $23.01 \pm 1.15^{aB}$ | $8.13 \pm 1.84^{abA}$ | $67.48 \pm 3.43^{abB}$ | $0.22 \pm 0.26^{aA}$ | $16.76 \pm 0.29^{aA}$ | $3.96 \pm 2.01^{bA}$ |
| | 2% HFD | $52.67 \pm 1.39^{cA}$ | $11.83 \pm 0.81^{aA}$ | $22.83 \pm 4.07^{aB}$ | $9.84 \pm 2.72^{aA}$ | $67.90 \pm 0.90^{aB}$ | $0.30 \pm 0.03^{aA}$ | $16.80 \pm 0.52^{aA}$ | $3.66 \pm 1.30^{bA}$ |

Data are expressed as mean ± standard deviations of triplicate determinations. Means with different lowercase letters in the same column indicate significant differences ($p < 0.05$) for the samples at different casein savinase hydrolysate levels (CSH). Different uppercase letters show statistically significant effects of frozen storage among the same bread types ($p < 0.05$). CD: control dough; CFD: control-frozen dough, HD: dough containing CSH, HFD: frozen dough containing CSH. The addition levels of CSH are 1%, 1.5% and 2%.

## 4. Conclusions

In this study, casein hydrolysate prepared with savinase (CSH) was used to improve the quality of frozen dough. The frozen dough with CSH did not demonstrate any notable alteration in the structure of α-helices following frozen storage at −30 °C for 28 days. The results obtained from the dynamic rheological measurement and freezable water content showed that the incorporation of CSH into frozen wheat dough slowed down the ice crystals' growth. These findings resulted in an amelioration in the bread quality of 1.5% and 2% CSH-containing samples since the lowest specific volume reduction and hardness increment were exhibited. In summary, casein hydrolysate obtained from savinase can be considered as an alternative bakery ingredient for frozen dough systems.

**Author Contributions:** H.B.: Investigation, data curation, formal analysis, methodology and writing—original draft. G.O.: Conceptualization, investigation, data curation and writing—original draft. O.S.: Supervision and writing—review and editing. All authors have read and agreed to the published version of the manuscript.

**Funding:** This research received no external funding.

**Data Availability Statement:** The data that support the findings of this study will be made available upon request.

**Conflicts of Interest:** The authors declare that there are no known conflict of interest.

## References


1. Gani, A.; Broadway, A.A.; Masoodi, F.A.; Wani, A.A.; Maqsood, S.; Ashwar, B.A.; Shah, A.; Rather, S.A.; Gani, A. Enzymatic hydrolysis of whey and casein protein-effect on functional, rheological, textural and sensory properties of breads. *J. Food Sci. Technol.* **2015**, *52*, 7697–7709. [CrossRef] [PubMed]
2. Abd El-Salam, M.H.; El-Shibiny, S. Preparation, properties, and uses of enzymatic milk protein hydrolysates. *Crit. Rev. Food Sci. Nutr.* **2017**, *57*, 1119–1132. [CrossRef]
3. Ryan, G.; Nongonierma, A.B.; O'Regan, J.; FitzGerald, R.J. Functional properties of bovine milk protein isolate and associated enzymatic hydrolysates. *Int. Dairy J.* **2018**, *81*, 113–121. [CrossRef]
4. Tang, T.; Wu, N.; Tang, S.; Xiao, N.; Jiang, Y.; Tu, Y.; Xu, M. Industrial application of protein hydrolysates in food. *J. Agric. Food Chem.* **2023**, *71*, 1788–1801. [CrossRef]
5. Hidalgo, M.E.; Côrrea, A.P.F.; Canales, M.M.; Daroit, D.J.; Brandelli, A.; Risso, P. Biological and physicochemical properties of bovine sodium caseinate hydrolysates obtained by a bacterial protease preparation. *Food Hydrocoll.* **2015**, *43*, 510–520. [CrossRef]
6. Clemente, A. Enzymatic protein hydrolysates in human nutrition. *Trends Food Sci. Technol.* **2000**, *11*, 254–262. [CrossRef]
7. Bekiroglu, H.; Bozkurt, F.; Karadag, A.; Ahhmed, A.M.; Sagdic, O. The effects of different protease treatments on the techno-functional, structural, and bioactive properties of bovine casein. *Prep. Biochem. Biotechnol.* **2022**, *52*, 1097–1108. [CrossRef]
8. Phimolsiripol, Y.; Siripatrawan, U.; Tulyathan, V.; Cleland, D.J. Effects of freezing and temperature fluctuations during frozen storage on frozen dough and bread quality. *J. Food Eng.* **2008**, *84*, 48–56. [CrossRef]
9. Berglund, P.T.; Shelton, D.R.; Freeman, T.P. by Duration of Frozen Storage and Freeze-Thaw Cycles'. *Cereal Chem.* **1991**, *68*, 105–107.
10. Naito, S.; Fukami, S.; Mizokami, Y.; Ishida, N.; Takano, H.; Koizumi, M.; Kano, H. Effect of freeze-thaw cycles on the gluten fibrils and crumb grain structures of breads made from frozen doughs. *Cereal Chem.* **2004**, *81*, 80–86. [CrossRef]
11. Aibara, S.; Nishimura, K.; Esaki, K. Effects of shortening on the loaf volume of frozen dough bread. *Food Sci. Biotechnol.* **2001**, *10*, 54–61.
12. Shon, J.; Yun, Y.; Shin, M.; Chin, K.B.; Eun, J.B. Effects of milk proteins and gums on quality of bread made from frozen dough. *J. Sci. Food Agric.* **2009**, *89*, 1407–1415. [CrossRef]
13. Yu, W.; Xu, D.; Zhang, H.; Guo, L.; Hong, T.; Zhang, W.; Jin, Y.; Xu, X. Effect of pigskin gelatin on baking, structural and thermal properties of frozen dough: Comprehensive studies on alteration of gluten network. *Food Hydrocoll.* **2020**, *102*, 105591. [CrossRef]
14. Chen, J.; Xiao, J.; Tu, J.; Yu, L.; Niu, L. The alleviative effect of sweet potato protein hydrolysates on the quality deterioration of frozen dough bread in comparison to trehalose. *LWT* **2023**, *175*, 114505. [CrossRef]
15. Zhou, B.; Dai, Y.; Guo, D.; Zhang, J.; Liang, H.; Li, B.; Sun, J.; Wu, J. Effect of desalted egg white and gelatin mixture system on frozen dough. *Food Hydrocoll.* **2022**, *132*, 107889. [CrossRef]
16. Cao, H.; Zheng, X.; Liu, H.; Yuan, M.; Ye, T.; Wu, X.; Yin, F.; Li, Y.; Yu, J.; Xu, F. Cryo-protective effect of ice-binding peptides derived from collagen hydrolysates on the frozen dough and its ice-binding mechanisms. *LWT* **2020**, *131*, 109678. [CrossRef]
17. Adler-Nissen, J. *Enzymic Hydrolysis of Food Proteins*; Elsevier Applied Science Publishers: Amsterdam, The Netherlands, 1986.
18. Shazly, A.B.; He, Z.; Abd El-Aziz, M.; Zeng, M.; Zhang, S.; Qin, F.; Chen, J. Fractionation and identification of novel antioxidant peptides from buffalo and bovine casein hydrolysates. *Food Chem.* **2017**, *232*, 753–762. [CrossRef]

19. Rao, P.S.; Mayur, A.; Harisha, N.B.; Bajaj, R.; Mann, B. Comparison of OPA and pH stat methods for measurement of degree of hydrolysis of alcalase and flavourzyme digested casein. *Indian J. Dairy Sci.* **2018**, *71*, 107–109.
20. Laemmli, U.K. Cleavage of Structural Proteins during the Assembly of the Head of Bacteriophage T4. *Nature* **1970**, *227*, 680–685. [CrossRef]
21. Wang, J.; Su, Y.; Jia, F.; Jin, H. Characterization of casein hydrolysates derived from enzymatic hydrolysis. *Chem. Cent. J.* **2013**, *7*, 62. [CrossRef] [PubMed]
22. Zhu, T.W.; Zhang, X.; Li, B.; Wu, H. Effect of interesterified blend-based fast-frozen special fat on the physical properties and microstructure of frozen dough. *Food Chem.* **2019**, *272*, 76–83. [CrossRef] [PubMed]
23. Khatkar, B.S.; Bell, A.E.; Schofield, J.D. The dynamic rheological properties of glutens and gluten sub-fractions from wheats of good and poor bread making quality. *J. Cereal Sci.* **1995**, *22*, 29–44. [CrossRef]
24. Lu, P.; Guo, J.; Fan, J.; Wang, P.; Yan, X. Combined effect of konjac glucomannan addition and ultrasound treatment on the physical and physicochemical properties of frozen dough. *Food Chem.* **2023**, *411*, 135516. [CrossRef] [PubMed]
25. Wang, X.Y.; Guo, X.N.; Zhu, K.X. Polymerization of wheat gluten and the changes of glutenin macropolymer (GMP) during the production of Chinese steamed bread. *Food Chem.* **2016**, *201*, 275–283. [CrossRef]
26. Wang, Q.; Li, Y.; Sun, F.; Li, X.; Wang, P.; Sun, J.; Zeng, J.; Wang, C.; Hu, W.; Chang, J.; et al. Tannins improve dough mixing properties through affecting physicochemical and structural properties of wheat gluten proteins. *Food Res. Int.* **2015**, *69*, 64–71. [CrossRef]
27. Wang, P.; Lee, T.C.; Xu, X.; Jin, Z. The contribution of glutenin macropolymer depolymerization to the deterioration of frozen steamed bread dough quality. *Food Chem.* **2016**, *211*, 27–33. [CrossRef]
28. Wang, P.; Chen, H.; Mohanad, B.; Xu, L.; Ning, Y.; Xu, J.; Wu, F.; Yang, N.; Jin, Z.; Xu, X. Effect of frozen storage on physico-chemistry of wheat gluten proteins: Studies on gluten-, glutenin-and gliadin-rich fractions. *Food Hydrocoll.* **2014**, *39*, 187–194. [CrossRef]
29. Lou, X.; Yue, C.; Luo, D.; Li, P.; Zhao, Y.; Xu, Y.; Wang, L.; Bai, Z. Effects of natural inulin on the rheological, physicochemical and structural properties of frozen dough during frozen storage and its mechanism. *LWT* **2023**, *184*, 114973. [CrossRef]
30. Yang, Z.; Xu, D.; Zhou, H.; Wu, F.; Xu, X. Rheological, microstructure and mixing behaviors of frozen dough reconstituted by wheat starch and gluten. *Int. J. Biol. Macromol.* **2022**, *212*, 517–526. [CrossRef]
31. Mao, X.; Liu, Z.; Ma, J.; Pang, H.; Zhang, F. Characterization of a novel β-helix antifreeze protein from the desert beetle *Anatolica polita*. *Cryobiology* **2011**, *62*, 91–99. [CrossRef]
32. Sharadanant, R.; Khan, K. Effect of hydrophilic gums on the quality of frozen dough: II. Bread characteristics. *Cereal Chem.* **2003**, *80*, 773–780. [CrossRef]

foods

MDPI

Article

# Vinegar-Preserved Sea Fennel: Chemistry, Color, Texture, Aroma, and Taste

**Sanja Radman [1], Petra Brzović [1], Mira Radunić [2,3], Ante Rako [4], Mladenka Šarolić [1], Tonka Ninčević Runjić [2], Branimir Urlić [4] and Ivana Generalić Mekinić [1,*]**

1 Department of Food Technology and Biotechnology, Faculty of Chemistry and Technology, University of Split, Ruđera Boškovića 35, HR-21000 Split, Croatia; sanja.radman@ktf-split.hr (S.R.); pbrzovic@ktf-split.hr (P.B.); msarolic@ktf-split.hr (M.Š.)

2 Department of Plant Science, Institute for Adriatic Crops and Karst Reclamation, Put Duilova 11, HR-21000 Split, Croatia; mira.radunic@krs.hr (M.R.); tonka.nincevic@krs.hr (T.N.R.)

3 Centre of Excellence for Biodiversity and Molecular Plant Breeding, Svetošimunska Cesta 25, HR-10000 Zagreb, Croatia

4 Department of Applied Science, Institute for Adriatic Crops and Karst Reclamation, Put Duilova 11, HR-21000 Split, Croatia; ante.rako@krs.hr (A.R.); branimir.urlic@krs.hr (B.U.)

* Correspondence: gene@ktf-split.hr; Tel.: +385-21-329-458

**Abstract:** The aim of this study was to produce non-fermented preserved sea fennel leaves in different pickle juices prepared with apple cider vinegar, wine vinegar and alcoholic vinegar, and to compare their chemical parameters (pH, titratable acidity and salt content), organoleptic properties (color and texture parameters; volatile aromatic compound profiles) and sensory attributes. The pH of the samples ranged from 3.49 to 3.64, the lowest being in the alcoholic vinegar sample and the highest being in the wine vinegar sample, while the titratable acidity and salinity were higher in the alcoholic vinegar pickle juice than those in the other two samples. The volatile aromatic compounds of the samples were also detected. The reddish color of the wine vinegar negatively affected the sea fennel color parameters (L* and b*), and was also negatively evaluated by the panelists, while the alcoholic vinegar maximally preserved the green tones of the leaf (a*). Firmness influences the quality perceived by consumers and was therefore also tested as one of the most important parameters for evaluating the textural and mechanical properties of the different products. All sensory parameters of the sea fennel preserved in alcoholic vinegar, namely color, texture, taste, aroma and overall impression, were given the highest scores, while the sample preserved in wine vinegar received the lowest scores. The intense aroma of the wine vinegar was described as a negative characteristic (off-flavor) of the sample.

**Keywords:** *Crithmum maritimum*; apple cider vinegar; wine vinegar; alcoholic vinegar; pickles; chemical characteristics; sensory attributes; organoleptic properties

## 1. Introduction

Vegetables have high nutritional value, but are highly perishable and have a short shelf life. Therefore, their preservation has long been practiced via various methods. Pickling in pickle juice is one of the oldest and most widely used traditional methods, which ensures preservation through high acidity. The production of pickles is simple, it requires relatively simple equipment, and the final products are relatively safe and long-lasting if properly preserved.

Pickling can be performed in several ways: via anaerobic fermentation with lactic acid bacteria, via acidification with vinegar with/without the addition of salt or via a combination of these processes [1–4]. The first method is based on the activity of fermentative microorganisms that produce various metabolites that create an environment that prevents the growth and multiplication of undesirable, spoilage and pathogenic microorganisms.

These metabolites lower the pH of the medium, and thus preserve the products. The second method results in non-fermented products and is based on the direct addition of acetic acid (an acidifier, usually in the form of vinegar) and other flavoring agents, followed by heat treatment (pasteurization) or cold storage in combination with the addition of preservatives [2–5]. For most products, the main preservatives used during processing are organic acids (with lactic acid and acetic acid being the most commonly used) and salt, but commercial preservatives such as sorbates, benzoates or sulfites may also be added. For non-fermented pickles, the addition of vinegar ensures a high acetic acid content, and in most cases this step is followed by pasteurization to ensure safety and prevent the product from spoiling. On the other hand, pickles produced via fermentation are not heated [1,3,4].

Salt and sugar are usually added to the pickles. Salt enhances the flavor, affects the texture and water content, and acts as a preservative by affecting water activity and reducing the solubility of oxygen [2,6]. It also helps prevent softening of tissue and assists in breaking the membranes of the raw material, allowing the diffusion/extraction of various components from the plant material into the pickle juice and vice versa [4]. For pasteurized products semi-preserved in weaker pickle juice or vinegar, salt provides some protection against microbial spoilage after the product has been opened at home [6]. Sugar, on the other hand, is added to accelerate fermentation or to make the product sweeter [1]. The addition of sugar and salt also allows the use of lower concentrations of acetic acid [3].

Sea fennel (*Crithmum maritimum* L., Apiaceae) is one of the most widespread halophyte species on the Adriatic coast. Although its leaves have been used for culinary and medicinal purposes for centuries, this valuable plant has been recently rediscovered and recognized as the star of coastal cuisine, but also as a "cash" plant for saline agriculture [7–9]. It stands out due to its sensory properties in terms of taste, odor and color, but also contains numerous valuable phytochemicals such as vitamin C, phenolics, unsaturated fatty acids, organic acids, minerals, etc. [10–17]. Sea fennel is used raw, cooked or preserved in various foods and dishes [8,9,18,19]. Pickled sea fennel is a product widely used in the cuisine of Mediterranean countries and usually can be found in local shops selling authentic and local traditional products. The aim of this study was to produce non-fermented pickled sea fennel using apple cider, wine and alcoholic vinegar, to evaluate the main chemical parameters of the products (pH, total acidity, salinity and volatile compounds), color and texture parameters, and to study the sensory attributes and general consumer acceptance of the products.

## 2. Materials and Methods

### 2.1. Pickling

Fresh, young and undamaged leaves of wild sea fennel collected in the area of the city of Split (Croatia) in late June 2023 were washed three times with tap water, drained and carefully dried with a paper towel. The fresh plant material was put into clean jars and immersed in pickle juice. In accordance with traditional recipes and the results of the preliminary study, three types of pickle juices were prepared using different types of vinegar: (a) apple cider vinegar (5% acetic acid; 1:5, *v*/*v*), (b) wine vinegar (6% acetic acid, 1:4, *v*/*v*) and (c) alcoholic vinegar (9% acetic acid, 1:5, *v*/*v*). All pickle juices contain salt (1%, *w*/*v*) and sucrose (granulated sugar) (1.5%, *w*/*v*). The filled and sealed jars were subjected to a pasteurization treatment (at 95 °C, for 10 min). After cooling (4 h), they were checked via lid inspection (if the center of the lid is depressed, the jar is sealed). The jars were stored at room temperature for 8 weeks before analysis.

### 2.2. Chemical Composition

Regarding the chemical composition, the pH, titratable acidity and salinity of the pickle juices were analyzed. The pH of the samples was measured using a digital pH meter (Hanna HI -2002-02 Edge, Hanna instruments Ltd., Leighton Buzzard, UK). Titratable acidity (g/100 mL) was determined using the modified AOAC method via the titration of pickle juices (10 mL) with standard sodium hydroxide (0.1 M) using phenolphthalein as the endpoint color indicator and expressed in acetic acid equivalents [20]. The salt content (g/100 mL) was determined in accordance with the modified method of Mohr by determining the chloride ion concentration by titrating the pickle juice sample (1 mL) diluted with 50 mL of distilled water with silver nitrate solution (0.1 M) [21]. All measurements were carried out in triplicate and the average values were reported.

### 2.3. Volatiles

The aromatic or volatile organic compounds (VOCs) of the pickled sea fennel were extracted via headspace–solid-phase microextraction (HS-SPME) and detected via GC-MS. HS-SPME was performed using a DVB/CAR/PDMS (divinylbenzene/carboxene/polydimethylsiloxane) SPME fiber (Agilent Technologies, Palo Alto, Santa Clara, CA, USA) conditioned in accordance with the manufacturer's instructions before analysis. Samples (1 g) were placed in 20 mL glass vials sealed with a stainless steel lid with a polytetrafluorethylene (PTFE)/silicon septum. First, the sample was equilibrated at a temperature of 40 °C for 15 min, after which extraction was continued for 45 min. Subsequent thermal desorption lasted 6 min in the inlet set at 250 °C, from where the compounds were passed directly into the GC column. GC-MS analysis was performed using a gas chromatograph (model 8890A, Agilent Technologies) and a tandem mass spectrometer (MS) model 7000D GC/TQ. Compounds were separated on a HP -5MS UI column (30 m × 0.25 mm × 0.25 μm; Agilent Technologies). The following operating conditions were used for the gas chromatograph: a 250 °C injector temperature; a 300 °C detector temperature; a split ratio of 1:5; a column temperature program of 2 min isothermal conditions at 70 °C, followed by a temperature gradient from 70 °C to 200 °C at 3 °C/min and a further retention of 15 min. The carrier gas was helium at a flow rate of 1.0 mL/min; the ion voltage was 70 eV; the mass range was set to 30 to 300 m/z. The compounds were identified by comparing their retention indices (RI), which were based on the retention times of n-alkanes (C8-C30), with those reported in the literature (National Institute of Standards and Technology) and by comparing their mass spectra with those from the Wiley 9 (Wiley, New York, NY, USA) and NIST 17 (Gaithersburg, MD, USA) mass spectral libraries. Percent composition was calculated using the normalization method (without correction factors). Analyses were performed in duplicate and expressed as an average percentage of peak area ± standard deviation.

### 2.4. Color and Texture Parameters

The color analysis of the pickled sea fennel leaves was determined using a CIELAB color system (CR-400 Chroma Meter, Konica, Tokyo, Japan) and expressed in terms of the parameters lightness, L*; a*; b*; C and h. Standard illuminant D65 was used as a reference. The color parameter L* indicates the lightness and varies from completely opaque (0) to completely transparent (100). The value a* represents the redness and ranges from negative values for green to positive values for red. The value b* is the yellowness ranging from negative values for blue to positive values for yellow. Saturation (C) is the color intensity and hue angle (h) is the color purity, and both describe the visual impression of a color. The value C was calculated as $C = [a \times 2 + b \times 2]^{1/2}$ and the hue (h) value as h = arctang b*/a*. The hue angle evaluates the different colors as follows: 0–90° red-violet, 90–180° yellow, 180–270° blue-green, and 270–360° blue [16,22]. All measurements were performed in triplicate, and the results are expressed as mean ± standard deviation.

The texture of sea fennel leaves was analyzed using a texture analyzer (TA Plus; Lloyd Instruments, Fareham, UK) equipped with a 500 N load cell and a Warner–Bratzler blade set with a rectangular slot blade. Prior to testing, the leaves were drained and brought to room temperature (25 °C). After reaching the pretest load of 0.01 N, the blade cut through the sample at a crosshead speed of 10 mm/min until the blade reached an extension of 5 mm and was then retracted to the zero position. Texture testing was carried out using Nexygen Plus 3 software. The parameters measured were hardness (N) and work of cutting (N/mm). In the software manual, hardness was defined as the maximum force during the test, while work of cutting was defined as the total work during the test. The analyses were performed in 12 replicates and the results are expressed as mean ± standard deviation.

*2.5. Sensory Evaluation*

Sensory evaluation was performed by 12 untrained panelists (aged between 28 and 48, including 6 men and 6 women) familiar with pickled sea fennel, using two methods. In the first method, samples were scored on a 5-point scale for color, texture, taste, aroma and overall impression, which includes visual, textural, taste and flavor attributes, as described by Amouroso et al. [16]. The scores were ranked as follows: 5: excellent, 4: good, 3: moderate, 2: poor, and 1: extremely poor. In addition, defects such as off-odor and mechanical damage were rated (5: none, 4: slight, 3: moderate, 2: severe, and 1: extreme). The coded samples were served at room temperature, and water was used as a palate cleanser. The arithmetic mean of the scores ($n = 12$) for each sensory attribute was used for further data processing. Although for reliable information on consumer acceptability the number of assessors should be higher, for the purpose of this study the same panelists were used for another method of sensory evaluation of the preserved sea fennel samples using a 9-point rating in a hedonic scale (the test of consumers' preference). In this method, the scores were ordered as follows: 9: like extremely, 8: like very much, 7: like moderately, 6: like slightly, 5: neither like nor dislike, 4: dislike slightly, 3: dislike moderately, 2: dislike very much, and 1: dislike extremely.

*2.6. Statistical Analysis*

Statistical procedures were carried out in SPSS Statistics (2012; version 21.0, IBM Corp., Armonk, NY, USA). A one-way ANOVA and the general linear model (GLM) were used to investigate the significance of physicochemical, aromatic, color and texture changes in sea fennel leaves placed in three different types of pickle juice for 8 weeks of storage at room temperature. Pairwise comparisons of means were made using the least significant difference (LSD) test.

## 3. Results and Discussion

While preservation methods for various vegetables via fermentation or the direct addition of acidulant have been extensively studied, there are few studies on sea fennel products. Fermented canned sea fennel was studied by Maoloni et al. [15], while artificially acidified, pasteurized sea fennel in pickle juice was studied by Maoloni et al. [23].

Organic acids have been used for decades as additives for food preservation; acetic acid (in the form of vinegar) is commonly used in pickles at a concentration of 0.5–2% [2,5]. In this study, we used three types of commercial vinegar, apple cider vinegar, red wine vinegar and alcoholic vinegar, to prepare pickling juice solutions (at a ratio of 1:5 $v/v$ in the case of apple cider vinegar and alcoholic vinegar and 1:4 $v/v$ for wine vinegar), which were used to pickle sea fennel leaves.

*3.1. Chemistry*

The results for the pH, salinity and total acidity of pickle juice of sea fennel pickles after two months of storage, which were investigated in the present study, are shown in Table 1.

**Table 1.** pH values, salt content and total acidity of the sea fennel samples preserved via the addition of three types of vinegars.

| | Apple Cider Vinegar (1:5, $v/v$) | Wine Vinegar (1:4, $v/v$) | Alcoholic Vinegar (1:5, $v/v$) | $p$-Value |
|---|---|---|---|---|
| pH value | 3.55 ± 0.05 | 3.64 ± 0.09 | 3.49 ± 0.06 | 0.09 |
| Titratable acidity (g/100 mL) | $1.62^{a} \pm 0.01$ | $1.65^{a} \pm 0.01$ | $2.48^{b} \pm 0.01$ | <0.01 |
| Salt (g/100 mL) | $1.16^{a} \pm 0.02$ | $1.16^{a} \pm 0.02$ | $1.25^{b} \pm 0.01$ | <0.01 |

Mean values within a row with different letters in superscript differ significantly ($p < 0.05$).

pH is a crucial parameter for the shelf life and safety of food products. According to Aljahani [5], the accepted standard for pickles is 4.6 or lower, which prevents the survival of most spoilage and pathogenic organisms and effectively preserves the final products. The pH of canned pickles is higher than that of fresh pickles, while the acidity and salt concentration of fresh pickles are higher. The acidity and acetic acid content of a pickled product are also important for its flavor, aroma and texture [2,24]. From the results presented in Table 1, the pH of the samples ranged from 3.49 to 3.64, being lowest in alcoholic vinegar and highest in wine vinegar. In the study by Maoloni et al. [15], the pH of the samples in most sea fennel lactic acid fermentation experiments of sea fennel shoots reached values below 3.80 after 24 h. On the other hand, Maoloni et al. [23] reported pH values of 3.85 for pickled sea fennel after 4 weeks of storage.

However, the titratable acidity of the sample preserved in alcoholic vinegar was significantly higher (2.5 g/100 mL) than that in two other samples, where it was 1.6 g/100 mL, which was expected due to the use of different acetic acid concentrations in pickle juice preparation. This was expected since the initial concentrations of acetic acid in prepared pickle juices differ (10 g/L in apple cider vinegar, 12.5 g/L in wine vinegar and 18 g/L in alcoholic vinegar). The salt content of the samples prepared with apple cider and wine vinegar was the same, while it was slightly higher (1.25 g/100 mL) when alcoholic vinegar was used.

### 3.2. Volatiles

The headspace profile of volatile organic compounds (VOCs) of the preserved sea fennel samples is presented in Table 2. The compounds were extracted via headspace–solid-phase microextraction (HS-SPME) and analyzed via gas chromatography–mass spectrometry (GC-MS).

In order to reach conclusions about the volatiles originating from the plant material and those from the vinegars, the sea fennel leaves before processing and pure vinegars were also analyzed. Briefly, 95.73% (apple cider vinegar sample) to 97.11% (alcoholic vinegar sample) of VOCs were identified. The majority of compounds originated from sea fennel (88.54% in the alcoholic vinegar sample; 90.01% in wine vinegar sample) (Figure 1) but vinegars were the source of only seven compounds detected (Table 2).

There is a lack of information on the HS-SPME analysis of sea fennel leaves; only GC-MS analyses of EO or hydrolate are available. Terpenes dominate in sea fennel essential oils. Politeo et al. [25] reported that sabinene is the most abundant volatile compound in sea fennel leaves (51.47%) followed by limonene (36.28%) and terpinen-4-ol (5.35%). Alves-Silva et al. [26] analyzed the essential oil of the flowering aerial parts of sea fennel and the major compounds were $\gamma$-terpinene (33.6%), sabinene (32.02%) and thymol methyl oxide (15.7%).

The monoterpene limonene was the major compound in the HS of preserved sea fennel leaf in all of vinegars (from 43.31% in the alcoholic vinegar sample to 52.64% in the apple cider vinegar sample) followed by the monoterpene alcohol terpinen-4-ol (from 9.40% in the alcoholic vinegar sample to 12.85% in the wine vinegar sample) (Table 2). Three other terpenes, $\gamma$-terpinene, (*E*)-$\beta$-ocimene, and $\alpha$-pinene, were more abundant than were other compounds. Among the compounds originating from vinegar, acetic acid was the most abundant (6.10% in the apple cider vinegar sample; 8.40% in the alcoholic vinegar sample). In addition to acetic acid, vinegar contains a variety of other secondary constituents that contribute to its flavor and that are a result of used raw material or vinegar processing steps [2]. These compounds were not detected, which is probably due to the predominance of compounds derived from the sea fennel leaves.

**Table 2.** The volatile organic compounds (VOCs, %) of preserved sea fennel leaf in apple cider, wine and alcoholic vinegar.

| No. | RT (min) | Compound | RI | RI Library | Apple Cider Vinegar (1:5, *v/v*) | Wine Vinegar (1:4, *v/v*) | Alcoholic Vinegar (1:5, *v/v*) | *p*-Value |
|---|---|---|---|---|---|---|---|---|
| 1. | 1.84 | Acetic acid * | <800 | 600 | $6.10^{a} \pm 0.30$ | $6.41^{a} \pm 0.08$ | $8.40^{b} \pm 0.54$ | <0.01 |
| 2. | 4.99 | α-Pinene | 947 | 939 | 4.69 ± 0.26 | 5.17 ± 0.34 | 5.04 ± 0.35 | 0.36 |
| 3. | 5.34 | Camphene | 962 | 961 | 0.06 ± 0.03 | 0.07 ± 0.03 | 0.10 ± 0.05 | 0.46 |
| 4. | 5.41 | (*E*)-Hept-2-enal | 965 | 956 | 0.07 ± 0.02 | 0.04 ± 0.01 | - | 0.21 |
| 5. | 5.56 | Benzaldehyde * | 971 | 961 | - | - | 0.02 ± 0.00 | - |
| 6. | 6.01 | Sabinene | 987 | 978 | $0.20^{a} \pm 0.00$ | $0.14^{b} \pm 0.01$ | $0.10^{c} \pm 0.01$ | <0.01 |
| 7. | 6.19 | 6-Methylhept-5-en-2-one | 992 | 987 | 0.01 ± 0.00 | - | - | - |
| 8. | 6.31 | β-Myrcene | 996 | 994 | $1.16^{a} \pm 0.09$ | $1.69^{b} \pm 0.10$ | $2.06^{c} \pm 0.13$ | <0.01 |
| 9. | 6.47 | (*E,Z*)-Hepta-2,4-dienal * | 1002 | 999 | 0.01 ± 0.00 | - | - | - |
| 10. | 6.64 | Octanal | 1008 | 1005 | 0.09 ± 0.02 | 0.05 ± 0.02 | 0.04 ± 0.02 | 0.10 |
| 11. | 6.74 | α-Phellanderene | 1012 | 1105 | $0.40^{a} \pm 0.05$ | $0.38^{a} \pm 0.01$ | $0.48^{b} \pm 0.02$ | <0.05 |
| 12. | 7.10 | α-Terpinene | 1023 | 1020 | $0.17^{a} \pm 0.00$ | $0.43^{b} \pm 0.03$ | $0.90^{c} \pm 0.06$ | <0.01 |
| 13. | 7.60 | Limonene | 1037 | 1035 | $52.64^{a} \pm 2.21$ | $43.75^{b} \pm 2.22$ | $43.31^{b} \pm 2.58$ | <0.01 |
| 14. | 7.76 | (*E*)-β-Ocimene | 1045 | 1040 | $5.99^{a} \pm 0.04$ | $8.30^{b} \pm 0.41$ | $8.50^{b} \pm 0.42$ | <0.01 |
| 15. | 8.05 | (*Z*)-β-Ocimene | 1056 | 1051 | $0.59^{a} \pm 0.04$ | $0.94^{b} \pm 0.01$ | $1.23^{c} \pm 0.01$ | <0.01 |
| 16. | 8.45 | γ-Terpinene | 1066 | 1064 | $6.26^{a} \pm 0.08$ | $8.04^{b} \pm 0.953$ | $7.81^{b} \pm 0.52$ | <0.01 |
| 17. | 9.44 | α-Terpinolene | 1097 | 1089 | $0.28^{a} \pm 0.00$ | $0.44^{b} \pm 0.02$ | $0.86^{c} \pm 0.05$ | <0.01 |
| 18. | 9.83 | Linalool | 1103 | 1104 | 0.02 ± 0.00 | 0.02 ± 0.00 | 0.01 ± 0.00 | 0.71 |
| 19. | 10.00 | (3*E*)-6-Methylhepta-3,5-dien-2-one * | 1108 | 1107 | $0.08^{a} \pm 0.00$ | $0.06^{b} \pm 0.00$ | $0.07^{b} \pm 0.00$ | <0.01 |
| 20. | 10.28 | *p*-Mentha-1,3,8-triene | 1117 | 1119 | - | - | 0.04 ± 0.01 | - |
| 21. | 10.31 | 2-Phenylethanol * | 1117 | 1117 | $0.05^{a} \pm 0.03$ | $0.32^{b} \pm 0.01$ | $0.07^{a} \pm 0.02$ | <0.01 |
| 22. | 10.60 | (3*E*,5*E*)-2,6-dimethylocta-1,3,5,7-tetraene | 1126 | 1134 | $0.29^{a} \pm 0.04$ | $0.21^{b} \pm 0.00$ | $0.17^{b} \pm 0.00$ | <0.02 |
| 23. | 10.92 | (*Z*)-Alloocimene | 1136 | 1130 | $1.22^{a} \pm 0.08$ | $2.04^{b} \pm 0.14$ | $2.23^{b} \pm 0.15$ | <0.03 |
| 24. | 11.52 | β-Terpineol | 1151 | 1159 | $0.09^{a} \pm 0.00$ | $0.12^{b} \pm 0.01$ | $0.11^{c} \pm 0.00$ | <0.04 |
| 25. | 11.94 | 4-prop-1-en-2-ylcyclohexene * | 1161 | 1161 | 0.33 ± 0.01 | 0.16 ± 0.00 | - | - |
| 26. | 12.08 | (*Z*)-Non-2-enal | 1161 | 1162 | 0.04 ± 0.01 | 0.02 ± 0.00 | 0.02 ± 0.00 | 0.23 |
| 27. | 12.36 | *p*-Mentha-1,5-dien-8-ol | 1171 | 1171 | 0.02 ± 0.01 | 0.02 ± 0.00 | 0.02 ± 0.00 | 1.00 |
| 28. | 12.88 | Terpinen-4-ol | 1183 | 1178 | $10.03^{a} \pm 0.32$ | $12.85^{b} \pm 0.95$ | $9.40^{a} \pm 0.64$ | <0.01 |
| 29. | 13.34 | α-Terpineol | 1193 | 1190 | $1.18^{a} \pm 0.02$ | $1.68^{b} \pm 0.07$ | $1.37^{c} \pm 0.06$ | <0.02 |
| 30. | 14.47 | (*E*)-Carveol | 1222 | 1222 | $0.36^{a} \pm 0.03$ | $0.21^{b} \pm 0.01$ | $0.16^{b} \pm 0.00$ | <0.03 |
| 31. | 14.94 | (*Z*)-Carveol | 1234 | 1232 | $0.09^{a} \pm 0.02$ | $0.06^{ab} \pm 0.01$ | $0.03^{b} \pm 0.01$ | <0.05 |
| 32. | 15.13 | Thymol methyl ether | 1239 | 1237 | 0.01 ± 0.00 | 0.01 ± 0.00 | 0.01 ± 0.00 | 0.42 |
| 33. | 15.49 | Carvone | 1248 | 1246 | $1.16^{a} \pm 0.12$ | $0.58^{b} \pm 0.04$ | $0.60^{b} \pm 0.04$ | <0.01 |

**Table 2.** *Cont.*

| No. | RT (min) | Compound | RI | RI Library | Apple Cider Vinegar (1:5, *v*/*v*) | Wine Vinegar (1:4, *v*/*v*) | Alcoholic Vinegar (1:5, *v*/*v*) | *p*-Value |
|---|---|---|---|---|---|---|---|---|
| 34. | 15.84 | α-Ionene | 1256 | 1255 | $0.02^{a} \pm 0.00$ | $0.05^{b} \pm 0.00$ | $0.02^{a} \pm 0.00$ | <0.05 |
| 35. | 16.03 | 2-Phenylethyl acetate * | 1261 | 1260 | $0.02^{a} \pm 0.00$ | $0.06^{b} \pm 0.01$ | $0.02^{a} \pm 0.00$ | <0.01 |
| 36. | 16.21 | (*E*)-Dec-2-enal | 1265 | 1260 | $0.07^{a} \pm 0.01$ | $0.04^{b} \pm 0.00$ | $0.03^{c} \pm 0.00$ | <0.01 |
| 37. | 17.54 | *o*-Thymol | 1295 | 1294 | $0.06^{a} \pm 0.00$ | $0.05^{a} \pm 0.00$ | $0.03^{b} \pm 0.00$ | <0.01 |
| 38. | 22.70 | β-Caryophyllene | 1421 | 1428 | $0.13^{a} \pm 0.00$ | $0.35^{b} \pm 0.02$ | $0.35^{b} \pm 0.02$ | <0.01 |
| 39. | 23.29 | γ-Elemene | 1436 | 1433 | $0.10^{a} \pm 0.04$ | $0.15^{a} \pm 0.05$ | $0.44^{b} \pm 0.05$ | <0.01 |
| 40. | 23.38 | α-Bergamotene | 1438 | 1436 | $0.14 \pm 0.03$ | $0.17 \pm 0.03$ | $0.20 \pm 0.04$ | 0.30 |
| 41. | 23.48 | Aromadendrene | 1441 | 1442 | $0.04 \pm 0.01$ | $0.07 \pm 0.01$ | $0.06 \pm 0.01$ | 0.50 |
| 42. | 25.27 | α-Curcumene | 1484 | 1485 | $0.18 \pm 0.02$ | $0.19 \pm 0.00$ | $0.18 \pm 0.00$ | 0.58 |
| 43. | 25.38 | (*E*)-β-Ionone | 1487 | 1485 | $0.02 \pm 0.00$ | $0.02 \pm 0.00$ | $0.02 \pm 0.00$ | 0.87 |
| 44. | 25.76 | α-Zingiberene | 1496 | 1496 | $0.17^{a} \pm 0.00$ | $0.38^{b} \pm 0.01$ | $0.79^{c} \pm 0.03$ | <0.01 |
| 45. | 26.30 | β-Bisabolene | 1510 | 1509 | $0.26^{a} \pm 0.02$ | $0.32^{b} \pm 0.00$ | $0.44^{c} \pm 0.01$ | <0.01 |
| 46. | 26.88 | β-Sesquiphellanderene | 1526 | 1526 | $0.41^{a} \pm 0.03$ | $0.49^{a} \pm 0.03$ | $0.70^{b} \pm 0.04$ | <0.01 |
| 47. | 27.52 | Selina-3,7(11)-dien | 1543 | 1543 | $0.02^{a} \pm 0.00$ | $0.03^{b} \pm 0.00$ | $0.09^{c} \pm 0.00$ | <0.01 |
| 48. | 28.11 | Germacrene B | 1558 | 1558 | $0.04^{a} \pm 0.00$ | $0.07^{b} \pm 0.00$ | $0.22^{c} \pm 0.02$ | <0.01 |
| 49. | 29.14 | Caryophyllene oxide | 1584 | 1583 | $0.15^{a} \pm 0.01$ | $0.15^{a} \pm 0.01$ | $0.11^{b} \pm 0.00$ | <0.01 |
| 50. | 29.44 | α-Guaiol | 1591 | 1595 | $0.21 \pm 0.02$ | $0.23 \pm 0.00$ | $0.20 \pm 0.01$ | 0.14 |

* originated from vinegar; RI—retention index. Results are expressed as mean ± standard deviation (SD); the average of two results was used as the third repetition in the ANOVA. Mean values within a row with different letters in superscript differ significantly ($p < 0.05$).

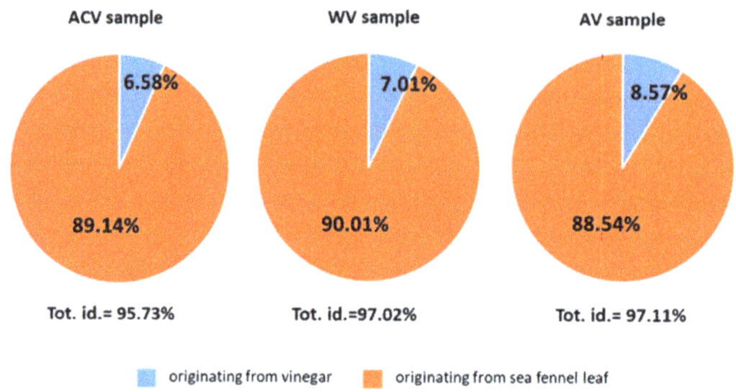


**Figure 1.** Distribution of origin of compounds in preserved sea fennel samples. ACV—apple cider vinegar; WV—wine vinegar; AV—alcoholic vinegar; tot. id.—total identified.

### 3.3. Color and Texture

Color is among the first and most important food quality parameters that attract consumers' attention and influence their choice and opinion of the product [22]. Since the human eye can only detect minimal changes in color, the use of the CIELAB color system in the color evaluation of different samples and products is widespread, especially in industry. Therefore, the color variability of the preserved sea fennel leaves in terms of lightness (L*), greenness/redness (a*), blueness/yellowness (b*), color intensity or saturation (C) and hue angle (h) was measured and the results are shown in Table 3. The tested samples are presented in Figure 2.

**Table 3.** Color and texture parameters of the sea fennel persevered via the addition of apple cider, wine and alcoholic vinegar.

| | Apple Cider Vinegar (1:5, *v*/*v*) | Wine Vinegar (1:4, *v*/*v*) | Alcoholic Vinegar (1:5, *v*/*v*) | *p*-Value |
|---|---|---|---|---|
| Color parameters | | | | |
| L* | $38.48^{a} \pm 0.33$ | $35.30^{b} \pm 0.32$ | $36.21^{b} \pm 0.33$ | <0.01 |
| a* | $-0.27^{a} \pm 0.01$ | $-0.84^{ab} \pm 0.01$ | $-1.45^{b} \pm 0.02$ | <0.05 |
| b* | $20.62^{a} \pm 0.60$ | $23.03^{b} \pm 0.59$ | $15.28^{c} \pm 0.61$ | <0.01 |
| C | $20.64^{a} \pm 0.58$ | $23.05^{b} \pm 0.57$ | $15.36^{c} \pm 0.56$ | <0.01 |
| h | $90.88^{a} \pm 0.43$ | $91.22^{a} \pm 0.43$ | $95.03^{b} \pm 0.42$ | <0.01 |
| Texture parameters | | | | |
| Load at Maximum Load (N) | $8.22 \pm 1.79$ | $10.47 \pm 0.45$ | $8.56 \pm 1.58$ | 0.21 |
| Work of shear (N/mm) | $14.19^{a} \pm 3.52$ | $19.62^{b} \pm 1.72$ | $13.91^{a} \pm 1.05$ | <0.05 |

Results are expressed as mean ± standard deviation; mean values within a row with different letters in superscript differ significantly ($p < 0.05$).

In our study, significantly higher lightness was recorded for sea fennel leaves preserved in apple cider vinegar (38.48) than for sea fennel leaves preserved in wine and alcoholic vinegar (35.30 and 36.21, respectively). Sea fennel leaves preserved in alcoholic vinegar had the highest green content (a* −1.45), and the lowest content of yellow (b* 15.28) pigment than those under the other two treatments. The color intensity (C) of sea fennel leaves preserved in alcoholic vinegar was the lowest (15.36), while the h (95.03) of those was the highest, indicating that the sea fennel leaves were generally lighter, i.e., they had a better-preserved natural green color. When treated with apple cider vinegar, there was a significant loss of green but also of yellow pigment (lowest a* and b* values). The color differences between the treatments could be closely related to the vinegar used. While the alcoholic vinegar was completely colorless, the apple cider vinegar was pale yellow and the wine vinegar was slightly reddish.

(a) (b) (c)

**Figure 2.** Samples of pickled sea fennel leaves in (**a**) apple cider vinegar, (**b**) wine vinegar and (**c**) alcoholic vinegar.

One of the key factors in the consumer acceptance and value of food, besides appearance and taste, is texture. Hardness is one of the most important parameters for evaluating the physical and mechanical properties of various foods and often influences their quality and acceptance by consumers.

It is well known that texture is determined via the moisture and fat content of the food, but also via the type and amount of structural components such as carbohydrates, proteins, hydrocolloids, etc. Thermal processes such as pasteurization, apart from their preservative effect, are often used to modify product texture, but the latter is also influenced by many other factors such as pH (with a minimum softening in the range of pH 4), salts and acids [27,28].

Therefore, when processing sea fennel leaves, special attention should be paid to the processing steps that could influence the nature of the final products, such as pasteurization (duration and temperature) and/or the composition of the soaking pickle juice (acidity and salinity). The results of our preliminary studies on preserved sea fennel were decisive for the choice of parameters used to find the optimal recipe for processing pickled sea fennel.

The textural properties of foods can be evaluated via descriptive sensory tests or instrumental analyses [28]. In this study, they were tested using a texture analyzer, and the results obtained, such as hardness and work of cutting of the product, are shown in Table 2. The work of cutting, described as the total work achieved during the test, was significantly ($p < 0.05$) higher for the sea fennel leaves preserved in wine vinegar than it was for the other two samples. The hardness of the pickled sea fennel leaves was not affected by the type of pickle juice.

Texture is one of the most important parameters for pickled sea fennel, because leaves that are too soft, tender and tough are undesirable. The fennel samples pickled in alcoholic vinegar had an average hardness and were characterized by a lower work of cutting, which was also positively evaluated by the panelists in the sensory evaluation, while the hardness of the samples pickled in wine vinegar was considered a negative attribute of the product. According to Brenes et al. [27], fruit softening is induced via the addition of acid, the effect being stronger with lactic acid than with acetic acid. In their study, the effect of acid on fruit hardness was greater at lower pHs and consequently at higher acid concentrations. According to the LSD test, the lowest values for work of cutting were found in the alcoholic vinegar samples, which had, significantly ($p < 0.01$), the highest titratable acidity. On the other hand, MacFeeters et al. [29] reported that salt softens low-acidity vegetables such as cucumbers. The LSD test also showed that the alcoholic vinegar sample had significantly ($p < 0.01$) the highest salt content in the pickle juice, indicating that the salt was probably extracted from the plant material, which could have an effect on softening and reducing the chopping values.

### 3.4. Sensory Evaluation

Descriptive sensory analysis was used to describe the sensory attributes of preserved sea fennel with three different vinegars (Figure 3 and Table 4).

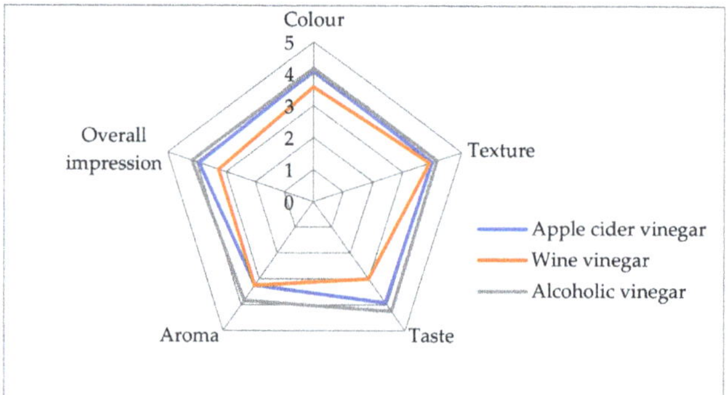


**Figure 3.** Sensory profiles of sea fennel preserved in pickle juices prepared using different vinegars (5: excellent, 4: good, 3: moderate, 2: poor, and 1: extremely poor).

**Table 4.** Sensory parameters of the sea fennel preserved via the addition of apple cider, wine and alcoholic vinegar.

| | **Apple Cider Vinegar (1:5, *v*/*v*)** | **Wine Vinegar (1:4, *v*/*v*)** | **Alcoholic Vinegar (1:5, *v*/*v*)** | ***p*-Value** |
|---|---|---|---|---|
| Color | 4.08 ± 0.38 | 3.58 ± 0.29 | 4.17 ± 0.52 | 0.25 |
| Texture | 4.00 ± 0.66 | 3.92 ± 0.29 | 4.17 ± 0.28 | 0.81 |
| Taste | $3.92^{a}$ ± 0.80 | $3.00^{b}$ ± 0.50 | $4.25^{a}$ ± 0.25 | <0.05 |
| Aroma | 3.25 ± 0.87 | 3.25 ± 0.25 | 3.83 ± 0.52 | 0.44 |
| Overall impression | 3.92 ± 0.63 | 3.25 ± 0.50 | 4.17 ± 0.38 | 0.16 |

Results are expressed as mean ± standard deviation; mean values within a row with different letters in superscript differ significantly ($p < 0.05$).

In this test, panelists rated the color, texture, taste, aroma and overall impression of the samples using a five-point scale and also noted any noticeable defects (off-odors and mechanical damage). As can be seen from the results, the sea fennel preserved with alcoholic vinegar received the highest values for all tested attributes. The highest value was given for taste (4.25 ± 0.25), color, texture and overall impressions, which were evaluated with an average score of 4.17, while the aroma of the product was evaluated to have a slightly lower score (3.83 ± 0.52). Sea fennel preserved with apple cider vinegar received high scores for color and texture, while the score for the aroma of the sample was significantly lower (3.25 ± 0.87). Few panelists described this sample as too acidic. Sea fennel preserved in wine vinegar, on the other hand, had the lowest values for all parameters (all below 4.00). The highest was for texture (3.92 ± 0.29), and the lowest was for the taste (3.00 ± 0.50) of the sample. The greatest differences between the samples were found for color, alcoholic (4.17) > apple (4.08) > wine vinegar (3.58), which was probably due to the coloration of the pickle juice samples. The color of the wine vinegar affected the color of the fennel leaves, so the split parts of the leaves and shoots were colored red, which was rated negatively by the panelists. The intense aroma of the wine vinegar was also described as a negative characteristic (off-flavor) of the sea fennel samples.

This study also included a test of consumer preference using the hedonic scale rating (Figure 4). The results show that only the sea fennel preserved in alcoholic vinegar received the highest rating (extremely like), while the sample in red wine vinegar was the only one with a low rating (dislike very much). In addition, the sample in alcoholic vinegar was evaluated mostly via high grades on a hedonic scale (like extremely, like very much and like moderately). These results are in accordance with the descriptive sensory tests of the preserved sea fennel samples.

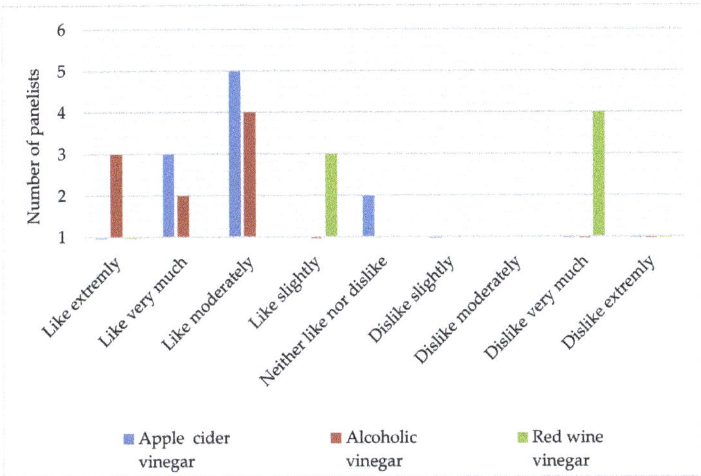


**Figure 4.** Hedonic scale rating of sea fennel preserved in pickle juices prepared using different vinegars.

## 4. Conclusions

Sea fennel is one of the most widespread halophytes on the Adriatic coast, with a valuable nutritional composition and a number of health-promoting ingredients. Its leaves, harvested in the summer months, have been used for centuries in the culinary tradition and in the food industry due to their special sensory properties (taste, smell and color) and are used in various foods raw, cooked or preserved. However, in the Croatian culinary tradition they are mainly prepared as fresh-packed pickles and used in salads. This study investigated the chemical, color, texture and organoleptic properties of sea fennel preserved with different vinegars, as well as overall consumer acceptance. The results showed that alcoholic vinegar received the highest score in the sensory evaluation, while the sample preserved with wine vinegar, which is mainly used in the traditional preservation of sea fennel in the Croatian coastal region, received the lowest scores and the negative perception of consumers, mainly due to the strong aromatic vinegar notes and negative attributes for color and firmness.

**Author Contributions:** Conceptualization, B.U. and I.G.M.; methodology, S.R., M.R., B.U. and I.G.M.; software, S.R., M.R. and A.R.; validation, S.R., P.B., M.R. and A.R.; formal analysis, S.R., P.B., M.R., A.R., M.Š., T.N.R., B.U. and I.G.M.; investigation, S.R., P.B., M.R., A.R. and I.G.M.; resources, S.R., P.B., M.R., A.R., M.Š., T.N.R., B.U. and I.G.M.; data curation, S.R., P.B., M.R., A.R., M.Š., T.N.R., B.U. and I.G.M.; writing—original draft preparation, S.R. and I.G.M.; writing—review and editing, P.B., M.R., A.R., M.Š., T.N.R. and B.U.; visualization, S.R., M.R., A.R., M.Š. and T.N.R.; supervision B.U. and I.G.M.; project administration, S.R., B.U. and I.G.M.; funding acquisition, I.G.M. All authors have read and agreed to the published version of the manuscript.

**Funding:** This work is part of the PRIMA program supported by the European Union. Project title: "Innovative sustainable organic sea fennel (*Crithmum maritimum* L.)-based cropping systems to boost agrobiodiversity, profitability, circularity, and resilience to climate changes in Mediterranean small farms" (acronym: SEAFENNEL4MED) (https://seafennel4med.com/, accessed on 17 October 2023).

**Institutional Review Board Statement:** Not applicable.

**Informed Consent Statement:** Informed consent was obtained from all subjects involved in the study.

**Data Availability Statement:** The data presented in this study are available on request from the corresponding author.

**Acknowledgments:** The authors are also thankful for the scientific-research equipment financed via the EU grant "Functional integration of the University of Split, PMFST, PFST and KTFST through the development of the scientific and research infrastructure" (KK.01.1.1.02.0018).

**Conflicts of Interest:** The authors declare no conflict of interest.

## References

1. FAO-AGS, Pickles—Instructional Manual, Nutritional Improvement of Food Products. 2007. Available online: https://www.fao.org/3/a-au116e.pdf (accessed on 12 September 2023).
2. Montaño, A.; Sánchez, A.H.; Beato, V.M.; López-López, A.; de Castro, A. Pickling. In *Encyclopedia of Food and Health*; Caballero, B., Finglas, P.M., Toldrá, F., Eds.; Academic Press: Cambridge, MA, USA, 2016; pp. 369–374. [CrossRef]
3. Ghnimi, S.; Guizani, N. Vegetable fermentation and pickling. In *Handbook of Vegetables and Vegetable Processing*, 2nd ed.; Siddiq, M., Uebersax, M.A., Eds.; John Wiley & Sons Ltd.: Hoboken, NJ, USA, 2018; pp. 407–427. [CrossRef]
4. Pérez-Díaz, I.M.; Breidt, F.; Buescher, R.W.; Arroyo-López, F.N.; Jiménez-Díaz, R.; Garrido-Fernández, A.; Bautista-Gallego, J.; Yoon, S.S.; Johanningsmeier, S.D. Fermented and Acidified Vegetables. In *Compendium of Methods for the Microbiological Examination of Foods*; American Journal of Public Health: Washington, DC, USA, 2015. [CrossRef]
5. Aljahani, A.H. Microbiological and Physicochemical Quality of Vegetable Pickles. *J. Saudi Soc. Agric. Sci.* **2020**, *19*, 415–421. [CrossRef]
6. Man, C.M.D. Technological functions of salt in food products. In *Woodhead Publishing Series in Food Science, Technology and Nutrition, Reducing Salt in Foods*; Kilcast, D., Angus, F., Eds.; Woodhead Publishing: Sawston, UK, 2007; pp. 157–173. [CrossRef]
7. Atia, A.; Barhoumi, Z.; Mokded, R.; Abdelly, C.; Smaoui, A. Enviromental eco-physiology and economical potential of the halophtye *Crithmum maritimum* L. (Apiaceae). *J. Med. Plants Res.* **2011**, *5*, 3564–3571.
8. Renna, M. Reviewing the prospects of sea fennel (*Crithmum maritimum* L.) as emerging vegetable crop. *Plants* **2018**, *7*, 92. [CrossRef] [PubMed]
9. Kraouia, M.; Nartea, A.; Maoloni, A.; Osimani, A.; Garofalo, C.; Fanesi, B.; Ismaiel, L.; Aquilanti, L.; Pacetti, D. Sea Fennel (*Crithmum maritimum* L.) as an Emerging Crop for the Manufacturing of Innovative Foods and Nutraceuticals. *Molecules* **2023**, *28*, 4741. [CrossRef]
10. Franke, W. Vitamin C in sea fennel (*Crithmum maritimum*), an edible wild plant. *Econ. Bot.* **1982**, *36*, 163–165. [CrossRef]
11. Generalić Mekinić, I.; Blažević, I.; Mudnić, I.; Burčul, F.; Grga, M.; Skroza, D.; Katalinić, V. Sea fennel (*Crithmum maritimum* L.): Phytochemical profile, antioxidative, cholinesterase inhibitory and vasodilatory activity. *J. Food Sci. Tech.* **2016**, *53*, 3104–3112. [CrossRef]
12. Generalić Mekinić, I.; Šimat, V.; Ljubenkov, I.; Burčul, F.; Grga, M.; Mihajlovski, M.; Lončar, R.; Katalinić, V.; Skroza, D. Influence of the vegetation period on sea fennel phenolic composition, antioxidant and anticholinesterase activities. *Ind. Crops. Prod.* **2018**, *124*, 947–953. [CrossRef]
13. Sarrou, E.; Siomos, A.S.; Riccadona, S.; Aktsoglou, D.C.; Tsouvaltzis, P.; Angeli, A.; Franceschi, P.; Chatzopoulou, P.; Vrhovsek, U.; Martens, S. Improvement of sea fennel (*Crithmum maritimum* L.) nutritional value through iodine biofortification in a hydroponic floating system. *Food Chem.* **2019**, *296*, 150–159. [CrossRef] [PubMed]
14. Sánchez-Faure, A.; Calvo, M.M.; Pérez-Jiménez, J.; Martín-Diana, A.B.; Rico, D.; Montero, M.P.; Gómez-Guillén, M.C.; López-Caballero, M.E.; Martínez-Alvarez, O. Exploring the potential of common iceplant, seaside arrowgrass and sea fennel as edible halophytic plants. *Food Res. Int.* **2020**, *137*, 109613. [CrossRef] [PubMed]
15. Maoloni, A.; Milanović, V.; Osimani, A.; Cardinali, F.; Garofalo, C.; Belleggia, L.; Foligni, R.; Mannozzi, C.; Mozzon, M.; Cirlini, M.; et al. Exploitation of sea fennel (*Crithmum maritimum* L.) for manufacturing of novel high-value fermented preserves. *Food Bioprod. Process.* **2021**, *127*, 174–197. [CrossRef]
16. Amoruso, F.; Signore, A.; Gómez, P.A.; Martínez-Ballesta, M.d.C.; Giménez, A.; Franco, J.A.; Fernández, J.A.; Egea-Gilabert, C. Effect of Saline-Nutrient Solution on Yield, Quality, and Shelf-Life of Sea Fennel (*Crithmum maritimum* L.) plants. *Horticulturae* **2022**, *8*, 127. [CrossRef]
17. Pedreiro, S.; Figueirinha, A.; Cavaleiro, C.; Cardoso, O.; Donato, M.M.; Salgueiro, L.; Ramos, F. Exploiting the *Crithmum maritimum* L. Aqueous Extracts and Essential Oil as Potential Preservatives in Food, Feed, Pharmaceutical and Cosmetic Industries. *Antioxidants* **2023**, *12*, 252. [CrossRef] [PubMed]
18. Renna, M.; Gonnella, M. The use of the sea fennel as a new spice-colorant in culinary preparations. *Inter. J. Gastron. Food Sci.* **2012**, *1*, 111–115. [CrossRef]
19. Renna, M.; Gonnella, M.; Caretto, S.; Mita, G.; Serio, F. Sea fennel (*Crithmum maritimum* L.): From underutilized crop to new dried product for food use. *Genet. Resour. Crop. Evol.* **2017**, *64*, 205–216. [CrossRef]
20. AOAC. Official Method 942.15. In *Official Methods of Analysis of AOAC INTERNATIONAL*, 22nd ed.; Latimer, G.W., Ed.; AOAC Publishing: New York, NY, USA, 2023. [CrossRef]
21. AOAC. Official Method 941.18. In *Official Methods of Analysis of AOAC INTERNATIONAL*, 22nd ed.; Latimer, G.W., Ed.; AOAC Publishing: New York, NY, USA, 2023. [CrossRef]
22. Montesano, F.F.; Gattullo, C.E.; Parente, A.; Terzano, R.; Renna, M. Cultivation of Potted Sea Fennel, an Emerging Mediterranean Halophyte, Using a Renewable Seaweed-Based Material as a Peat Substitute. *Agriculture* **2018**, *8*, 96. [CrossRef]

23. Maoloni, A.; Cardinali, F.; Milanović, V.; Osimani, A.; Verdenelli, M.C.; Coman, M.M.; Aquilanti, L. Exploratory Study for Probiotic Enrichment of a Sea Fennel (*Crithmum maritimum* L.) Preserve in Brine. *Foods* **2022**, *11*, 2219. [CrossRef]
24. Garden-Robinson, J. *Food Preservation: Making Pickled Products (FN189, Revised Aug. 2019)*; North Dakota State University: Fargo, ND, USA, 2019. Available online: https://www.ndsu.edu/agriculture/extension/publications/food-preservation-making-pickled-products (accessed on 2 September 2023).
25. Politeo, O.; Popović, M.; Veršić Bratinčević, M.; Koceić, P.; Ninčević Runjić, T.; Mekinić, I.G. Conventional vs. Microwave-Assisted Hydrodistillation: Influence on the Chemistry of Sea Fennel Essential Oil and Its By-Products. *Plants* **2023**, *12*, 1466. [CrossRef] [PubMed]
26. Alves-Silva, J.M.; Guerra, I.; José Gonçalves, M.; Cavaleiro, C.; Teresa Cruz, M.; Figueirinha, A.; Salgueiro, L. Chemical composition of Crithmum maritimum L. essential oil and hydrodistillation residual water by GC-MS and HPLC-DAD-MS/MS, and their biological activities. *Ind. Crops Prod.* **2020**, *149*, 112329. [CrossRef]
27. Brenes, M.; Garcia, P.; Garrido, A. Influence of salts and pH on the firmness of olives in acid conditions. *J. Food Qual.* **1994**, *17*, 335–346. [CrossRef]
28. Alemu, T. Texture Profile and Design of Food Product. *J. Agric. Horti. Res.* **2023**, *6*, 272–281.
29. MacFeeters, R.F.; Senter, M.M.; Fleming, H.P. Softening Effects of Monovalent Cations in Acidified Cucumber Mesocarp Tissue. *J. Food Sci.* **1989**, *54*, 366–370. [CrossRef]

foods

MDPI

Article

# Cooking Methods for Preserving Isothiocyanates and Reducing Goitrin in *Brassica* Vegetables

**Thanaporn Panduang [1], Pakkapong Phucharoenrak [2], Weeraya Karnpanit [3] and Dunyaporn Trachootham [2,*]**

[1] Master of Science Program in Toxicology and Nutrition for Food Safety, Institute of Nutrition, Mahidol University, Nakhon Pathom 73170, Thailand; thanaporn.pad@student.mahidol.ac.th
[2] Institute of Nutrition, Mahidol University, Nakhon Pathom 73170, Thailand; pakkapong.phu@mahidol.ac.th
[3] School of Science, Western Sydney University, Locked Bag 1797, Penrith, NSW 2751, Australia; w.karnpanit@westernsydney.edu.au
* Correspondence: dunyaporn.tra@mahidol.ac.th or dunyaporn.tra@mahidol.edu; Tel.: +66-02-800-2380 (ext. 326)

**Abstract:** Glucosinolates in *Brassica* vegetables can be hydrolyzed into various products, e.g., chemopreventive agents, isothiocyanates (ITCs) and anti-thyroid substance, goitrin. Cooking can reduce goitrin but destroy isothiocyanates. This study aimed to optimize cooking conditions for reducing goitrin while preserving isothiocyanates in *Brassica* vegetables. Cabbage and Chinese kale samples were divided evenly into raw, blanched, steamed, and water-based stir-fried samples. Cooking temperature and time were varied at 60, 80, or 100 °C for 2, 4, or 6 min. The levels of goitrin, benzyl isothiocyanate (BITC), and sulforaphane (SFN) were measured using LC-MS/MS. Response surface model (RSM) was used to identify the optimal cooking conditions to reduce goitrin but preserve ITCs. Results showed that goitrin content in cabbage depended on the cooking methods, temperature, and time, while that of Chinese kale only depended on the methods. In contrast, the concentrations of SFN in cabbage and BITC in kale depended on the cooking temperature and time but not methods. Based on RSM analysis, the suggested household cooking methods for preserving isothiocyanates and reducing goitrin are steaming cabbage at 80–100 °C for 4 min and stir-frying Chinese kale at 60–100 °C for 2 min. Such methods may preserve the bioactive compounds while reducing food hazards.

**Keywords:** cooking; *Brassica* vegetables; goitrogen; goitrin; isothiocyanates; benzyl isothiocyanate; sulforaphane; LC-MS/MS

## 1. Introduction

*Brassica* vegetables, or cruciferous vegetables, e.g., cabbage, Chinese kale, cauliflower, brussels sprouts, watercress, red cabbage, bok choy, and broccoli have two main characteristics: a bitter taste and a sulfurous smell [1]. Such unique properties are derived from a family of phytochemicals, isothiocyanates (ITCs). ITCs are metabolic products obtained from the hydrolysis of various types of glucosinolates. Once the vegetables are chopped, cooked with heat, or chewed in the mouth, the myrosinase enzyme will be activated, catalyzing the generation of ITCs via a hydrolysis reaction [2]. ITCs including beta-phenethyl isothiocyanate (PEITC), sulforaphane, and benzyl isothiocyanate (BITC) were well documented for cancer chemopreventive effects [3,4]. Several previous studies have found that ITCs can inhibit the development of cancer in a variety of organs in animal models, including oral, prostate, breast, colon, lung, gastric, and liver [5–7]. Clinical trials showed that ITCs can promote detoxification and urinary excretion of tobacco-derived and grilled meat-derived carcinogens [8–10]. Case–control and cohort studies also found that consumption of *Brassica* vegetables is associated with decreased risk of several types of cancer [11–13]. However, these vegetables also contain 2-hydroxy-3-butenyl glucosinolate (progoitrin) and indole glucosinolate, which can be converted to goitrin and thiocyanate, which act as goitrogens in animal models. The conversion can occur either via spontaneous cyclization or activated by the myrosinase enzyme [14,15]. These goitrogens may block iodide transport to the thyroid gland and inhibit the incorporation of iodine into thyroglobulin, impairing the synthesis of the active thyroid hormone. To compensate for the deficiency, thyroid-stimulating hormone (TSH) has been increasingly produced in the pituitary gland to stimulate more thyroid hormone production [16]. The increased TSH secretion was found to be associated with an increased risk of thyroid cancer [16]. A previous meta-analysis reported that the consumption of *Brassica* vegetables may be associated with an increased risk of thyroid cancer, especially in areas where iodine deficiency is commonly found [17,18]. A previous review suggested that the level of thiocyanate from human consumption of *Brassica* vegetables was extremely low with negligible adverse effects [14,19]. In contrast, the level of goitrin derived from the vegetable can be detectable and still be able to decrease iodine uptake by the thyroid gland [19]. Therefore, goitrin is still a goitrogenic substance of concern. Previous studies have shown that uncooked Chinese kale and cabbage contain high amounts of progoitrin, compared to other types of *Brassica* vegetables [19,20]. Since progoitrin can be converted to goitrin and cabbage and Chinese kale are popularly consumed worldwide, strategies to reduce goitrin in Chinese kale and cabbage are crucial.

Some food preparation methods such as washing, soaking, and boiling were shown to help reduce goitrogens in food by destroying myrosinase and preventing the liberation of goitrin from progoitrin [14,15,20]. Unfortunately, heat can also destroy glucosinolate and isothiocyanates [21–23]. To date, no studies have been conducted to identify which cooking condition is the best for reducing goitrin, while preserving isothiocyanates in *Brassica* vegetables. A previous study showed that moist heat methods can reduce progoitrin with blanching being the best one, compared with boiling and steaming [22]. However, cooking with high heat for a long duration such as boiling can destroy ITCs [22,23]. Therefore, we hypothesize that the best method to reduce goitrin while preserving ITCs should be moist heat with short duration. This study aimed to optimize household cooking conditions to preserve functional substances like ITCs while also reducing anti-thyroid substances like goitrin. The effect of cooking methods via moist heat including steaming, blanching, and water-based stir-frying, cooking time, and temperature were studied. Among the isothiocyanates, sulforaphane and BITC have been found in significant amounts in cabbage and Chinese kale, respectively. Furthermore, these compounds' chemopreventive effects against various types of cancer are well documented [24,25]. Thus, in this paper, we explore the cooking conditions to reduce goitrin while preserving SFN and BITC.

The insight from this work may be used for the recommendation of household cooking methods to help reduce hazardous chemicals and preserve the beneficial substances in *Brassica* vegetables.

## 2. Materials and Methods

### 2.1. Materials, Reagents, and Standards

Cabbage (*Brassica oleracea* L. var. *capitata* L.) and Chinese kale (*Brassica alboglabra* L.H. Bailey) were purchased from Salaya market, Phutthamonthon District, Nakhon Pathom Province, Thailand. Benzyl isothiocyanate standard (Purity: 98%) and (R)-sulforaphane standard (Purity: 98%) were products of Toronto Research Chemicals (Toronto, ON, Canada), which were purchased from Phoenix Scientific Co., Ltd. (Bangkok, Thailand). Goitrin standard (Purity: 98%) was purchased from Wuhan ChemFaces Biochemical Co., Ltd. (Wuhan, China). Ammonia (2 M) in methanol was a product of AcroSeal™ (Thermo Scientific Chemicals, Somerville, NJ, USA). Formic acid, dimethyl sulfoxide (DMSO), and liquid chromatography-mass spectrometry (LC-MS) grade of methanol and acetonitrile (ACN) were products of Merck Millipore (Merck KGaA, Darmstadt, Germany. Hexane was a product of J.T. Baker Chemical (Phillipsburg, NJ, USA). LC-MS grade of acetonitrile was a product of RCI Labscan Ltd. (Bangkok, Thailand). Type I grade water was obtained freshly from a Purelab Quest UV water purifying system (Elga LabWater, High Wycombe, UK).

### 2.2. Cooking Conditions and Sample Collections

Sample preparation was performed according to previous studies [20,22] with some modifications. Cabbage and Chinese kale were purchased from the Salaya market, Nakhon Pathom, Thailand. Each vegetable was randomly selected. The cabbage was rinsed with clean water and chopped into four equal parts, as shown in Figure 1. Each part was used for uncooked, blanching, steaming, and stir-frying samples. For Chinese kale, both leaf and stalk were used, rinsed with clean water, and cut into four equal parts for uncooked, blanching, steaming, and stir-frying samples (Figure 1).

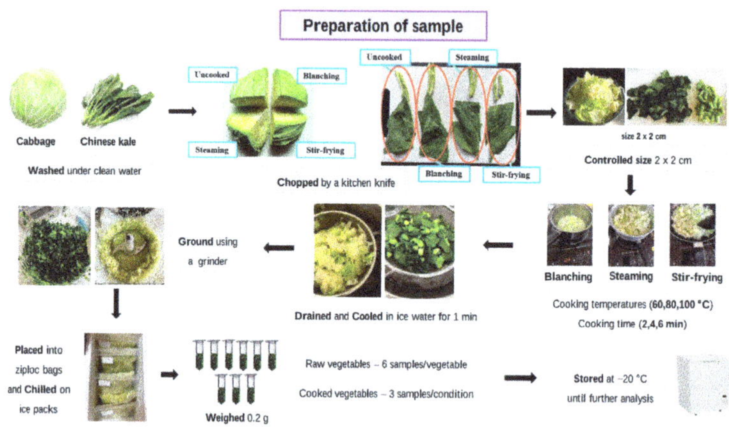


**Figure 1.** Preparation of cabbage and Chinese kale samples using various cooking conditions.

Each divided part of the vegetable was chopped into 2 × 2 cm pieces before cooking. The samples were processed using different cooking methods, i.e., blanching, steaming, and stir-frying under specific cooking temperatures (60, 80, and 100 °C) and cooking times (2, 4, and 6 min) conditions. All cooking conditions were performed using the same induction cooker (Touch type Induction Cooker, SEAGULL touch set, SEAGULL®, Bangkok, Thailand) and the same saucepan and pot. The detailed cooking condition for each cooking method is shown in Table 1.

**Table 1.** Cooking water, cooking temperature, and cooking time under different cooking methods for cabbage and Chinese kale.

| | Cooking Methods | Weight (g) | Water Volume (mL) | Cooking Time (min) | | | Cooking Temp. (°C) | | |
|---|---|---|---|---|---|---|---|---|---|
| Cabbage | Blanching | 150 | 400 | 2 | 4 | 6 | 60 | 80 | 100 |
| | Steaming | 150 | 400 | 2 | 4 | 6 | 60 | 80 | 100 |
| | Stir-frying | 150 | 400 | 2 | 4 | 6 | 60 | 80 | 100 |
| Chinese kale | Blanching | 150 | 400 | 2 | 4 | 6 | 60 | 80 | 100 |
| | Steaming | 150 | 400 | 2 | 4 | 6 | 60 | 80 | 100 |
| | Stir-frying | 150 | 400 | 2 | 4 | 6 | 60 | 80 | 100 |

Uncooked (raw) vegetables were used as control samples. After cooking, the cooked vegetables were drained and cooled in ice water (with 1000 g ice and 2000 g water) for 1 min to stop the heat-induced hydrolysis reaction. Each cooked or uncooked sample was ground separately using a grinder (OTTO®, Bangkok, Thailand). After that, each sample was placed immediately into Ziploc bags and chilled on ice packs in polystyrene containers for transport from the kitchen to the lab. Three samples (0.2 g each) of each cooking condition were randomly collected into a 2 mL micro-centrifuge tube. For the uncooked vegetables, 6 samples were collected. All samples were stored at −20 °C in the freezer until further analysis. Each sample has been prepared separately before analysis. Therefore, the mean ± standard deviation displayed in the table is derived from n = 6 for uncooked vegetables and n = 3 for each cooking condition.

### 2.3. Sample Preparation for Determination of Goitrin, BITC, and Sulforaphane in Cooked and Uncooked Vegetables

The sample preparation was performed according to previous studies [9,26,27] with some modifications. Hydrolysis and hexane extraction were performed similarly for all substances. Three randomly selected samples of cooked vegetables and six randomly selected samples of uncooked vegetables were collected. Frozen vegetable samples (0.2 g) were mixed with 1 mL of PBS (1:5) and mixed on a high-speed vortex for 30 s. To stimulate hydrolysis and release goitrins, BITC, and SFN, the mixtures were incubated for three hours in a water bath at 37 °C. Then, 300 μL hexane was added to each sample using a positive displacement pipette (Mettler-Toledo Rainin LLC, Oakland, CA, USA), and mixed on a high-speed vortex for 30 s. Subsequently, 2000 g samples were centrifuged at 25 °C for 3 min, and 200 μL of the clear supernatants of samples were collected into a new 2 mL microcentrifuge tube. Hexane extraction was performed twice and combined with the first extract. Ammonia derivatization was performed before the analysis of goitrin and BITC. Briefly, 400 μL of 2 M ammonia in methanol was added to the hexane extract and mixed for 30 s on a vortex. Then, the samples were incubated at room temperature on a shaker for 3 h followed by 4 °C overnight. The samples were dried in a SpeedVac concentrator (Labconco, Kansas City, MO, USA) at 55 °C for 40 min to remove the solvents and stored at −20 °C until further analysis. The dried samples were dissolved with 250 μL of 60% ACN (acetonitrile: water in a 3:2 ratio) and mixed on vortex for 1 min. Then, the samples were filtered through a 0.2 μm nylon filter before LC-MS/MS analysis. For the analysis of SFN, a non-derivatized method was used. For SFN, samples were prepared without derivatization. Briefly, the hexane extract was dried in a SpeedVac concentrator at 55 °C for 40 min to remove the solvents and stored at −20 °C until further analysis. The dried samples were dissolved with 200 μL of 60% ACN (acetonitrile: water in a 3:2 ratio) and mixed on vortex for 1 min. Then, the samples were filtered through a 0.2 μm of nylon filter for LC-MS/MS analysis.

### 2.4. Preparation of Standard Concentrations of Goitrin-$NH_3$, BITC-$NH_3$, and SFN for Standard Calibration Curves

Preparation of the standard of goitrin-$NH_3$, BITC-$NH_3$, and SFN for calibration curves was performed according to a previous study [9,26,27] with some modifications. Calibration curves of BITC-$NH_3$, goitrin-$NH_3$, and SFN were used for calculation. The concentrations used to generate the calibration curve for BITC included 0.1, 0.5, 1, 5, 10, and 50 ng/mL. Those of goitrin-$NH_3$ and SFN included 0.1, 0.5, 1, 5, 10, 50, 100, 250, and 500 ng/mL. The standard stock solution of goitrin (4 mg/mL) was prepared by dissolving 20 mg of goitrin in 5 mL of DMSO. Then, the solution was further diluted with ACN to the working concentrations. The standard stock solution of BITC (1 mg/mL) was prepared by dissolving 1 μL of BITC standard in 1.125 mL of methanol. The solution was further diluted with methanol to the working concentrations. All standard solutions of goitrin and BITC had an equal final volume of 200 μL. For each standard of BITC and SFN, 140 μL of 2 M ammonia in methanol was added to derivatize and mixed for 30 s on a vortex. The standard mixtures were incubated at room temperature on a shaker for 3 h, followed by 4 °C overnight. The mixtures were dried in a SpeedVac concentrator at 55 °C for 40 min to remove the solvents. The dried samples were stored at −20 °C. On the day of analysis, the dried samples were dissolved with 200 μL of 60% ACN (ACN: water in a 3:2 ratio) and mixed on vortex for 1 min. Then, the standard solution at each concentration of goitrin-$NH_3$ or BITC-$NH_3$ was placed in a glass vial with an insert for LC-MS/MS analysis.

The standard stock solution of SFN (10 mg/mL) was prepared by dissolving 10 mg of SFN in 1 mL of methanol. The solution was further diluted with methanol to the working concentrations at the volume of 200 μL. Then, the standard was put in a glass vial with an insert for the LC-MS/MS analysis.

### 2.5. Determination of the Goitrin, BITC, and SFN in Samples Using LC-MS/MS Assay

LC-MS/MS analysis was performed according to a previous study [9] with some modifications. Measurement of goitrin and BITC was performed simultaneously, while measurement of SFN was performed in different injections but in the same batch of run as those of goitrin and BITC. The extracts were analyzed via the Ultimate 3000 Ultra High-performance liquid chromatography (UHPLC) system (Thermo Scientific, Waltham, MA, USA) using Hypersil GOLD$^{TM}$ C18 column with a 100 × 2.1 mm, 1.9 m particle size, at a flow rate of 0.3 mL/min. The injection volume was 5 μL. The mobile phase consisted of 100% ACN (A) and 5 mM formic acid (B). For the determination of goitrin-$NH_3$ and BITC-$NH_3$, a gradient running program of 5.00 min was set as follows: 0–1.20 min, 50–50% B; 1.20–2.00 min, 50–75% B; 2.00–3.50 min, 75–75% B; 3.50–5.00 min, and 75–50% B. The retention time of BITC-$NH_3$ and goitrin-$NH_3$ were 1.246 and 1.019 min, respectively. For SFN analysis, an isocratic running program at 50:50 of the mobile phase A: B was run for 3.50 min. The retention time of SFN was 1.179 min.

Tandem mass spectrometric analysis was conducted using positive ion mode (ESI+) combined with an electron spray ionization probe at a spray voltage of 3500 V under the $N_2$ sheath. Sheath gas had an arbitrary unit (Arb) value of 50 while auxiliary gas had a value of 10.0. Temperatures for the vaporizing and ion transfer tubes (ITT) were 325 °C and 350 °C, respectively. For simultaneous analysis of multiple masses, the mass spectrometer analysis was conducted via the selected reaction monitoring (SRM) mode. The mass-to-charge ratio of goitrin-$NH_3$ precursor, quantified, and confirmation product ions were 146.85 $m/z$, 126.49 $m/z$ (transition collision energy at 5.25 V), and 105.97 $m/z$ (transition collision energy at 5.34 V), respectively. The mass-to-charge ratio of BITC-$NH_3$ precursor, quantified, and confirmation product ions were 167 $m/z$, 91.125 $m/z$ (transition collision energy at 17.43 V), and 65.042 $m/z$ (transition collision energy at 39.5 V), respectively. The mass-to-charge ratio of SFN precursor, quantified, and confirmation product ions were 178 $m/z$, 144.07 $m/z$ (transition collision energy at 10.48 V), and 72.04 $m/z$ (transition collision energy at 26.57 V), respectively. Chromeleon™ Chromatography Data System (CDS) was used to acquire and analyze the LC-MS/MS data.

### 2.6. Method Validation

The validation of analytical methods was conducted according to the International Council for Harmonisation of Technical Requirements for Pharmaceuticals for Human Use (ICH) guidelines [28] and guidelines for standard method performance requirements of The Association of Official Analytical Chemists (AOAC) 2016 [29]. Performance characteristics of the method validation of the determination of goitrin, BITC, and SFN included linearity, precision, and accuracy. Calibration curves were created for linearity by plotting the peak area versus calibration standard concentration. The acceptable coefficient of determination ($R^2$) is 0.995 or higher [30]. To estimate the inter-day precision, 10 replicates of vegetable samples spiked with goitrin, BITC, and SFN at a concentration of 1 ng/mL were analyzed on 10 different days. The percent of relative standard deviation (% RSD) was calculated. The acceptance criteria of % RSD are <21% at a concentration of 10 ng/mL and <30% at a concentration of 1 ng/mL [29]. The accuracy of the proposed method was examined as percentages of recovery. The standard solutions of goitrin, BITC, and SFN (10 ng/mL) were spiked to vegetable sample solutions in triplicate. The concentrations of the analytes were determined for the spiked samples. The acceptable recovery percentages were 60–115% at a concentration of 10 ng/mL [29].

### 2.7. Data Analysis

Graphing and statistical analyses were performed using GraphPad Prism version 9.5.1 (GraphPad Software, Inc., Boston, MA, USA) and Minitab Statistical Software version 21.1.0 (Minitab, Inc., State College, PA, USA). All data were shown as the mean ± standard deviation. The concentrations of goitrin, SFN, and BITC in uncooked and cooked vegetables were calculated using a linear regression equation generated from the respective standard curves. The dilution factor was applied when it was needed. Comparison of the goitrin, BITC, and SFN levels between cooked vegetables via different cooking methods under controlled different cooking temperatures and time (blanching, steaming, and stir-frying) and uncooked vegetables were analyzed via one-way analysis of variance (ANOVA), followed by Tukey post hoc multiple comparison test.

Response surface methodology (RSM) was performed using Design-Expert version 13 (Stat-Ease, Inc., Minneapolis, MN, USA). The RSM model was analyzed using the central composite design (CCD). Our study used RSM with CCD to study the effects of three independent variables (cooking time, cooking temperature, and cooking methods) on the dependent variables (the yield of goitrin, BITC, and SFN). The program could help to create the predicted model of relationship among variables and the three-dimension graphs of RSM were generated. Optimization for the best cooking conditions was performed by weighting to minimize goitrin (+++++) as the priority for both cabbage and Chinese kale. For the second priority, we set different weights of BITC and SFN for the two kinds of vegetables. For cabbage, we assigned a higher weight to maximize SFN (++++) than that of BITC (+) because of a slightly higher amount of SFN than BITC in raw cabbage. For Chinese kale, more weight was given to maximize BITC (++++) than SFN (+) due to a much higher amount of BITC than SFN in raw Chinese kale.

## 3. Results and Discussion

### 3.1. *Chromatograms and Mass Spectra of Goitrin (GN), Benzyl Isothiocyanate (BITC), and Sulforaphane (SFN)*

The mass-to-charge ratio ($m/z$) of the GN-$NH_3$ precursor and quantified product mass were 146.85 and 126.49, respectively. A collision energy of 5.25 V was used for the transition. The confirmation product mass for GN-$NH_3$ was $m/z$ 105.97, with a collision energy of 5.34 V for the transition. The retention time of GN-$NH_3$ was 1.019 min. The Chromatogram and mass spectra of GN-$NH_3$ are shown in Figure 2A.

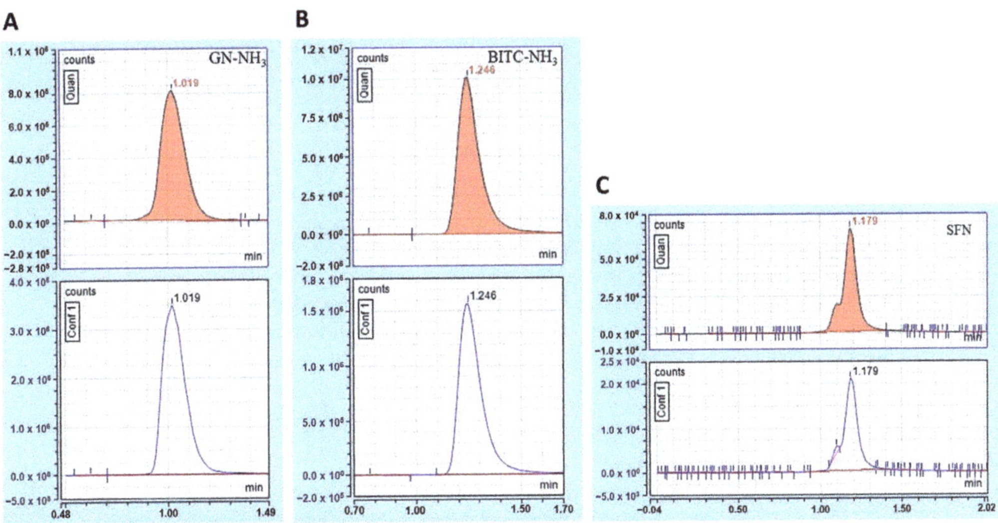


**Figure 2.** Reference compounds analyzed via LC-MS/MS. Chromatogram of GN-$NH_3$ (**A**), BITC-$NH_3$ (**B**), and SFN (**C**), for quantified (upper) and confirmation product masses (lower).

The mass-to-charge ratio ($m/z$) of the BITC-$NH_3$ precursor and quantified product mass were 167 and 91.125, respectively. A collision energy of 17.43 V was used for the transition. The confirmation product mass for BITC-$NH_3$ was $m/z$ 65.042 with a collision energy of 39.5 V for the transition. The retention time of BITC was 1.246 min. The Chromatogram and mass spectra of BITC-$NH_3$ are shown in Figure 2B.

The mass-to-charge ratio ($m/z$) of the SFN precursor and quantified product mass were 178 and 114.07, respectively. A collision energy of 10.48 V was used for the transition. The confirmation product mass for SFN was $m/z$ 72.04 with a collision energy of 26.57 V for the transition. The retention time of SFN was 1.179 min. The Chromatogram and mass spectra of SFN are shown in Figure 2C. The shouldered peak was found in both standards and samples. Sulforaphane has two isomers: (R)- and (S)-. (R)-Sulforaphane is the major isomer with R configuration at the sulfinyl group. The S-form is an enantiomer of (R)- form with (S)-configuration. The main peak is likely the (R)-form, while the shoulder is likely the S-form. For quantitation, we used the whole-shouldered peak.

Calibration curves of the three compounds showed good linearity with an average $R^2$ of 0.9997 for GN-$NH_3$, 0.9993 for BITC-$NH_3$, and 0.9998 for SFN (Figure S1). Inter-day precision as %RSD for the analyses of GN-$NH_3$, BITC-$NH_3$, and SFN was 2.28–9.56%, 1.00–2.72%, and 2.44–2.48%, respectively. The mean recovery for the analyses of GN-$NH_3$, BITC-$NH_3$, and SFN was 85–88%, 82–93%, and 90–94%, respectively. Satisfactory linearity, precision, and recovery were noted when compared to guidelines for standard method performance requirements of AOAC (2016) and ORA Laboratory Manual Volume II of the U.S. FDA [29,30]. The validation data are shown in Table 2.

**Table 2.** Validation data for LC-MS/MS analysis of goitrin, BITC, and SFN in cabbage and Chinese kale.

| Analyte | $R^2$ | Mean Recovery (%) | | | Inter-Day Precision (%RSD) | | | |
|---|---|---|---|---|---|---|---|---|
| | | Cabbage | Chinese Kale | Expected Criteria [a] | Vegetable Samples | Expected Criteria [b] | Standard 1 ng/mL | Expected Criteria [c] |
| Goitrin | 0.9997 | 88 | 85 | 60–115 | 2.28 | 21 | 9.56 | 30 |
| BITC | 0.9993 | 93 | 82 | 60–115 | 1.00 | 21 | 2.72 | 30 |
| SFN | 0.9998 | 94 | 90 | 60–115 | 2.44 | 21 | 2.48 | 30 |

[a] Expected recovery at 10 ng/mL from AOAC Peer-Verified Methods Program (2016). [b] Expected precision (repeatability) at 10 ng/mL from AOAC Peer-Verified Methods Program (2016). [c] Expected precision (repeatability) at 1 ng/mL from AOAC Peer-Verified Methods Program (2016).

For measurement of goitrin and BITC, ammonia derivatization was performed with the hexane extract (not with the vegetable samples) before LC/MS-MS analysis. Therefore, the chance of the occurrence of decomposition of glucosinolates by a strong base in this step is low. Or even if it happens, it would happen with all samples in a similar fashion since all samples were treated with ammonia in exactly similar conditions. Therefore, ammonia derivatization should not interfere with the result. We performed a pretest to measure goitrin, BITC, and SFN with or without ammonia derivatization. The result of the pretest showed that goitrin and BITC also can be measured without ammonia derivatization. However, without ammonia derivatization, their ionization patterns in mass spectrometric analysis are not as stable as the ones with ammonia derivatization. This is the reason why we used ammonia derivatization for goitrin and BITC. Unlike BITC and goitrin, the non-derivatized method provided better linearity for SFN than the derivatized method.

### 3.2. Effect of Different Cooking Methods on the Content of Goitrin, BITC, and SFN in Cabbage and Chinese Kale

The yield of goitrin, BITC, and SFN in cabbage after being cooked in different conditions are shown in Table 3 and Figure S2. Uncooked cabbage contained a total of 251.50 ± 59.57 ng/mL of goitrin. A significant decrease in goitrin ($p < 0.0001$) was observed after certain conditions of cooking, i.e., a decrease of 61–81% after blanching at 80 °C for 4–6 min and 100 °C for 2–6 min, a significant decrease of 57–87% after steaming under all tested conditions, and a significant decrease of 58–84% after stir-frying under all test conditions (Table 5). Moreover, the effects of blanching, steaming, and stir-frying in goitrin reduction are not significantly different ($p > 0.05$; Table 3). As shown in Table 5, the best cooking condition for reducing goitrin in cabbage was steaming at the temperature of 80 °C for 4 min, decreasing the percentage change by 87%. Uncooked cabbage contains 108.70 ± 1.08 ng/mL of BITC (Table 3). Blanching, steaming, and stir-frying significantly decreased ($p < 0.01$) BITC content by 6–46%, 12–41%, and 20–38%, respectively when compared with raw samples (Table 5). Based on Table 5, the best cooking condition for preserving BITC in cabbage was blanching at the temperature of 60 °C for 2 min, with the lowest percent loss of 6%. Uncooked cabbage contained 202.08 ± 60.09 ng/mL of SFN. In cabbage, almost all cooking conditions increased SFN by 1.2–16.9-fold (Table 5). Based on Table 5, the best cooking condition for preserving SFN in cabbage was blanching at 60 °C for 2 min, with a 16.9-fold increase ($p < 0.0001$).

**Table 3.** The yield of goitrin, BITC, and SFN in cabbage under different cooking conditions.

| Cooking Methods | Cooking Condition | | Goitrin (ng/mL) | BITC (ng/mL) | SFN (ng/mL) |
|---|---|---|---|---|---|
| | Cooking Time (min) | Cooking Temp. (°C) | | | |
| Uncooked | – | – | 251.5 ± 59.6 $^{a}$ | 108.7 ± 1.1 $^{a}$ | 202.8 ± 60.1 $^{f}$ |
| Blanching | 2 | 60 | 296.5 ± 128.4 $^{a}$ | 102.6 ± 1.1 $^{ab}$ | 3423.0 ± 1014.0 $^{a}$ |
| | 4 | 60 | 209.8 ± 7.6 $^{ab}$ | 88.8 ± 0.8 $^{bcdefgh}$ | 698.9 ± 162.6 $^{bcdef}$ |
| | 6 | 60 | 152.1 ± 85.1 $^{bcd}$ | 86.0 ± 1.4 $^{bcdefghi}$ | 350.5 ± 101.0 $^{def}$ |
| Blanching | 2 | 80 | 199.4 ± 40.7 $^{abc}$ | 94.8 ± 9.0 $^{abcde}$ | 853.0 ± 231.7 $^{bcd}$ |
| | 4 | 80 | 72.5 ± 13.4 $^{de}$ | 97.8 ± 10.7 $^{abc}$ | 395.5 ± 24.5 $^{cdef}$ |
| | 6 | 80 | 97.6 ± 27.5 $^{bcde}$ | 58.5 ± 0.9 $^{m}$ | 510.1 ± 46.7 $^{bcdef}$ |
| Blanching | 2 | 100 | 53.2 ± 2.4 $^{de}$ | 84.4 ± 0.4 $^{cdefghij}$ | 509.6 ± 31.9 $^{bcdef}$ |
| | 4 | 100 | 48.2 ± 5.6 $^{de}$ | 60.8 ± 20.9 $^{lm}$ | 68.0 ± 15.7 $^{f}$ |
| | 6 | 100 | 76.3 ± 8.5 $^{de}$ | 71.9 ± 0.6 $^{hijklm}$ | 104.6 ± 49.5 $^{f}$ |
| Steaming | 2 | 60 | 108.5 ± 26.9 $^{bcde}$ | 74.3 ± 0.4 $^{ghijklm}$ | 242.9 ± 59.5 $^{def}$ |
| | 4 | 60 | 80.9 ± 5.8 $^{de}$ | 89.9 ± 0.2 $^{bcdefg}$ | 276.7 ± 26.8 $^{def}$ |
| | 6 | 60 | 53.4 ± 4.5 $^{de}$ | 86.0 ± 9.7 $^{bcdefghi}$ | 1033.0 ± 208.1 $^{bc}$ |
| Steaming | 2 | 80 | 62.8 ± 11.7 $^{de}$ | 91.8 ± 6.3 $^{bcdef}$ | 150.8 ± 7.3 $^{ef}$ |
| | 4 | 80 | 33.7 ± 0.6 $^{e}$ | 72.2 ± 0.2 $^{hijklm}$ | 802.6 ± 46.4 $^{bcde}$ |
| | 6 | 80 | 45.5 ± 8.1 $^{de}$ | 64.3 ± 0.2 $^{klm}$ | 440.5 ± 87.3 $^{cdef}$ |
| Steaming | 2 | 100 | 45.2 ± 15.3 $^{de}$ | 77.8 ± 0.0 $^{fghijkl}$ | 130.6 ± 16.5 $^{ef}$ |
| | 4 | 100 | 45.0 ± 1.7 $^{de}$ | 95.5 ± 0.2 $^{abcd}$ | 679.3 ± 36.6 $^{bcdef}$ |
| | 6 | 100 | 55.3 ± 4.8 $^{de}$ | 77.4 ± 0.4 $^{fghijkl}$ | 314.6 ± 62.8 $^{def}$ |
| Stir-frying | 2 | 60 | 67.9 ± 16.3 $^{de}$ | 86.6 ± 0.2 $^{bcdefghi}$ | 1164.0 ± 21.5 $^{b}$ |
| | 4 | 60 | 40.8 ± 6.2 $^{de}$ | 76.6 ± 0.6 $^{fghijkl}$ | 508.2 ± 134.1 $^{bcdef}$ |
| | 6 | 60 | 39.6 ± 14.0 $^{de}$ | 75.4 ± 7.9 $^{fghijklm}$ | 462.2 ± 66.3 $^{cdef}$ |
| Stir-frying | 2 | 80 | 57.6 ± 0.4 $^{de}$ | 78.1 ± 0.0 $^{efghijk}$ | 253.0 ± 37.8 $^{def}$ |
| | 4 | 80 | 52.9 ± 1.4 $^{de}$ | 67.5 ± 0.2 $^{jklm}$ | 519.5 ± 247.9 $^{bcdef}$ |
| | 6 | 80 | 48.1 ± 14.2 $^{de}$ | 79.8 ± 0.2 $^{defghijk}$ | 561.5 ± 81.5 $^{bcdef}$ |
| Stir-frying | 2 | 100 | 102.4 ± 15.6 $^{bcde}$ | 78.4 ± 0.4 $^{efghijk}$ | 319.7 ± 26.4 $^{def}$ |
| | 4 | 100 | 86.3 ± 2.1 $^{cde}$ | 73.4 ± 0.0 $^{ghijklm}$ | 478.8 ± 37.4 $^{cdef}$ |
| | 6 | 100 | 105.1 ± 2.8 $^{bcde}$ | 69.9 ± 0.2 $^{ijklm}$ | 458.5 ± 195.6 $^{cdef}$ |

The table shows mean ± SD (n = 3 for each cooking condition, n = 6 for uncooked). The superscript letter $^{a-m}$ indicates the results of statistical analyses using One-way ANOVA with Tukey's multiple comparison test. The same letter represents no significant difference for pair-wise comparison between conditions ($p \geq 0.05$). The different letter represents a significant difference for pair-wise comparison between conditions ($p < 0.05$).

The yield of goitrin, BITC, and SFN in Chinese kale are shown in Table 4 and Figure S3. Uncooked Chinese kale had lower levels of goitrin than that of cabbage, with a total amount of goitrin at 82.27 ± 25.23 ng/mL. Blanching of Chinese kale at 60–80 °C for 2–6 min was found to increase goitrin. The result may be explained by an increase in myrosinase activity when heated at 60–80 °C, as reported in a previous study [20]. In contrast, blanching

Chinese kale at 100 °C for 6 min significantly decreased goitrin by 73% ($p < 0.05$) when compared with the raw samples (Table 6). However, blanching at 100 °C 4 min, steaming, and stir-frying Chinese kale decreased goitrin by 30%, 1–67%, and 3–66%, respectively, but no significant decrease ($p > 0.05$) was observed when compared with the raw samples (Table 6). Based on Table 6, the best cooking condition for reducing goitrin in Chinese kale was blanching at the temperature of 100 °C for 6 min, with a 73% decrease. Uncooked Chinese kale contains 107.20 ± 2.11 ng/mL of BITC (Table 4). Blanching, steaming, and stir-frying in some conditions did not significantly decrease BITC content ($p > 0.05$) when compared with the raw samples. Instead, some conditions such as stir-frying at 80 °C for 2 min and stir-frying at 100 °C for 2 min significantly increased BITC amount by 36% ($p < 0.0001$) and 19% ($p < 0.01$), respectively (Table 6). The best cooking condition for preserving BITC in Chinese kale was stir-frying at the temperature of 80 °C for 2 min with a 36% increase. Chinese kale contains 9.74 ± 3.86 ng/mL of SFN. While certain conditions increased SFN by 1.1–15.2–fold, some conditions decreased SFN by around 70–80%. The best cooking condition for preserving SFN in Chinese kale was steaming at 60 °C for 6 min, with a 15.2-fold increase (Table 6).

Previously published works only reported changes in progoitrin, the precursor of goitrin, after cooking. Thus, our current study is the first to report the changes in goitrin after cooking. Volden et al. found that the progoitrin in red cabbage was significantly reduced after blanching (78% decrease), boiling (65% decrease), and steaming (54% decrease) [22]. Compared to the previous study, our current work revealed a higher percent decrease of goitrin in cabbage after cooking (81–87%). This distinction may be due to different cooking conditions, analytical methods, and types of cabbage. Wennberg et al. also showed varied percentages of progoitrin reduction after blanching in different cabbage types (47% and 87% decrease in Hekla and Predikant cabbage, respectively [31].

For isothiocyanate (ITCs), our results are consistent with a previous study by Wang et al. that showed an increase in ITCs after steaming and stir-frying. Nevertheless, in our current study, the magnitude of SFN increase in cabbage after steaming (15 folds) is higher than those reported in the previous study (1.7 fold) [21]. This may be due to different cooking conditions, analytical methods, and types of vegetables. Consistently, two studies demonstrated that high cooking temperatures can enhance the antioxidant effect of SFN [32,33]. Therefore, heat-induced cooking can also be beneficial for some bioactive compounds.

In our study, the loss of goitrin, BITC, and SFN during different cooking conditions could be explained by various mechanisms: (i) the high-temperature cooking could denature the myrosinase enzyme in *Brassica* vegetables, resulting in the slower conversion of glucosinolates to hydrolysis products (goitrin, BITC, and SFN) during cooking and after mastication [34]; (ii) the loss of glucosinolates at high temperature cooking or leaching into the cooking water would reduce the amount of goitrin and isothiocyanates found in cooked vegetables [21]; (iii) the longer the cooking time, the greater the breakdown of glucosinolates [35]. Moreover, we can indicate that other factors have affected the yield of goitrin and isothiocyanates in our study such as cooking water, sampling samples, food matrix of vegetables, plant species and genotype, plant organ, pre-and -post-harvest factors [36]. These variables may have an impact on the broad SD value in sample vegetables.

### 3.3. Optimization of Cooking Methods for Cabbage and Chinese Kale

The results of this research demonstrated that the concentrations of goitrin, BITC, and SFN were influenced by cooking temperature, cooking time, and cooking methods. The interactions between these factors were found via ANOVA analysis. To determine which cooking method is the best for reducing goitrin while preserving isothiocyanates in *Brassica* vegetables, response surface methodology (RSM) models were used. The RSM graphs of goitrin (Figures 3A–C and 4A–C), BITC (Figures 3D–F and 4D–F), and SFN (Figures 3G–I and 4G–I) are shown. The warm tone color zones (lowest in blue and maximum in red) represent the components' optimal values.

**Table 4.** The yield of goitrin, BITC, and SFN in Chinese kale under different cooking conditions.

| Cooking Condition | | | Goitrin (ng/mL) | BITC (ng/mL) | SFN (ng/mL) |
|---|---|---|---|---|---|
| Cooking Methods | Cooking Time (min) | Cooking Temp. (°C) | | | |
| Uncooked | – | – | 82.3 ± 25.2 [abc] | 107.2 ± 2.1 [cde] | 9.7 ± 3.9 [g] |
| Blanching | 2 | 60 | 84.0 ± 1.0 [abcd] | 108.4 ± 0.4 [bcde] | 55.1 ± 8.7 [cde] |
| | 4 | 60 | 109.5 ± 2.8 [a] | 102.5 ± 0.2 [cdefg] | 19.6 ± 1.6 [fg] |
| | 6 | 60 | 112.4 ± 3.7 [a] | 94.5 ± 0.4 [defghi] | 10.7 ± 0.6 [g] |
| Blanching | 2 | 80 | 104.5 ± 34.6 [a] | 94.7 ± 0.0 [defghi] | 61.5 ± 5.1 [cd] |
| | 4 | 80 | 96.6 ± 1.7 [ab] | 87.7 ± 0.4 [fghi] | 2.7 ± 0.2 [g] |
| | 6 | 80 | 86.3 ± 1.9 [abcd] | 82.1 ± 0.4 [hi] | 2.4 ± 0.1 [g] |
| Blanching | 2 | 100 | 90.4 ± 1.5 [abc] | 93.7 ± 0.2 [defghi] | 26.0 ± 2.9 [efg] |
| | 4 | 100 | 57.4 ± 12.1 [abcd] | 90.0 ± 0.4 [efghi] | 3.6 ± 0.2 [g] |
| | 6 | 100 | 22.4 ± 4.3 [d] | 85.3 ± 0.4 [ghi] | 2.8 ± 0.2 [g] |
| Steaming | 2 | 60 | 33.3 ± 1.1 [bcd] | 118.0 ± 33.3 [bc] | 4.9 ± 0.2 [g] |
| | 4 | 60 | 85.3 ± 1.5 [abcd] | 92.2 ± 0.0 [efghi] | 26.8 ± 0.7 [efg] |
| | 6 | 60 | 58.9 ± 26.3 [abcd] | 81.6 ± 0.2 [i] | 148.4 ± 6.5 [a] |
| Steaming | 2 | 80 | 93.2 ± 19.0 [ab] | 106.1 ± 0.2 [cdef] | 5.4 ± 0.7 [g] |
| | 4 | 80 | 81.8 ± 38.0 [abcd] | 106.3 ± 4.9 [cdef] | 104.6 ± 7.2 [b] |
| | 6 | 80 | 113.2 ± 2.2 [a] | 95.8 ± 0.0 [defghi] | 3.5 ± 0.5 [g] |
| Steaming | 2 | 100 | 56.1 ± 17.3 [abcd] | 95.9 ± 0.2 [defghi] | 23.3 ± 30.6 [fg] |
| | 4 | 100 | 62.2 ± 26.5 [abcd] | 85.4 ± 0.2 [ghi] | 4.4 ± 0.2 [g] |
| | 6 | 100 | 27.4 ± 2.9 [cd] | 82.7 ± 0.2 [ghi] | 3.5 ± 0.1 [g] |
| Stir-frying | 2 | 60 | 27.6 ± 3.5 [cd] | 102.4 ± 0.4 [cdefg] | 31.4 ± 11.0 [defg] |
| | 4 | 60 | 60.8 ± 54.9 [abcd] | 113.1 ± 0.4 [bcd] | 14.9 ± 1.9 [fg] |
| | 6 | 60 | 91.8 ± 4.6 [abc] | 109.1 ± 0.0 [bcde] | 16.8 ± 9.5 [fg] |
| Stir-frying | 2 | 80 | 33.1 ± 4.8 [bcd] | 145.6 ± 0.4 [a] | 43.7 ± 2.2 [cdef] |
| | 4 | 80 | 39.9 ± 38.0 [bcd] | 81.1 ± 0.2 [i] | 3.4 ± 0.3 [g] |
| | 6 | 80 | 55.3 ± 9.1 [abcd] | 81.6 ± 0.4 [i] | 1.9 ± 0.1 [g] |
| Stir-frying | 2 | 100 | 66.1 ± 31.4 [abcd] | 127.3 ± 0.2 [ab] | 63.7 ± 39.1 [c] |
| | 4 | 100 | 79.7 ± 3.0 [abcd] | 101.7 ± 0.2 [cdefgh] | 6.0 ± 0.1 [g] |
| | 6 | 100 | 66.3 ± 3.2 [abcd] | 87.0 ± 0.2 [fghi] | 3.0 ± 0.2 [g] |

The table shows mean ± SD (n = 3 for each cooking condition, n = 6 for uncooked). The superscript letter [a–i] indicates the results of statistical analyses using One-way ANOVA with Tukey's multiple comparison test. The same letter represents no significant difference for pair-wise comparison between conditions ($p \geq 0.05$). The different letter represents a significant difference for pair-wise comparison between conditions ($p < 0.05$).

**Table 5.** The mean fold changes and percentage change (%) of goitrin, BITC, and SFN in cabbage under different cooking conditions compared with uncooked vegetables.

| Cooking Condition | | | Mean of Fold Change/Percent Change (%), Compared to Uncooked Vegetables | | |
|---|---|---|---|---|---|
| Cooking Methods | Cooking Time (min) | Cooking Temp. (°C) | Goitrin | BITC | SFN |
| Uncooked | – | – | 100% | 100% | 100% |
| Blanching | 2 | 60 | 1.2/+18% | 0.9/−6% | 16.9/+1588% |
| | 4 | 60 | 0.8/−17% | 0.8/−18% | 3.4/+245% |
| | 6 | 60 | 0.6/−40% | 0.8/−21% | 1.7/+73% |
| Blanching | 2 | 80 | 0.8/−21% | 0.9/−13% | 4.2/+321% |
| | 4 | 80 | 0.3/−71% | 0.9/−10% | 2.0/+95% |
| | 6 | 80 | 0.4/−61% | 0.5/−46% | 2.5/+152% |

**Table 5.** *Cont.*

| Cooking Condition | | | Mean of Fold Change/Percent Change (%), Compared to Uncooked Vegetables | | |
|---|---|---|---|---|---|
| **Cooking Methods** | **Cooking Time (min)** | **Cooking Temp. (°C)** | **Goitrin** | **BITC** | **SFN** |
| Blanching | 2 | 100 | 0.2/−79% | 0.8/−22% | 2.5/+151% |
| | 4 | 100 | 0.2/−81% | 0.6/−44% | 0.3/−66% |
| | 6 | 100 | 0.3/−70% | 0.7/−34% | 0.5/−48% |
| Steaming | 2 | 60 | 0.4/−57% | 0.7/−32% | 1.2/+20% |
| | 4 | 60 | 0.3/−68% | 0.8/−17% | 1.4/+36% |
| | 6 | 60 | 0.2/−79% | 0.8/−21% | 5.1/+409% |
| Steaming | 2 | 80 | 0.2/−75% | 0.8/−16% | 0.7/−26% |
| | 4 | 80 | 0.1/−87% | 0.7/−34% | 4.0/+296% |
| | 6 | 80 | 0.2/−82% | 0.6/−41% | 2.2/+117% |
| Steaming | 2 | 100 | 0.2/−82% | 0.7/−28% | 0.6/−36% |
| | 4 | 100 | 0.2/−82% | 0.9/−12% | 3.3/+235% |
| | 6 | 100 | 0.2/−78% | 0.7/−29% | 1.6/+55% |
| Stir-frying | 2 | 60 | 0.3/−73% | 0.8/−20% | 5.7/+474% |
| | 4 | 60 | 0.2/−84% | 0.7/−30% | 2.5/+151% |
| | 6 | 60 | 0.2/−84% | 0.7/−31% | 2.3/+128% |
| Stir-frying | 2 | 80 | 0.2/−77% | 0.7/−28% | 1.2/+25% |
| | 4 | 80 | 0.2/−79% | 0.6/−38% | 2.6/+156% |
| | 6 | 80 | 0.2/−81% | 0.7/−27% | 2.8/+177% |
| Stir-frying | 2 | 100 | 0.4/−59% | 0.7/−28% | 1.6/+58% |
| | 4 | 100 | 0.3/−66% | 0.7/−32% | 2.4/+136% |
| | 6 | 100 | 0.4/−58% | 0.6/−36% | 2.3/+126% |

The table shows mean fold changes and percentage change (%), which have the minus sign (−) and the plus sign (+) denoting decreasing and increasing levels, respectively, in each substance under different cooking conditions compared with uncooked. Minus sign (−) under cooking time and temperature represents no cooking.

**Table 6.** The mean fold changes and percentage change (%) of goitrin, BITC, and SFN in Chinese kale under different cooking conditions compared with uncooked condition.

| Cooking Condition | | | Mean of Fold Change/Percent Change (%), Compared to Uncooked Vegetables | | |
|---|---|---|---|---|---|
| **Cooking Methods** | **Cooking Time (min)** | **Cooking Temp. (°C)** | **Goitrin** | **BITC** | **SFN** |
| Uncooked | − | − | − | − | − |
| Blanching | 2 | 60 | 1.0/+2% | 1.0/+1% | 5.7/+466% |
| | 4 | 60 | 1.3/+33% | 1.0/−4% | 2.0/+101% |
| | 6 | 60 | 1.4/+37% | 0.9/−12% | 1.1/+10% |
| Blanching | 2 | 80 | 1.3/+27% | 0.9/−12% | 6.3/+531% |
| | 4 | 80 | 1.2/+17% | 0.8/−18% | 0.3/−72% |
| | 6 | 80 | 1.0/+5% | 0.8/−23% | 0.2/−75% |
| Blanching | 2 | 100 | 1.1/+10% | 0.9/−13% | 2.7/+167% |
| | 4 | 100 | 0.7/−30% | 0.8/−16% | 0.4/−63% |
| | 6 | 100 | 0.3/−73% | 0.8/−20% | 0.3/−72% |
| Steaming | 2 | 60 | 0.4/−60% | 1.1/+10% | 0.5/−50% |
| | 4 | 60 | 1.0/+4% | 0.9/−14% | 2.8/+176% |
| | 6 | 60 | 0.7/−28% | 0.8/−24% | 15.2/+1424% |

**Table 6.** *Cont.*

| Cooking Condition | | | Mean of Fold Change/Percent Change (%), Compared to Uncooked Vegetables | | |
|---|---|---|---|---|---|
| **Cooking Methods** | **Cooking Time (min)** | **Cooking Temp. (°C)** | **Goitrin** | **BITC** | **SFN** |
| Steaming | 2 | 80 | 1.1/+13% | 1.0/−1% | 0.6/−45% |
| | 4 | 80 | 1.0/−1% | 1.0/−1% | 10.7/+974% |
| | 6 | 80 | 1.4/+38% | 0.9/−11% | 0.4/−64% |
| Steaming | 2 | 100 | 0.7/−32% | 0.9/−11% | 2.4/+139% |
| | 4 | 100 | 0.8/−24% | 0.8/−20% | 0.5/−55% |
| | 6 | 100 | 0.3/−67% | 0.8/−23% | 0.4/−64% |
| Stir-frying | 2 | 60 | 0.3/−66% | 1.0/−4% | 3.2/+222% |
| | 4 | 60 | 0.7/−26% | 1.1/+6% | 1.5/+53% |
| | 6 | 60 | 1.1/+12% | 1.0/+2% | 1.7/+73% |
| Stir-frying | 2 | 80 | 0.4/−60% | 1.4/+36% | 4.5/+349% |
| | 4 | 80 | 0.5/−52% | 0.8/−24% | 0.4/−65% |
| | 6 | 80 | 0.7/−33% | 0.8/−24% | 0.2/−80% |
| Stir-frying | 2 | 100 | 0.8/−20% | 1.2/+19% | 6.5/+554% |
| | 4 | 100 | 1.0/−3% | 0.9/−5% | 0.6/−39% |
| | 6 | 100 | 0.8/−19% | 0.8/−19% | 0.3/−69% |

The table shows mean fold changes and percentage change (%), which have the minus sign (−) and the plus sign (+) denoting decreasing and increasing in each substance under different cooking conditions compared with uncooked. Minus sign (−) under cooking time and temperature represents no cooking.

In this study, the weights of BITC and SFN in the analyses for cabbage and Chinese kale were set differently. Uncooked cabbage contains a higher amount of SFN than BITC. In contrast, raw Chinese kale has a higher amount of BITC than SFN. Therefore, we provided a higher weight in SFN for cabbage and a higher weight in BITC for Chinese kale.

In cabbage, the response surface analysis (RSA) using ANOVA showed that the concentration of goitrin depended on the cooking temperature, cooking time, and cooking methods, but the concentration of BITC did not significantly depend on those factors. Furthermore, the concentration of SFN in cabbage depended on the cooking temperature and cooking time but not on cooking methods. When the concentrations were weighed after minimizing goitrin (+++++), maximizing BITC (+), and maximizing SFN (++++), the data showed that the best cooking method to reduce goitrin and preserve ITCs in cabbage was steaming at a temperature of 90.92 °C for 4.26 min. (desirability 0.38) (Table 7 and Figure S4).

In Chinese kale, the response surface analysis (RSA) using ANOVA showed that the concentrations of goitrin and SFN depended on cooking methods but not cooking temperature and time. In contrast, the concentration of BITC depended on the cooking temperature and cooking time but not on cooking methods. When the concentrations were weighed after by minimizing goitrin (+++++), maximizing BITC (++++), and maximizing SFN (+), the data showed that the best cooking method to reduce goitrin and preserve ITCs in Chinese kale was stir-frying at a temperature of 84.56 °C for 2 min. (desirability 0.63) (Table 8 and Figure S4). It is worth noting that although blanching Chinese kale at 100 °C for 6 min was the best method to significantly decrease goitrin, such a method also caused a significant loss of BITC and SFN. When RSM analysis was performed by weighing both the reduction in goitrin and preservation of BITC and SFN in Chinese kale, the results suggest that stir-frying was a better method with higher desirability than that of blanching (Tables 7 and 8).

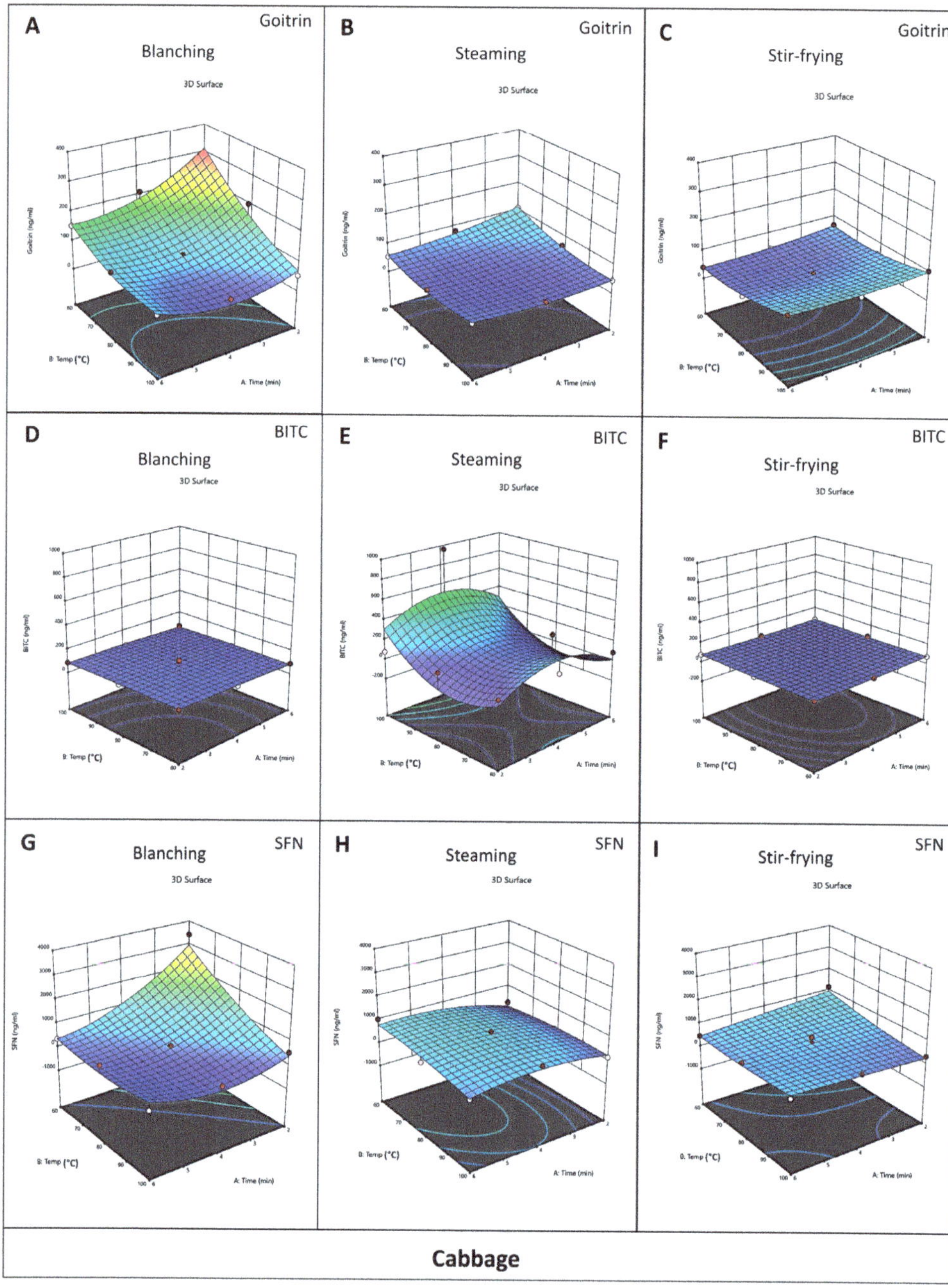


**Figure 3.** Relationship between cooking factors in cabbage. Three-dimensional (3D) response surface methodology (RSM) diagrams show concentrations of goitrin (**A–C**), BITC (**D–F**), and SFN (**G–I**) in different cooking conditions of cabbage as the function of cooking time, cooking temperature, and cooking methods. Red, yellow, green, and blue areas depict the highest to lowest goitrin, BITC, and SFN concentrations.

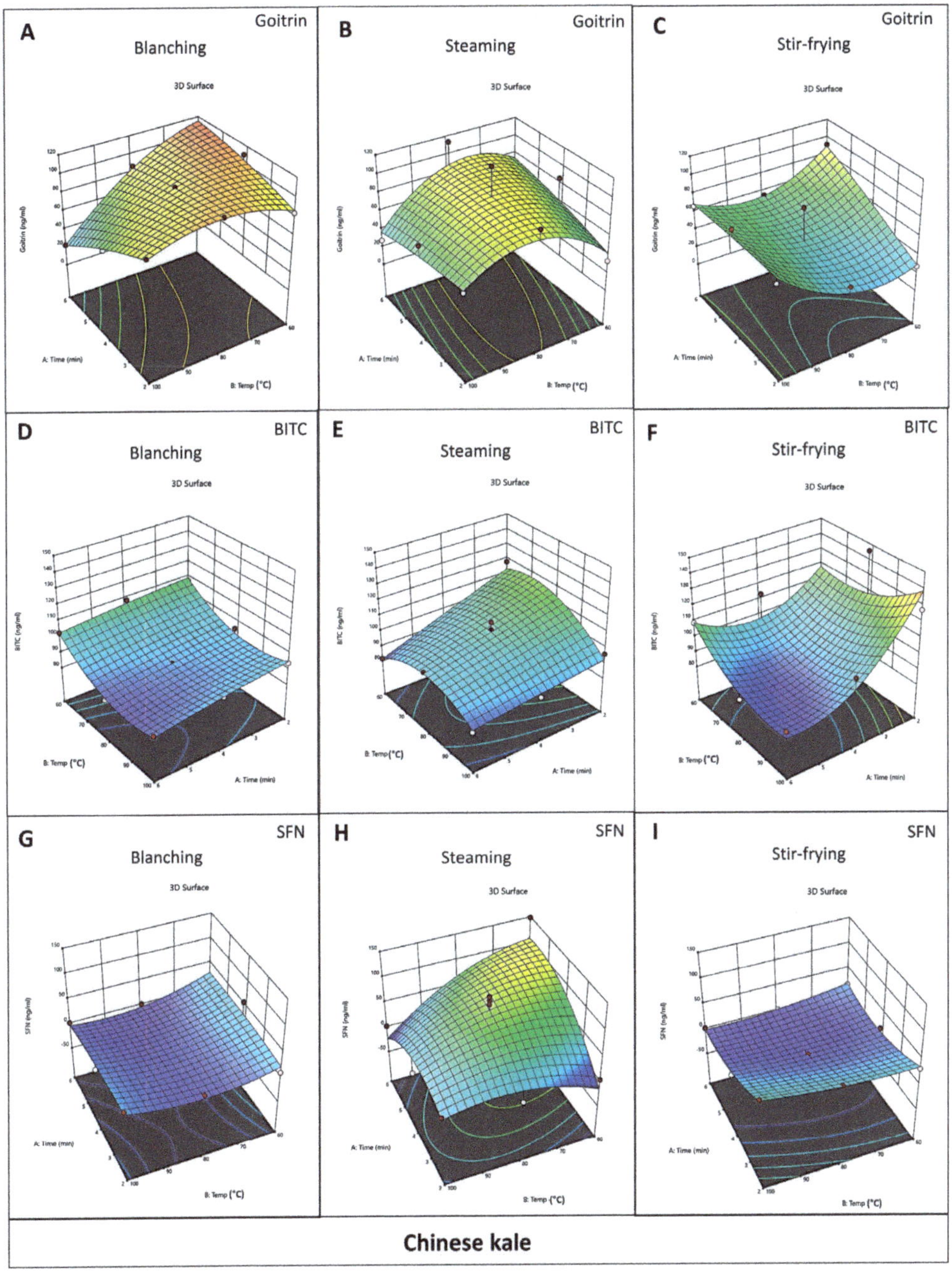


**Figure 4.** Relationship between cooking factors in Chinese kale. Three-dimensional (3D) response surface methodology (RSM) diagrams show concentrations of goitrin (**A–C**), BITC (**D–F**), and SFN (**G–I**) in different cooking conditions of Chinese kale as the function of cooking time, cooking temperature, and cooking methods. Red, yellow, green, and blue areas depict the highest to lowest goitrin, BITC, and SFN concentrations.

**Table 7.** The rank of conditions with the best desirability values of cabbage.

| Number | Time (min) | Temp. (°C) | Cooking Method | Goitrin (ng/mL) | BITC (ng/mL) | SFN (ng/mL) | Desirability |
|---|---|---|---|---|---|---|---|
| 1 | 4.26 | 90.92 | Steaming | 33.45 | 278.72 | 681.39 | 0.38 |
| 2 | 2.71 | 60.00 | Stir-frying | 56.30 | 80.10 | 875.54 | 0.29 |
| 3 | 2.68 | 60.00 | Stir-frying | 56.71 | 80.39 | 879.95 | 0.29 |
| 4 | 2.63 | 60.00 | Stir-frying | 57.29 | 80.67 | 885.96 | 0.29 |
| 5 | 2.87 | 60.00 | Stir-frying | 54.50 | 79.11 | 855.86 | 0.29 |
| 6 | 2.96 | 60.00 | Stir-frying | 53.54 | 78.60 | 844.90 | 0.29 |
| 7 | 4.09 | 60.00 | Stir-frying | 43.96 | 74.60 | 704.61 | 0.28 |
| 8 | 4.30 | 60.00 | Stir-frying | 42.66 | 74.35 | 677.52 | 0.27 |
| 9 | 4.70 | 60.00 | Stir-frying | 40.68 | 74.32 | 626.40 | 0.27 |
| 10 | 2.42 | 92.81 | Blanching | 78.21 | 88.59 | 472.44 | 0.18 |
| 11 | 5.89 | 98.65 | Blanching | 73.32 | 63.37 | 343.13 | 0.14 |

The table shows 11 solutions found for 3 combinations of categoric factor levels in cabbage from Design-Expert version 13. Grey shade represents condition with the best desirability value.

**Table 8.** The rank of conditions with the best desirability values of Chinese kale.

| Number | Time (min) | Temp. (°C) | Cooking Method | Goitrin (ng/mL) | BITC (ng/mL) | SFN (ng/mL) | Desirability |
|---|---|---|---|---|---|---|---|
| 1 | 2.00 | 84.56 | Stir-frying | 31.90 | 121.22 | 45.26 | 0.63 |
| 2 | 2.00 | 84.78 | Stir-frying | 32.26 | 121.39 | 45.44 | 0.63 |
| 3 | 2.00 | 84.95 | Stir-frying | 32.54 | 121.51 | 45.57 | 0.63 |
| 4 | 2.00 | 85.10 | Stir-frying | 32.80 | 121.63 | 45.70 | 0.63 |
| 5 | 2.00 | 84.01 | Stir-frying | 31.02 | 120.81 | 44.82 | 0.63 |
| 6 | 2.00 | 83.77 | Stir-frying | 30.65 | 120.65 | 44.64 | 0.63 |
| 7 | 2.00 | 85.86 | Stir-frying | 34.12 | 122.23 | 46.33 | 0.63 |
| 8 | 2.00 | 87.39 | Stir-frying | 37.01 | 123.54 | 47.67 | 0.63 |
| 9 | 2.00 | 62.12 | Stir-frying | 26.49 | 117.22 | 35.37 | 0.61 |
| 10 | 2.00 | 61.75 | Stir-frying | 26.92 | 117.37 | 35.34 | 0.61 |
| 11 | 2.00 | 61.11 | Stir-frying | 27.70 | 117.64 | 35.31 | 0.61 |
| 12 | 2.00 | 60.25 | Stir-frying | 28.84 | 118.04 | 35.28 | 0.61 |
| 13 | 2.00 | 97.00 | Stir-frying | 61.74 | 134.44 | 57.77 | 0.59 |
| 14 | 2.72 | 60.00 | Steaming | 49.52 | 107.39 | 30.35 | 0.46 |
| 15 | 2.74 | 60.00 | Steaming | 49.72 | 107.24 | 31.40 | 0.46 |
| 16 | 2.78 | 60.00 | Steaming | 50.13 | 106.93 | 33.55 | 0.46 |
| 17 | 2.66 | 60.00 | Steaming | 48.91 | 107.85 | 27.12 | 0.46 |
| 18 | 2.80 | 60.00 | Steaming | 50.39 | 106.73 | 34.89 | 0.46 |
| 19 | 2.00 | 60.00 | Blanching | 86.26 | 109.12 | 21.77 | 0.31 |
| 20 | 6.00 | 60.00 | Blanching | 86.45 | 109.10 | 21.90 | 0.31 |
| 21 | 2.32 | 60.29 | Blanching | 86.73 | 108.72 | 21.26 | 0.31 |
| 22 | 6.00 | 63.22 | Stir-frying | 80.98 | 104.01 | 15.86 | 0.30 |
| 23 | 2.32 | 60.00 | Blanching | 89.59 | 108.80 | 24.01 | 0.30 |
| 24 | 4.11 | 60.00 | Stir-frying | 62.49 | 97.11 | 8.11 | 0.29 |
| 25 | 3.80 | 100.00 | Blanching | 62.41 | 90.65 | 15.34 | 0.26 |
| 26 | 3.82 | 100.00 | Blanching | 62.22 | 90.62 | 15.28 | 0.26 |
| 27 | 3.84 | 100.00 | Blanching | 61.81 | 90.56 | 15.14 | 0.26 |
| 28 | 3.91 | 100.00 | Blanching | 60.67 | 90.39 | 14.76 | 0.26 |
| 29 | 4.35 | 100.00 | Blanching | 53.14 | 89.23 | 12.00 | 0.25 |
| 30 | 4.30 | 60.00 | Blanching | 106.45 | 105.99 | 32.20 | 0.20 |

The table shows 30 solutions found for three combinations of categoric factor levels in Chinese kale from Design-Expert version 13. Grey shade represents condition with the best desirability value.

The optimized conditions for reducing goitrin and preserving isothiocyanates are in decimal digits, and it might be difficult to achieve such values in household cooking. Moreover, due to the limitation of the central composite design in this research, the exact optimized condition obtained from the analysis may not be suitable for recommendation. Thus, a range of values was further identified. Tables 7 and 8 illustrates the top-ranked cooking conditions with the best desirability values for reducing goitrin and preserving isothiocyanates in Chinese kale and cabbage. For Chinese kale, the best range of optimized cooking conditions with the best desirability values (0.59–0.63) is stir-frying at 60–100 °C for 2 min (Table 8). For cabbage, since the best desirability value is still quite low (0.38), we considered the data analyzed via RSM (Table 7) together with the data analyzed with ANOVA (Table 5). The results showed that steaming cabbage at 80 °C and 100 °C for 4 min still could reduce goitrin and also preserve BITC and SFN. The best cooking of cabbage obtained from RSM was steaming at 90.92 °C for 4.26 min.

Based on these findings, we suggest that the optimized household cooking methods for preserving isothiocyanates and reducing goitrin are steaming cabbage at 80–100 °C for 4 min, and stir-frying Chinese kale at 60–100 °C for 2 min.

### 3.4. *Strengths and Limitations of This Research*

This study has several strengths. First, several affecting factors have been controlled. Cooking was performed in a home kitchen using the very same induction cooker for all cooking methods (blanching, steaming, and stir-frying), ensuring real household cooking conditions and fair control of the cooking apparatus. The temperature of the cooker is also controlled with a digital system, allowing the user to adjust the temperature precisely. The water used in all conditions was from the same source. In addition, the same bulb of vegetables has been divided evenly into four parts for all cooking conditions and raw samples. Since our sample contains a mixture of leaf and stem, we ground the cooked or uncooked samples before collection for analysis. This is to ensure the homogeneity of collected samples. Second, this is the first study to report the changes in goitrin after cooking, while previous studies only reported changes in progoitrin, the precursor of goitrin, after cooking. Thus, the findings of this study can guide which cooking conditions may help to reduce goitrin. Third, this work studied two *Brassica* vegetables (cabbage and Chinese kale), which are commonly consumed in many countries around the world. Therefore, the insight from this work may have broad application.

There were some limitations of this study. First, the yield of goitrin, BITC, and SFN in vegetable samples had wide standard deviation (SD) values. Though we tried our best to control the possible confounding factors, there are still some variables we cannot control. Since we used manual household cooking in the open air, the whole pieces of vegetables may not be evenly cooked (unlike the factory cooking in the closed system). The second limitation is that we set the temperature of cooking from an induction cooker and did not directly measure the temperature on the vegetables during cooking. Therefore, the insight from this work is only applicable for household cooking, and may not be practical for industrial use. Future studies for closed-system cooking are warranted for industrial purposes. The third limitation is the design of RSM analysis. Since the cooking method is not in numerical or ordinal scales and cannot make three levels of factorial experiments, the Box–Behnken design (BBD) cannot be used in this study. A central composite design was used instead. For the last limitation of this study, we used cooking water for the stir-frying method in our study because other cooking methods (blanching and steaming) also used water. Nevertheless, water stir-frying is not as commonly used as oil stir-frying. Future works are warranted to study the best condition of oil-based methods, such as stir-frying, deep frying, and grilling with oil, to reduce goitrin while preserving ITCs. In addition, cooking methods to reduce goitrin while preserving other ITCs with chemo-preventive effects such as allyl isothiocyanate (AITC), phenethyl isothiocyanate (PEITC), 3-Butenyl Isothiocyanate (MTBITC), warrants further studies.

## 4. Conclusions

Using LC-MS/MS for the determination of goitrin, BITC, and SFN together with the response surface methodology, we found that cooking temperature, cooking time, and cooking methods were the factors that influenced the levels of goitrin and isothiocyanates (BITC and SFN) in cabbage and Chinese kale. The optimum cooking method for preserving isothiocyanates and reducing goitrin is steaming cabbage at 80–100 °C for 4 min and stir-frying Chinese kale at 60–100 °C for 2 min. Such cooking methods may help to maintain cancer chemopreventive compounds (ITCs) and remove the anti-thyroid compound (goitrin). Eventually, it may help to reduce the risk of cancer in the long term.

**Supplementary Materials:** The following supporting information can be downloaded at: https://www.mdpi.com/article/10.3390/foods12193647/s1. Figure S1: Average calibration curves of GN-$NH_3$ (A), BITC-$NH_3$ (B), and SFN (C). Figure S2. Goitrin and isothiocyanate content in cabbage samples after cooking conditions: (A) goitrin content; (B) benzyl-isothiocyanate (BITC) content; (C) sulforaphane (SFN) content. Figure S3. Goitrin and isothiocynate content in Chinese kale samples after cooking conditions: (A) goitrin content; (B) benzyl-isothiocyanate (BITC) content; (C) sulforaphane (SFN) content. Figure S4. The optimal condition of cabbage (A) and Chinese kale (B) for reducing goitrin while preserving isothiocyanates (BITC and SFN) in *Brassica* vegetables through the response surface methodological (RSM).

**Author Contributions:** Conceptualization, T.P., W.K. and D.T.; Data curation, T.P., P.P. and D.T.; Formal analysis, T.P.; Investigation, T.P., P.P. and D.T.; methodology, T.P., W.K. and D.T.; Writing—original draft preparation, T.P.; Writing—review and editing, P.P., W.K. and D.T.; Supervision, W.K. and D.T. All authors have read and agreed to the published version of the manuscript.

**Funding:** This research received no external funding.

**Data Availability Statement:** Data is contained within the article or Supplementary Material.

**Acknowledgments:** The authors are grateful to the Institute of Nutrition for providing the opportunity to do this study and for all the resources throughout this study. The assistance of all staff of the Institute of Nutrition in this study is also appreciated.

**Conflicts of Interest:** The authors declare no conflict of interest.

## References

1. Fahey, J.W. Brassica: Characteristics and Properties. In *Encyclopedia of Food and Health*; Caballero, B., Finglas, P.M., Toldrá, F., Eds.; Academic Press: Cambridge, MA, USA, 2016; pp. 469–477. [CrossRef]
2. Abellán, Á.; Domínguez-Perles, R.; Moreno, D.A.; García-Viguera, C. Sorting out the Value of Cruciferous Sprouts as Sources of Bioactive Compounds for Nutrition and Health. *Nutrients* **2019**, *11*, 429. [CrossRef] [PubMed]
3. Melim, C.; Lauro, M.R.; Pires, I.M.; Oliveira, P.J.; Cabral, C. The Role of Glucosinolates from Cruciferous Vegetables (Brassicaceae) in Gastrointestinal Cancers: From Prevention to Therapeutics. *Pharmaceutics* **2022**, *14*, 190. [CrossRef] [PubMed]
4. Hudlikar, R.; Wang, L.; Wu, R.; Li, S.; Peter, R.; Shannar, A.; Chou, P.J.; Liu, X.; Liu, Z.; Kuo, H.D.; et al. Epigenetics/Epigenomics and Prevention of Early Stages of Cancer by Isothiocyanates. *Cancer Prev. Res.* **2021**, *14*, 151–164. [CrossRef]
5. Rekha, K.; Venkidasamy, B.; Govindasamy, R.; Neralla, M.; Thiruvengadam, M. Isothiocyanates (AITC & BITC) bioactive molecules: Therapeutic potential for oral cancer. *Oral Oncol.* **2022**, *133*, 106060. [CrossRef] [PubMed]
6. Elkashty, O.A.; Tran, S.D. Sulforaphane as a Promising Natural Molecule for Cancer Prevention and Treatment. *Curr. Med. Sci.* **2021**, *41*, 250–269. [CrossRef] [PubMed]
7. Lam-Ubol, A.; Fitzgerald, A.L.; Ritdej, A.; Phonyiam, T.; Zhang, H.; Myers, J.N.; Huang, P.; Trachootham, D. Sensory acceptable equivalent doses of β-phenylethyl isothiocyanate (PEITC) induce cell cycle arrest and retard the growth of p53 mutated oral cancer in vitro and in vivo. *Food Funct.* **2018**, *9*, 3640–3656. [CrossRef] [PubMed]
8. Soundararajan, P.; Kim, J.S. Anti-Carcinogenic Glucosinolates in Cruciferous Vegetables and Their Antagonistic Effects on Prevention of Cancers. *Molecules* **2018**, *23*, 2983. [CrossRef] [PubMed]
9. Kaewsit, N.; Winuprasith, T.; Trachootham, D. Detoxification of heterocyclic aromatic amines from grilled meat using a PEITC-rich vegetable sauce: A randomized crossover controlled trial. *Food Funct.* **2021**, *12*, 10411–10422. [CrossRef]
10. Phikulkhao, N.; Tanaviyutpakdee, P.; Lam-ubol, A.; Trachootham, D. Nutri-phenethyl isothiocyanate Jelly Promotes Detoxification of a Tobacco-specific Oral Carcinogen in Male Active Cigarette Smokers. *Cancer Screen. Prev.* **2023**, *2*, 30–41. [CrossRef]
11. Liu, X.; Lv, K. Cruciferous vegetables intake is inversely associated with risk of breast cancer: A meta-analysis. *Breast* **2013**, *22*, 309–313. [CrossRef]
12. Long, J.; Liu, Z.; Liang, S.; Chen, B. Cruciferous Vegetable Intake and Risk of Prostate Cancer: A Systematic Review and Meta-Analysis. *Urol. Int.* **2023**, *107*, 723–733. [CrossRef] [PubMed]
13. Tse, G.; Eslick, G.D. Cruciferous vegetables and risk of colorectal neoplasms: A systematic review and meta-analysis. *Nutr. Cancer* **2014**, *66*, 128–139. [CrossRef] [PubMed]
14. Petroski, W.; Minich, D.M. Is There Such a Thing as "Anti-Nutrients"? A Narrative Review of Perceived Problematic Plant Compounds. *Nutrients* **2020**, *12*, 2929. [CrossRef] [PubMed]

15. Bajaj, J.K.; Salwan, P.; Salwan, S. Various Possible Toxicants Involved in Thyroid Dysfunction: A Review. *J. Clin. Diagn. Res.* **2016**, *10*, FE01–FE03. [CrossRef] [PubMed]
16. Boelaert, K. The association between serum TSH concentration and thyroid cancer. *Endocr. Relat. Cancer* **2009**, *16*, 1065–1072. [CrossRef] [PubMed]
17. Cho, Y.A.; Kim, J. Dietary Factors Affecting Thyroid Cancer Risk: A Meta-Analysis. *Nutr. Cancer* **2015**, *67*, 811–817. [CrossRef] [PubMed]
18. Truong, T.; Baron-Dubourdieu, D.; Rougier, Y.; Guénel, P. Role of dietary iodine and cruciferous vegetables in thyroid cancer: A countrywide case-control study in New Caledonia. *Cancer Causes Control* **2010**, *21*, 1183–1192. [CrossRef] [PubMed]
19. Felker, P.; Bunch, R.; Leung, A.M. Concentrations of thiocyanate and goitrin in human plasma, their precursor concentrations in brassica vegetables, and associated potential risk for hypothyroidism. *Nutr. Rev.* **2016**, *74*, 248–258. [CrossRef]
20. Oloyede, O.O.; Wagstaff, C.; Methven, L. The Impact of Domestic Cooking Methods on Myrosinase Stability, Glucosinolates and Their Hydrolysis Products in Different Cabbage (*Brassica oleracea*) Accessions. *Foods* **2021**, *10*, 2908. [CrossRef]
21. Wang, Z.; Kwan, M.L.; Pratt, R.; Roh, J.M.; Kushi, L.H.; Danforth, K.N.; Zhang, Y.; Ambrosone, C.B.; Tang, L. Effects of cooking methods on total isothiocyanate yield from cruciferous vegetables. *Food Sci. Nutr.* **2020**, *8*, 5673–5682. [CrossRef]
22. Volden, J.; Borge, G.I.A.; Bengtsson, G.B.; Hansen, M.; Thygesen, I.E.; Wicklund, T. Effect of thermal treatment on glucosinolates and antioxidant-related parameters in red cabbage (*Brassica oleracea* L. ssp. *capitata* f. *rubra*). *Food Chem.* **2008**, *109*, 595–605. [CrossRef]
23. Baenas, N.; Marhuenda, J.; García-Viguera, C.; Zafrilla, P.; Moreno, D.A. Influence of Cooking Methods on Glucosinolates and Isothiocyanates Content in Novel Cruciferous Foods. *Foods* **2019**, *8*, 257. [CrossRef] [PubMed]
24. Dinh, T.N.; Parat, M.O.; Ong, Y.S.; Khaw, K.Y. Anticancer activities of dietary benzyl isothiocyanate: A comprehensive review. *Pharmacol. Res.* **2021**, *169*, 105666. [CrossRef] [PubMed]
25. Mordecai, J.; Ullah, S.; Ahmad, I. Sulforaphane and Its Protective Role in Prostate Cancer: A Mechanistic Approach. *Int. J. Mol. Sci.* **2023**, *24*, 6979. [CrossRef]
26. Ji, Y.; Morris, M.E. Determination of phenethyl isothiocyanate in human plasma and urine by ammonia derivatization and liquid chromatography–tandem mass spectrometry. *Anal. Biochem.* **2003**, *323*, 39–47. [CrossRef] [PubMed]
27. Bell, L.; Yahya, H.N.; Oloyede, O.O.; Methven, L.; Wagstaff, C. Changes in rocket salad phytochemicals within the commercial supply chain: Glucosinolates, isothiocyanates, amino acids, and bacterial load increase significantly after processing. *Food Chem.* **2017**, *221*, 521–534. [CrossRef]
28. International Conference on Harmonization of Technical Requirements for Registration of Pharmaceuticals for Human Use (ICH). Q2 (R1), A Validation of Analytical Procedures: Text and Methodology. 2005. Available online: https://database.ich.org/sites/default/files/Q2%28R1%29%20Guideline.pdf (accessed on 23 June 2023).
29. AOAC. International Guidelines for Standard Method Performance Requirements AOAC Official Methods of Analysis. 2016. Appendix F, 1–18. Available online: https://www.aoac.org/wp-content/uploads/2019/08/app_f.pdf (accessed on 23 June 2023).
30. Food and Drug Administration Office of Regulatory Affairs ORA Laboratory Manual Volume II: U.S. Food and Drug Administration. Available online: https://www.fda.gov/media/73920/download (accessed on 23 June 2023).
31. Wennberg, M.; Ekvall, J.; Olsson, K.; Nyman, M. Changes in carbohydrate and glucosinolate composition in white cabbage (*Brassica oleracea* var. *capitata*) during blanching and treatment with acetic acid. *Food Chem.* **2006**, *95*, 226–236. [CrossRef]
32. Cedrowski, J.; Dąbrowa, K.; Przybylski, P.; Krogul-Sobczak, A.; Litwinienko, G. Antioxidant activity of two edible isothiocyanates: Sulforaphane and erusin is due to their thermal decomposition to sulfenic acids and methylsulfinyl radicals. *Food Chem.* **2021**, *353*, 129213. [CrossRef]
33. Cedrowski, J.; Dąbrowa, K.; Krogul-Sobczak, A.; Litwinienko, G. A Lesson Learnt from Food Chemistry-Elevated Temperature Triggers the Antioxidant Action of Two Edible Isothiocyanates: Erucin and Sulforaphane. *Antioxidants* **2020**, *9*, 1090. [CrossRef]
34. Nugrahedi, P.Y.; Verkerk, R.; Widianarko, B.; Dekker, M. A Mechanistic Perspective on Process-Induced Changes in Glucosinolate Content in *Brassica* Vegetables: A Review. *Crit. Rev. Food Sci. Nutr.* **2015**, *55*, 823–838. [CrossRef]
35. Jones, R.B.; Faragher, J.D.; Winkler, S. A review of the influence of postharvest treatments on quality and glucosinolate content in broccoli (*Brassica oleracea* var. *italica*) heads. *Postharvest Biol. Technol.* **2006**, *41*, 1–8. [CrossRef]
36. Gu, Z.-X.; Guo, Q.-H.; Gu, Y.-J. Factors Influencing Glucoraphanin and Sulforaphane Formation in *Brassica* Plants: A Review. *J. Integr. Agric.* **2012**, *11*, 1804–1816. [CrossRef]

foods

MDPI

Article

# Determining the Optimal Vacuum Frying Conditions for Silver Herring (*Spratelloides gracilis*) Using the Response Surface Methodology

**Hung-I Chien †, Chiu-Chu Hwang †, Yi-Chen Lee, Chun-Yung Huang , Shu-Chuan Chen, Chia-Hung Kuo and Yung-Hsiang Tsai ***

Department of Seafood Science, National Kaohsiung University of Science and Technology, Kaohsiung 811213, Taiwan; aaoc1018@yahoo.com.tw (H.-I.C.); omics1@gmail.com (C.-C.H.); lionlee@nkust.edu.tw (Y.-C.L.); cyhuang@nkust.edu.tw (C.-Y.H.); litengren273@gmail.com (S.-C.C.); kuoch@nkust.edu.tw (C.-H.K.)

* Correspondence: yht@nkust.edu.tw; Tel.: +886-7-3617141 (ext. 23609); Fax: +886-7-3640634

† These authors contributed equally to this work.

**Abstract:** Vacuum frying (VF) is a selective technique for producing high-quality fried food that is mostly used on vegetables, fruits, and potato chips. It is rarely applied to the production of aquatic (especially fish) products. The purpose of this study is to explore whether VF technology can be applied to the preparation of dried silver herring products and to obtain the optimal VF conditions. The response surface methodology (RSM) was used to examine the factors affecting the quality of silver herring (*Spratelloides gracilis*) products after VF, namely temperature (75, 90, and 105 °C), duration (25, 35, and 45 min), and concentration (0, 15, and 30%) of maltose solution used to immerse the samples during pre-processing. The results indicated that VF temperatures had significant impacts on water activity (Aw), moisture content, yield, oil content, lightness ($L^*$ value), and colour difference (*ÄE*). The higher the VF temperature, the lower the Aw, moisture content, yield, and oil content of the product, but the higher the $L^*$ value and $\Delta E$. Next, a longer VF duration resulted in higher oil content of the product. Maltose concentration was significantly and positively correlated with the yield and fracturability of the product. RSM analysis indicated that the optimal combination of processing conditions was a VF temperature of 105 °C, VF duration of 25 min, and maltose concentration of 27%. Under these VF conditions, the silver herring products had a moisture content of 3.91%, Aw of 0.198, oil content of 21.69%, $L^*$ value of 28.19, $\Delta E$ of 27.31, and fracturability of 359 g/s. In summary, vacuum frying technology is suitable for the preparation of dried silver herring products, and this study can provide the optimal processing conditions for seafood processors to obtain better quality.

**Keywords:** vacuum frying; response surface methodology; oil content; colour; fracturability

## 1. Introduction

The silver herring (*Spratelloides gracilis*) is a small pelagic and migratory fish that is widely distributed in the Indo-West Pacific Ocean, and can be caught in the offshore areas surrounding Taiwan [1]. It is a fast-growing and abundant fish species. Being small in size (average body length of approximately 5–6 cm) and rich in calcium, the entire fish is usually sold for consumption in a fresh, dried, or salted state [1]. In general, the traditional method of making dried silver herring in Taiwan is to pour the fresh fish into boiling 10% NaCl water and cook it for 3 min, then expose the fish body to the sun for 8–12 h until dry, or dry the fish with hot air at 50–60 °C for 3–6 h [1]. The dried fish then serves as a snack food or cooking ingredient [1].

Traditional frying (TF), as a drying method, involves heating oil under normal pressure to 160–180 °C before the moisture in the ingredients is dried by boiling and vaporising upon coming into contact with air [2]. The physicochemical qualities of the food change during the frying process, causing the products to have shortcomings such as a rancid odour, dark brown colouration, and shrunken shape [3]. In addition, the quality of frying oil deteriorates with excessive heating, which adversely affects the flavour, stability, colour, and texture of the food. Such products may also be harmful to human health [4–6].

Vacuum frying (VF) is another method of frying and drying. VF is different from TF in that the frying and heating processes occur inside a decompressed and airtight container [3]. Because the boiling point of water decreases when in a decompressed state, moisture is rapidly vaporised and removed in the vacuum system even with lower oil temperatures. Consequently, the food is dehydrated and dried to become crisp, while the original colour and flavour are maintained [7–9]. Moreover, decompression reduces the boiling point of the oil (that is, a drop in the temperature required for frying), which effectively prevents its oxidation and deterioration, thereby prolonging the service life of the frying oil and saving money [8,9]. Currently, common VF products include chips made from fruits, mushrooms, sweet potatoes, yams, carrots, and potatoes [4,8,10].

Generally, food ingredients undergo pre-processing prior to VF for the purpose of promoting dehydration, reducing oil content, maintaining colour, and increasing crispness [3,4]. Common pre-processing methods before VF include sugaring, salting, freezing, and blanching. Sugaring refers to immersing the ingredients in a sugar solution (maltose is commonly used) before VF. The sugar molecules enter the ingredients owing to osmotic pressure, whereas water molecules flow out to cause dehydration. This process reduces oil absorption during subsequent frying [4]. By reducing the oil content of VF products, sugaring helps to better maintain their colour and crispness [4,10]. The principle of salting is similar to that of sugaring, except that it will make the products taste salty. It is generally less used in the production of vegetable, fruit, and starch (potato) chips. Freezing converts the moisture content within the tissues of food ingredients into ice crystals. These sublimate and accelerate dehydration when the temperature rises and the pressure drops during frying. Concurrently, the tissues become hollow and porous, thereby increasing the crispness of the product [3,4,10]. During blanching, the ingredients are heated with boiling water for at least 30 s, which fixes their shape and maintains the colour and appearance of the finished product. For example, green vegetables must first be blanched before VF to better maintain their green lustre [3,4,10].

The response surface methodology (RSM) is used to examine the impact of various factors on the response during a multi-factor experiment. It involves appropriate experimental design, mathematical inference, and systematic analysis, and aims to determine the lowest number of experimental groups to obtain the optimal processing conditions [11,12]. Specifically, RSM predicts the corresponding results produced under different control factor settings. It does so by establishing the functional relationship between the factors and response, thereby arriving at the optimal response or operating conditions [11,12].

Most of the foods used for VF research are starches (such as potatoes) or fruits and vegetables (such as carrots) [13–15]. However, VF research using fish as the ingredient are rare [16,17]. The purpose of this study is to explore whether VF technology can be applied to the preparation of dried silver herring products and to obtain the optimal VF conditions. RSM was used to design experimental conditions involving three factors and three levels per factor, namely (i) VF temperature (75, 90, and 105 °C), (ii) VF duration (25, 35, and 45 min), and (iii) maltose concentration used for pre-processing (0, 15, and 30%). The selection of the above factors and levels is mainly based on our previous preliminary experimental results in the laboratory and the research results of Yang (2004) [18]. The aim was to examine the impacts of these factors and levels on the quality of the finished products. RSM calculation and analysis were also used to arrive at the optimal processing conditions.

## 2. Materials and Methods

### 2.1. Preparation of Silver Herring Samples

Fresh silver herring were purchased from a supermarket in Kaohsiung City, Taiwan, placed in crushed ice, and transported back to the National Kaohsiung University of Science and Technology laboratory. Upon arrival, the fish were immediately boiled in 90 °C salt water (1.0% NaCl) for 30 s, according to Yang (2004) [18]. After removal, the fish were drained dry, cooled to room temperature, and then stored in a freezer for subsequent use. Palm oil was used for frying the samples in the experiments.

### 2.2. VF Experiments

The vacuum fryer used in the study was a Type VF-5 (Chin Ying Fa Co., Ltd., Changhua, Taiwan). The VF level was set to 6.65 kPa, the centrifugal de-oiling speed was 1000 rpm, and the de-oiling time was 1 min. RSM using a Box–Behnken design (BBD) (a factorial design) was used to investigate the response patterns and also to determine the optimum synergy of variables. The BBD model of RSM was applied to establish a tri-factor, tri-level design, namely VF temperature ($X_1$; 75, 90, and 105 °C), VF duration ($X_2$; 25, 35, and 45 min), and maltose concentration during pre-processing ($X_3$; 0, 15, and 30%). A total of 15 sets of experiments were designed (Table 1). The selection of the above factors and levels is mainly based on our previous preliminary experimental results in the laboratory and the research results of Yang (2004) [18]. Three replicate experiments were conducted in each treatment group.

**Table 1.** The levels and ratios of the response surface methodology design with different operation conditions.

| Treatment | $X_1$ | $X_2$ | $X_3$ |
|---|---|---|---|
| 1 | 75 (−1) | 25 (−1) | 15 (0) |
| 2 | 105 (+1) | 25 (−1) | 15 (0) |
| 3 | 75 (−1) | 45 (+1) | 15 (0) |
| 4 | 105 (+1) | 45 (+1) | 15 (0) |
| 5 | 75 (−1) | 35 (0) | 0 (−1) |
| 6 | 75 (−1) | 35 (0) | 30 (+1) |
| 7 | 105 (+1) | 35 (0) | 0 (−1) |
| 8 | 105 (+1) | 35 (0) | 30 (+1) |
| 9 | 90 (0) | 25 (−1) | 0 (−1) |
| 10 | 90 (0) | 25 (−1) | 30 (+1) |
| 11 | 90 (0) | 45 (+1) | 0 (−1) |
| 12 | 90 (0) | 45 (+1) | 30 (+1) |
| 13 (Centre point) | 90 (0) | 35 (0) | 15 (0) |
| 14 (Centre point) | 90 (0) | 35 (0) | 15 (0) |
| 15 (Centre point) | 90 (0) | 35 (0) | 15 (0) |

$X_1$ = VF temperature (°C), $X_2$ = VF duration (min), $X_3$ = Maltose concentration (%).

For the sugaring step, the boiled fish samples were immersed in maltose solutions of various concentrations at 4 °C for 1 h. The impacts of the three processing factors of VF, namely temperature ($X_1$), VF duration ($X_2$), and maltose concentration ($X_3$), on the quality of the fried silver herring products were then examined. The use of these three processing factors is based on Yang's (2004) research [18]. The responses for sample testing included water activity (Aw), moisture content (%), yield (%), oil content (%), lightness ($L^*$ value), colour difference ($\Delta E$), and fracturability (g/s).

### 2.3. *Quality Analysis of the Fried Products*

#### 2.3.1. Water Activity (Aw)

After 10 g of the sample was finely ground, 1.0 g of the finely ground sample was placed into a plastic container for Aw determination using the AquaLab Model Series 3 (Decagon, Pullman, WA, USA) water activity meter [19]. The Aw measurement of each treatment sample was performed in duplicate.

#### 2.3.2. Moisture Content

Approximately 5 g of the finely ground sample was placed in an aluminium dish, and the moisture content was determined using an infrared moisture analyser (Model HE53, Mettler Toledo GmbH, Greifensee, Switzerland). The moisture content of the sample was recorded when a constant weight was reached [20]. The moisture content measurement of each treatment sample was performed in duplicate.

#### 2.3.3. Yield

The yield is the ratio of the weight of the sample after frying to its weight before frying and is calculated as a percentage (%). The weight of the sample was measured with an electronic balance (Mettler Toledo GmbH, Greifensee, Switzerland) [21]. The yield measurement of each treatment sample was performed in duplicate.

#### 2.3.4. Oil Content

Oil content was extracted from the sample with petroleum ether using Soxhlet extraction equipment (Soxtec System 2055 Tecator, FOSS, Hillerod, Denmark) and then gravimetrically determined [21]. The oil content measurement of each treatment sample was performed in duplicate.

#### 2.3.5. Colour Analysis

A colour meter (CR-300 Chroma meter, Konica Minolta, Inc., Tokyo, Japan) was used to measure the colour values of the sample. The indicators included lightness ($L^*$), +red to −green ($a^*$), and +yellow to −blue ($b^*$). $\Delta E$ was calculated based on the $L^*$, $a^*$, and $b^*$ values using the following equation [21]:

$$\Delta E = \sqrt{(L^* - L_0^*)^2 + (a^* - a_0^*)^2 + (b^* - b_0^*)^2} \quad (1)$$

where $L_0^*$, $a_0^*$, and $b_0^*$ refer to the colour values of the sample before VF; and $L^*$, $a^*$, and $b^*$ are the colour values after VF. The colour measurement of each treatment sample was performed in duplicate.

#### 2.3.6. Texture Measurement

The texture profile analysis of the samples was measured with a Brookfield CT3 texture analyser [16]. The measurement item was fracturability (g/s), with the conditions set as follows: (i) probe: TA 39 cylinder, 2 mm D, 20 mm L; (ii) target value: 4.00 mm; (iii) predicted speed: 2 mm/s; (iv) test and return speeds: 1.5 mm/s; and (v) trigger point load: 2 g. In this study, the parameters and conditions used for texture analysis are mainly modified from the research of Su et al. [21]. The fracturability measurement of each treatment sample was performed in duplicate.

### 2.4. *Optimization Procedure*

Optimization of processing conditions is based on responses of moisture content, oil content, texture (fracturability), and $L^*$ value. The numerical optimization techniques are used to optimize multiple responses simultaneously. The desired goal for each processing parameter and response is chosen, and all processing parameters are maintained within the specified parameter range. The target values on oil content and moisture content were minimized, while target values on $L^*$ value and fracturability were maximized. The response surface plot was prepared after the optimal regression model was obtained using the numerical optimization tool of Design-Expert 7.0 software.

### 2.5. Statistical Analysis

The Design-Expert 7.0 (Stat-Ease Inc., Minneapolis, MN, USA) statistical software was used to analyse the results of the various factor combinations and determine the optimal conditions. Data obtained from sample analysis were using an analysis of variance (ANOVA) to study the effect of processing conditions on the responses of vacuum fried silver herring. The probability level $p < 0.05$ indicated that the difference was significant. In addition, a regression analysis was performed on the experimental data and expressed using an empirical second-order polynomial model, as follows:

$$Y = \beta_0 + \sum_{i=1}^{3} \beta_i X_i + \sum_{i=1}^{3} \beta_{ii} X_i^2 + \sum_{i=1}^{2} \sum_{j=i+1}^{3} \beta_{ij} X_i X_j \quad (2)$$

where $Y$ is the predicted response, $\beta_0$ is a constant, $\beta_i$ is the linear coefficient, $\beta_{ii}$ is the quadratic coefficient, $\beta_{ij}$ is the interaction coefficient of variables $i$ and $j$, and $X_i$ and $X_j$ are independent variables.

RSM was used to analyse the impacts of VF frying on the samples' parameters, including Aw, moisture content, yield, oil content, $L^*$ value, $\Delta E$, and fracturability. BBD was used to analyse the response ($Y$) of these qualities and obtain the optimal VF conditions. Long et al., (2014) [22] concluded that the larger the F value of the model and the smaller the P value, the more significant the impact of this factor on the response ($Y$). Jentzer et al., (2015) [23] stated that the coefficient of determination ($R^2$) represents the goodness-of-fit that the prediction model of the regression method has on the experimental data. The closer the value of $R^2$ to 1, the greater the reliability of the model in explaining the experiment. Lack of fit can be used to analyse parameters of the experiments to determine the suitability of the model. It is usually recommended to be higher than 0.1.

## 3. Results

The regression coefficients and analysis of variance (ANOVA) of the response as a function of the independent variables on vacuum-fried silver herring are presented in Table S1. In addition, the experimental and predicted results of the central composite rotatable design for VF silver herring are shown in Tables S2 and S3, respectively. VF uses lower temperature than traditional frying to remove moisture from the food and reduce the oil content in the final product. The results indicated that VF temperatures had significant impacts on Aw, moisture content, yield, oil content, lightness ($L^*$ value), and colour difference ($\ddot{A}E$). Higher VF temperatures resulted in lower Aw, moisture content, yield, and oil content of the product, but higher $L^*$ value and $\Delta E$. Next, the longer the VF duration, the higher oil content of the product. Maltose concentration was significantly and positively correlated with the yield and fracturability of the product.

### 3.1. Changes in Aw

Following a statistical regression analysis of the Aw data, the prediction model of the established quadratic polynomial regression equation was as follows:

$$Y = 0.16 - 0.12X_1 - 0.051X_2 + 0.024X_3 + 0.057X_1X_2 - (3.250E - 003)X_1X_3 - (6.833E - 0003)X_2X_3 + 0.092{X_1}^2 + 0.077{X_2}^2 - 0.040{X_3}^2 \quad (3)$$

where $Y$ is the Aw, $X_1$ is the VF temperature (°C), $X_2$ is the VF duration (min), and $X_3$ is the maltose concentration (%). Regression coefficients and analysis of variance (ANOVA) of the Aw as a function of the independent variables on vacuum fried silver herring are shown in Table S1.

Based on the BBD variance analysis (Table S1), the $p$ value of this model was 0.0428 < 0.05, which meant that under the interactive effects of the three factors, the model of the quadratic polynomial regression equation had significant impacts on Aw. For the one-order analysis, the $F$ value of VF temperature ($X_1$) was the highest at 25.21, indicating that the impact of VF temperature ($X_1$) on Aw was higher than that of VF duration ($X_2$) and maltose concentration ($X_3$). The impact was also significant ($p < 0.05$). $R^2$ was 0.9028, indicating that the prediction model could explain 90.28% of the total variance. The $p$ value of 0.0526 for the lack-of-fit test was not significant ($p > 0.05$), indicating that the established regression equation actually reflected the relationship between the factors and Aw ($Y$).

For the one-order analysis, VF temperature ($X_1$) and duration ($X_2$) had relatively large impacts on Aw. Hence, the Aw response surface and contour plots were prepared after these two factors were optimized (Figure 1). When the maltose concentration was fixed at 15%, the RSM analysis showed that Aw decreased with increasing VF temperature ($X_1$) and duration ($X_2$). The contour plot indicated that VF must be carried out at a high temperature or for a long time to reach the lowest Aw. In this experiment, the lowest Aw was obtained when the VF temperature and duration were 105 °C and 45 min, respectively. Our findings were similar to those of Reis et al., (2008) [24], namely, the higher the VF temperature and the longer the VF duration, the lower the Aw of the fried food product. The interaction between the variable factors is significant if the contour plot presents an elliptical shape, but not significant if the plot is circular [25,26]. The interaction between the two factors of VF temperature ($X_1$) and duration ($X_2$) was significant because the Aw contour plot of our results tended to be elliptical.

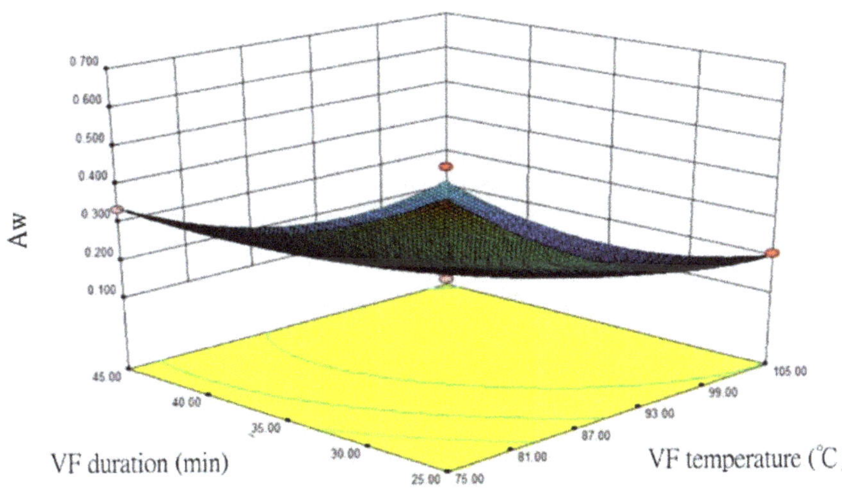


**Figure 1.** *Cont.*

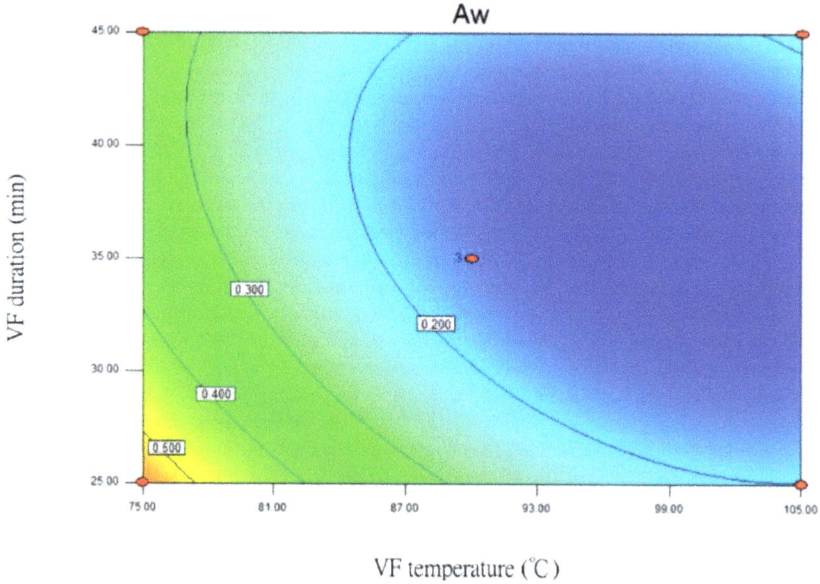


**Figure 1.** The response surface and contour plots for the impacts of vacuum frying (VF) temperature and duration on the Aw in fried silver herring (maltose concentration at 15%).

*3.2. Changes in Moisture Content*

The prediction model of the established quadratic polynomial regression equation following a statistical regression analysis of the data on moisture content was as follows:

$$Y = 2.67 - 0.9X_1 - 0.34X_2 + 0.35X_3 - 0.42X_1X_2 + 0.12X_1X_3 - 0.28X_2X_3 + 0.44{X_1}^2 + 0.25{X_2}^2 + 0.14{X_3}^2 \quad (4)$$

where $Y$ is moisture content, $X_1$ is VF temperature (°C), $X_2$ is VF duration (min), and $X_3$ is maltose concentration (%). Regression coefficients and analysis of variance (ANOVA) of the moisture content as a function of the independent variables on vacuum fried silver herring are shown in Table S1.

Based on the BBD variance analysis (Table S1), the $p$ value of this model was 0.0917 > 0.05, indicating that the quadratic polynomial regression equation had no significant impact on the moisture content following the interactions of these three factors. The $F$ value of 17.63 for VF temperature ($X_1$) was the highest in the one-order analysis, indicating that its impact on moisture content was higher than that of VF duration ($X_2$) and maltose concentration ($X_3$). Furthermore, the impact was significant ($p < 0.05$). $R^2$ was 0.9034, meaning that the prediction model could explain 90.34% of the total variance. The value of $p$ in the lack-of-fit test was 0.1554, which was not significant ($p > 0.1$). This indicated that the established regression equation actually reflected the relationship between the factors and moisture content ($Y$).

During the one-order analysis, VF temperature ($X_1$) and maltose concentration ($X_3$) had relatively large impacts on moisture content. Hence, the response surface and contour plots for moisture content were prepared after these two factors were optimized (Figure 2). When the VF duration ($X_2$) was fixed at 35 min, the RSM analysis showed that the higher the VF temperature ($X_1$), the lower the moisture content. The moisture content was the lowest when the VF temperature ($X_1$) was the highest (105 °C). The contour plot tended to be circular, indicating that the interaction between the two factors of VF temperature and maltose concentration was not significant. The impact of maltose concentration on moisture content was also not significant.

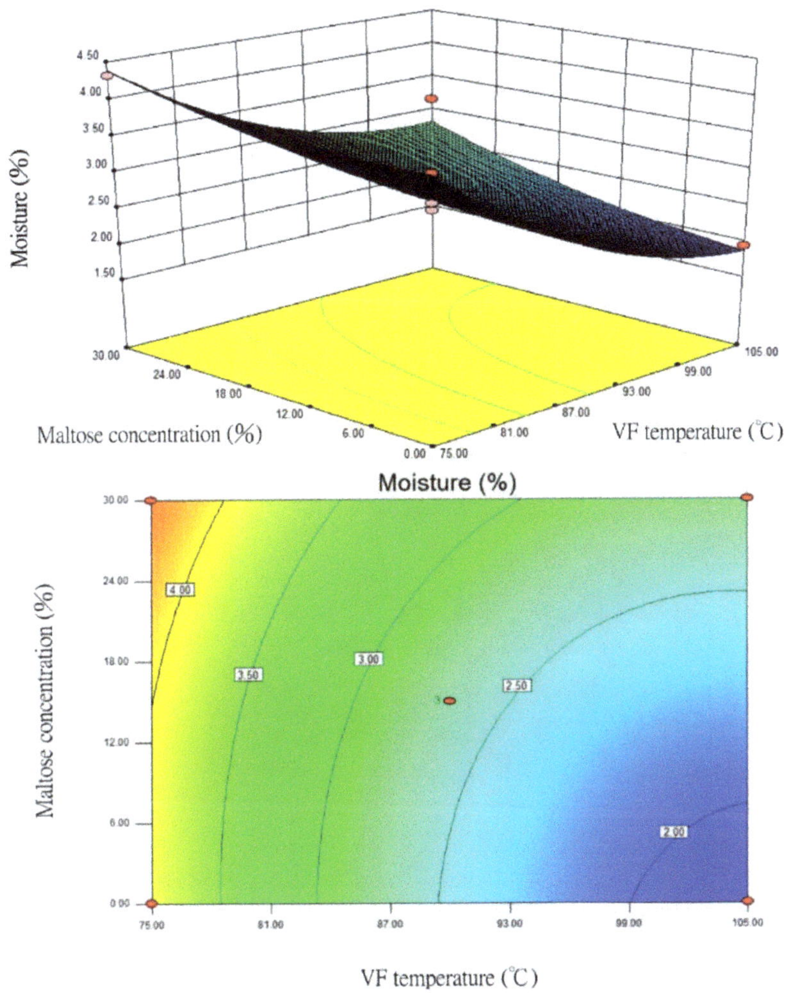


**Figure 2.** The response surface and contour plots for the impacts of VF temperature and maltose concentration on the moisture content in fried silver herring (VF duration of 35 min).

### 3.3. *Changes in Yield*

Following a statistical regression analysis of the yield data, the prediction model of the established quadratic polynomial regression equation was as follows:

$$Y = 22.96 - 2.09X_1 + 7.500E - 003X_2 + 3.53X_3 + 0.64X_1X_2 - 0.33X_1X_3 + 0.79X_2X_3 + 0.35X_1^2 + 0.79X_2^2 + 0.33X_3^2 \quad (5)$$

where $Y$ is the yield, $X_1$ is the VF temperature (°C), $X_2$ is the VF duration (min), and $X_3$ is the maltose concentration (%). Regression coefficients and analysis of variance (ANOVA) of the yield as a function of the independent variables on vacuum fried silver herring are shown in Table S1.

Based on the BBD variance analysis (Table S1), the $p$ value of this model was 0.0575 > 0.05, indicating that the impact of the quadratic polynomial regression equation on the yield was not significant following the interaction of these three factors. For the one-order analysis, the $F$ value of 28.07 for maltose concentration ($X_3$) was the highest, showing that its impact on the yield was very significant ($p < 0.01$) and higher than that of VF temperature ($X_1$) and duration ($X_2$). The VF temperature ($X_1$) also had a significant impact on the yield ($0.05 > p > 0.01$). However, the $p$ value of the VF duration ($X_2$) was 0.9914, showing that its impact on the yield was minimal. $R^2$ was 0.8890, indicating that the prediction model could explain 88.90% of the total variance. The value of $p$ in the lack-of-fit test was 0.0585, which was not significant ($0.1 > p > 0.05$). This indicated that the established regression equation accurately reflected the relationship between the factors and the yield ($Y$).

The one-order analysis revealed that VF temperature ($X_1$) and maltose concentration ($X_3$) had relatively large impacts on the yield. Therefore, the response surface and contour plots of the yield were prepared after these two factors were optimized (Figure 3). When the VF duration ($X_2$) was fixed at 35 min, the RSM analysis showed that the yield increased with increasing maltose concentrations ($X_3$). However, the higher the VF temperature ($X_1$) resulted in the lower the yield. When the VF temperature was 105 °C, the yield of the sample was lower than 20%. Because the purpose of this study was not to improve the yield, the results were not listed in the follow-up optimization conditions.

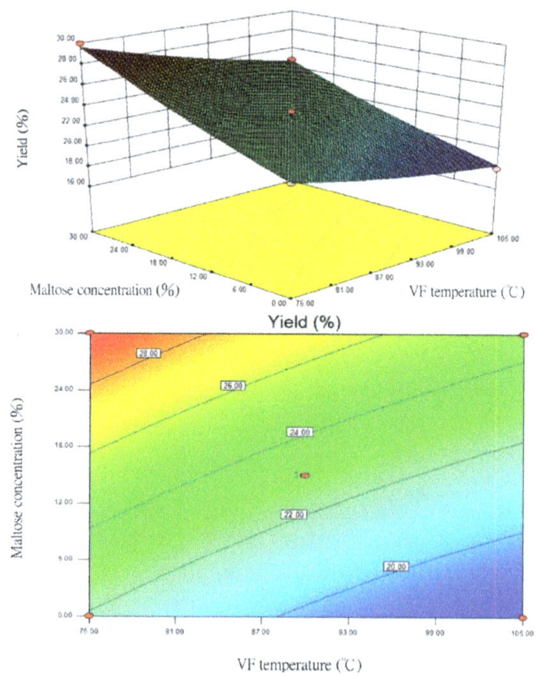


**Figure 3.** The response surface and contour plots for the impacts of VF temperature and maltose concentration on the yield of fried silver herring (VF duration of 35 min).

### 3.4. Changes in Oil Content

After a statistical regression analysis of the oil content data, the prediction model established by the quadratic polynomial regression equation was as follows:

$$Y = 24.95 - 1.1X_1 + 1.47X_2 - 0.88X_3 + 1.06X_1X_2 - 1.15X_1X_3 - 0.19X_2X_3 + 1.17X_1^2 + 1.59X_2^2 - 1.4X_3^2 \quad (6)$$

where $Y$ is oil content, $X_1$ is VF temperature (°C), $X_2$ is VF duration (min), and $X_3$ is maltose concentration (%). Regression coefficients and analysis of variance (ANOVA) of the oil content as a function of the independent variables on vacuum fried silver herring are shown in Table S1.

Based on the BBD variance analysis (Table S1), the $p$ value of this model was $0.0319 < 0.05$, indicating that the quadratic polynomial regression equation had a significant impact on oil content following the interaction of these three factors. The $F$ value for VF duration ($X_2$) was the highest at 14.02, indicating that the impact on oil content was higher than that of VF temperature ($X_1$) and maltose concentration ($X_3$). The impact was also significant ($p < 0.05$). The coefficient of determination $R^2$ was 0.9146, meaning that the prediction model could explain 91.46% of the total variance. The $p$ value of 0.7440 for the lack-of-fit model was not significant ($p > 0.1$). This showed that the established regression equation reflected the relationship between the factor and oil content ($Y$). In summary, the regression model was reasonable and the equation could be used to predict oil content.

Under the one-order analysis, the impacts of VF temperature ($X_1$) and duration ($X_2$) on oil content were relatively large. Therefore, after these two factor variables were optimized, the response surface and contour plots of oil content were drawn (Figure 4). When the maltose concentration ($X_3$) was fixed at 15%, the RSM analysis results showed that the higher the VF temperature had the lower the oil content of the sample. In contrast, the longer the VF duration resulted in the higher the oil content. In the contour plot, the relationship between VF temperature and duration was close to an ellipse, indicating that the two factors had significant impacts on oil content. Samples with the lowest oil content were produced when the VF temperature was between 99 °C and 104 °C and the VF duration was between 26 and 28 min.

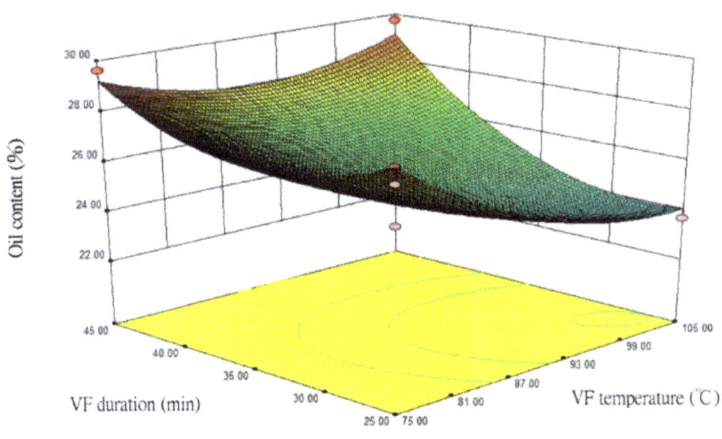


**Figure 4.** *Cont.*

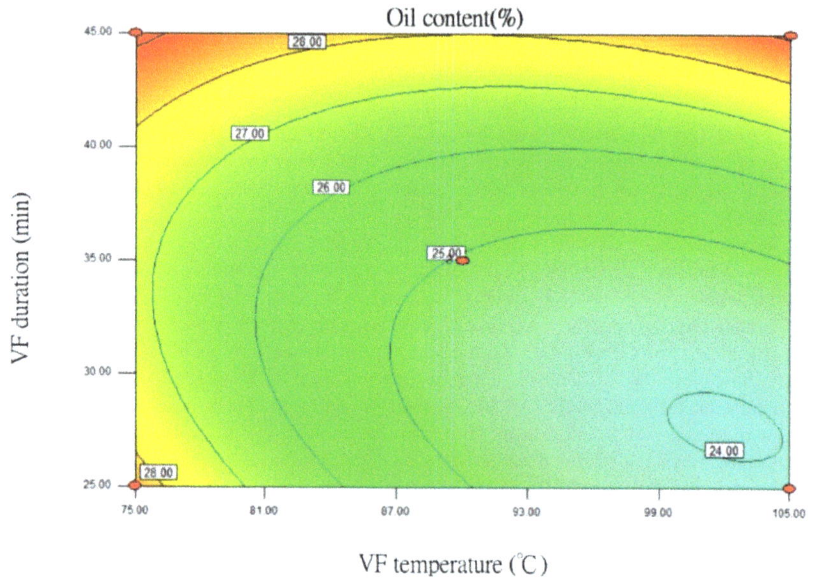


**Figure 4.** The response surface and contour plots for the impacts of VF temperature and duration on the oil content of fried silver herring (maltose concentration at 15%).

### 3.5. Changes in the *L** Value

The prediction model of the established quadratic polynomial regression equation, following the statistical regression analysis of the *L** value data, was as follows:

$$Y = 26.99 + 1.73X_1 - 0.3X_2 - 0.5X_3 - 0.91X_1X_2 - 0.42X_1X_3 + 0.36X_2X_3 - 0.062X_1{}^2 + 0.39X_2{}^2 - 1.11X_3{}^2 \quad (7)$$

where $Y$ is the $L^*$ value, $X_1$ is the VF temperature (°C), $X_2$ is VF duration (min), and $X_3$ is the maltose concentration (%). Regression coefficients and analysis of variance (ANOVA) of the $L^*$ as a function of the independent variables on vacuum fried silver herring are shown in Table S1.

Based on the BBD variance analysis (Table S1), the $p$ value of this model was 0.0554 > 0.05, indicating that the quadratic polynomial regression equation did not have any significant impact on the $L^*$ value after the interaction of these three factors. However, in the one-order analysis, the VF temperature ($X_1$) had the highest F value of 26.68. This indicated that the impact of $X_1$ on the $L^*$ value was higher than that of VF duration ($X_2$) and maltose concentration ($X_3$), and that the impact was very significant ($p < 0.01$). The $R^2$ value of 0.8908 indicated that the prediction model could explain 89.08% of the total variance. The $p$ value of 0.8873 for the lack-of-fit test was not significant ($p > 0.1$), indicating that the established regression equation could reflect the relationship between the factors and the $L^*$ value. Based on the above, the regression model was reasonable and the equation could be used to predict the $L^*$ value.

For the one-order analysis, VF temperature ($X_1$) and maltose concentration ($X_3$) had relatively large impacts on the $L^*$ value. Therefore, after optimizing these two factor variables, the response surface and contour plots of the $L^*$ value were prepared (Figure 5). When the VF duration ($X_2$) was fixed at 35 min, the RSM analysis showed that as the VF temperature ($X_1$) increased, the $L^*$ values of the samples also increased significantly. Although increasing maltose concentrations ($X_3$) caused the $L^*$ value to rise, the $L^*$ inversed and decreased when the concentration exceeded 16%. The surface of the samples became darker, which might be related to the browning caused by the Maillard reaction of excess maltose and amino acids. The contour plot shows that the relationship between VF temperature ($X_1$) and maltose concentration ($X_3$) tended to be elliptical, which meant that the two factors had significant impacts on the $L^*$ value.

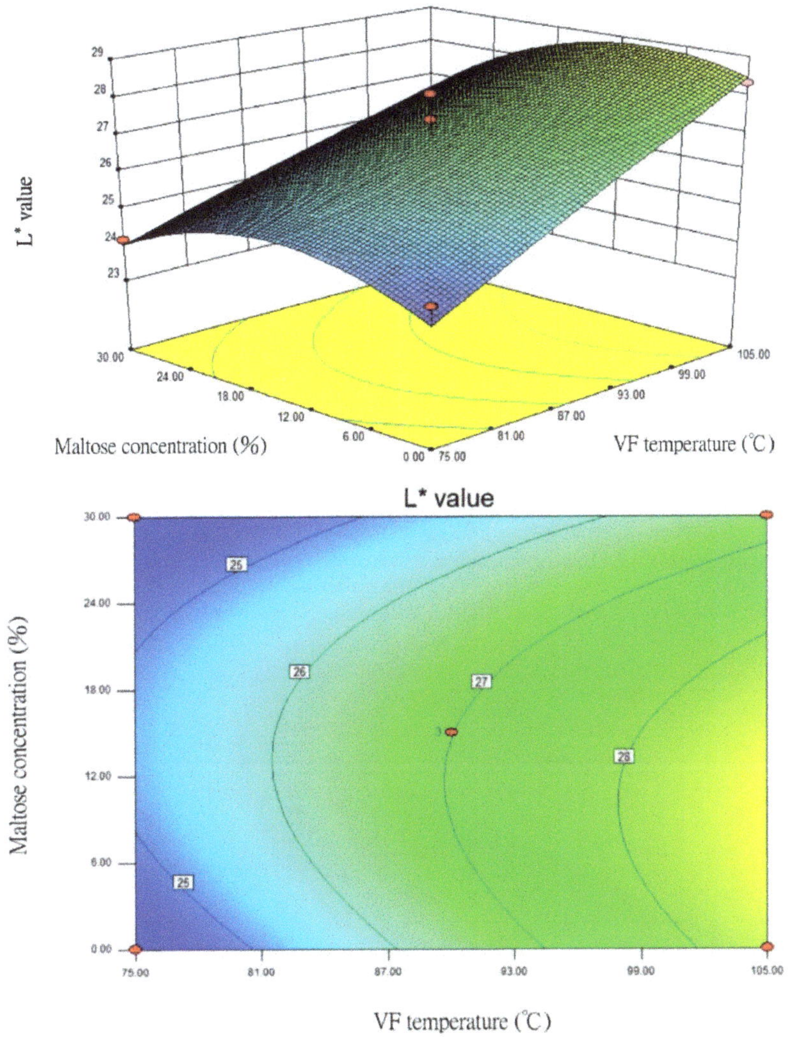


**Figure 5.** The response surface and contour plots for the impacts of VF temperature and maltose concentration on the $L^*$ value of fried silver herring (VF duration of 35 min).

### 3.6. Changes in *ÄE*

Following a statistical regression analysis of the *ÄE* data, the prediction model of the quadratic polynomial regression equation was as follows:

$$Y = 27.2 + 1.72X_1 - 0.3X_2 - 0.53X_3 - 0.92X_1X_2 - 0.41X_1X_3 + 0.36X_2X_3 - 0.094X_1^2 + 0.4X_2^2 - 1.1X_3^2 \quad (8)$$

where $Y$ is *ÄE*, $X_1$ is VF temperature (°C), $X_2$ is VF duration (min), and $X_3$ is maltose concentration (%). Regression coefficients and analysis of variance (ANOVA) of the *ÄE* as a function of the independent variables on vacuum fried silver herring are shown in Table S1.

Based on the BBD variance analysis (Table S1), the values of *ÄE* and *L** had the same trend under the BBD variable analysis. The $p$ value of this *ÄE* model was 0.0576 > 0.05, indicating that the impact of the quadratic polynomial regression equation model was not significant. The $R^2$ was 0.8888, which meant that the prediction model could explain 88.88% of the total variance. The value of $p$ in the lack-of-fit test was 0.9056. This was not significant ($p > 0.1$), indicating that the established regression equation reflected the relationship between the factors and *ÄE*.

For the one-order analysis, the $F$ value of VF temperature ($X_1$) was the highest at 25.81 and the impact was significant ($p < 0.01$). This $F$ value was much higher than that of VF duration ($X_2$) and maltose concentration ($X_3$), indicating that $X_1$ had a significant impact on *ÄE* and that the impact was much higher than that of $X_2$ and $X_3$. The regression model was deemed to be reasonable based on the above, and the equation could be used to predict *ÄE*.

The impacts of VF temperature ($X_1$) and maltose concentration ($X_3$) on the *ÄE* of silver herring were relatively large during the one-order analysis. Hence, the response surface and contour plots of *ÄE* were drawn after these two factor variables were optimized (Figure 6). This figure presents similar results to that for the *L** values (Figure 5). In other words, when the VF duration ($X_2$) was fixed at 35 min, the hotter the VF temperature ($X_1$) resulted in the higher the *ÄE*. Although the *ÄE* increased with an increasing maltose concentration ($X_3$), the *ÄE* inversed and decreased when the concentration exceeded 16%. This meant that the colour difference on the surface of the samples diminished and they tended to become darker. This might be related to the browning caused by the Maillard reaction arising from excess maltose and amino acids. The contour plot showed that the relationship between VF temperature ($X_1$) and maltose concentration ($X_3$) tended to be elliptical, indicating that both factors had significant impacts on the *ÄE* value.

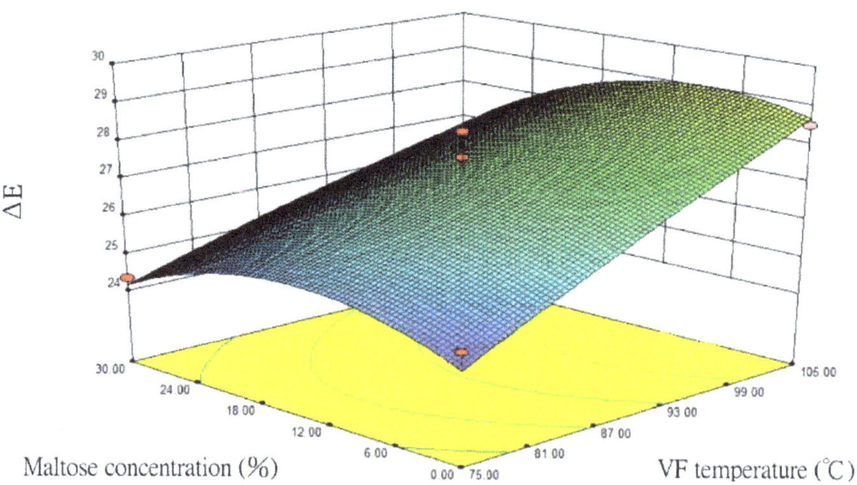


**Figure 6.** *Cont.*

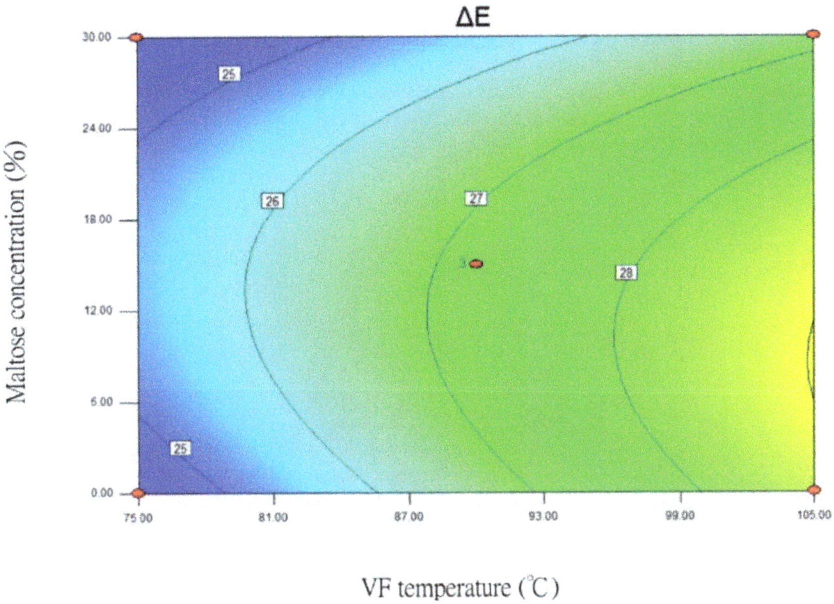


**Figure 6.** The response surface and contour plots for the impacts of VF temperature and maltose concentration on the *ÄE* value of fried silver herring (VF duration of 35 min).

### 3.7. Changes in Fracturability

After the statistical regression analysis of the fracturability data, the prediction model of the quadratic polynomial regression equation was established as follows:

$$Y = 284.69 - 16.13X_1 + 27.27X_2 + 76.29X_3 + 5.79X_1X_2 + 74.32X_1X_3 + 9.37X_2X_3 + 36.2X_1^2 + 29.27X_2^2 + 41.85X_3^2 \quad (9)$$

where $Y$ is fracturability, $X_1$ is VF temperature (°C), $X_2$ is VF duration (min), and $X_3$ is maltose concentration (%). Regression coefficients and analysis of variance (ANOVA) of the fracturability as a function of the independent variables on vacuum fried silver herring are shown in Table S1.

Based on the BBD variance analysis (Table S1), the $p$ value of this model was $0.1300 > 0.05$, which meant that the quadratic polynomial regression equation had no significant impact on fracturability following the interaction of these three factors. Nevertheless, the prediction model could explain 83.73% of the total variance because the value of the coefficient of determination $R^2$ was 0.8373. The $p$ value of the lack-of-fit model was 0.0524, which was not significant ($p > 0.05$). This indicated that the established regression equation truly reflected the relationship between the factors and fracturability. The $F$ value of maltose concentration ($X_3$) during the one-order analysis was the highest at 13.34 and the impact was significant ($0.01 < p < 0.05$). This $F$ value was much higher than those for VF temperature ($X_1$) and duration ($X_2$), which were 0.60 and 1.71, respectively. This indicated that the impact of maltose concentration on fracturability was higher than those of VF temperature and duration.

Under the one-order analysis, VF duration ($X_2$) and maltose concentration ($X_3$) had relatively large impacts on the fracturability of the fried silver herring. After those two factors were optimized, the response surface and contour plots were prepared (Figure 7). When the VF temperature ($X_1$) was fixed at 90 °C, the RSM analysis showed that the higher the maltose concentration had the higher the fracturability. The products had less fracturability when the maltose concentration was between 0 and 6%. Next, the longer the VF duration resulted in the lower the fracturability. When the duration was between 28 and 35 min, fracturability was relatively poor. The contour plot showed that the relationship between the VF temperature ($X_1$) and maltose concentration ($X_3$) was close to an ellipse, indicating that the two factors had significant impacts on fracturability. Overall, the findings showed a strong relationship between maltose concentration and fracturability.

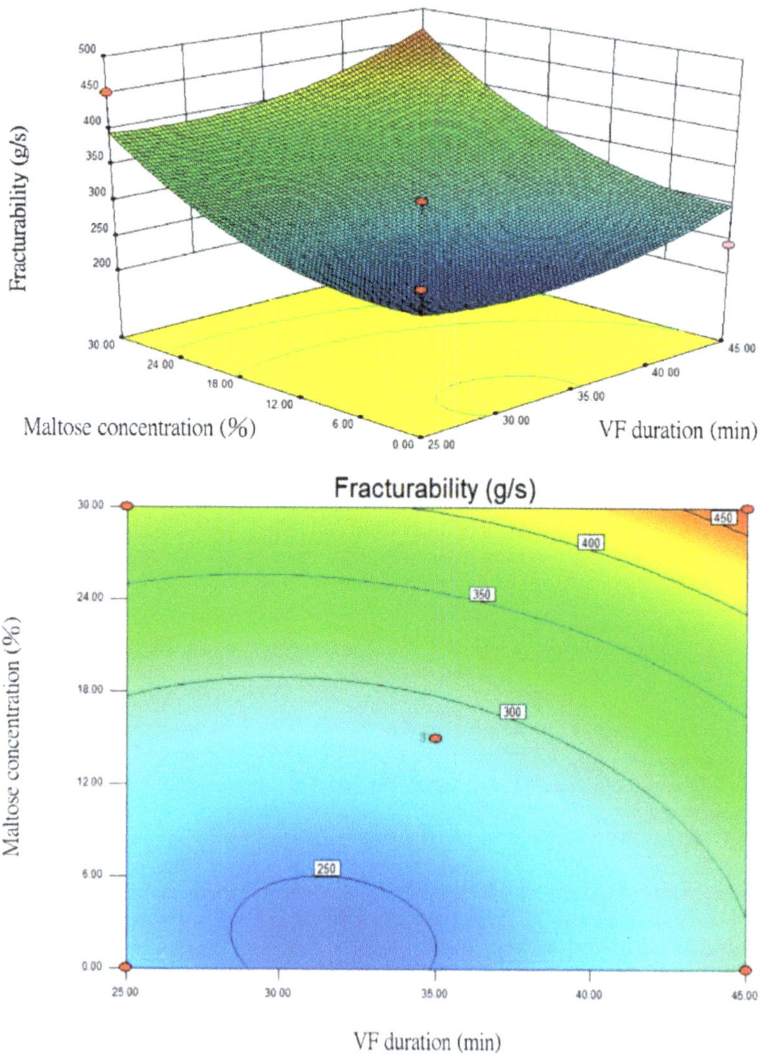


**Figure 7.** The response surface and contour plots for the impacts of VF duration and maltose concentration on the fracturability of fried silver herring (VF temperature at 90 °C).

### *3.8. Analysis of Optimal VF Conditions for Silver Herring*

The Design-Expert 7.0 (Stat-Ease Inc., Minneapolis, MN, USA) statistical software was used to analyse the results of the various combinations of samples and optimal conditions. The optimal regression model was arrived at after preparing a response surface plot (Figure 8). Specifically, the optimal combination of processing conditions was a VF temperature ($X_1$) of 105 °C, VF duration ($X_2$) of 25 min, and maltose concentration ($X_3$) of 27%. The resultant quality values of the fried silver herring products were as follows: moisture content at 3.91%, Aw at 0.198, oil content at 21.69%, *L** value at 28.19, *ÄE* at 27.31, and fracturability at 359 g/s.

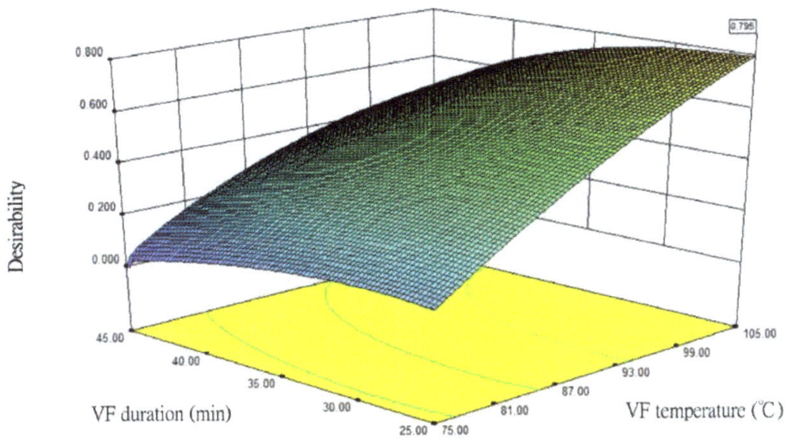


**Figure 8.** The response surface plot for the optimal VF conditions for fried silver herring. VF temperature of 105 °C, VF duration of 25 min, and maltose concentration of 27%.

## 4. Discussion

According to the results of this study, VF temperature had a significant impact on Aw and moisture content: the higher the VF temperature, the lower the Aw and moisture content of the products. These results were similar to those of Pan et al. [19] and Andres-Bello et al. [20], who stated that the moisture content of VF breaded shrimps and sea bream fillets decreased with rising VF temperatures. Also, the oil content reduced with increasing temperatures. Andres-Bello et al. [20] reported similar results for VF of sea bream fillets: the oil absorption of the samples reduced when the VF temperature increased (90–110 °C), and fish fillets prepared at the highest temperature of 110 °C had the lowest oil content. In contrast, the longer the VF duration, the higher the oil content tended to be. The results of this study were also similar to the reports of Pan et al. [19] and Andres-Bello et al. [20], who concluded that the oil content of VF breaded shrimps and sea bream fillets increased with longer VF durations. Most studies on VF fruit, vegetable, and potato chips had similar findings to ours [10,12,24–26]. Akinpelu et al. [12] proposed that the oil absorption amount was positively correlated with VF duration in a VF system.

In terms of yield, the higher the VF temperature resulted in the lower the yield of the samples. Andres-Bello et al. [20] similarly reported that the weight loss of VF sea bream fillets increased with higher VF temperatures. This study further found a significantly positive correlation between maltose concentration and yield: the stronger the concentration, the higher the yield. Shyu et al. [10] immersed carrot chips in fructose before VF and found that the longer the duration of immersion, the higher the sugar content in the products. When fried in this state, the products retained more moisture but had a lower oil content. Consequently, the weight loss of the samples was reduced. Jayaraman et al. [27] proposed that immersing cauliflower in a salt and sucrose solution before dehydration could reduce shrinkage and preserve the integrity of the cell walls. It was deduced that the maltose molecules entered the tissues of the ingredients due to osmotic pressure. This led to the retention of additional water molecules and reduced oil absorption during subsequent frying. As a result, the appearance and cell integrity of the products were maintained, weight loss was reduced, and the yield increased.

In terms of the colour value, this study found that the higher the VF temperature had the higher the $L^*$ value of the samples. This result was contrary to the findings of previous research on VF vegetable, fruit, and potato chips; sea bream fillets; and breaded shrimps, most of which reported that the higher the VF temperature had the lower the $L^*$ value and the darker the appearance [10,12,19,20,24,25]. The main cause for this difference was the skin of the silver herring, which is covered in small, silvery-white scales. The skin became brighter and whiter when heated at high temperatures, resulting in an increase in the $L^*$ value [1]. Next, the higher the VF temperature resulted in the higher the $\ddot{A}E$ of the samples, which was similar to the results of previous research [10,12,19,20,24,25]. The main reason was that high-temperature thermal treatment increased the rate of non-enzymatic browning reaction and products [10,19,20].

In general, dehydration caused by the frying process leads to textural changes in the samples, such as increased crispness of the product. In this study, fracturability was used as an indicator of the crispness of the fried sliver herring. Specifically, the higher the fracturability had the crisper the product. This study found that the stronger the maltose concentration, the higher the fracturability. The possible reason was that malt sugar molecules entered the tissues of the ingredients due to osmotic pressure, whereas water molecules flowed out, leading to dehydration. The result was greater crispness [4] after VF.

## 5. Conclusions

After analysing the results of fried silver herring using RSM, we found that VF temperature, VF duration, and maltose concentration all affect the main quality characteristics of fried silver herring, especially in terms of oil content. The optimal combination of conditions was found to be a VF temperature of 105 °C, VF duration of 25 min, and maltose concentration of 27%. This was the first study in which the maltose concentration for immersion of the ingredients during pre-processing was included as one of the parameters for VF of fish and combined with other factors to examine the impacts on product quality. Therefore, vacuum frying is a viable option applied to produce fried silver herring products with better quality.

**Supplementary Materials:** The following supporting information can be downloaded at: https://www.mdpi.com/article/10.3390/foods12193533/s1, Table S1: Regression coefficients and analysis of variance (ANOVA) of the response as a function of the independent variables on vacuum fried silver herring. Table S2: The experimental results of the central composite rotatable design for VF silver herring. Table S3: The predicted results of the central composite rotatable design for VF silver herring.

**Author Contributions:** Conceptualization, H.-I.C., C.-C.H. and Y.-H.T.; methodology, Y.-C.L., C.-Y.H. and Y.-H.T.; formal analysis, C.-Y.H. and S.-C.C.; data curation, C.-Y.H., C.-H.K. and Y.-C.L.; writing—original draft preparation, C.-C.H. and H.-I.C.; writing—review and editing, S.-C.C. and Y.-H.T.; supervision, C.-C.H. and H.-I.C.; project administration, Y.-C.L. and C.-H.K.; funding acquisition, Y.-C.L. and Y.-H.T. All authors have read and agreed to the published version of the manuscript.

**Funding:** This research was funded by NKUST grants (111KK017 and 111KK020). It was also supported by the National Science and Technology Council, Taiwan (Contract No. MOST 111-2637-E-992-011).

**Institutional Review Board Statement:** Not applicable.

**Informed Consent Statement:** Not applicable.

**Data Availability Statement:** The data presented in this study are available on request from the corresponding author. The data are not publicly available due to privacy and ethical reasons.

**Conflicts of Interest:** The authors declare no conflict of interest.

## References


1. Shao, K.T.; Taiwan Fish Database. WWW Web Electronic Publication. Available online: http://fishdb.sinica.edu.tw (accessed on 5 July 2023).
2. Sharanabasava, M.R. Vacuum processing of food–a mini review. *MOJ Food Process. Technol.* **2018**, *6*, 283–290.
3. Diamante, L.M.; Shi, S.; Hellmann, A.; Busch, J. Vacuum frying foods: Products, process and optimization. *Int. Food Res. J.* **2015**, *22*, 15–22.
4. Ramadan, M.F.; Amer, M.M.A.; Sulieman, A.E.R.M. Correlation between physicochemical analysis and radical-scavenging activity of vegetable oil blends as affected by frying of french fries. *Eur. J. Lipid Sci. Technol.* **2006**, *108*, 670–678. [CrossRef]
5. Sosa-Morales, M.E.; Solares-Alvarado, A.P.; Aguilera-Bocanegra, S.P.; Muñoz-Roa, J.F.; Cardoso-Ugarte, G.A. Reviewing the effects of vacuum frying on frying medium and fried foods properties. *Int. J. Food Sci. Technol.* **2022**, *57*, 3278–3291. [CrossRef]
6. Manzoor, S.; Masoodi, F.A.; Rashid, R.; Wani, S.M.; Naqash, F.; Ahmad, M. Advances in vacuum frying: Recent developments and potential applications. *J. Food Eng.* **2023**, *46*, e14219. [CrossRef]
7. Moreira, R.G.; Da Silva, P.F.; Gomes, C. The effect of a de-oiling mechanism on the production of high quality vacuum fried potato chips. *J. Food Eng.* **2009**, *92*, 297–304. [CrossRef]
8. Granda, C.; Moreira, R.G.; Tichy, S.E. Reduction of acrylamide formation in potato chips by low-temperature vacuum frying. *J. Food Sci.* **2004**, *69*, E405–E411. [CrossRef]
9. Fan, L.P.; Zhang, M.; Mujumdar, A.S. Vacuum frying of carrot chips. *Dry. Technol.* **2005**, *23*, 645–656. [CrossRef]
10. Shyu, S.L.; Hau, L.B.; Hwang, L.S. Effects of processing conditions on the quality of vacuum-fried carrot chips. *J. Sci. Food Agric.* **2005**, *85*, 1903–1908. [CrossRef]
11. Chen, J.; Lei, Y.; Zuo, J.; Guo, Z.; Miao, S.; Zheng, B.; Lu, X. The Effect of vacuum deep frying technology and *Raphanus sativus* on the quality of surimi cubes. *Foods* **2021**, *10*, 2544. [CrossRef] [PubMed]
12. Akinpelu, O.R.; Idowu, M.A.; Sobukola, O.P.; Henshaw, F.; Sanni, S.A.; Bodunde, G.; Agbonlahor, M.; Munoz, L. Optimization of processing conditions for vacuum frying of high quality fried plantain chips using response surface methodology (RSM). *Food Sci. Biotech.* **2014**, *23*, 1121–1128. [CrossRef]
13. Garayo, J.; Moreira, R. Vacuum frying of potato chips. *J. Food Eng.* **2002**, *55*, 181–191. [CrossRef]
14. Dueik, V.; Robert, P.; Bouchon, P. Vacuum frying reduces oil uptake andimproves the quality parameter of carrots crisps. *Food Chem.* **2010**, *119*, 1143–1149. [CrossRef]
15. Da Silva, P.F.; Moreira, R.G. Vacuum frying of high-quality fruit and vegetable-based snacks. *LWT-Food Sci. Technol.* **2008**, *41*, 1758–1767. [CrossRef]
16. Pei, Z.; Ma, T.; Wen, P.; Xue, C.; Feng, A.; Li, C.; Xu, Y.; Shen, X. Effect of vacuum frying and atmospheric frying on the quality and protein oxidation of squid (*Loligo chinensis*). *J. Food Sci.* **2021**, *86*, 4316–4329. [CrossRef] [PubMed]
17. Fang, M.C.; Chin, P.S.Y.; Sung, W.C.; Chen, T.Y. Physicochemical and volatile flavor properties of fish skin under conventional frying, air frying and vacuum frying. *Molecules* **2023**, *28*, 4376. [CrossRef] [PubMed]
18. Yang, I.C. The Effect of Syrup-Soaking Concentration, Frying Temperature and Frying Time on the Quality of Vacuum Frying Carambola and Silver Anchovy. Master's Thesis, National Pingtung University of Science and Technology, Pingtung, Taiwan, 2004. Available online: https://hdl.handle.net/11296/cw4ty5 (accessed on 5 July 2023).
19. Pan, G.; Ji, H.; Liu, S.; He, X. Vacuum frying of breaded shrimps. *LWT-Food Sci. Technol.* **2015**, *62*, 734–739. [CrossRef]
20. Andrés-Bello, A.; García-Segovia, P.; Martinez-Monzo, J. Vacuum frying process of gilthead sea bream (*Sparus aurata*) fillets. *Innov. Food Sci. Emerg. Technol.* **2010**, *11*, 630–636. [CrossRef]
21. Su, Y.; Zhang, M.; Zhang, W.; Adhikari, B.; Yang, Z. Application of novel microwave-assisted vacuum frying to reduce the oil uptake and improve the quality of potato chips. *LWT* **2016**, *73*, 490–497. [CrossRef]
22. Long, D.; Ye, F.; Zhao, G. Optimization and characterization of wheat bran modified by in situ enhanced $CO_2$ blasting extrusion. *LWT-Food Sci. Technol.* **2014**, *59*, 605–611. [CrossRef]
23. Jentzer, J.B.; Alignan, M.; Vaca-Garcia, C.; Rigal, L.; Vilarem, G. Response surface methodology to optimise Accelerated Solvent Extraction of steviol glycosides from *Stevia rebaudiana* bertoni leaves. *Food Chem.* **2015**, *166*, 561–567. [CrossRef] [PubMed]
24. Reis, F.R.; Masson, M.L.; Waszczynskyj, N. Influence of a blanching pretreatment on color, oil uptake and water activity of potato sticks, and its optimization. *J. Food Process. Eng.* **2008**, *31*, 833–852. [CrossRef]
25. Ma, Z.; Wang, W.; Gong, Z.; Wang, Y.; Cui, W.; Jia, F. A study on vacuum fried *Pleurotus eryngii* chips and optimization technology. *Hans J. Food Nutr. Sci.* **2016**, *5*, 114–127. [CrossRef]
26. Ma, R.; Gao, T.; Song, L.; Zhang, L.; Jiang, Y.; Li, J.; Zhang, X.; Gao, F.; Zhou, G. Effects of oil-water mixed frying and pure-oil frying on the quality characteristics of soybean oil and chicken chop. *Food Sci. Technol.* **2016**, *36*, 329–336. [CrossRef]
27. Jayaraman, K.S.; Gupta, D.D.; Rao, N.B. Effect of pretreatment with salt and sucrose on the quality and stability of dehydrated cauliflower. *Int. J. Food Sci. Technol.* **1990**, *25*, 47–60. [CrossRef]

*foods*

MDPI

*Article*

# Quality Evaluation of Shiitake Blanched and Centrifuged Broths as Functional Instant Drinks

**Shin-Yu Chen [1,†], Jim Tseng [2,†], Cheng-Rong Wu [2] and Sheng-Dun Lin [2,*]**

1. Department of Food Science, National Pingtung University of Science and Technology, Pingtung 912301, Taiwan; sychen@mail.npust.edu.tw
2. Department of Food Science and Technology, Hungkuang University, 1018, Sec. 6, Taiwan Boulevard, Shalu District, Taichung 433304, Taiwan; tsengxjim85@gmail.com (J.T.); sam2763366@gmail.com (C.-R.W.)

* Correspondence: lin54@hk.edu.tw; Tel.: +886-4-26318652 (ext. 5038); Fax: +886-4-26319176

† These authors contributed equally to this work.

**Abstract:** In the process of making mushrooms into vacuum-fried crisps, the resulting blanched broth (BB) and centrifuged broth (CB) are often discarded, thereby increasing the amount of wastewater and treatment costs. This study measured the proximate compositions, bioactive components, taste components, and minerals of freeze-dried BB and CB and then used functional indigestible dextrin (Fibersol-2) as a carrier to make these two broths into instant drinks. The solids of the BB and CB contained protein (16.88–19.21%), fat (0.01–0.23%), ash (12.89–13.50%), carbohydrate (67.28–70.00%), sugars and polyols (40.55–45.68%), free amino acids (6.58–6.69%), 5′-nucleotides (0.98–1.47%), and bioactive components, especially polysaccharides (4.53–7.45%), ergothioneine (both 0.19%), and total phenols (0.15–0.36%). The equivalent umami concentration of BB was 2.77-fold higher than that of the CB. Both BB and CB showed compositions and essential minerals that are rich in taste. Using a nine-point hedonic test, it was found that the solid contents of BB and CB in the instant drink affected the consumer's preference. The flavor and overall preference of instant drinks with 2.5% BB or CB were the best amongst consumers. Overall, the BB and CB were rich in nutrients and bioactive and taste components and could be developed as a functional food in the form of a drink.

**Keywords:** shiitake; blanched broth; centrifuged broth; indigestible dextrin; sensory evaluation

## 1. Introduction

*Lentinula edodes* (Berk.) Pegler, or shiitake in Japanese and xianggu in Chinese, is an edible and medicinal mushroom native to East Asia. It is the second most widely cultivated mushroom around the world because of its unique flavor, texture, and medicinal and nutritional value [1–4]. Shiitake is rich in polysaccharides, proteins, flavor compounds, amino acids, sterols, dietary fibers and vitamins, and is low in fat and sodium [2,5,6]. The protein, ash, and carbohydrate of fresh shiitake cap and stipe are 15.5–18.5%, 4.3–6.2%, and 72.3–77.6%, respectively [5]. In Taiwan, fresh shiitake is divided into special grade, premium grade, and excellent grade, and in three different size classes according to the diameter of the caps: small (3–5 cm), medium (5–7 cm), and large (>7 cm) [5]. Fresh small-sized shiitake of good grade are very perishable, and they have a low commodity value in the market. Therefore, they are commonly used as raw materials for processing mushrooms, such as vacuum-fried crisps and prepared foods [3,5,7,8]. Nevertheless, a few studies have utilized shiitake as an ingredient to produce alcoholic and non-alcoholic drinks by fermentation [9–11].

Hot water blanching is one of the most common traditional pretreatment methods for vegetable processing and is often used to reduce or eliminate the bitterness of vegetables. The main purpose of blanching shiitake is to destroy enzyme activity and reduce the deterioration rate [5]. Blanching is also an important step in the process of preparing shiitake as vacuum-fried crisps and prepared foods. It can improve the color of the vacuum-fried samples and reduce oil absorption [7]. After hot water blanching, water is retained in the shiitake, and the moisture level is increased. Then, centrifugation is utilized to remove some of the moisture from the mushrooms, so that the mushrooms can absorb the seasoning liquid more easily [3,8].

However, the blanching process requires large amounts of water and reduces the nutrients in the shiitake [4,12]. Most of the proximate composition, minerals, sugar and polyols, amino acids, 5′-nucleotides, and bioactive compounds of shiitake leak into the blanching water during the process, but carbohydrates remain [5]. The blanched broth (BB) and centrifuged broth (CB) produced are often discarded, resulting in increased waste water and treatment costs [8,12,13]. The BB and CB were subjected to preliminary quality analyses, and they were found to contain many taste-related and bioactive components. Therefore, finding methods for reusing processing by-products in order to increase added value and manufacturers' profits and reduce waste water represents important research [14,15].

Indigestible dextrin is often used in the manufacturing of foods and drugs because of its many applications and benefits. Fibersol-2 (FS-2) is a highly digestion-resistant maltodextrin fiber supplement, and it is used as a bulking agent [16–18]. The process of making FS-2 includes the heating and enzymatic treatment of corn starch, generating a random distribution of α- and β-(1,4), (1,6), (1,2), and (1,3) linkages with an average molecular weight of 2000 Da [16]. Only α-(1,4) and (1,6) linkages in the FS-2 can be hydrolyzed and digested in the human digestive system, and they have an energy value of 0.5 kcal/g, found through a series of in vivo and in vitro experiments [17]. FS-2 has no known acute toxicity or mutagenicity [19]. FS-2 also has many good processing characteristics, such as low viscosity, water solubility, acid stability, freeze/thaw stability, and heat stability, making it suitable for use in products such as juices, sauces, puddings, and drinks. Several studies have reported that FS-2 has the following beneficial effects: improves intestinal regularity [20], moderates postprandial blood glucose levels [21], lowers serum cholesterol and triglyceride levels [22], and regulates satiety and hunger hormones to enhance satiety after a meal [23]. Therefore, it is important to explore the use of FS-2 as a carrier to make blanched and centrifuged broths into instant drinks with health benefits. Additionally, Nutriose is also a commercial resistant maltodextrin [24,25], and it may be used for further comparative study.

To the best of our knowledge, there are few studies on the quality characteristics of BB and CB, especially on their development into functional instant drinks. As a result, the purpose of this study was to utilize the BB and CB from shiitake to develop a novel functional product in the form of an instant drink, with FS-2 as the carrier. The contents of proximate compositions, sugars and polyols, free amino acids, 5′-nucleotides, bioactive components, and minerals in BB and CB were determined. The physical and hedonic sensory characteristics of instant drinks containing them were also evaluated.

## 2. Materials and Methods

### 2.1. Materials and Reagents

Small-sized, good-grade fresh shiitake were purchased from a mushroom farm in Puli, Nantou, Taiwan. After collecting the shiitake, they were packed in polyethylene bags and then transported to Hungkuang University in a refrigerated truck (4–7 °C). The FS-2 was purchased from Matsutani Chemical Industrial Co. (Hyogo, Japan).

Gallic acid, Folin–Ciocalteu phenol reagent, sugars and polyols, amino acids and 5′-nucleotides, γ-aminobytyric acid, ergosterol, ergothioneine, sodium dodecyl sulfate, trimethylamine, betaine, 2-mercapto-1-methylimidazole, sodium acetate, tetrahydrofuran, and 1,4-dithiothreitol were purchased from Sigma-Aldrich (St. Louis, MO, USA). Minerals were purchased from AccuStandard, Inc. (New Haven, CT, USA). Ethanol (95%) was purchased from Echo Chemical (Miaoli, Taiwan). Nitric acid was purchased from J.T. Baker (Radnor, PA, USA). Hydrogen peroxide was purchased from Honeywell International (Detroit, MI, USA). Hydrochloric acid, sulfuric acid, and sodium phosphate were purchased from Union Chemical Works (Hsinchu, Taiwan). Methanol and acetonitrile were purchased from Avantor Performance Materials (Radnor, PA, USA). Hexane was purchased from Tedia (Fairfield, OH, USA). Cupric sulfate, anhydrous sodium carbonate, and sodium hydroxide were purchased from Shimakyu's Pure Chemicals (Osaka, Japan). Phenol reagent was purchased from Wako Pure Chemical Industries (Osaka, Japan). Potassium sulfite was purchased from Nihon Shiyaku Reagent (Tokyo, Japan). Methyl red pure was purchased from Koch Light Research Laboratories (Gauteng, South Africa). Methylene blue was purchased from Katayama Chemical Industries (Osaka, Japan).

### 2.2. Blanched and Centrifuged Broths

#### 2.2.1. Preparation

After washing fresh shiitake (running water, 30 s), we made up a ratio of washed shiitake (2.5 kg) to boiling water of 1:5 ($w/w$) and blanched the mushrooms for 5 min. After blanching 5 batches of shiitake with the same boiling water, they were cooled and filtered with a Mini Jet Filter (5–7 mm, Buon Vino Manufacturing, Cambridge, ON, Canada), taking the filtrate as the blanched broth (BB). We immediately put the above blanched shiitake into an ice bath for 1.5 min to cool. The cooled samples were put into plastic baskets, and we placed them in a freezer at −25 °C for 24 h. The frozen shiitake were then left to defrost in the refrigerator at 4–7 °C. The thawed shiitake were centrifuged in a stainless steel centrifugal separator (TF-16, Tung Fu Machinery, Taichung, Taiwan) at 1400 rpm for 2 min. The centrifuged broth thus collected was filtered through a Mini Jet Filter, taking the filtrate as the centrifuged broth (CB). The process of preparing BB and CB was repeated 3 times. Parts of the BB and CB were subjected to physicochemical quality characteristic analyses after freeze-drying (FD). The remaining BB and CB were prepared for use in instant drinks. Shiitake is edible and treated with regular processing steps; no food safety concerns exist.

#### 2.2.2. Determination of Quality Characteristics of Blanched and Centrifuged Broths

The blanched broth yield (%) = (filtered weight of blanched broth after blanching 5 batches of washed shiitake/weight of initial blanching water) × 100. Centrifuged broth yield (%) = (weight of centrifuged broth/weight of blanched shiitake) × 100. The moisture in BB and CB was determined by American Association of Cereal Chemists Approved Methods 44-40.01 [26]. The soluble solids and pH values of BB and CB were measured with a hand refractometer (Atago Co., Tokyo, Japan) and a pH meter (MP225, Mettler-Toledo, Switzerland), respectively.

#### 2.2.3. Determination of Proximate Compositions, Sugars, and Polyols

The crude ash, crude fat, moisture, and crude protein of the freeze-dried broths were determined by the methods 08-01.01, 30-25.01, 44-40.01, and 46-11.02, respectively, as approved by the American Association of Cereal Chemists [26]. The nitrogen conversion factor of the sample used to calculate the crude protein content was 4.38 [27]. On a dry basis, the carbohydrate content (%) was calculated by subtracting the protein, fat, and ash content from 100% dry matter. The sugars and polyols of freeze-dried broths were determined according to the method of Tsai et al. [28]. We identified and quantified each of the sugars and polyols based on a calibration curve of authentic compounds.

#### 2.2.4. Determination of Free Amino Acids, 5′-Nucleotides, and Equivalent Umami Concentration

The free amino acids of the freeze-dried broths were determined according to the method used by Hou et al. [29]. The 5′-nucleotides were measured according to the method of Tsai et al. [26]. Each compound was identified and quantified using the calibration curve of the authentic compound. The equivalent umami concentration (EUC, mg sodium glutamate (MSG)/100 g) was calculated according to the addition equation established by Yamaguchi et al. [30,31]. EUC gives the concentration of MSG, which is equivalent to the relative umami intensity of a mixture of MSG and 5′-nucleotides.

2.2.5. Determination of Minerals

Minerals were determined following the method described in CNS [32]. The powdered sample (0.5 g) was added to 2 mL of 30% $H_2O_2$ and 10 mL of 65% $HNO_3$. The mixture was heated on a pre-heated hot plate, and 5 mL of $H_2O_2$ was added. The following mixture was heated to 200 °C for 10 min and then cooled to room temperature. After cooling, the solution was put into a microwave digester (SINEO Microwave Chemistry Technology, Shanghai, China) to perform a digestion reaction at 700 W. After digestion, the solution was quantified to 20 mL with deionized water and filtered through a 0.22 μm syringe filter. The filtered solution was analyzed by high-resolution ICP-MS (Thermo Fisher Scientific, Waltham, MA, USA). Each mineral was identified and quantified on the basis of the calibration curve of the authentic compound.

2.2.6. Determination of Bioactive Compounds

According to the method previously described [33], ergothioneine and γ-aminobutyric acid were extracted and determined, and they were quantified using the calibration curve of the corresponding authentic compound. The measurement of crude polysaccharides was performed according to the method of Mau et al. [34]. The determination of total phenols was made according to the method of Mau et al. [5]. The analysis of ergosterol was based on the method of Qian and Sheng [35].

2.3. *Instant Powder and Drink*

2.3.1. Preparation

The BB and CB were vacuum-concentrated (70 cmHg) with an outer pot temperature of 95 ± 2 °C and an inner pot of 70 ± 2 °C in a vacuum extraction concentrator (GV-800, Amos Instruments, Taipei, Taiwan). When the solid contents of the blanched broth concentrate (BBC) or the centrifuged broth concentrate (CBC) had reached 6%, the concentrate was mixed with FS-2 in a dry matter ratio of 1:3, 1:4, or 1:5, and then, the solid content of the mixed solution was adjusted to 20% with water. The mixed solution was spray-dried in a spray-dryer (SD-1000, Eyela, Tokyo Rikakikai, Tokyo, Japan). The working conditions for the spray-drying were an inlet temperature of 180 °C, an outlet temperature of 85–90 °C, a dry air volume of 0.65 $m^3$/min, a spraying air pressure of 250 kPa, and a pumping volume of 4.2 mL/min. After being spray-dried, the instant drink powder was packed in aluminum foil laminated bags (PET/Al/PE), and ratios of 1:3, 1:4, and 1:5 were made up, designated as SD13, SD14, and SD15, respectively.

In terms of brewing conditions, the instant powders were brewed with boiling water into drinks containing 1.0%, 1.5%, 2.0%, 2.5%, 3.0%, and 3.5% of dried broth, designated as 1.0, 1.5, 2.0, 2.5, 3.0, and 3.5 instant drinks, respectively.

2.3.2. Determination of Physical Quality Characteristics

The moisture of the instant powder was determined according to method 44-40.01 of AACC [20]. The water activity was determined using an AquaLab CX-2 water activity meter (Decagon Devices, Pullman, WA). The reflective color of the powder and the transmittance color of the drink were determined using a spectrophotometer (CM-5, Konica Minolta, Tokyo, Japan). The spectrophotometer was set for CIE $L^*$, $a^*$, and $b^*$ values with a D65 illuminant at 10°. The hue angle ($h^*$, arctan ($b^*/a^*$), first quadrant ($+a^*$, $+b^*$); 180 + arctan ($b^*/a^*$), second quadrant ($-a^*$, $+b^*$)) [36,37], chroma ($c^*$, $\sqrt{a^{*2}+b^{*2}}$), and total color difference ($\Delta E$) values were calculated from the $L^*$, $a^*$ and $b^*$ values. The $\Delta E$ of the powder and drink in comparison to the control was calculated using the equation $\Delta E = \sqrt{(L_c{}^* - L_s{}^*)^2 + (a_c{}^* - a_s{}^*)^2 + (b_c{}^* - b_s{}^*)^2}$, where $L^*$, $a^*$, and $b^*$ are the color coordinates in the control (c) and the powder or drink (s). Each analysis was carried out in triplicate.

2.3.3. Sensory Evaluation

A hedonic scale test was used to measure the color of instant powder, as well as the color, flavor, and overall enjoyableness of the drink. Powder samples were placed on white dishes. Drinks were placed in white cups. Each of the consumers received three samples and was asked to rate them based on the degree of liking them on a nine-point hedonic scale (1 = dislike extremely, 5 = neither like nor dislike, 9 = like extremely). All samples were identified with random three-digit numbers, and the consumers were not informed of the ingredients. Consumers were provided with warm water to rinse after each sample. The evaluation was conducted in a classroom with white light, a temperature of 25 ± 1 °C, and a relative humidity of 60 ± 5% RH. Untrained consumers ($n$ = 90, 19–56 years old, 37 males and 53 females) were recruited from the students, staff, and faculty of Hungkuang University.

### 2.4. Statistical Analysis

Each measurement was conducted in triplicate, except for the hedonic test (n = 90). The experimental data were subjected to an analysis of variance using the Statistical Analysis System software package version 9.4 (SAS Institute, Cary, NC, USA). When a significant difference was found among treatments, Duncan's multiple range test was used to determine the differences among the mean values at the level of $\alpha = 0.05$.

## 3. Results and Discussion

### 3.1. Quality Characteristics of Blanched and Centrifuged Broths

The washed shiitake (2.5 kg) was placed in boiling water (12.5 kg) and blanched for 5 min. After blanching five batches of shiitake with the same boiling water, the BB obtained after filtration weighed 12.04 kg, and the yield was 96.32% (Table 1). After centrifuging the thawed shiitake, the CB obtained after filtration comprised 404.7 g/kg of blanched shiitake, and the yield was 40.47%. The above results suggest that much of the BB and CB recovered from the processing process represents a heavy burden for sewage treatment. The moisture (%), soluble solids (°Brix), and pH values of the BB and CB were 99.20 and 97.67, 0.90 and 2.90, and 6.14 and 6.40, respectively (Table 1). On a dry basis, the ash and protein levels were significantly higher in freeze-dried BB than in freeze-dried CB, while crude fat and carbohydrate showed the reverse (Table 1). The dry matter of BB and CB contained 16.88–19.21% of protein, 12.89–13.50% of ash, and 67.28–70.00% of carbohydrate. The protein, fat, and ash of the shiitake leak into the blanching water to a higher degree than carbohydrate during hot blanching process [5]. Indeed, when comparing our results with the proximate compositions of fresh shiitake described in Mau et al. [5], the ratios of protein and ash in BB and CB are here increased, while carbohydrates show the reverse.

Mushrooms contain large amounts of functional compounds, and these components remain in their by-products. Therefore, the reuse of mushroom by-products has been widely discussed [14,37]. The solids also contain substantial amounts of bioactive components, such as crude polysaccharides, ergothioneine, and total phenols (Table 1). However, $\gamma$-aminobutyric acid was not detected in either broth, and ergosterol was not detected in the BB, while the CB contained 0.01 g of ergosterol per 100 g dry matter. Further, when comparing the bioactive components between BB and CB, the content of crude polysaccharides in BB was obviously higher than that in CB, whereas the contents of ergothioneine and total phenols in CB were significantly higher than in BB.

Amongst these bioactive components, polysaccharides are considered the most frequently recycled component, due to their health benefits and high presence in mushroom by-products [14]. Ultrasound was used to extract the polysaccharides from the by-products generated from *Agaricus bisporus*, and the highest extraction yield was 4.7% [38]. The contents of polysaccharides in BB and CB were 7.45% and 4.53%, respectively. Ergothioneine was found at high levels in mushrooms and demonstrated a high antioxidant ability. The dry matter of *L. edodes* contained 0.92 mg/g of ergothioneine [39]. The ergothioneine contents of BB and CB were calculated at 1.28 and 1.92 mg/g dry matter, respectively. The total phenolic content of the water extract of *L. edodes* and *Volvariella volvacea* showed 1.33 and 1.34 mg of GAEs/g of dry mushroom, which suggests higher phenol contents and antioxidant activities compared to the methanol extract [40]. The total phenols of BB and CB were 0.152 and 0.362 mg GAE/100 g dry matter, respectively. According to the above results, the blanched and centrifuged broths maintained a large amount of polysaccharides, ergothioneine, and total phenols when compared to the fruiting bodies of the mushrooms. These components are important for maintaining health, and so BB and CB show a high reuse value. The freeze-dried solids of BB and CB contained high amounts of non-volatile taste components, including sugars and polyols (40.56–45.69%), free amino acids (6.58–6.69%), and 5′-nucleotides (0.98–1.47%) (Table 1). The sugars and polyols were significantly higher in CB than in freeze-dried BB, while 5′-nucleotides showed the reverse.

**Table 1.** The physicochemical quality characteristics of blanched and centrifuged broths from shiitake.

| | BB | CB |
|---|---|---|
| Yield (%) [1] | 96.32 ± 0.63 [3] | 40.47 ± 0.76 |
| Moisture (g/100 g) | 99.20 ± 0.04 | 97.67 ± 0.01 |
| Soluble solids (°Brix) | 0.90 ± 0.10 | 2.90 ± 0.10 |
| pH | 6.14 ± 0.04 | 6.40 ± 0.06 |
| Proximate composition (g/100 g dry matter) | | |
| Crude protein | 19.21 ± 0.73 [a] | 16.88 ± 0.04 [b] |
| Crude fat | 0.01 ± <0.01 [b] | 0.23 ± 0.01 [a] |
| Crude ash | 13.50 ± 0.06 [a] | 12.89 ± 0.05 [b] |
| Carbohydrate | 67.28 ± 0.77 [b] | 70.00 ± 0.45 [a] |
| Bioactive composition (g/100 g dry matter) | | |
| γ-Aminobutyric acid | nd | nd |
| Crude polysaccharide | 7.45 ± 0.36 [a] | 4.53 ± 0.17 [b] |
| Ergosterol | nd [4] | 0.01 ± <0.01 |
| Ergothioneine | 0.19 ± <0.01 [b] | 0.19 ± <0.01 [a] |
| Total phenols [2] | 0.15 ± 0.01 [b] | 0.36 ± 0.02 [a] |
| Taste composition (g/100 g dry matter) | | |
| Sugars and polyols | 40.56 ± 0.86 [b] | 45.69 ± 1.01 [a] |
| Free amino acids | 6.58 ± 0.11 [a] | 6.69 ± 0.04 [a] |
| 5′-Nucleotides | 1.47 ± 0.02 [a] | 0.98 ± 0.02 [b] |
| EUC (g MSG/100 g sample) | 858.06 ± 7.63 [a] | 309.66 ± 39.15 [b] |

[1] BB yield (%) = (filtered weight of blanched broth after blanching 5 batches of shiitake/weight of initial blanching water) × 100. CB yield (%) = (filtered weight of centrifuged broth/weight of blanched shiitake) × 100. [2] Unit: expressed as mg GAE (gallic acid equivalent)/100 g dry matter. [3] Each value is expressed as a mean ± standard deviation ($n$ = 3). Values with different lowercase letters within a row differ significantly ($p < 0.05$). [4] nd: not detected.

The main sugars and polyols in both freeze-dried broths were mannitol (33.155–34.315 g/100 dry matter), arabitol (4.684–7.476 g/100 dry matter), and trehalose (2.715–3.888 g/100 dry matter) (Table 2). The total sugar and polyol content of freeze-dried CB was significantly higher than that of BB. In terms of taste-related components, arabinose, fructose, and glucose were not detected in either of the freeze-dried broths. There was no significant difference in total free amino acid content between the two freeze-dried broths (Table 3). The main amino acids in the two freeze-dried broths were glutamine, followed by arginine and glutamic acid. All essential amino acids (EAA) were found in both freeze-dried broths, and all non-essential amino acids (NEAA) were detected, except cystine. Except for the higher content of phenylalanine in freeze-dried BB, there was no significant difference in EAA between the freeze-dried broths. Of the NEAAs, the aspartic acid, glutamic acid, glycine, and proline levels were significantly higher in the BB than in the CB, while the glutamine level was significantly higher in the CB than in the BB, and the rest of the NEAAs showed no significant differences between the two freeze-dried broths. There were no significant differences in the concentrations of total amino acids (TAA), EAAs, NEAAs, and branched-chain amino acids (BCAAs) in the two freeze-dried broths. The MSG-like amino acids in BB were significantly more prevalent than in CB, while sweet amino acids showed the reverse. The bitter and tasteless amino acids did not differ significantly between the two freeze-dried broths. In both freeze-dried broths, the highest 5′-nucleotide was 5′-CMP (Table 4). The BB was significantly higher in all 5′-nucleotides than in the CB, except for 5′-UMP, which showed no significant difference. The contents of flavor 5′-nucleotides were 1.97-fold higher in BB than in CB and 1.50-fold higher in total 5′-nucleotides. The sugars and polyols showed lower molecular weights and higher water solubility compared to 5′-nucleotides. This may explain the CB maintaining a higher ratio of arabitol, mannitol, and trehalose but a lower ratio of 5′-nucleotides after centrifugation.

**Table 2.** Sugars and polyols of blanched and centrifuged broths from shiitake.

| | Content (g/100 g Dry Matter) | |
|---|---|---|
| | **BB** | **CB** |
| Arabinose | nd [1] | nd |
| Arabitol | 4.684 ± 0.306 [b,2] | 7.476 ± 0.188 [a] |
| Fructose | nd | nd |
| Glucose | nd | nd |
| Mannitol | 33.155 ± 0.447 [b] | 34.315 ± 0.804 [a] |
| Trehalose | 2.715 ± 0.178 [b] | 3.888 ± 0.203 [a] |
| Total | 40.554 ± 0.862 [b] | 45.679 ± 1.008 [a] |

[1] nd: Not detected. [2] Each value is expressed as mean ± standard deviation ($n$ = 3). Values with different lowercase letters within a row differ significantly ($p < 0.05$).

**Table 3.** Free amino acids of blanched and centrifuged broths from shiitake.

| Amino Acids | Content (g/100 g Dry Matter) | |
|---|---|---|
| | **BB** | **CB** |
| Essential amino acid | | |
| Histidine | 0.099 ± 0.006 [a,2] | 0.109 ± 0.010 [a] |
| Isoleucine | 0.142 ± 0.002 [a] | 0.134 ± 0.003 [a] |
| Leucine | 0.241 ± 0.006 [a] | 0.222 ± 0.006 [a] |
| Lysine | 0.276 ± 0.017 [a] | 0.279 ± 0.015 [a] |
| Methionine | 0.294 ± 0.022 [a] | 0.280 ± 0.006 [a] |
| Phenylalanine | 0.135 ± 0.002 [a] | 0.122 ± 0.001 [b] |
| Threonine | 0.231 ± 0.012 [a] | 0.237 ± 0.023 [a] |
| Tryptophan | 0.031 ± 0.001 [a] | 0.028 ± 0.003 [a] |
| Valine | 0.305 ± 0.007 [a] | 0.299 ± 0.003 [a] |
| Non-essential amino acid | | |
| Alanine | 0.521 ± 0.020 [a] | 0.457 ± 0.008 [a] |
| Arginine | 0.803 ± 0.040 [a] | 0.834 ± 0.019 [a] |
| Asparagine | 0.198 ± 0.005 [a] | 0.184 ± 0.003 [a] |
| Aspartic acid | 0.215 ± 0.004 [a] | 0.175 ± 0.004 [b] |
| Cystine | nd [3] | nd |
| Glutamic acid | 0.730 ± 0.017 [a] | 0.638 ± 0.016 [b] |
| Glutamine | 1.676 ± 0.014 [b] | 2.110 ± 0.042 [a] |
| Glycine | 0.256 ± 0.013 [a] | 0.190 ± 0.012 [b] |
| Proline | 0.093 ± 0.001 [a] | 0.074 ± 0.001 [b] |
| Serine | 0.232 ± 0.006 [a] | 0.213 ± 0.003 [a] |
| Tyrosine | 0.105 ± 0.008 [a] | 0.103 ± 0.008 [a] |
| Total amino acids (TAA) | 6.583 ± 0.112 [a] | 6.688 ± 0.042 [a] |
| Essential amino acids (EAA) | 1.754 ± 0.001 [a] | 1.710 ± 0.034 [a] |
| Non-essential amino acids (NEAA) | 4.829 ± 0.111 [a] | 4.978 ± 0.076 [a] |
| Branched-chain amino acid (BCAA) [1] | 0.688 ± 0.014 [a] | 0.655 ± 0.011 [a] |
| Taste characteristic | | |
| MSG-like | 0.945 ± 0.022 [a] | 0.813 ± 0.020 [b] |
| Sweet | 3.009 ± 0.040 [b] | 3.281 ± 0.018 [a] |
| Bitter | 2.050 ± 0.071 [a] | 2.028 ± 0.023 [a] |
| Tasteless | 0.381 ± 0.025 [a] | 0.382 ± 0.023 [a] |

[1] Branched-chain amino acid = leucine + isoleucine + valine. MSG-like (monosodium glutamate-like) = aspartic acid + glutamic acid. Sweet amino acids = alanine + glutamine + glycine + proline + serine + threonine. Bitter = arginine + histidine + isoleucine + leucine + methionine + phenylalanine + tryptophan + valine. Tasteless = lysine + tyrosine. [2] Each value is expressed as mean ± standard deviation ($n$ = 3). Values with different lowercase letters within a row differ significantly ($p < 0.05$). [3] nd: Not detected.

**Table 4.** 5′-Nucleotides of blanched and centrifuged broths from shiitake.

| | Content (g/100 g Dry Matter) | |
|---|---|---|
| | **BB** | **CB** |
| 5′-AMP [1] | 0.179 ± 0.012 [a,3] | 0.128 ± 0.001 [b] |
| 5′-CMP | 0.684 ± 0.012 [a] | 0.486 ± 0.002 [b] |
| 5′-GMP | 0.383 ± 0.008 [a] | 0.151 ± 0.017 [b] |
| 5′-IMP | 0.018 ± 0.002 [a] | 0.007 ± 0.001 [b] |
| 5′-UMP | 0.182 ± 0.002 [a] | 0.190 ± 0.003 [a] |
| 5′-XMP | 0.022 ± 0.001 [a] | 0.019 ± 0.001 [b] |
| Flavor 5′-Nucleotides [2] | 0.602 ± 0.006 [a] | 0.305 ± 0.017 [b] |
| Total 5′-Nucleotides | 1.468 ± 0.020 [a] | 0.981 ± 0.022 [b] |

[1] 5′-AMP, 5′-adenosine monophosphate; 5′-CMP, 5′-cytidine monophosphate; 5′-GMP, 5′-guanosine monophosphate; 5′-IMP, 5′-inosine monophosphate; 5′-UMP, 5′-uridine monophosphate; 5′-XMP, 5′-xanthosine monophosphate. [2] Flavor 5′-Nucleotides = 5′-IMP + 5′-GMP + 5′-XMP + 5′-AMP. [3] Each value is expressed as mean ± standard deviation ($n$ = 3). Values with different lowercase letters within a row are significantly different ($p < 0.05$).

Using the addition equation established via sensory evaluation by Yamaguchi et al. [30], the EUC values of the BB and CB were found to be 858.06 and 309.66 g MSG/100 g dry matter, respectively (Table 1). The EUC was significantly higher in BB than in CB. Compared with CB, the EUC of BB is 2.77-fold higher. EUC values can be grouped into four levels: (1) >1000 g MSG/100 g, (2) 100–1000 g MSG/100 g, (3) 10–100 g MSG/100 g, and (4) <10 g MSG/100 g [41]. The BB and CB both reached level 3, indicating that our blanched and centrifuged broths were rich sources of umami compounds.

Potassium, magnesium, sodium, calcium, phosphorus, zinc, manganese, selenium, iron, and copper are essential minerals for the human body and play important roles in human cell metabolism, biosynthesis, and physiological functions [2]. The blanching process damages the cell structure, causing the release of minerals into the blanching water [5]. Indeed, BB and CB contain a variety of essential mineral elements (Table 5). The most prevalent major mineral in both freeze-dried broths was potassium (3387.6–3708.2 mg/100 g dry matter), followed by magnesium (183.2–230.2 mg/100 g dry matter) and sodium (129.5–132.3 mg/100 g dry matter). Potassium intake can reduce the risk of cardiovascular and renal disease. However, most populations around the world consume less than the recommended amount of potassium, whereas the intake of sodium is generally double the recommended amount [42,43]. Moreover, the BB and CB samples were shown to be high in potassium and low in sodium. Potassium can reduce the absorption of sodium and has been used as a low-sodium salt. It could act as a good food source for people with high blood pressure. Due to their advantage of having lower sodium, the mushrooms were mixed into a beef taco filling to reduce the overall sodium [6].

In regard to the trace minerals, both BB and CB had the highest amounts of zinc (1.811–7.042 mg/100 g dry matter), followed by manganese (0.479–3.26 mg/100 g dry matter) and selenium (1.811–7.042 mg/100 g dry matter). In particular, the CB contained more manganese and zinc than BB, while selenium was the reverse. The trace minerals are important to maintaining human health. In particular, selenium enhances some of the biological activities of mushrooms [44]. As such, blanched and centrifuged broths are good sources of essential mineral elements.

**Table 5.** Minerals of blanched and centrifuged broths from shiitake.

| Mineral | Content (mg/100 g Dry Matter) | |
|---|---|---|
| | BB | CB |
| Major mineral | | |
| Calcium (Ca) | 56.7 ± 2.3 [a,1] | 33.7 ± 1.5 [b] |
| Magnesium (Mg) | 183.2 ± 7.7 [b] | 230.2 ± 3.0 [a] |
| Phosphorus (P) | 2.0 ± 0.1 [b] | 16.1 ± 0.5 [a] |
| Potassium (K) | 3387.6 ± 59.8 [b] | 3708.2 ± 12.3 [a] |
| Sodium (Na) | 132.3 ± 4.9 [a] | 129.5 ± 4.1 [b] |
| Trace mineral | | |
| Aluminum (Al) | 0.117 ± 0.005 [a] | 0.080 ± <0.001 [b] |
| Chromium (Cr) | 0.008 ± <0.001 [b] | 0.029 ± 0.001 [a] |
| Iron (Fe) | 0.285 ± 0.013 [a] | 0.168 ± 0.005 [b] |
| Manganese (Mn) | 0.479 ± 0.016 [b] | 3.260 ± 0.132 [a] |
| Selenium (Se) | 0.474 ± 0.016 [a] | 0.055 ± 0.001 [b] |
| Zinc (Zn) | 1.811 ± 0.082 [b] | 7.042 ± 0.342 [a] |

[1] Each value is expressed as mean ± standard deviation ($n$ = 3). Values with different lowercase letters within a row differ significantly ($p < 0.05$).

### 3.2. Quality Characteristics of Instant Powders

The concentrated broth (BBC and CBC) was visually darker than the original broth (BB and CB), and the BBC was visually darker than the CBC (Figure S1). In order to produce an instant drink powder with more health benefits, FS-2 was utilized as a carrier. Figure 1 shows the color of spray-dried instant powders prepared from mixtures of BBC or CBC with FS-2. The BB instant powder was yellower than the CB powder. The moisture contents of the BB and CB instant powders (4.41–7.06%, 1.72–3.55%) were, in descending order, SD13 > SD14 > SD15 (Table S1). As the level of FS-2 increased, the water activity (0.315–0.365, 0.226–0.307) also decreased significantly. This finding is related to changes in the moisture level in instant powder. It seems that FS-2 could retain less water and reduce the water activity, which means that the shelf-life can be extended. Among the three ratios of FS-2, 1:5 is recommended, because a higher ratio of microcapsule packaging helps to reduce the risk of core components interacting with the external environment.

**Figure 1.** The appearance color of Fibersol-2, BB, and CB instant powders.

In BB and CB instant powders, the $L^*$ and $WI$ values increased significantly with the increase in FS-2, while the $a^*$ (except CB) and $b^*$ values decreased significantly (Table S1), which means that with the increase in FS-2, the instant powder became brighter, whiter, less red, and less yellow. The CB instant powder was brighter, whiter, less red, and less yellow than the BB instant powder. These two instant powders were less bright, less white, redder (except CB), and yellower than FS-2, because the instant powders contained BB or CB. The colors of the BB and CB instant powders were compared with FS-2, and their $\Delta E$ values decreased with the increase in FS-2, a difference which could even be observed with the naked eye. Song et al. [45] also reported an $\Delta E > 3$ and a color difference that was obvious to the human eye. The CB instant powders were whiter than the BB instant powder, and their color was closer to that of FS-2. However, according to the sensory evaluation, the color of the powder was insignificant, and the scores were around 7.4–7.6.

### 3.3. Sensory Evaluation of Instant Drink

All instant drinks were made by brewing instant powder with boiling water. Figure 2 shows the color of the instant drinks in which instant powders were brewed in boiling water. As the solid concentrations of BB or CB increased, the drinks became darker. For all instant drinks, as the solid concentration of BB or CB increased, the $L^*$ and $h^*$ values decreased significantly, while $a^*$, $b^*$, $c^*$, and $\Delta E$ showed the reverse (Table S2). The $h^*$ values of all instant drinks were 80.77–98.50° (Table S2), which implies that they were between yellow (60°) and green (120°) [46], as is consistent with the color of the instant drinks shown in Figure 2.

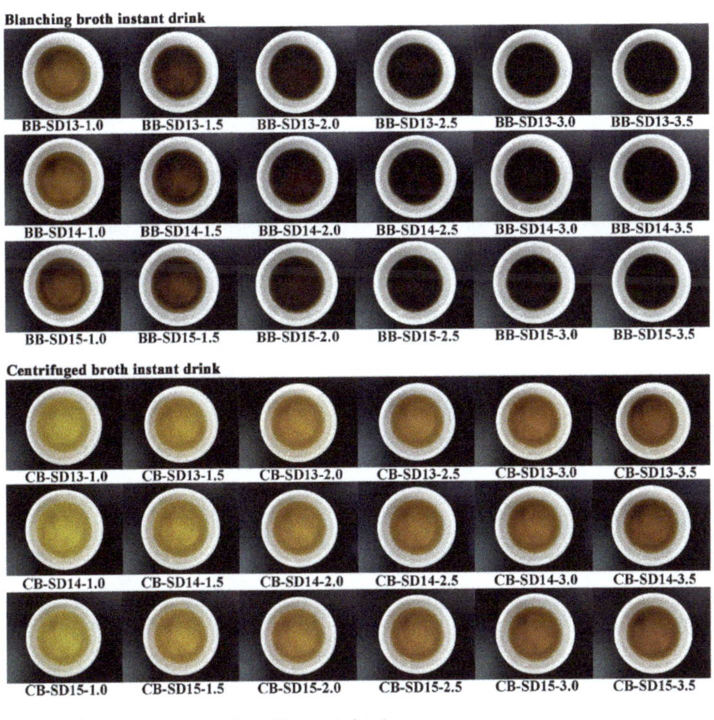


**Figure 2.** The appearance color of instant drinks.

Table 6 shows the results of the hedonic sensory evaluation of instant drinks. As the concentration of BB in the drinks increased, the color scores were lower, but the highest color score of the CB instant drink was 2.0–2.5%. The flavor and overall preference scores of all instant drinks increased from 1.0% to 2.5% and then decreased from 2.5% to 3.5%, indicating that the drink with 2.5% instant powder was the most popular. Consumers believed that in terms of the flavor intensity, 1.0% to 2.0% was too weak, but more than 3.0% was too strong and had a slightly bitter taste; 2.5% was the best. The flavor and overall preference were similar between BB and CB. It seems that the significant differences in taste components between BB and CB, such as in sugars and polyols, 5′-nucleotides, EUC, and MSG-like, and sweet amino acids, did not affect the hedonic sensory results.

**Table 6.** Hedonic sensory evaluation of instant drinks prepared from instant drink powders with hot water.

| | BB | | | CB | | |
|---|---|---|---|---|---|---|
| | Color | Flavor | Overall | Color | Flavor | Overall |
| SD13-1.0 [1] | $7.2 \pm 0.9^{A,2}$ | $4.5 \pm 0.9^{E}$ | $4.4 \pm 0.6^{F}$ | $6.9 \pm 0.8^{E}$ | $4.4 \pm 0.9^{E}$ | $4.3 \pm 0.6^{F}$ |
| SD13-1.5 | $7.1 \pm 0.9^{A}$ | $5.5 \pm 1.1^{D}$ | $5.4 \pm 0.8^{E}$ | $6.8 \pm 0.7^{E}$ | $5.4 \pm 1.0^{D}$ | $5.3 \pm 0.8^{E}$ |
| SD13-2.0 | $6.6 \pm 1.0^{B}$ | $6.6 \pm 0.8^{C}$ | $6.8 \pm 0.8^{B}$ | $7.4 \pm 0.8^{ABC}$ | $6.7 \pm 0.8^{C}$ | $6.9 \pm 0.7^{B}$ |
| SD13-2.5 | $6.5 \pm 1.0^{B}$ | $7.3 \pm 0.8^{A}$ | $7.3 \pm 0.8^{A}$ | $7.3 \pm 0.9^{ABC}$ | $7.4 \pm 0.7^{A}$ | $7.5 \pm 0.9^{A}$ |
| SD13-3.0 | $5.5 \pm 1.1^{CD}$ | $4.3 \pm 0.5^{EF}$ | $3.9 \pm 0.7^{G}$ | $6.0 \pm 0.8^{F}$ | $4.3 \pm 0.5^{EF}$ | $3.9 \pm 0.7^{HI}$ |
| SD13-3.5 | $5.3 \pm 1.0^{CD}$ | $3.7 \pm 0.5^{H}$ | $3.3 \pm 0.5^{IJ}$ | $5.9 \pm 0.8^{F}$ | $3.8 \pm 0.6^{H}$ | $3.4 \pm 0.8^{K}$ |
| SD14-1.0 | $7.1 \pm 0.7^{A}$ | $4.4 \pm 1.4^{E}$ | $4.4 \pm 1.4^{F}$ | $7.0 \pm 0.7^{E}$ | $4.1 \pm 0.8^{G}$ | $4.1 \pm 0.7^{GH}$ |
| SD14-1.5 | $7.4 \pm 1.0^{A}$ | $5.6 \pm 1.0^{D}$ | $5.2 \pm 0.7^{E}$ | $7.3 \pm 1.0^{CD}$ | $5.7 \pm 1.0^{D}$ | $5.2 \pm 0.8^{E}$ |
| SD14-2.0 | $6.4 \pm 1.3^{B}$ | $6.7 \pm 1.1^{C}$ | $6.4 \pm 1.0^{C}$ | $7.4 \pm 0.9^{ABC}$ | $6.9 \pm 1.0^{BC}$ | $6.9 \pm 1.0^{C}$ |
| SD14-2.5 | $6.6 \pm 1.4^{B}$ | $7.1 \pm 0.9^{A}$ | $7.3 \pm 1.0^{A}$ | $7.6 \pm 0.7^{A}$ | $7.2 \pm 0.9^{AB}$ | $7.3 \pm 0.8^{A}$ |
| SD14-3.0 | $5.6 \pm 0.8^{C}$ | $4.0 \pm 0.9^{G}$ | $3.7 \pm 0.9^{GH}$ | $5.8 \pm 0.7^{F}$ | $4.0 \pm 1.0^{G}$ | $3.7 \pm 1.0^{IJ}$ |
| SD14-3.5 | $5.3 \pm 0.5^{CD}$ | $3.4 \pm 0.6^{I}$ | $3.2 \pm 0.9^{J}$ | $5.5 \pm 0.7^{G}$ | $3.4 \pm 0.7^{I}$ | $3.1 \pm 0.6^{L}$ |
| SD15-1.0 | $7.3 \pm 1.1^{A}$ | $4.2 \pm 0.8^{EFG}$ | $4.4 \pm 0.7^{F}$ | $7.0 \pm 0.9^{E}$ | $4.1 \pm 0.8^{FG}$ | $4.3 \pm 0.6^{FG}$ |
| SD15-1.5 | $7.3 \pm 1.0^{A}$ | $5.7 \pm 1.1^{D}$ | $5.6 \pm 0.7^{D}$ | $7.1 \pm 1.0^{DE}$ | $5.6 \pm 1.1^{D}$ | $5.6 \pm 0.6^{D}$ |
| SD15-2.0 | $6.7 \pm 1.0^{B}$ | $6.8 \pm 1.2^{BC}$ | $6.6 \pm 0.7^{BC}$ | $7.3 \pm 0.8^{BCD}$ | $6.9 \pm 1.2^{BC}$ | $6.7 \pm 0.5^{BC}$ |
| SD15-2.5 | $6.4 \pm 1.3^{B}$ | $7.0 \pm 1.1^{AB}$ | $7.1 \pm 1.2^{A}$ | $7.6 \pm 1.0^{AB}$ | $7.3 \pm 1.2^{A}$ | $7.3 \pm 1.1^{A}$ |
| SD15-3.0 | $5.3 \pm 0.5^{CD}$ | $4.0 \pm 0.6^{FG}$ | $3.5 \pm 0.9^{HI}$ | $5.4 \pm 0.6^{G}$ | $4.1 \pm 0.9^{FG}$ | $3.6 \pm 0.9^{JK}$ |
| SD15-3.5 | $5.2 \pm 1.0^{D}$ | $3.2 \pm 0.9^{I}$ | $2.8 \pm 0.5^{K}$ | $5.3 \pm 1.1^{G}$ | $3.3 \pm 1.2^{I}$ | $2.9 \pm 0.6^{L}$ |

[1] SD: Spray drying. 13, 14, and 15: dry solid of blanched broth (or centrifuged broth) concentrate/indigestible dextrin (*w*/*w*). 1.0, 1.5, 2.0, 2.5, 3.0, and 3.5: instant drinks containing 1.0%, 1.5%, 2.0%, 2.5%, 3.0%, and 3.5% dried broth, respectively. [2] Each value is expressed as mean ± standard deviation ($n$ = 90). Means with different capital letters within a column differ significantly ($p < 0.05$). Nine-point hedonic scale with 1, 5, and 9 representing extreme dislike, neither like nor dislike, and extreme like, respectively.

A heatmap based on the correlation between color parameters and the hedonic sensory scores of the instant drinks has been constructed (Figure 3). The color scores of the hedonic sensory evaluation show high positive correlations between $L^*$ and $h$ (°)* and negative ones between $a^*$, $b^*$, and $c^*$. Interestingly, the BB drink showed stronger relations when compared to CB. This indicates that the dark color of BB impacts its color hedonic score more strongly than in the case of CB.

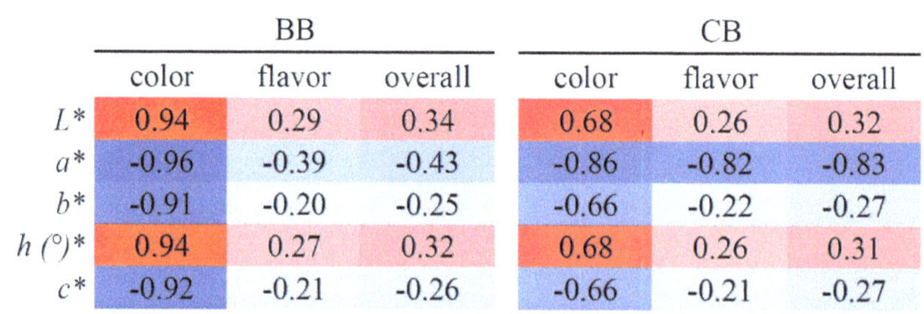


**Figure 3.** The correlation heatmap of instant drink and color property.

## 4. Conclusions

Blanched and centrifuged shiitake broths are rich in polysaccharides, total phenols, ergothioneine, essential minerals, and taste compounds and are low in fat and sodium. This indicates that the blanching water of shiitake is not only tasty but also has health benefits. The BB and CB were made into functional instant drinks using functional FS-2 as a carrier, which was dried via spray-drying. The color became whiter as more FS-2 was added. However, in the sensory evaluation, the color of the powder was insignificant. The solid contents of BB and CB in the instant drinks caused significant differences in the degrees to which the consumers liked them, with 2.5% solid content being the most popular. According to the data obtained in this study, when BB and CB were mixed with FS-2 to produce an instant drink, it will not only be more acceptable for the consumers but will also offer greater health benefits. We suggest that manufacturers should use functional FS-2 to make the blanched and centrifuged broths produced during the production of vacuum-fried shiitake crisps into functional instant drinks, which can not only reduce wastewater treatment costs but also increase economic benefit.

**Supplementary Materials:** The following supporting information can be downloaded at: https://www.mdpi.com/article/10.3390/foods12152925/s1, Figure S1: The color of blanched and centrifuged broths and their concentrates; Table S1: Moisture, water activity and color property of instant drink powders; Table S2: Color property of instant drinks prepared from instant drink powders with hot water.

**Author Contributions:** Conceptualization, S.-D.L.; data curation, S.-Y.C., J.T., C.-R.W. and S.-D.L.; formal analysis, S.-Y.C., J.T. and C.-R.W.; funding acquisition, S.-D.L.; investigation, S.-Y.C., J.T., C.-R.W. and S.-D.L.; methodology, S.-Y.C., J.T. and S.-D.L.; project administration, S.-D.L.; resources, S.-D.L.; software, S.-Y.C. and J.T.; supervision, S.-D.L.; validation, S.-Y.C., J.T. and C.-R.W.; visualization, S.-Y.C.; writing—original draft, S.-Y.C. and J.T.; writing—review and editing, S.-D.L. All authors have read and agreed to the published version of the manuscript.

**Funding:** This research was funded by Ministry of Science and Technology of Taiwan, ROC, for the financial support under Grant No. MOST 107-2622-E-241-001-CC3.

**Informed Consent Statement:** Untrained consumers consented to participate in a hedonic scale test on behalf of themselves.

**Data Availability Statement:** All available data are contained within the article.

**Acknowledgments:** We would like to pay our gratitude and deep respects to Jeng-Leun Mau. He passed away on 21 June 2022. Mau was a dedicated professor in the Department of Food Science and Biotechnology at National Chung Hsing University, with a passion for mushroom study. We greatly appreciate your suggestions for this study. We hope we made you proud.

**Conflicts of Interest:** The authors declare no conflict of interest.

## References

1. Chen, W.; Li, W.; Yang, Y.; Yu, H.; Zhou, S.; Feng, J.; Li, X.; Liu, Y. Analysis and evaluation of tasty components in the pileus and stipe of *Lentinula edodes* at different growth stages. *J. Agric. Food Chem.* **2015**, *63*, 795–801. [CrossRef]
2. Li, S.; Wang, A.; Liu, L.; Tian, G.; Wei, S.; Xu, F. Evaluation of nutritional values of shiitake mushroom (*Lentinus edodes*) stipes. *J. Food Meas. Charact.* **2018**, *12*, 2012–2019. [CrossRef]
3. Ren, A.; Pan, S.; Li, W.; Chen, G.; Duan, X. Effect of various pretreatments on quality attributes of vacuum-fried shiitake mushroom chips. *J. Food Qual.* **2018**, *2018*, 4510126. [CrossRef]
4. Ahmad, I.; Arif, M.; Xu, M.; Zhang, J.; Ding, Y.; Lyu, F. Therapeutic values and nutraceutical properties of shiitake mushroom (*Lentinula edodes*): A review. *Trends Food Sci. Technol.* **2023**, *134*, 123–135. [CrossRef]
5. Mau, J.L.; Tseng, J.; Wu, C.R.; Chen, C.H.; Lin, S.D. Chemical, nutritional, and bioactive compositions of fresh, washed, and blanched shiitake. *Czech. J. Food Sci.* **2021**, *39*, 426–434. [CrossRef]
6. Wong, K.M.; Decker, E.A.; Autio, W.R.; Toong, K.; DiStefano, G.; Kinchla, A.J. Utilizing mushrooms to reduce overall sodium in taco filling using physical and sensory evaluation. *J. Food Sci.* **2017**, *82*, 10. [CrossRef] [PubMed]
7. García-Segovia, P.; Urbano-Ramos, A.M.; Fiszman, S.; Martínez-Monzó, J. Effects of processing conditions on the quality of vacuum fried cassava chips (*Manihot esculenta* Crantz). *LWT-Food Sci. Technol.* **2016**, *69*, 515–521. [CrossRef]

8. Mau, J.L.; Tseng, J.; Wu, C.R.; Chen, C.H.; Lin, S.D. Instrumental texture and sensory preference of vacuum-fried shiitake crisps as affected by isomalto-oligosaccharide pretreatment. *Int. J. Food Prop.* **2021**, *24*, 859–870. [CrossRef]
9. Lin, P.H.; Huang, S.Y.; Mau, J.L.; Liou, B.K.; Fang, T.J. A novel alcoholic beverage developed from shiitake stipe extract and cane sugar with various Saccharomyces strains. *LWT-Food Sci. Technol.* **2010**, *43*, 971–976. [CrossRef]
10. Qiu, L.; Zhang, M.; Chang, L. Effects of lactic acid bacteria fermentation on the phytochemicals content, taste and aroma of blended edible rose and shiitake beverage. *Food Chem.* **2023**, *405*, 134722. [CrossRef]
11. Zhang, Y.; Hartung, N.M.; Fraatz, M.A.; Zorn, H. Quantification of key odor-active compounds of a novel nonalcoholic beverage produced by fermentation of wort by shiitake (*Lentinula edodes*) and aroma genesis studies. *Food Res. Int.* **2015**, *70*, 23–30. [CrossRef]
12. Li, B.; Kimatu, B.M.; Pei, F.; Chen, S.; Feng, X.; Hu, Q.; Zhao, L. Non-volatile flavour components in *Lentinus edodes* after hot water blanching and microwave blanching. *Int. J. Food Prop.* **2017**, *20*, S2532–S2542. [CrossRef]
13. Lin, S.D.; Wu, Y.T.; Lo, Y.C.; Mau, J.L. Quality characteristics of centrifuged broth from blanched *Pleurotus eryngii* and its application as instant drink. *J. Food Process. Preserv.* **2018**, *42*, e13356. [CrossRef]
14. Wang, C.Y. A review on the potential reuse of functional polysaccharides extracted from the by-products of mushroom processing. *Food Bioprocess Technol.* **2020**, *13*, 217–228. [CrossRef]
15. De Corato, U.; De Bari, I.; Viola, E.; Pugliese, M. Assessing the main opportunities of integrated biorefining from agro-bioenergy co/by-products and agroindustrial residues into high-value added products associated to some emerging markets: A review. *Renew. Sust. Energy Rev.* **2018**, *88*, 326–346. [CrossRef]
16. Ohkuma, K.; Wakabayashi, S. Chapter 44: Fibersol-2: A soluble, nondigestible, starch-derived dietary fibre. In *Advanced Dietary Fibre Technology*; McCleary, B.V., Prosky, L., Eds.; Blackwell Science: Oxford, UK, 2001; pp. 509–523. [CrossRef]
17. Tsuji, K.; Gordon, D.T. Energy value of a mixed glycosidic linked dextrin determined in rats. *J. Agric. Food Chem.* **1998**, *46*, 2253–2259. [CrossRef]
18. McCleary, B.V. Measurement of dietary fiber: Which AOAC official method of analysisSM to use. *J. AOAC Int.* **2023**, qsad051. [CrossRef]
19. Wakabayashi, S.; Satouchi, M.; Ueda, Y.; Ohkuma, K. Acute toxicity and mutagenicity studies of indigestible dextrin and its effect on bowel movement of the rat. *Food Hyg. Saf. Sci. (Shokuhin Eiseigaku Zasshi)* **1992**, *33*, 557–562. [CrossRef]
20. Watanabe, N.; Suzuki, M.; Yamaguchi, Y.; Egashira, Y. Effects of resistant maltodextrin on bowel movements: A systematic review and meta-analysis. *Clin. Exp. Gastroenterol.* **2018**, *11*, 85–96. [CrossRef] [PubMed]
21. Wakabayashi, S.; Kishimoto, Y.; Nanbu, S.; Matsuoka, A. Effects of indigestible dextrin on postprandial rise in blood glucose levels in man. *J. Jpn. Assoc. Diet. Fiber Res.* **1999**, *3*, 13–19. (In Japanese) [CrossRef]
22. Kishimoto, Y.; Wakabayashi, S.; Tokunaga, K. Effects of long-term administration of indigestible dextrin on visceral fat accumulation. *J. Jpn. Assoc. Diet. Fiber Res.* **2000**, *4*, 59–65. (In Japanese) [CrossRef]
23. Ye, Z.; Arumugam, V.; Haugabrooks, E.; Williamson, P.; Hendrich, S. Soluble dietary fiber (Fibersol-2) decreased hunger and increased satiety hormones in humans when ingested with a meal. *Nutr. Res.* **2015**, *35*, 393–400. [CrossRef] [PubMed]
24. Włodarczyk, M.; Śliżewska, K. Efficiency of resistant starch and dextrins as prebiotics: A review of the existing evidence and clinical trials. *Nutrients* **2021**, *13*, 3808. [CrossRef] [PubMed]
25. Bojarczuk, A.; Skąpska, S.; Khaneghah, A.M.; Marszałek, K. Health benefits of resistant starch: A review of the literature. *J. Funct. Foods* **2022**, *93*, 105094. [CrossRef]
26. AACC International. *Approved Methods of the AACC*, 10th ed.; American Association of Cereal Chemists: St. Paul, MN, USA, 2000.
27. Crisan, E.V.; Sands, A. Edible mushrooms: Nutritional value. In *The Biology and Cultivation of Edible Mushrooms*; Chang, S.T., Hayes, W.A., Eds.; Academic Press: New York, NY, USA, 1978; pp. 137–165.
28. Tsai, S.Y.; Wu, T.P.; Huang, S.J.; Mau, J.L. Nonvolatile taste components of *Agaricus bisporus* harvested at different stages of maturity. *Food Chem.* **2007**, *103*, 1457–1464. [CrossRef]
29. Hou, J.H.; Wu, P.W.; Liao, C.D.; Kao, Y.M.; Wang, D.Y.; Chen, H.F. Simultaneously determination of glucosamine, taurine, and 20 amino acids in foods and the surveillance report. *Annu. Rep. Food Drug Res.* **2018**, *9*, 1–6. Available online: https://www.fda.gov.tw/tc/includes/GetFile.ashx?id=f636812673608919855 (accessed on 20 June 2023).
30. Yamaguchi, S.; Yoshikawa, T.; Ikeda, S.; Ninomiya, T. Measurement of the relative taste intensity of some $\alpha$-amino acid and 5′-nucleotides. *J. Food Sci.* **1971**, *36*, 846–849. [CrossRef]
31. Yamamoto, T.; Inui-Yamamoto, C. The flavor-enhancing action of glutamate and its mechanism involving the notion of kokumi. *npj Sci. Food* **2023**, *7*, 3. [CrossRef]
32. *CNS 12869:2018*; Methods of Test for Minerals in Infant Formula -Test of Copper, Iron, Magnesium, Manganese, Potassium, Sodium and Zinc. National Standards of the Republic of China: Taiwan, China, 2018.
33. Chen, S.Y.; Ho, K.J.; Hsieh, Y.J.; Wang, L.T.; Mau, J.L. Contents of lovastatin, $\gamma$-aminobutyric acid and ergothioneine in mushroom fruiting bodies and mycelia. *LWT-Food Sci. Technol.* **2012**, *47*, 274–278. [CrossRef]
34. Mau, J.L.; Chen, Y.L.; Chien, R.C.; Lo, Y.C.; Lin, S.D. Taste quality of the hot water extract from *Flammulina velutipes* and its application in umami seasoning. *Food Sci. Technol. Res.* **2018**, *24*, 201–208. [CrossRef]
35. Qian, H.; Sheng, M. Simultaneous determination of fat-soluble vitamins A, D and E and pro-vitamin $D_2$ in animal feeds by one-step extraction and high-performance liquid chromatography analysis. *J. Chromatogr. A* **1998**, *825*, 127–133. [CrossRef] [PubMed]

36. Mclellan, M.R.; Lind, L.R.; Kime, R.W. Hue angle determinations and statistical analysis for multiquadrant Hunter L,a,b data. *J. Food Qual.* **1995**, *18*, 235–240. [CrossRef]
37. Beraldo, I.M.; Assunção Botelho, R.B.; Romão, B.; de Alencar, E.R.; Zandonadi, R.P. Dried apples replacing sugar in pound cakes: Physicochemical composition and sensory analysis. *Int. J. Gastron. Food Sci.* **2023**, *32*, 100731. [CrossRef]
38. Aguiló-Aguayo, I.; Walton, J.; Viñas, I.; Tiwari, B.K. Ultrasound assisted extraction of polysaccharides from mushroom by-products. *LWT-Food Sci. Technol.* **2017**, *77*, 92–99. [CrossRef]
39. Kalaras, M.D.; Richie, J.P.; Calcagnotto, A.; Beelman, R.B. Mushrooms: A rich source of the antioxidants ergothioneine and glutathione. *Food Chem.* **2017**, *233*, 429–433. [CrossRef]
40. Cheung, L.M.; Cheung, P.C.K.; Ooi, V.E.C. Antioxidant activity and total phenolics of edible mushroom extracts. *Food Chem.* **2003**, *81*, 249–255. [CrossRef]
41. Ginos, B.N.R.; Engberink, R.H.G.O. Estimation of sodium and potassium intake: Current limitations and future perspectives. *Nutrients* **2020**, *12*, 3275. [CrossRef] [PubMed]
42. Mau, J.L. The umami taste of edible and medicinal mushrooms. *Int. J. Med. Mushrooms* **2005**, *7*, 119–126. [CrossRef]
43. Weaver, C.M. Potassium and Health. *Adv. Nutr.* **2013**, *4*, 368S–377S. [CrossRef]
44. Hossain, A.; Skalicky, M.; Brestic, M.; Maitra, S.; Sarkar, S.; Ahmad, Z.; Vemuri, H.; Garai, S.; Mondal, M.; Bhatt, R.; et al. Selenium biofortification: Roles, mechanisms, responses and prospects. *Molecules* **2021**, *26*, 881. [CrossRef]
45. Song, K.Y.; O, H.; Joung, K.Y.; Shin, S.Y.; Kim, Y.S. Effects of basil (*Ocimum basilicum* L.) seed mucilage substituted for fat source in sponge cake: Physicochemical, structural, and retrogradation properties. *Ital. J. Food Sci.* **2017**, *29*, 681–696. [CrossRef]
46. Zhao, Z.; Yan, C.; Xu, F.; Liu, J. Study on dyeing properties and color characteristics of wool fabrics dyed with *Geranium caespitosum* L. extract—A new natural yellow dye. *Coatings* **2023**, *13*, 1125. [CrossRef]

*foods* MDPI

*Article*

# Effect of Ultrasonic Pretreatment on the Far-Infrared Drying Process and Quality Characteristics of Licorice

**Jianwei Shang, Qian Zhang, Tongxun Wang, Yanrui Xu, Zepeng Zang, Fangxin Wan, Yuanman Yue and Xiaopeng Huang ***

College of Mechanical and Electronical Engineering, Gansu Agricultural University, Lanzhou 730070, China; zangzp@st.gsau.edu.cn (Z.Z.)
* Correspondence: huangxp@gsau.edu.cn

**Abstract:** In this paper, the effects of different ultrasonic pretreatment processes on the far-infrared drying characteristics, quality indexes, and microstructure of licorice are evaluated. The results showed that ultrasonic pretreatment, combined with far-infrared drying, significantly reduced the drying time and moisture content of licorice compared with those of the control group. The highest total flavonoid content was obtained at an ultrasound power of 80 W. The total phenolic content (0.686 mg gallic acid equivalent/g) was higher than that in the control group, the increase was 19.4%, and its content was the highest at the sonication frequency of 20 kHz. The antioxidant capacity tended to increase and then decrease with the increase in sonication time, sonication power, and sonication frequency, and was the highest at 30 min of sonication. The soluble sugar content (31.490 mg glucose equivalent/g) was the highest at 30 kHz and 30 min. Observation of the microstructure revealed that the surface structure of the ultrasonic pretreated licorice slices changed significantly, forming more micropore channels, which facilitated the mass heat transfer during the drying process. In conclusion, ultrasonic pretreatment can significantly improve the quality of licorice tablets and significantly reduce the time required for subsequent drying. The combination of pretreatment parameters of 60 W ultrasonic power and 40 kHz ultrasonic frequency for 30 min was found to be an optimal combination of pretreatment parameters; therefore, this study may provide a technical reference for the industrialization of licorice drying.

**Keywords:** licorice; far-infrared drying; ultrasonic pretreatment; drying kinetics; quality

## 1. Introduction

Licorice is the dried root of *Glycyrrhiza uralensis* Frich, *Glycyrrhiza inflate* Bat, or *Glycyrrhiza glabra* L. [1]. Licorice is internationally popular as a medicinal plant due to its high value multipurpose medicinal components, including triterpene glycyrrhizic acid and the flavonoids liquiritin, isoliquiritin, liquiritigenin, and osoliquiritigenin [2,3], which have anti-allergic, anti-viral, and immunomodulatory pharmacological effects and are often used in the treatment of coughs, oral diseases, and liver damage [4]. Licorice boasts high medicinal and economic value. However, due to the large water content of fresh licorice, it is prone to mildew and corruption during long-term storage and transportation, which reduces its edible value. As an important food preservation technology, drying can effectively reduce the activity of water in food and inhibit the growth of microorganisms, thus reducing its storage time at room temperature.

The traditional drying method is sun drying, but the drying time is long and subject to the weather, which is not conducive to the retention of the medicinal components of heat-sensitive medicinal plants. In recent years, some scholars have applied hot air drying, vacuum drying, microwave drying, vacuum freeze drying, and other methods to licorice to extend its storage period and reduce the loss of active ingredients [4–6]. Icier et al. studied the drying characteristics, energy/combustion efficiency, and quality properties (color and the total phenolic content) of licorice samples subjected to infrared drying (ID) and carbon fiber assisted drying (CFACD). The results show that the drying time of ID is shorter, but the CFACD method is better for maintaining color characteristics and improving energy/combustion efficiency [7].

The energy released by the infrared radiation source penetrates the surface of the material and can simultaneously heat the surface and interior of the material to achieve uniform internal heating and rapid dehydration [8]. In the study of kiwifruit drying, Zeng et al. found that a higher retention rate of total phenolic content (TPC), total flavonoid content (TFC), and vitamin content (VC) can be obtained in a certain range of infrared temperatures, but it will accelerate the Maillard reaction in the drying process of kiwifruit, resulting in serious color browning [9]. Huang et al. found that far-infrared drying increased the number of surface and internal micropores of stevia and improved the quality of dried products [10]. In the study of Cistanche, Jiang et al. found that far-infrared dried samples had more pores and better boundary integrity than did those subjected to natural drying [11]. Since food drying is mainly controlled by internal diffusion, internal resistance prevents water from being removed and diffused to the surface [12]. In order to increase the drying rate, it is necessary to use a pretreatment method to reduce the internal resistance as much as possible, reduce the initial water content of the raw material, or change the organizational structure of the raw material to shorten the drying time [13]. Ultrasonic waves (US), as a non-thermal radiation technology, interacts with the medium to produce mechanical effects, cavitation effects and thermal effects [14], which can change the organizational structure, reduce the resistance of water migration in the material, and promote internal water diffusion [15]. Ultrasonic pretreatment can effectively enhance heat transfer, improve drying efficiency, and inhibit the deterioration of material quality. Studies have shown that ultrasonic pretreatment, combined with far-infrared drying, can better retain the chemical components, particularly antioxidant activity and flavonoid content, of saffron [16]. In the study of *Lycium barbarum*, Zhang et al. found that the use of ultrasound can improve the quality of far-infrared dried *Lycium barbarum* products, increase the micropores on the surface of the material, increase the capillaries, reduce the mass transfer resistance, and facilitate the diffusion of water in the material [17]. Xi et al. used ultrasound, combined with far-infrared drying, to dry potato slices, which significantly improved the porous structure of potatoes, enhanced the heat and mass transfer process, and helped to retain the TPC and TFC content [18].

In this study, we intend to investigate the effects of different ultrasonic action times, ultrasonic power, and ultrasonic frequency on the drying characteristics, active ingredients, and microstructure of licorice slices subjected to far-infrared drying, with a view to providing theoretical references for the industrial processing of licorice.

## 2. Materials and Methods

### 2.1. Test Materials and Reagents

The licorice used in the experiment was purchased from the licorice planting base in Gongjing Township, Yuzhong County, Gansu Province. In order to prevent the water loss of fresh licorice after excavation, after purchase, it was placed in a 4 °C constant temperature fresh-keeping cabinet. The test materials were selected to reflect a uniform appearance and size in the samples. The average moisture content of fresh licorice held at 105 °C for 24 h, as determined by the official AOAC method, was 52.15% [19].

### 2.2. Chemicals

Liquiritin apioside (standard), glycyrrhizic acid (standard), isoliquiritin apioside (standard), neoisoliquiritin (standard), liquiritigenin (standard), retrochalcone (standard), liquiritin (standard), isoliquiritin (standard), phenol, 1,1-diphenyl-2-trinitrophenylhydrazine (DPPH), catechin, ascorbic acid, and forinol were used. The rest of the reagents, such as methanol, were analytically pure. All reagents were purchased from Chengdu Desite Biotechnology Co., Ltd., Chengdu, China.

### 2.3. Ultrasonic Pretreatment

Licorice samples of uniform size and water content were selected as test materials. The samples were cleaned, placed in distilled water, and pretreated by an ultrasonic device (KQ-300VDE three-frequency CNC ultrasonic cleaner, Kunshan Ultrasonic Instrument Co., Ltd., Shanghai, China). Based on the preliminary testing, the ultrasonic power (40 W, 60 W, 80 W), ultrasonic frequency (20 kHz, 40 kHz, 60 kHz), and ultrasonic treatment time (20 min, 30 min, 40 min) were selected as the adjustment parameters. After the sample pretreatment was completed, the sample was removed, and the excess water on the surface of the sample was absorbed with hygroscopic paper. The arrangement of the ultrasonic one-way test for licorice slices is shown in Table 1.

**Table 1.** Ultrasonic single-factor test of cistanche slices.

| Experimental Number | Ultrasonic Power (W) | Ultrasonic Frequency (kHz) | Ultrasonic Treatment Time (min) |
|---|---|---|---|
| 1 | 60 | 20 | 30 |
| 2 | 60 | 40 | 30 |
| 3 | 60 | 60 | 30 |
| 4 | 40 | 40 | 30 |
| 5 | 80 | 40 | 30 |
| 6 | 60 | 40 | 20 |
| 7 | 60 | 40 | 40 |

### 2.4. Drying Test Method

The drying test was carried out in a far-infrared drying oven. Before the test, the equipment was adjusted to the preset parameters (drying power 900 W, temperature 50 °C) for preheating. After ultrasonic pretreatment, the samples were drained from the surface, sliced, weighed (120.0 ± 0.5) in a single layer, placed on trays, and dried in the far-infrared drying equipment (the optimal process for the far-infrared drying of licorice was obtained from the pre-test at a drying temperature of 50 °C and a slice thickness of 4 mm, so these were set as the fixed drying parameters). During the drying process, the sample trays were removed every 20 min, weighed on a digital balance with an accuracy of ±0.001 g (AUW-220D, Shimadzu, Japan), and then placed back in the drying oven to continue drying until the moisture content of the sample decreased by less than 12% [1]. Each group of tests was repeated three times, and the average value was taken as the test value. To verify the result of pretreatment on the of far-infrared drying effect on licorice, far-infrared dried licorice without ultrasonic pretreatment was selected as a control group for comparison. The experimental procedure is shown in Figure 1.

### 2.5. Calculation of Drying Parameters

#### 2.5.1. Moisture Content of the Dry Base

The formula for calculating the moisture content of the dry base is as follows [20].

$$M_t = \frac{(M_a - M_0)}{M_0} \times 100\% \quad (1)$$

where $M_t$ is the dry basis moisture content ($g \cdot g^{-1}$), $M_a$ is the mass of the licorice slice at time $t$ (g), and $M_0$ is the mass of the dry matter of the licorice slice (g).

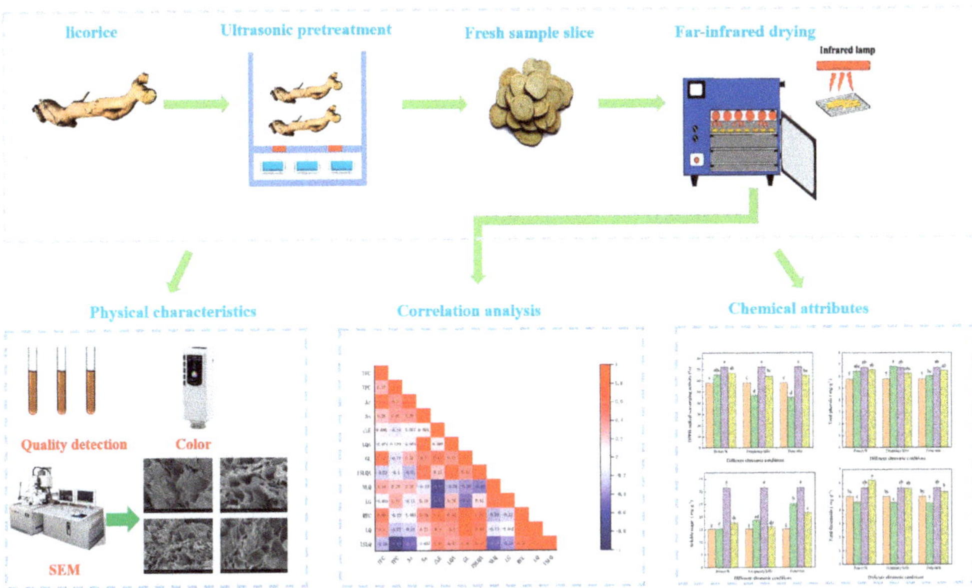


**Figure 1.** Schematic diagram of far-infrared drying and quality detection of licorice under different ultrasonic pretreatment conditions. The vertical bars indicate the standard deviation from the mean. The letters reveal significant differences ($p < 0.05$) according to the Duncan test.

#### 2.5.2. Moisture Ratio

The moisture ratio of the licorice slices under different drying conditions was calculated by the following equation [20].

$$MR = \frac{M_t}{M_0} \tag{2}$$

where $MR$ is the moisture ratio of the licorice slices, $M_t$ is the dry basis moisture content of the licorice slices at moment $t$ ($g \cdot g^{-1}$), and $M_0$ is the initial dry basis moisture content of the licorice slices ($g \cdot g^{-1}$).

#### 2.5.3. Drying Rate

The drying rate of licorice slices was calculated using the following equation [21].

$$DR = \frac{M_{t1} - M_{t2}}{t_2 - t_1} \tag{3}$$

where $t_1$ and $t_2$ are arbitrary drying times, and $M_{t1}$ and $M_{t2}$ are the dry basis moisture content ($g \cdot g^{-1}$) of the licorice slices at moments $t_1$ and $t_2$, respectively.

### *2.6. Determination of Quality Indicators*

#### 2.6.1. Preparation of Sample Extracts

The licorice slices were first crushed through an 80 mesh sieve [22]. A total of 1.0 g of powder was precisely weighted and placed it in a 25 mL 80% absolute ethanol stopper flask, rotated, vibrated (120 r/min) for 48 h under shaded room temperature, and then centrifuged for 10 min (rotation speed: 5000 r/min). The supernatant was diluted with absolute ethanol to 25 mL and stored in a refrigerator at 4 °C until subsequent determination and analysis of total phenolic content, antioxidant activity, soluble sugars, and total flavonoids [23]. Each test was repeated three times, and the average was then taken.

### 2.6.2. Determination of Total Phenolic Content

The total phenol content was determined by the Folin–Ciocalteu method [24,25]. A total of 50 μL of extract solution was added to 2.0 mL of 10% Folin-Ciocalteu solution and 1.0 mL of 7.5% sodium carbonate solution, mixed and shaken for 5 min, and placed in a water bath at 37 °C for 1 h. The absorbance (A) was measured at 765 nm with a spectrophotometer, with the addition of no extract solution as a blank control. The standard curve of total phenol content was customized with gallic acid as the reference standard. The calculation formula of total phenol content is:

$$TPC = \frac{V_2 C_1}{V_1 M} \quad (4)$$

where $C_1$ represents the mass concentration of gallic acid, mg·g$^{-1}$; $V_1$ represents the volume of the sample extract, mL; $V_2$ represents the total volume of the sample extract, mL; and $M$ represents sample mass, g.

### 2.6.3. Determination of Antioxidant Activity

The total antioxidant activity of organic active substances was determined using the DPPH method [26]. A total of 300 μL of extract solution was added to 3.0 mL of 0.1 mmol/L DPPH methanol solution under shaking at room temperature in darkness for 30 min, followed by the determination of the absorbance ($A$) value at 515 nm. According to the above operation, 75% ethanol was used as a blank control, and 500 μmol/L 90% ascorbic acid (ASA) methanol solution was used as a positive control. The antioxidant calculation formula is as follows:

$$\text{Inhibition rate} = \frac{A_0 - A}{A_0} \times 100\% \quad (5)$$

where $A$ is the absorbance value of the sample solution, and $A_0$ is the absorbance of the solution without sample.

### 2.6.4. Determination of Soluble Sugar

The soluble sugar content was measured using the anthrone sulfuric acid method [27]. To 120 μL of the extract, 1.0 mL of 9% phenol solution was added and mixed thoroughly; 3.0 mL of concentrated sulfuric acid was subsequently added and the mixture was shaken for 5 min, reacted at room temperature for 30 min, and the A value at 485 nm was then measured. The standard curve of soluble sugar content was customized with glucose as a reference substance. The soluble sugar content calculation formula is:

$$\text{Soluble sugars} = \frac{V_2 C_2}{V_1 M} \quad (6)$$

where $C_2$ represents the mass concentration of soluble sugar, g · g$^{-1}$; $V_1$ represents the volume of the sample extract, mL; $V_2$ represents the total volume of the sample extract, mL; and $M$ represents sample mass, g.

### 2.6.5. Determination of Total Flavonoids

The content of total flavonoids was determined using $NaNO_2$-$Al(NO_2)_3$-NaOH [28]. To 800 μL of the extract, 2.0 mL of distilled water and 0.3 mL of 5% sodium nitrate solution were added, the mixture was shaken for 5 min, 0.3 mL of 10% aluminum chloride solution was added, the mixture was mixed and shaken for 5 min, and finally, 2.0 mL of 1 mol/L sodium hydroxide solution was added. The sample was allowed to fully react, using no extract as blank control, and the absorbance (A) value was measured at 510 nm. The standard curve of total flavonoids content was customized with catechin as reference substance. The total flavonoid content calculation formula is:

$$TFC = \frac{V_2 C_3}{V_1 M} \quad (7)$$

where $C_3$ represents the mass concentration of catechin, $mg \cdot g^{-1}$; $V_1$ represents the volume of sample extract used, mL; $V_2$ represents the total volume of sample extract, mL; and $M$ represents the mass of sample, g.

2.6.6. Color and Luster

The color of fresh licorice slices and licorice slices under different drying conditions were measured using a colorimeter (CR-10 type, Konica Minolta Ltd., Tokyo, Japan). The difference in color between the samples was represented by $\Delta E$. The smaller the $\Delta E$, the better the quality of the dried product. $\Delta E$ is calculated as follows:

$$\triangle E = \sqrt{(L^* - L_0)^2 + (a^* - a_0)^2 + (b^* - b_0)^2} \quad (8)$$

where $L^*$, $a^*$, and $b^*$ are the brightness, red-green value, and blue-yellow value of the dried licorice product. $L_0$, $a_0$, and $b_0$ are the brightness, red-green value, and blue-yellow value of the fresh sample.

2.6.7. Microstructure

The surface microstructure of the licorice slices under different drying conditions was observed using electron microscopy [29]. Before the test, each sample was cut into 5 mm × 5 mm pieces and then immediately fixed with 2.5% glutaraldehyde solution to stabilize the structure and composition of the biological system. After spraying, the samples were observed by SEM (magnification of ×300), and the acceleration voltage was 5.0 KV.

## *2.7. Determination of Bioactive Constituent Content in Licorice*

Preparation of reference substances: a total of 3 mg of liquiritin apioside (LA), glycyrrhizic acid (GA), isoliquiritin apioside (ILA), neoisoliquiritin (NIL), liquiritigenin (LG), retrochalcone (RC), liquiritin (LQ), and isoliquiritin (ILQ) standards were accurately weighed to prepare a reference substance with a mass concentration of 1 $mg \cdot mL^{-1}$, which was then diluted into different concentration gradients for linear relationship investigation.

Preparation of test samples: The sample powder was sieved through an 80 mesh sieve, and 0.5 g was accurately weighed and placed in a 25 mL 80% anhydrous ethanol stopper triangular flask for ultrasonic treatment for 25 min (power 100 W, frequency 40 kHz). The constant volume with anhydrous ethanol was 25 mL, which was then centrifuged for 10 min, and the supernatant was filtered using a 0.22 μm filter membrane and injected for analysis.

The test materials used are as follows—chromatographic column: Agilent SB-C18 (250 mm × 4.6 mm, 5 μm); mobile phase: acetonitrile (A) −1% aqueous acetic acid (B); gradient elution program: 0–4 min (15–40% A), 4–8 min (40–65% A), 8–10 min (65–85% A), 10–12 min (85–15% A), 12–16 min (15–15% A); mobile phase flow rate: 1.0 mL/min; column temperature: 40 °C; sample volume: 2 μL; and detection wavelength: 250 nm.

## *2.8. Statistical Analysis*

Each group of tests was repeated three times and the average value was acquired. The obtained data were evaluated for moisture ratio and drying rate using Excel 2019. The curve and histogram were drawn by Origin 8.0. SPSS 24.0 was used for the analysis of variance (ANOVA). Tukey's multiple-range test was used to analyze the significance of the mean difference, and the significance level was 0.05.

## 3. Results and Analysis

### 3.1. Effects of Drying Characteristics

The effects of different ultrasonic powers on the moisture ratio and drying rate of licorice slices at an ultrasonic frequency of 40 kHz and an ultrasonic treatment time of 30 min are shown in Figure 2a,b. It can be seen from the figure that the drying time required for the moisture content to reach the safe level was shortened from 220 min to 180 min, saving 9.1–18.2%. It is worth noting that the drying rate curve of ultrasonic power 80 W is lower than that of 40 W and 60 W, and the drying time is the longest (200 min). This may be because the higher the ultrasonic power, the more water infiltrated inside the licorice [30]. Water can fill the intercellular space or enter the cells. The water in the intercellular space can be easily removed during the drying process. However, water that accumulates within the cells may hinder the drying process, as it must first diffuse through the cell membrane and then reach the outer layer of the sample in order to be vaporized [31]. It follows that sonication pretreatment at the optimal power allows the licorice sections to reach drying standards more quickly.

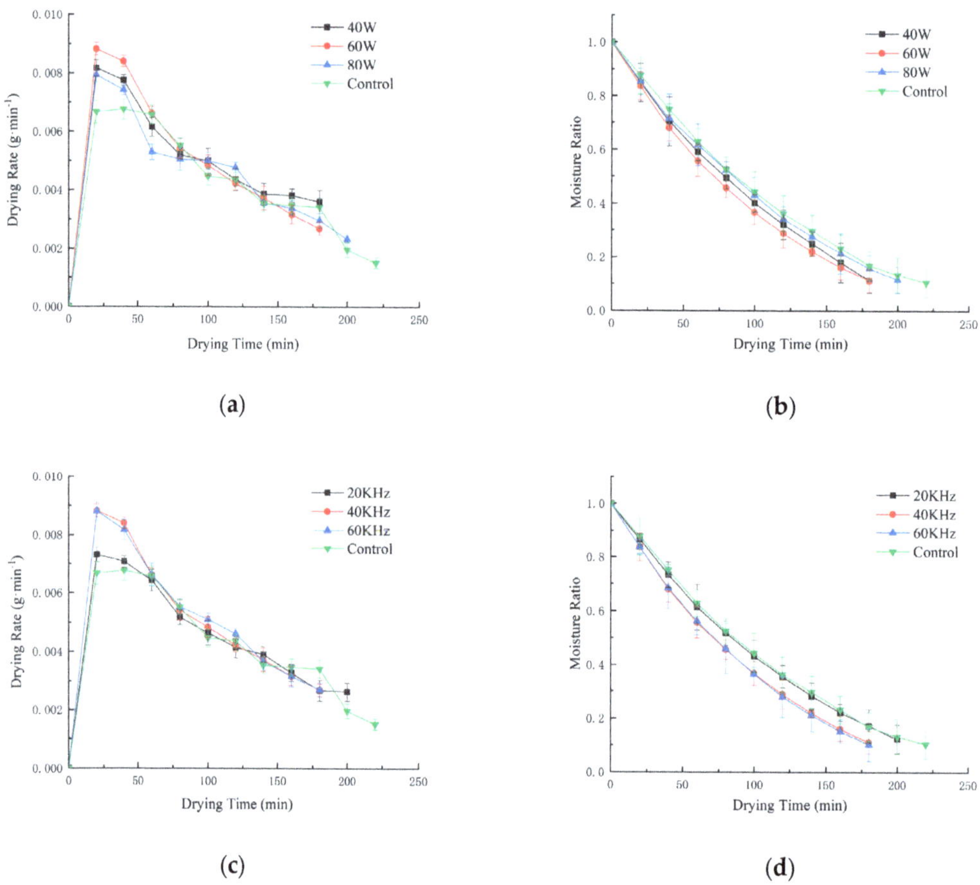


**Figure 2.** *Cont.*

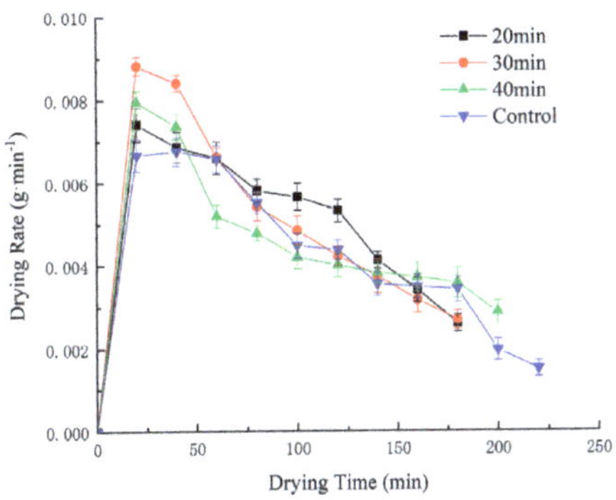


(e)

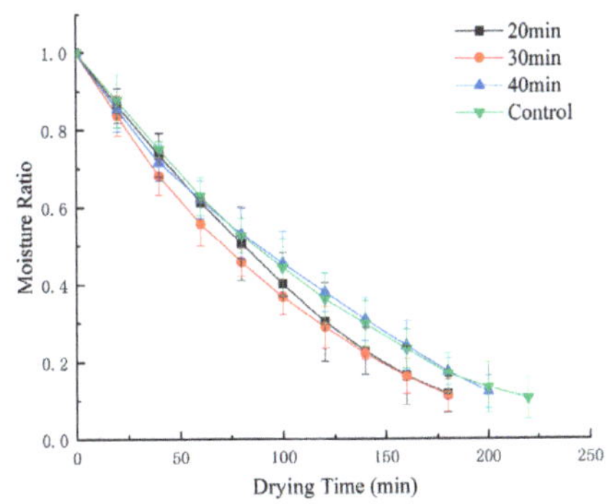


(f)

**Figure 2.** Drying rate curve (**a**) and moisture ratio curve (**b**) of licorice slices subjected to different ultrasonic power; drying rate curve (**c**) and moisture ratio curve (**d**) of licorice slices subjected to different ultrasonic frequency; drying rate curve (**e**) and moisture ratio curve (**f**) of licorice slices subjected to different ultrasonic time.

The effects of different ultrasonic frequencies on the moisture ratio and drying rate at 60 W ultrasonic power and 30 min ultrasonic treatment time are shown in Figure 2c,d. With increasing ultrasound frequency, the time required to dry licorice to concentrations below the safe moisture content level was 200, 180, and 180 min, respectively. Compared with the control group, the moisture ratio curve of licorice slices showed a greater decrease after pretreatment, and the drying time was shortened by 9.1%, 18.2%, and 18.2%, respectively. The higher the ultrasound frequency, the less time is required for drying, which is attributed to the "sponge effect" caused by ultrasound [32]. Under the high-frequency vibration of ultrasound, the internal structure of the licorice slices rapidly compresses and expands, generating a large number of microbubbles, which instantly generate strong kinetic and compression energy when they rupture. These energies can reduce the adhesion of water molecules tightly bound to the microtubule wall, thereby enhancing the fluidity of water and making it easier for water molecules to diffuse in licorice tablets. This energy will also change the structure of the liquid in licorice, as well as the force between water molecules, thereby promoting the internal diffusion rate of water. In addition, the higher the ultrasonic frequency, the stronger the perturbation effect generated by the ultrasonic wave, which is more conducive to enhancing the turbulence of water in the material inside the infrared radiation heat source area, increasing the degree of freedom of the hydrogen protons, improving the fluidity of water molecules, reducing the thickness and concentration difference of the solid–liquid mass transfer boundary, making it conducive to heat and mass transfer. Other studies have also shown that the "sponge effect" caused by ultrasound is beneficial to the absorption of infrared energy and the loss of water inside the material [33].

The effects of different ultrasonic treatment times on the moisture ratio and drying rate of licorice slices at an ultrasonic power of 60 W and an ultrasonic frequency of 40 kHz are shown in Figure 2e,f. It can be seen from the figure that the different ultrasonic treatment times affect the strengthening effect during the sample drying process. When the ultrasonic treatment was 30 min, the maximum drying speed and the minimum drying time could be achieved, but with the extension of the ultrasonic treatment time, the strengthening effect was no longer significant. This may occur because in the process of ultrasonic pretreatment, the osmotic exchange between the licorice slices and the ultrasonic medium is achieved by water loss and solids increase [34]. With the increase in ultrasonic treatment time, the solids content will increase accordingly. The cavitation effect of the ultrasonic waves will show a larger characteristic impedance in the solid, and its attenuation coefficient will be larger, which means that longer ultrasonic treatment times will weaken the strengthening effect [35]. Therefore, the more suitable range of ultrasonic treatment time is 20~30 min for the ultrasonic pretreatment of licorice prior to far-infrared drying.

The drying rate curve was analyzed. Under different drying conditions, the drying rate of each sample reached the maximum at about 20 min, and then decreased continuously. On the one hand, this is because the moisture content of the material is continuously reduced, which greatly increases the diffusion resistance of water in the sample; on the other hand, the attenuation coefficient of ultrasonic wave propagation in the material will increase with the decrease in the dry basis moisture content of the material, the mechanical effect and cavitation effect will be weakened, and the ultrasonic strengthening effect is not obvious [35].

### 3.2. *Influence of Quality*

#### 3.2.1. Influence of Total Flavonoid Content

The effect of different ultrasonic pretreatment conditions on the TFC of licorice slices is shown in Figure 3a. It can be seen from the figure that with the increase in ultrasonic power, the retention of flavonoids in licorice slices increased gradually. When the ultrasonic power was 80 W, the content of flavonoids was the highest (1.984 mg catechin equivalent/g), at 23.4% higher than that of the control group (1.608 mg catechin equivalent/g), which was basically consistent with the results obtained in the study of sweet potatoes by Muhammad Tayyyab Rashid et al. [36]. This may be because with the enhancement of ultrasonic action, the drying time of the material is shortened, which reduces the oxidative decomposition of flavonoids. In addition, the high-intensity ultrasonic action makes the pores inside the licorice more porous [37], and it is easier to penetrate the interior of the cell to break the covalent bond between the polymers during far-infrared radiation, which is helpful for the synthesis of flavonoids [38]. With the extension of ultrasonic time, the flavonoid content first increased and then decreased, indicating that an extended ultrasonic treatment time was not conducive to the synthesis of flavonoids. This may be because the ultrasonic action itself will produce a thermal effect. With the increase in ultrasonic treatment time, the temperature in licorice is also rising, but at higher temperatures, the related flavonoid synthase will undergo structural and activity changes, resulting in some flavonoid synthase cannot be synthesized [10]. The previous study of this subject showed that under this condition, the drying process was relatively long, and the oxidative degradation of flavonoids may exist during the drying process, so that the content of flavonoids was lower than that without ultrasonic treatment.

#### 3.2.2. Effect of Total Phenolic Content

The effects of different ultrasonic pretreatment conditions on the TPC of licorice slices are shown in Figure 3b. It can be seen that the total phenol content of the samples after ultrasonic treatment was higher than that of the control group, and with the increase in ultrasonic power and the extension of ultrasonic treatment time, the total phenol content in licorice samples first increased and then decreased, and with the increase in ultrasonic frequency, the total phenol content decreased. The main reason may be that the shock wave and high-speed jet generated by ultrasonic cavitation damaged the cell wall of the licorice, promoting the release of polyphenols (such as chlorogenic acid, caffeic acid, rutin, and gallic acid) in the sample and increased the content of polyphenols in the material [39]. Moreover, ultrasonic treatment can reduce the content of dissolved oxygen in the licorice slices and slow down the oxidative decomposition of phenols. Nevertheless, if the ultrasonic power, ultrasonic frequency, and ultrasonic treatment time are too high, the degradation rate of phenolic substances such as hydroxybenzoic acid will increase. The study of Joanna Kroehnke et al. also confirmed this. They found that the content of phenolic compounds in carrots was significantly higher after ultrasound pretreatment compared to that of the unsonicated samples [40]. Observing the overall trend, it was found that the highest total phenolic content (0.686 mg/g) was obtained when the sonication power was 60 W, the sonication frequency was 20 kHz, and the sonication time was 30 min, which increased by 19.4% relative to the control group (0.576 mg gallic acid equivalent/g). This is because the mechanical action of ultrasound breaks the bond between the macromolecules and the phenolic compounds, and the phenolic compounds are released [41]. However, an ultrasonic frequency that is too high will destroy the integrity of the cells and cause oxidative decomposition. On the whole, high ultrasonic power, ultrasonic frequency, and ultrasonic treatment time are not conducive to the retention of total phenolic content.

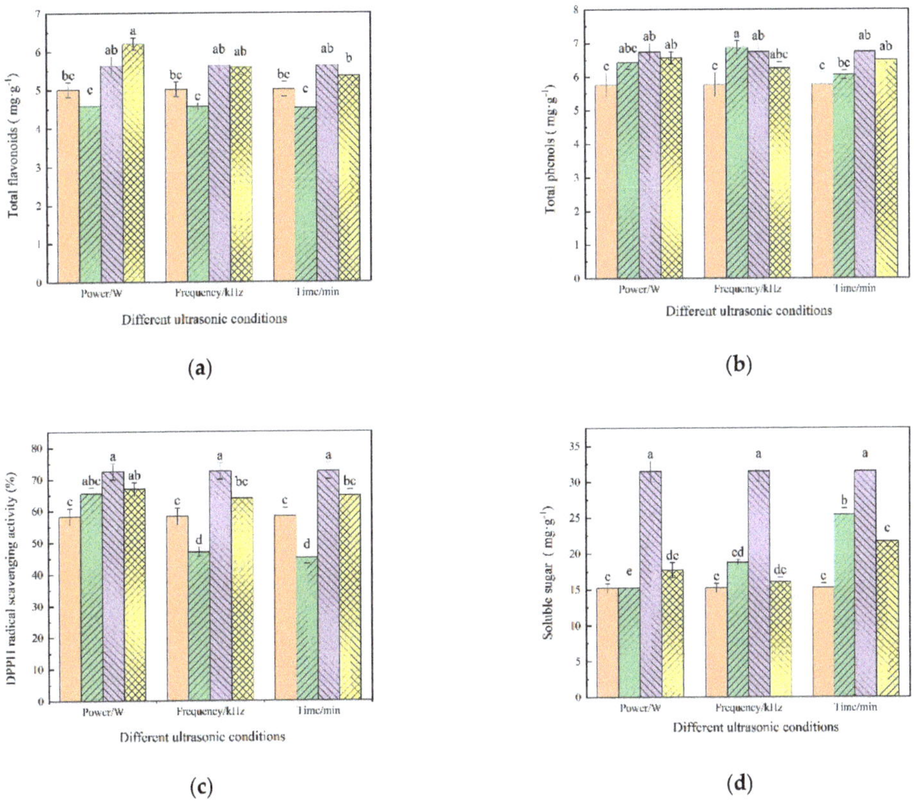


**Figure 3.** (**a**) Total flavonoids content; (**b**) total phenolic content; (**c**) DPPH radical scavenging activity; (**d**) Soluble sugar content. The total phenolic, antioxidant, soluble sugar, and total flavonoid content in licorice under different ultrasonic conditions. Control group ( ), 40 W ( ), 60 W ( ), 80 W ( ). The vertical bars indicate the standard deviation from the mean. The letters reveal significant differences ($p < 0.05$) according to the Duncan test.

3.2.3. Effect of Antioxidant Capacity (DPPH)

The effects of different ultrasonic conditions on the antioxidant activity of licorice slices are shown in Figure 3c. It can be seen that the highest antioxidant capacity of the material (72.54%) was achieved when the sonication pretreatment conditions were 60 W, 40 kHz, and 30 min of sonication time, which was 24.6% higher than that of the control group (58.22%). With the increase in ultrasonic power, the antioxidant activity of the material showed a trend of first increasing and then decreasing, and its change rule was basically consistent with that of the phenols. Studies have shown that polyphenol content is an important factor affecting antioxidant activity because it can affect the oxidation stability of substances [36], and the addition of a second hydroxyl group to the aromatic ring adjacent or opposite to the phenolic compound can enhance the antioxidant capacity of materials [42]. After ultrasonic pretreatment, the antioxidant capacity of the licorice slices was better than that of the control because ultrasonic pretreatment could alleviate the loss rate of phenolic substances and enhance the antioxidant capacity of the samples [22]. However, ultrasound, due to its high temperature and high pressure, will generate a large number of free groups and initiate an oxidation reaction, thus reducing the antioxidant activity of the body.

#### 3.2.4. Effect of Soluble Sugars

The effects of different ultrasonic conditions on the soluble sugar content of licorice slices are shown in Figure 3d. It can be seen that the content of soluble sugars was consistently higher than that of the control under all conditions, except for the ultrasound power of 40 W, the ultrasound frequency of 40 kHz, and the ultrasound treatment time of 30 min, under which the content (15.225 mg glucose equivalent/g) was comparable to that of the control. Ultrasonic pretreatment contributes to the retention of soluble sugar, which may be due to the mechanical effect produced by the shearing force generated by ultrasound, which enhances the penetration of the solvent into the sample matrix, increases the contact surface area between the solid and liquid phases, and improves the diffusion rate of the solvent such that the soluble sugar content increases. The soluble sugar content tended to increase and then decrease with increasing ultrasonic power, ultrasonic frequency, and ultrasonic treatment time. The highest soluble sugar content (31.490 mg glucose equivalent/g) was found in the samples at a treatment time of 30 min at 60 W ultrasonic power and 40 kHz ultrasonic frequency, which was a two-fold increase relative to the control group (15.207 mg glucose equivalent/g). This is due to the enhanced mechanical and cavitation effects of ultrasound, which reduce the mass transfer resistance, facilitate the diffusion of water, shorten the time required for drying, and reduce the oxidation of soluble sugars [43]. The Maillard reaction may occur between protein and polysaccharide, which is a typical chemical reaction. It can produce some products with a large molecular weight and a medium molecular weight. These products will change the structure of polysaccharide, thus reducing its content. However, excessive ultrasonic intensity or higher temperatures will affect the occurrence of this reaction [44].

### 3.3. *Effect of Ultrasonic Treatment on Bioactive Constituents Content*

The effect of the ultrasonic pretreatment of far-infrared radiation drying on the content of active ingredients in licorice is shown in Table 2. The influence of different ultrasonic conditions on the effective components is quite different. For example, at a sonication time of 40 min, GA decreased by 12.8%, and NIL and LG increased by 49.3% and 50%, respectively, compared to the control group, while the content of the other active ingredients did not change significantly. This may be because glycyrrhetinic acid is less stable under heat, and after prolonged heating, the ether glycoside bond breaks and glucuronide is removed to produce glycyrrhetinic acid and glycyrrhetinic acid 3-O-glucuronide [45,46]. When the ultrasonic power increased from 40 W to 80 W, the contents of GA, NIL, RC, LQ, and ILQ increased by 24.03%, 16.35%, 60.56%, 61.76%, and 16.59%, respectively, while the contents of LA and ILA increased first and then decreased, and the maximum contents were 11.403 mg/g and 2.588 mg/g at 60 W, respectively. According to the literature, LA and ILA are flavonoids, and flavonoids mainly contain 2-phenylchromanone as the basic parent nucleus, and -OH is the main reactive group. It is presumed that the 7′-OH contained in them can be oxidized to a quinone structure under high temperatures. The high ultrasonic power is more likely to destroy their molecular structures [47]. The effects of different ultrasonic frequencies on the active components were observed. With the increase in ultrasonic frequency, the content of LG decreased, the content of LA, ILA, and NIL increased first and then decreased, and the content of GA, RC, LQ, and ILQ increased gradually.

**Table 2.** Effective component content of licorice under different drying conditions (mg/g).

| Different Ultrasound Conditions | | LA | GA | ILA | NIL | LG | RC | LQ | LLQ |
|---|---|---|---|---|---|---|---|---|---|
| Control | | 7.96 ± 0.12 [cd] | 26.12 ± 1.50 [ab] | 2.10 ± 0.05 [bc] | 0.93 ± 0.14 [a] | 2.23 ± 0.18 [bc] | 1.89 ± 0.22 [bc] | 27.04 ± 0.27 [c] | 1.14 ± 0.19 [d] |
| 40 kHz 30 min | 40 W | 6.76 ± 0.18 [d] | 22.15 ± 1.05 [d] | 1.55 ± 0.12 [d] | 1.02 ± 0.20 [a] | 1.54 ± 0.05 [e] | 1.63 ± 0.11 [c] | 20.26 ± 0.75 [d] | 1.69 ± 0.14 [cd] |
| | 60 W | 11.40 ± 1.02 [ab] | 27.25 ± 1.60 [a] | 2.59 ± 0.07 [b] | 1.02 ± 0.12 [a] | 2.08 ± 0.14 [bcd] | 2.41 ± 0.23 [ab] | 29.15 ± 0.37 [b] | 1.84 ± 0.11 [c] |
| | 80 W | 6.80 ±0.25 [d] | 27.47 ± 1.05 [a] | 1.63 ± 0.11 [cd] | 1.18 ± 0.16 [a] | 1.75 ± 0.15 [cde] | 2.61 ± 0.13 [a] | 32.78 ± 0.91 [a] | 1.97 ± 0.13 [c] |
| 60 W 30 min | 20 kHz | 9.45 ± 0.13 [b] | 22.31 ± 0.42 [cd] | 2.34 ± 0.12 [b] | 1.02 ± 0.14 [a] | 2.57 ± 0.13 [b] | 2.19 ± 0.18 [abc] | 28.19 ± 0.63 [bc] | 1.66 ± 0.19 [cd] |
| | 40 kHz | 11.40 ± 1.02 [ab] | 27.25 ± 1.60 [a] | 2.59 ± 0.07 [b] | 1.02 ± 0.12 [a] | 2.08 ± 0.14 [bcd] | 2.41 ± 0.23 [ab] | 29.15 ± 0.37 [b] | 1.84 ± 0.11 [c] |
| | 60 kHz | 9.02 ± 0.33 [b] | 27.49 ± 0.54 [a] | 2.06 ± 0.19 [bcd] | 0.94 ± 0.14 [a] | 1.63 ± 0.21 [de] | 2.49 ± 0.09 [ab] | 31.90 ± 0.58 [a] | 3.03 ± 0.26 [a] |
| 60 W 40 kHz | 20 min | 12.15 ± 0.22 [a] | 27.25 ± 0.46 [a] | 3.19 ± 0.23 [b] | 0.90 ± 0.15 [a] | 1.89 ± 0.06 [bcd] | 2.63 ± 0.20 [a] | 33.66 ± 0.77 [a] | 4.29 ± 0.30 [a] |
| | 30 min | 11.40 ± 1.02 [ab] | 27.25 ± 1.60 [a] | 2.59 ± 0.07 [b] | 1.02 ± 0.12 [a] | 2.08 ± 0.14 [bcd] | 2.41 ± 0.23 [ab] | 29.15 ± 0.37 [b] | 1.84 ± 0.11 [c] |
| | 40 min | 9.77 ± 0.44 [bc] | 22.77 ± 1.65 [cd] | 2.08 ± 0.21 [bcd] | 1.39 ± 0.26 [a] | 3.35 ± 0.25 [a] | 2.02 ± 0.19 [abc] | 27.64 ± 0.83 [bc] | 1.65 ± 0.14 [cd] |

Note: Data are expressed as means ± standard deviation of triplicate samples. The letters in the same column reveal significant differences ($p < 0.05$) according to the Duncan test.

### 3.4. Color and Luster

Color difference ($\Delta E$) is an important parameter for assessing the quality of dried samples [48]. From Table 3, it can be seen that the $L^*$, $a^*$, and $b^*$ values of fresh licorice were 38.95, 39.30, and 25.63, respectively. In the case of ultrasonic pretreatment, the $b^*$ values of the dried licorice slices were all smaller than those of the fresh samples, indicating that ultrasonic treatment affects the yellowness values of licorice. The $a^*$ values increased with increasing ultrasonic time, suggesting that the length of ultrasonic action had a significant effect on the glycyrrhiza redness values, possibly because changes in $a^*$ can be linked to the formation of colored compounds [49], which undergo enzymatic and non-enzymatic browning reactions during treatment—the brown pigments are formed from colorless polyphenols and lead to changes in optical properties [50]. At the same time, longer ultrasonic action will increase the gap between cells, form more micropores, damage the surface of the material, and increase the effective area of contact with the ultrasonic medium, resulting in color changes. [51]. With the increase in ultrasonic frequency, the $\Delta E$ of licorice slices showed a declining trend, indicating that the higher the frequency, the better the color of licorice.

**Table 3.** Color value of far-infrared dried licorice slices under different pretreatment conditions.

| Different Ultrasound Factors | | $L^*$ | $a^*$ | $b^*$ | $\Delta E$ |
|---|---|---|---|---|---|
| Fresh Sample | | 78.10 ± 0.03 [b] | 1.91 ± 0.32 [a] | 28.56 ± 0.90 [a] | — |
| 40 kHz 30 min | 40 W | 81.39 ± 0.54 [a] | 2.23 ± 0.08 [a] | 24.60 ± 1.17 [c] | 4.93 ± 0.42 [c] |
| | 60 W | 75.71 ± 0.68 [c] | 1.79 ± 0.41 [a] | 25.00 ± 0.38 [bc] | 4.42 ± 0.67 [c] |
| | 80 W | 77.73 ± 0.02 [b] | 2.18 ± 0.09 [a] | 23.45 ± 0.67 [bc] | 4.64 ± 0.82 [c] |
| 60 W 30 min | 20 kHz | 78.83 ± 0.17 [b] | 2.00 ± 0.81 [a] | 22.93 ± 0.73 [bc] | 5.28 ± 0.81 [c] |
| | 40 kHz | 75.71 ± 0.68 [c] | 1.79 ± 0.41 [a] | 25.00 ± 0.38 [bc] | 4.42 ± 0.67 [c] |
| | 60 kHz | 77.49 ± 0.76 [b] | 2.47 ± 0.08 [a] | 25.94 ± 1.79 [abc] | 3.97 ± 2.07 [c] |
| 60 W 40 kHz | 20 min | 63.89 ± 0.51 [e] | 0.06 ± 0.16 [b] | 24.74 ± 0.37 [bc] | 14.87 ± 0.69 [a] |
| | 30 min | 75.71 ± 0.68 [c] | 1.79 ± 0.41 [a] | 25.00 ± 0.38 [bc] | 4.42 ± 0.67 [c] |
| | 40 min | 68.35 ± 0.28 [d] | 2.07 ± 0.59 [a] | 26.27 ± 0.06 [ab] | 10.29 ± 0.14 [b] |

Note: Data are expressed as means ± standard deviation of triplicate samples. The letters in the same column reveal significant differences ($p < 0.05$) according to the Duncan test.

### 3.5. Correlation Analysis

The correlation between the quality indexes of licorice slices is shown in the Figure 4. It can be seen that TFC is positively correlated with LQ, REC, NLQ, GL, $\Delta E$, Ss, Ac, and TPC. The higher the TFC content, the greater the correlation with Ac (r = 0.75). Deng et al. also revealed that the antioxidant activity was positively correlated with TPC [19]. There was a significant positive correlation between REC and LQ (r = 0.96). TPC was negatively correlated with ISLQ (r = −0.84). Ss was positively correlated with LQA and ISLQA, and the correlation was significant (r = 0.84, r = 0.72). By observing the correlation between $\Delta E$

and each component, it was found that the index was negatively correlated with NLQ and LG, and positively correlated with other components. The positive correlation coefficient between LG and ISLQA was 0.95, and the correlation was extremely significant. On the whole, there is a significant correlation between pretreatment conditions and the quality indexes of licorice.

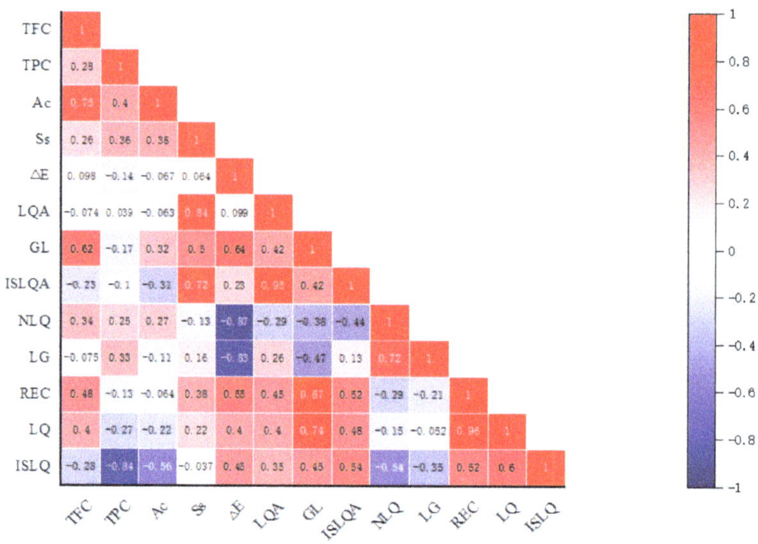


**Figure 4.** Correlation analysis between the indicators under different pretreatment conditions.

### 3.6. Microstructural Analysis

In the drying process, the microstructure of the material has an important influence on the heat and mass transfer and water diffusion characteristics [49]. The microstructure of the dried products dried under different ultrasonic pretreatment conditions is shown in Figure 5. It can be clearly seen that the licorice slice tissue will continue to shrink during the drying process, and the cavity will become smaller, while the cell tissue structure of the dried products resulting from natural drying (Figure 5a) without ultrasonic pretreatment (Figure 5b) is more compact and denser. When the ultrasonic frequency was 40 kHz and the treatment time was 30 min, the higher the ultrasonic power, the clearer the arrangement of the cell structure microchannels on the surface of the licorice slices (Figure 5c–e). When the ultrasonic power was 60 W and the time was 30 min, regardless of the frequency, the microchannel of the cell structure on the surface of the licorice slices showed little change, indicating that the ultrasonic frequency had an insignificant effect. When the ultrasonic power was 60 W and the frequency was 40 kHz, with the increase in ultrasonic action time, the structure of the material cells and the formed microchannels was gradually destroyed, and the tissue integrity of the cells was also reduced (Figure 5h,d,i). In addition, when the ultrasonic pretreatment power was 80 W and the frequency was 40 kHz for 30 min, the skeletal structure of the licorice slice cells remained in good condition, and the microchannel was clearly visible. Under this condition, the fiber network structure could be observed.

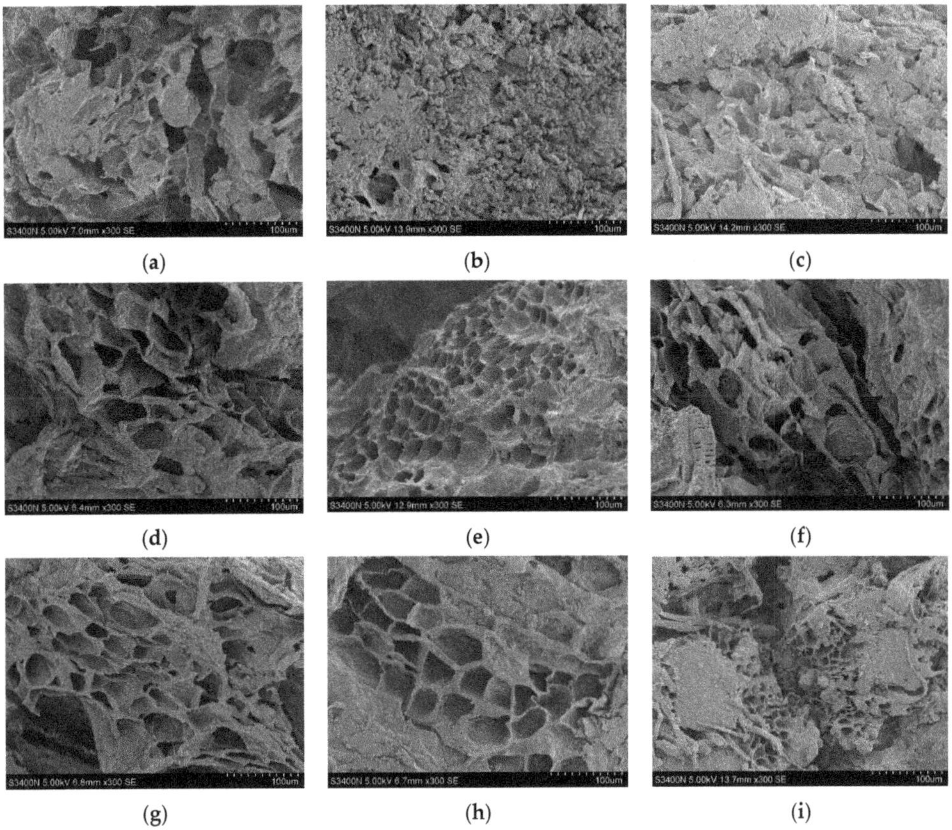

**Figure 5.** (**a**) Natural drying; (**b**) control; (**c**) 40 W, 40 kHz, 30 min; (**d**) 60 W, 40 kHz, 30 min; (**e**) 80 W, 40 kHz, 30 min; (**f**) 60 W, 20 kHz, 30 min; (**g**) 60 W, 60 kHz, 30 min; (**h**) 60 W, 40 kHz, 20 min; (**i**) 60 W, 40 kHz, 40 min. Microstructure of far infrared dried licorice under different pretreatment conditions.

## 4. Conclusions

In this study, far-infrared-dried licorice slices were pretreated using ultrasound technology, and the effects of different ultrasound conditions on drying characteristics, quality characteristics and microstructure were investigated. The results show that ultrasonic pretreatment mainly affects the microstructure changes, such as cell destruction and microchannel formation. When the ultrasonic pretreatment is carried out for a long period of time, the formation of the microchannel increases the porosity of the sample and enhances the subsequent drying process, but the ultrasonic effect will be weakened if the treatment time is excessive. In addition, the high frequency vibration is more favorable to the absorption of infrared energy and the loss of moisture within the material. Finally, it was confirmed that the extension of ultrasonic pretreatment time and excessive power would weaken the effect of ultrasonic pretreatment. The overall quality of the licorice slices was improved by ultrasonic pretreatment. The combination of ultrasonic power of 60 W, ultrasonic frequency of 40 kHz, and ultrasonic time of 30 min was found to be the best combination of pretreatment parameters. In addition, the retention of active ingredients in licorice is closely related to the integrity of its cell structure. By observing the microstructure, it was found that the surface cell structure changed to form microchannels, which provided more mass transfer paths for water migration. Therefore, ultrasonic pretreatment, combined with far-infrared drying, is an efficient method to obtain the retention of multiple components in licorice slices.

In this paper, the analysis of the drying characteristics and quality properties of licorice slices under ultrasonic pretreatment followed by far-infrared drying is investigated, but the more specific heat and mass transfer process inside the slices was still unclear. In future research, the heat and mass transfer mechanism of licorice slices should be studied in depth through simulation in order to provide a theoretical basis for the optimization of the drying process at a later stage.

**Author Contributions:** J.S.: writing—original draft, conceptualization, formal analysis, and writing—review and editing; Q.Z.: investigation, formal analysis, and conceptualization; T.W.: investigation and formal analysis; Y.X.: investigation and formal analysis; Z.Z.: investigation and formal analysis; Y.Y.: data curation and software; F.W.: investigation and funding acquisition; X.H.: methodology, investigation, and project administration. All authors have read and agreed to the published version of the manuscript.

**Funding:** This research was funded by the National Natural Science Foundation of China grant number 32160426 and the APC was funded by Fangxin wan.

**Data Availability Statement:** The datasets generated for this study are available on request to the corresponding author.

**Acknowledgments:** This research was financed with funds under the agricultural mechanization and automation research team of Gansu Agricultural University for their help and encouragement.

**Conflicts of Interest:** The authors declare that they have no known competing financial interests or personal relationships that could have appeared to influence the work reported in this paper.

## References

1. National Pharmacopoeia Committee. *Chinese Pharmacopoeia*; China Pharmaceutical Science and Technology Press: Beijing, China, 2020.
2. Wang, C.; Jin, Y.; Jin, Y. Promoting effect of licorice extract on spermatogonial proliferation and spermatocytes differentiation of neonatal mice in vitro. *Vitr. Cell. Dev. Biol. Anim.* **2016**, *52*, 149–155. [CrossRef] [PubMed]
3. Asl, M.; Hosseinzadeh, H. Review of pharmacological effects of *Glycyrrhiza* sp. and its bioactive compounds. *Phytother. Res.* **2008**, *22*, 709–724. [CrossRef] [PubMed]
4. Li, Y.H.; Qi, Y.R.; Wu, Z.F.; Wang, Y.Q.; Wang, X.C.; Wang, F.; Yang, M. Comparative study of microwave-vacuum and vacuum drying on the physicochemical properties and antioxidant capacity of licorice extract powder. *Powder Technol.* **2017**, *317*, 430–437. [CrossRef]
5. Asım, B.; Ömer, Ş. Microwave drying kinetics of a thinlayer liquorice root. *Dry. Technol.* **2012**, *30*, 859–864.
6. Icier, F.; Ozmen, D.; Cevik, M.; Cokgezme, O.F. Drying of Licorice Root by Novel Radiative Methods. *J. Food Process. Preserv.* **2021**, *45*, e15214. [CrossRef]
7. Sun, R.; Hikosaka, S.; Goto, E.; Sawa, H.; Saito, T.; Kudo, T.; Ohno, T.; Yoshimatsu, K.; Kawano, N.; Inui, T.; et al. Effects of post-harvest storage and drying temperatures on four medicinal compounds in the root of Chinese licorice (*Glycyrrhiza uralensis*). *Environ. Control. Biol.* **2013**, *51*, 149–155. [CrossRef]
8. Çakmak, R.; Tekeoğlu, O.; Bozkır, H.; Ergün, A.R.; Baysal, T. Effects of electrical and sonication pretreatments on the drying rate and quality of mushrooms. *LWT* **2016**, *69*, 197–202. [CrossRef]
9. Zeng, Y.; Liu, Y.; Zhang, J.; Xi, H.; Duan, X. Effects of far-infrared radiation temperature on drying characteristics, water status, microstructure and quality of kiwifruit slices. *J. Food Meas. Charact.* **2019**, *13*, 3086–3096. [CrossRef]
10. Huang, X.; Li, W.; Wang, Y.; Wan, F. Drying characteristics and quality of *Stevia rebaudiana* leaves by far-infrared radiation. *LWT* **2021**, *140*, 110638. [CrossRef]
11. Jiang, C.; Wan, F.; Zang, Z.; Zhang, Q.; Xu, Y.; Huang, X. Influence of far-infrared vacuum drying on drying kinetics and quality characteristics of *Cistanche slices*. *J. Food Process. Preserv.* **2022**, *11*, 866. [CrossRef]
12. Zhang, Z.; Liu, Z.; Liu, C.; Li, D.; Jiang, N.; Liu, C. Effects of ultrasound pretreat-ment on drying kinetics and quality parameters of button mushroom slices. *Dry. Technol.* **2016**, *34*, 1791–1800. [CrossRef]
13. Garcia-Perez, J.V.; Ortuño, C.; Puig, A.; Carcel, J.A.; Perez-Munuera, I. Enhancement of Water Transport and Microstructural Changes Induced by High-Intensity Ultrasound Application on Orange Peel Drying. *Food Bioprocess Technol.* **2012**, *5*, 2256–2265. [CrossRef]
14. Yuan, L.; He, X.; Lin, R.; Cheng, S. Effect of ultrasonic pretreatment on moisture state and hot air drying characteristics of kiwifruit. *J. Agric. Eng.* **2021**, *37*, 263–272.

15. Liu, Y.; Zeng, Y.; Hu, X.; Sun, X. Effect of Ultrasonic Power on Water Removal Kinetics and Moisture Migration of Kiwifruit Slices During Contact Ultrasound Intensified Heat Pump Drying. *Food Bioprocess Technol.* **2020**, *13*, 430–441. [CrossRef]
16. Pei, Y.; Li, Z.; Xu, W.; Song, C.; Li, J.; Song, F. Effects of ultrasound pretreatment followed by far-infrared drying on physicochemical properties, antioxidant activity and aroma compounds of saffron (*Crocus sativus* L.). *Food Biosci.* **2021**, *42*, 101186. [CrossRef]
17. Zhang, Q.; Wan, F.; Zang, Z.; Jiang, C.; Xu, Y.; Huang, X. Effect of ultrasonic far-infrared synergistic drying on the characteristics and qualities of wolfberry (*Lycium barbarum* L.). *Ultrason. Sonochem.* **2022**, *89*, 106134. [CrossRef]
18. Xi, H.; Liu, Y.; Guo, L.; Hu, R. Effect of ultrasonic power on drying process and quality properties of far-infrared radiation drying on potato slices. *Food Sci. Biotechnol.* **2020**, *29*, 93–101. [CrossRef]
19. AOAC. *Official Method of Analysis*; No. 934.06; Association of Official Analytical Chemists: Washington, DC, USA, 1990.
20. Zhang, Z.; Li, L.; Li, D. Characterization and kinetic modeling of thin-layer vacuum microwave drying of lotus root slices. *Jiangsu Agric. Sci.* **2019**, *47*, 201. [CrossRef]
21. Tao, Y.; Wang, P.; Wang, Y.; Kadam, S.U.; Han, Y.; Wang, J.; Zhou, J. Power ultra-sound as a pretreatment to convective drying of mulberry (*Morus alba* L.) leaves: Impact on drying kinetics and selected quality properties. *Ultrason. Sonochem.* **2016**, *31*, 310–318. [CrossRef]
22. Wan, F.; Luo, Y.; Li, W. Effects of different pretreatment methods on the far-infrared drying characteristics and quality of *Lycium barbarum*. *Chin. Herb. Med.* **2020**, *51*, 4183. [CrossRef]
23. Manuel, B.V.; Francisco, O.; Francisco, M.; Alfredo, M. Changes in phenolic compounds in garlic (*Allium sativum* L.) owing to the cultivar and location of growth. *Plant Foods Hum. Nutr.* **2011**, *66*, 218–223.
24. Chen, Z.; Jiang, J.; Ju, X. Active ingredients and their antioxidant activities in various edible rice. *Food Ind. Sci. Technol.* **2018**, *39*, 71. [CrossRef]
25. Cristina, N.; Andrea, M.; Gabriele, F.G.; Lucia, M. In vitro Antioxidant Activity of Aged Extracts of some Italian Allium Species. *Plant Foods Hum. Nutr.* **2011**, *66*, 11.
26. Feng, Y.; Cao, D.; Guo, Y. Determination of polysaccharides in *Panax notoginseng* by anthrone sulfuric acid colorimetric method. *J. Northwestern Univ. (Nat. Sci. Ed.)* **2013**, *43*, 921.
27. Ma, L.M.; Anuar, K.S.; Sadegh, M.; Nurestri, A.M.S. Phytochemical constituents, nutritional values, phenolics, flavonols, flavonoids, antioxidant and cytotoxicity studies on *Phaleria macrocarpa* (Scheff.) Boerl fruits. *BMC Complement. Altern. Med.* **2014**, *14*, 152.
28. Wang, Y.; Li, J.; Xuan, L.; Liu, Y.; Shao, L.; Ge, H.; Gu, J.; Wei, C.; Zhao, M. Astragalus Root dry extract restores connexin43 expression by targeting miR-1 in viral myocarditis. *Phytomedicine* **2018**, *46*, 32–38. [CrossRef] [PubMed]
29. Ricce, C.; Rojas, M.L.; Miano, A.C.; Siche, R.; Augusto, P.E.D. Ultrasound pretreatment enhances the carrot drying and rehydration. *Food Res. Int.* **2016**, *89*, 701–708. [CrossRef]
30. Rojas, M.L.; Leite, T.S.; Cristianini, M.; Alvim, I.D.; Augusto, P.E. Peach juiceprocessed by the ultrasound technology: Changes in its microstructure improve its physical properties and stability. *Food Res. Int.* **2016**, *82*, 22–33. [CrossRef]
31. Zhang, Y.; Abatzoglou, N. Review: Fundamentals, applications and potentials of ultrasound-assisted drying. *Chem. Eng. Res. Des.* **2020**, *154*, 21–46. [CrossRef]
32. Wu, B.; Guo, X.; Guo, Y.; Ma, H.; Zhou, C. Enhancing jackfruit infrared drying by combining ultrasound treatments: Effect on drying characteristics, quality properties and microstructure. *Food Chem.* **2021**, *358*, 129845. [CrossRef]
33. Xu, B.; Tiliwa, E.S.; Wei, B.; Wang, B.; Hu, Y.; Zhang, L.; Mujumdar, A.; Zhou, C.; Ma, H. Multi-frequency power ultrasound as a novel approach improves intermediate-wave infrared drying process and quality attributes of pineapple slices. *Ultrason. Sonchem.* **2022**, *88*, 106083. [CrossRef] [PubMed]
34. Wiktor, A.; Sledz, M.; Nowacka, M.; Rybak, K.; Witrowa-Rajchert, D. The influence of immersion and contact ultrasound treatment on selected properties of the apple tissue. *Appl. Acoust.* **2016**, *103*, 136–142. [CrossRef]
35. Tayyab, R.M.; Liu, K.; Ahmed, J.M.; Safdar, B.; Lv, D.; Wei, D. Developing ultrasound-assisted hot-air and infrared drying technology for sweet potatoes. *Ultrason. Sonochem.* **2022**, *86*, 106047. [CrossRef]
36. Hayat, K.; Zhang, X.; Chen, H.; Xia, S.; Jia, C.; Zhong, F. Liberation and separation of phenolic compounds from citrus mandarin peels by microwave heating and its effect on antioxidant activity. *Sep. Purif. Technol.* **2010**, *73*, 371–376. [CrossRef]
37. Chang, C.-H.; Lin, H.-Y.; Chang, C.-Y.; Liu, Y.-C. Comparisons on the antioxidant properties of fresh, freeze-dried and hot-air-dried tomatoes. *J. Food Eng.* **2006**, *77*, 478–485. [CrossRef]
38. Zhao, Y.; Zhu, H.; Xu, J.; Zhuang, W.; Zheng, B.; Lo, Y.M.; Tian, Y. Microwave vacuum drying of lotus (*Nelumbo nucifera* Gaertn.) seeds: Effects of ultrasonic pretreatment on color, antioxidant activity, and rehydration capacity. *LWT* **2021**, *149*, 111603. [CrossRef]
39. Kroehnke, J.; Szadzińska, J.; Stasiak, M.; Radziejewska-Kubzdela, E.; Biegańska-Marecik, R.; Musielak, G. Ultrasound- and microwave-assisted convective drying of carrots—Process kinetics and product's quality analysis. *Ultrason. Sonochem.* **2018**, *48*, 249–258. [CrossRef]
40. Bryan, G.G.; Guy, S.; Katleen, R.; John, V.C. Combined alkaline hydrolysis and ultra-sound-assisted extraction for the release of nonextractable phenolics from cauliflower (*Brassica oleracea* var. botrytis) waste. *J. Agric. Food Chem.* **2014**, *62*, 3371–3376.
41. Ashokkumar, M.; Sunartio, D.; Kentish, S.; Mawson, R.; Simons, L.; Vilkhu, K.; Versteeg, C. Modification of food ingredients by ultrasound to improve functionality: A preliminary study on a model system. *Innov. Food Sci. Emerg. Technol.* **2008**, *9*, 155–160. [CrossRef]

42. Firouz, M.S.; Farahmandi, A.; Hosseinpour, S. Recent advances in ultrasound application as a novel technique in analysis, processing and quality control of fruits, juices and dairy products industries: A review. *Ultrason. Sonochem.* **2019**, *57*, 73–88. [CrossRef]
43. McClements, D. Advances in the application of ultrasound in food analysis and processing. *Trends Food Sci. Technol.* **1995**, *6*, 293–299. [CrossRef]
44. Sun, J.; Zhao, J.; Fu, D.; Gu, S.; Wang, D. Extraction, Optimization and Antimicrobial Activity of IWSP from Oleaginous Microalgae *Chlamydomonas* sp. YB-204. *Food Sci. Technol. Res.* **2017**, *23*, 819–826. [CrossRef]
45. Ota, M.; Xu, F.; Li, Y.L.; Shang, M.Y.; Makino, T.; Cai, S.Q. Comparison of chemical constituents among licoriceroasted licorice, and roasted licorice with honey. *J. Nat. Med.* **2018**, *72*, 80–95. [CrossRef] [PubMed]
46. Sung, M.W.; Li, P.C. Chemical analysis of raw, dryroasted, and honey-roasted licorice by capillary electrophoresis. *Electrophoresis* **2004**, *25*, 3434–3440. [CrossRef]
47. Wen, W.; Li, L.; Li, D.; Zhang, L.; Zhou, D.; Yang, Y.; Yang, J.; Yu, B.; Ju, A. Differential analysis of chemical composition of licorice concoction products based on liquid mass spectrometry and plant metabolomics. *Chin. J. Exp. Formul.* **2020**, *26*, 104–110.
48. Vadivambal, R.; Jayas, D.S. Changes in quality of microwave-treated agricultural products—A review. *Biosyst. Eng.* **2007**, *98*, 1–16. [CrossRef]
49. Maskan, M. Kinetics of colour change of kiwifruits during hot air and microwave drying. *J. Food Eng.* **2001**, *48*, 169–175. [CrossRef]
50. Fratianni, A.; Niro, S.; Alam, M.D.R.; Cinquanta, L.; Di Matteo, M.; Adiletta, G.; Panfili, G. Effect of a physical pre-treatment and drying on carotenoids of goji berries (*Lycium barbarian* L.). *LWT* **2018**, *92*, 313–328. [CrossRef]
51. Deng, L.-Z.; Pan, Z.; Mujumdar, A.; Zhao, J.-H.; Zheng, Z.-A.; Gao, Z.-J.; Xiao, H.-W. High-humidity hot air impingement blanching (HHAIB) enhances drying quality of apricots by inactivating the enzymes, reducing drying time and altering cellular structure. *Food Control.* **2018**, *96*, 104–111. [CrossRef]

foods

MDPI

*Article*

# Physicochemical, Antioxidant, Sensory, and Starch Digestibility Properties of Steamed Bread Fortified with Tamarillo Powder

**Pei-Ci Syu, Qi-Fang Zhang and Sheng-Dun Lin ***

Department of Food Science and Technology, Hungkuang University, Taichung 433304, Taiwan; spc679163@gmail.com (P.-C.S.); 0712fa520@gmail.com (Q.-F.Z.)
* Correspondence: lin54@hk.edu.tw; Tel.: +886-4-26318652 (ext. 5038); Fax: +886-4-26319176

**Abstract:** The effects of lyophilized tamarillo powder (TP) on the physicochemical, antioxidant, sensory, and starch digestibility characteristics of steamed breads were studied. The TP was used to substitute 5–20% of wheat flour to make steamed breads, assigned as T5, T10, T15, and T20, respectively. The results showed that TP is rich in dietary fiber (36.45%). Its extract is rich in bioactive components, including phenolic compounds (28.90 mg GAE/g extract), ascorbic acid (3.25 mg/g extract), total anthocyanins (316.35 μg C3GE/g extract), and total carotenoids (12.68 μg βCE/g extract) and has good antioxidant capacity. As the level of TP increased, the color of steamed breads became darker, redder, and yellower; the texture became harder, and the overall consumption preference decreased. However, their bioactive components content and antioxidant activity increased. The starch hydrolysis percentage of T5 (43.82%), T10 (41.57%), T15 (37.41%), and T20 (35.63%) at 180 min was significantly lower than that of the control (49.80%) ($p < 0.05$). The in vitro predicted glycemic index (80.02) of T20 was categorized as a medium-GI food when bread was used as the reference. On a nine-point hedonic test, control and T5 had the highest overall preference scores (7.1–7.4). The T20 supplemented with extra 15–20% water improved its volume and specific volume, and the overall preference scores (7.4–7.5) were not significantly different from the control ($p > 0.05$). Overall, a partial replacement of wheat flour with TP in steamed bread making could be developed as a new type of medium-GI value food containing more bioactive components and effective antioxidant capacity.

**Keywords:** tamarillo; steamed bread; bioactive component; antioxidants; sensory evaluation; glycemic index

## 1. Introduction

Traditional steamed bread is one of the staple foods loved by Asians. Its basic ingredients are wheat flour, water, and yeast. After mixing the ingredients to form fermented dough, it is steamed for consumption. Compared with baked bread, the outer layer of traditional steamed bread is smooth and white, and because of the relatively low production temperature (<100 °C), the steamed bread contains fewer Maillard reaction products, such as acrylamide and furan [1]. However, several studies found that traditional steamed bread made from refined wheat flour, is a high-glycemic-index (GI, 88.1–98.3) product [2–4]. The intake of high-GI foods by the human body may cause a rapid increase in blood sugar levels and insulin secretion, which in turn increases the risk of cardiovascular disease [5]. Therefore, developing steamed bread with lower GI is beneficial for patients with diabetes and other diseases [6]. In view of this, many scholars have used plants, such as wholegrains and seeds, to improve the GI, physicochemical, antioxidant, and sensory qualities of the SB [7–9]. However, there is little information on the application of fruit to improve the GI of steamed bread.

Tamarillo (*Solanum betaceum* Cav.), also known as tree tomato, is a kind of Solanaceae fruit that grows in the subtropical climate of South America, especially the countries across the Andes Mountains. It is a subtropical fruit with unique color and flavor, rich in nutrients and bioactive components, including minerals; dietary fiber; non-starch polysaccharides;

vitamins A, C, $B_6$, E, and K, carotenoids; phenolic compounds; and anthocyanins [10–15]. Its potential health benefits comprise anti-obesity [16], anti-inflammatory [16], antibacterial [17], antioxidant [10,12,18,19], and anti-tumor [11,18] properties; inhibition of enzymes activity related to metabolic syndrome characteristics [19]; and a potential "multitarget" activity against Alzheimer's disease [20]. Because of its bioactive ingredients and novel eating experience, tamarillo has been used in a variety of products, including salads, sauces, jellies, ice cream, juices, liqueurs, yogurt, and effervescent fruit tablets [13,14,21,22]. Therefore, the tamarillo is an important source that has the potential to be developed into healthy products.

Increasing intake of dietary fiber may be one way to lower postprandial blood glucose [7,23]. The content of dietary fiber and bioactive components in traditional SB is insufficient. Therefore, it would be beneficial to develop a new formulation of steamed bread containing tamarillo powder (TP). The purpose of this study was to measure the physicochemical quality characteristics of TP and wheat flour and then use 5%, 10%, 15%, and 20% (*w*/*w*) of TP instead of wheat flour to make steamed breads to evaluate the effect on the physicochemical, antioxidant, sensory, and starch digestibility properties of the steamed breads. Furthermore, the effect of various amounts of water on the quality characteristics of tamarillo steamed breads was also studied.

## 2. Materials and Methods

### 2.1. Materials

The red variety of tamarillo fruit (*Solanum betaceum* Cav.) (Figure 1) was provided from the farmer (Nantou, Taiwan) and harvested at 22–24 weeks after anthesis at 1100–1200 m sea level. The plant was 5 years old, and all harvested fruits are usually consumed. Then, the fruits were packed in cartons and were transported to Hungkuang University (about 2 h). Fresh fruits were washed with tap water, drained, and then cut into two pieces with a knife. The samples were subjected to vacuum freeze-drying using a freeze-dryer (FD-20L-6S, Kingmech Co., Ltd., New Taipei, Taiwan) with drying conditions of <0.2 mmHg, chamber temperature of 25 °C, and condenser temperature of −50 °C. Dried samples were ground using a mill (RT-30HS, Rong Tsong Precision Technology Co., Taichung, Taiwan) and screened through a 0.7 mm sieve. The dried powders were sealed in PET/Al/PE bags and then stored at −25 °C before use. Low-protein flour (LPF, protein, 8%) of soft white wheat and middle protein flour (MPF, protein, 11.6%) of hard red winter wheat from US were purchased from Chia Fha Enterprise Co., Ltd. (Taichung, Taiwan). Sugar (Taiwan Sugar Co., Ltd., Tainan, Taiwan), soybean powder (Mauhshii Co., Ltd., New Taipei, Taiwan), Lesaffer Saf-Instant yeast (red) (Lesaffer, Marcq-en-Baroeul, France), and shortening (Namchow Oil and Fat Co., Ltd., Taoyuan, Taiwan) were purchased from the local market.

**Figure 1.** The appearance and color of tamarillo tree, fresh tamarillo, freeze-dried tamarillo, and tamarillo powder.

Gallic acid, Folin–Ciocalteu phenol reagent, 1,1-diphenyl-2-picrylhydrazyl (DPPH), potassium ferricyanide, ferric chloride, pepsin, α-amylase (heat stable), α-amylase (from porcine pancreas, type VI-B), α-amylase (from human saliva, type XIII-A), 3,5-dinitrosalicylic acid, maltose, protease (from *Bacillus licheniformis*), amyloglucosidase, celite filter aid, tichloroacetic acid, ascorbic acid, and acetic acid were purchased from Sigma-Aldrich Chemical (St. Louis, MO, USA). Hexane was purchased from Tedia (Fairfield, OH, USA).

Acetonitrile, methanol, and acetone were purchased from Avantor Performance Materials (Center Valley, PA, USA). Potassium chloride was purchased from Showa Chemical (Tokyo, Japan). Cupric sulfate, anhydrous sodium carbonate, and sodium hydroxide were purchased from Shimakyu's Pure Chemicals (Osaka, Japan). Methyl red pure was purchased from Koch Light Research Laboratories (Gauteng, South Africa). Methylene blue and anhydrous sodium acetate were purchased from Katayama Chemical Industries (Osaka, Japan). Potassium sulfite was purchased from Nihon Shiyaku Reagent (Tokyo, Japan). Potassium sodium (+)-tartrate tetrahydrate was purchased from Wako Pure Chemical Industries (Osaka, Japan). Dialysis membrane (molecular weight cut-off, 6000–8000) was purchased from Membrane Filtration Products (Seguin, TX, USA). Sodium dihydrogen phosphate (monobasic), sodium phosphate (dibasic, 12 hydrates), sulfuric acid, and hydrochloric acid were purchased from Union Chemical (Hsinchu, Taiwan). Ethanol was purchased from Taiwan Tobacco & Liquor (Tainan, Taiwan).

### 2.2. Steamed Bread Making

For the control group of steamed bread making, the method and formula (Table 1) of traditional Chinese wheat food was followed, and the dough was prepared by a straight dough method [24]. The formula of the control steamed bread contained middle protein flour (160 g), low-protein flour (40 g), soybean powder (4 g), yeast (4 g), sugar (10 g), shortening (2 g), and water (110 g) on the basis of the formulation. The TP was used to substitute 5%, 10%, 15%, and 20% (*w*/*w*) of low-protein flour to make various steamed breads, designated as T5, T10, T15, and T20, respectively. First of all, TP was mixed with water, and all the ingredients (except the shorting) were poured into the mixing tank in a mixer (SC-20L, Sun-Mate Machinery Co., Ltd., New Taipei, Taiwan) at a low speed (146 rpm) for 2 min. Melted shortening was added into the paste and mixed using a mixer at low speed for 1 min and then turned to a high speed (585 rpm) for 4 min to complete the dough. The dough was divided into 110 g each, 3 parts in total. After folding 3 times by hand, it was shaped into a strip of 17 cm, the head and tail were cut off (16 cm left), and the dough was divided into 4 equal parts (25 g). The dough was placed into an incubator (SP-18S, Sun-Mate Machinery Co., Ltd., New Taipei, Taiwan) at 35 °C and 75% RH for 20 min. Conventional steaming was performed for 7 min in a steaming basket with a boiling wok underneath. Afterwards, the steamed breads were taken out of the basket, cooled to room temperature for 30 min and weighed. The steamed breads were packed in polyethylene bags before analyses of physicochemical and sensory characteristics. Twelve batches (12 steamed breads per batch) were manufactured for each formula. To determine the effect of adding extra water on the quality characteristics of T20, the formula of T20 contained middle protein flour (160 g), TP (40 g), soybean powder (4 g), yeast (4 g), sugar (10 g), shortening (2 g), and water (110 g) on the basis of the formulation (Table 1). The extra 5%, 10%, 15%, 20%, 25%, and 30% (*w*/*w*) amounts of water were on the basis of the amount of wheat flour in the control as 100%. The total amounts of water added to SB of T20-05, T20-10, T20-15, T20-20, T20-25, and T20-30 were 120 g, 130 g, 140 g, 150 g, 160 g, and 170 g, respectively.

**Table 1.** Formulae of tamarillo steamed breads.

| Ingredient (%) | | C [1] | T5 | T10 | T15 | T20 | T20-05 | T20-10 | T20-15 | T20-20 | T20-25 | T20-30 |
|---|---|---|---|---|---|---|---|---|---|---|---|---|
| Wheat flour | middle protein flour | 80 | 80 | 80 | 80 | 80 | 80 | 80 | 80 | 80 | 80 | 80 |
| | low-protein flour | 20 | 15 | 10 | 5 | 0 | 0 | 0 | 0 | 0 | 0 | 0 |
| Tamarillo powder | | 0 | 5 | 10 | 15 | 20 | 20 | 20 | 20 | 20 | 20 | 20 |
| Soybean powder | | 2 | 2 | 2 | 2 | 2 | 2 | 2 | 2 | 2 | 2 | 2 |
| Yeast | | 2 | 2 | 2 | 2 | 2 | 2 | 2 | 2 | 2 | 2 | 2 |
| Sugar | | 5 | 5 | 5 | 5 | 5 | 5 | 5 | 5 | 5 | 5 | 5 |
| Shortening | | 1 | 1 | 1 | 1 | 1 | 1 | 1 | 1 | 1 | 1 | 1 |
| Water | | 55 | 55 | 55 | 55 | 55 | 60 | 65 | 70 | 75 | 80 | 85 |
| Total | | 165 | 165 | 165 | 165 | 165 | 170 | 175 | 180 | 185 | 190 | 195 |

[1] Control, T5, T10, T15, and T20: steamed bread prepared with 0%, 5%, 10%, 15%, and 20% (*w*/*w*) replacement of low-protein flour with tamarillo powder, respectively. T20-05, T20-10, T20-15, T20-20, T20-25, and T20-30: formula of T20 added with extra 5%, 10%, 15%, 20%, 25%, and 30% (*w*/*w*) water added, with the amount of wheat flour in the control as 100%, respectively.

### 2.3. Determination of Proximate Composition, Total Dietary Fiber, and Water Holding Capacity

The moisture, crude fat, crude protein, and crude ash of the samples were determined according to the method of American Association of Cereal Chemists [25]. The nitrogen conversion factor of wheat flour, TP, and steamed breads used for crude protein calculation was 5.7, 6.25, and 5.7, respectively. The total dietary fiber of the samples was analyzed according to the AOAC method 985.29 [26]. The carbohydrate content (g/100 g) and water holding capacity (g $H_2O$ absorbed/g sample) of the samples were measured according to the method of Mau et al. [27]. Each analysis was carried out in triplicate.

### 2.4. Determination of Physical Characteristics

The weight of steamed breads was measured using a scale. The maximum length, width, and height of the steamed breads were measured using a Vernier caliper (Mitutoyo, Kawasaki, Kanagawa, Japan). The spread ratio was calculated by dividing the maximum width with the maximum height of the steamed breads. The volume of the samples was determined by the rapeseed displacement method [28]. The specific volume was calculated by dividing the volume with the weight of the steamed breads. The water activity of the SB was determined using a water activity meter (LabSwift-aw, Novasina AG, Neuheimstrasse, Lachen, Switzerland). Each analysis was carried out with 10 replications.

The reflective color of the sample was measured using a color measurement spectrophotometer (ColorFlex-Diffuse, Hunter Associates Laboratory, Reston, VA, USA) set for CIE $L^*$, $a^*$, and $b^*$ values with a D65 illuminant at 10° [27]. The hue angle ($h^*$), chroma ($c^*$), and whiteness index ($WI^*$) of the steamed breads were calculated as arctan $(b^*/a^*)$ and $(a^{*2} + b^{*2})^{1/2}$ and $100 - [(100 - L^*)^2 + a^{*2} + b^{*2}]^{1/2}$, respectively. The total color difference ($\Delta E$) between the sample and the control was calculated as $\Delta E = [(L^*_c - L^*_t)^2 + (a^*_c - a^*_t)^2 + (a^*_c - a^*_t)^2]^{1/2}$, where $L^*$, $a^*$, and $b^*$ are the color coordinates in the control (c) and tamarillo bread (t), respectively. The results of each color property were averaged from 10 replications.

Texture profile analysis (TPA) of the midsection (2.5 cm × 2.5 cm × 2.5 cm) of the SB was measured using a texture analyzer (TA-XT2, 25 kg model; Stable Micro Systems, Surrey, UK) [27]. After the steamed breads were steamed and cooled to room temperature (25–27 °C), the upper and lower layers of the steamed breads were cut off, and then the crumb was cut into cubes (2.5 cm × 2.5 cm × 2.5 cm) for measurement. The recorded texture parameters were hardness, cohesiveness, springiness, gumminess, chewiness, and resilience. The results for each texture property were presented as the mean of 10 replicates.

### 2.5. Determination of Bioactive Components

Fifteen grams of the samples (wheat flour, TP, and lyophilized steamed bread powder) were extracted with 150 mL of 50% ($v/v$) aqueous ethanol in a water bath at 75 °C (150 rpm) for 30 min and then centrifuged at 4450× $g$ for 30 min and filtered with Advantec No. 1 filter paper. The residue was re-extracted with one extra 150 mL portions of solvent as described. The combined ethanolic extracts were rotary evaporated (40 °C) and freeze-dried. Total phenols of lyophilized extracts were determined following the method of Mau et al. [27]. The total phenols content of the extracts was calculated according to the standard curve of gallic acid [760 nm absorbance = 0.0008 $C_{\text{gallic acid}}$ (μg/mL) − 0.0123, $R^2$ = 0.9995]. Results were expressed as milligram of gallic acid equivalent (GAE) per gram of lyophilized extract. Each analysis was carried out in triplicate.

Total anthocyanins of the extracts were measured by the pH-differential method [29]. Total anthocyanins of the extract were expressed as micrograms of cyanidin-3-glucoside equivalent (C3GE) per gram of lyophilized extract. Each analysis was carried out in triplicate.

Total carotenoids of the lyophilized extracts were analyzed following the method of Knockaert et al. [30] with minor modification. Briefly, each extract (500 mg) was extracted with 20 mL of extraction solvent (hexane: acetone: ethanol, 2:1:1, $v/v$) and 1 mL of 10% KOH solution in a water bath at 25 °C (125 rpm) for 1 h and then centrifuged at 4000× $g$ for 5 min. The supernatant was diluted to 25 mL with the extraction solvent and then poured into a separatory funnel (protect from light). Next, 25 mL of deionized water was added, and the mixture was shaken vigorously to mix (to remove the cover to exhaust air) and left to stand for 30 min. The upper organic layer (yellow) was taken out, and the absorbance was measured at 450 nm. Analyses were performed three times. Total carotenoids of the extract were expressed as micrograms of β-carotene equivalent (βCE) per gram of lyophilized extract. Each analysis was carried out in triplicate.

The ascorbic acid content in the extract was measured based on the method of Attia [31] with minor modification. The extract (100 mg) was dissolved in methanol (3 mL) using an ultrasonic bath with 40 kHz for 15 min, and the volume was adjusted to 5 mL with methanol and then centrifuged at 1463× *g* for 10 min (3000 rpm). The solution was then filtered using a PVDF syringe filter (13 mm × 0.45 μm) prior to injection onto a HPLC. The HPLC system consisted of a Hitachi L-2130 pump, a Hitachi 5430 diode array detector, and a Luna C18 (2) column (250 mm × 4.6 mm, 5 μm particle size; Phenomenex, Torrance, CA, USA). The column temperature was maintained at 25 °C. The mobile phase consisted of 0.1% acetic acid (in water) and acetonitrile (75:25, *v/v*) at a flow rate of 1.0 mL/min. The sample injection volume was 10 μL. The ascorbic acid in the samples was identified by comparing its relative retention time and UV spectra with the authentic compound. Ascorbic acid content was calculated on the basis of the calibration curves of the corresponding authentic compound. Each analysis was carried out in triplicate.

### 2.6. Determination of Antioxidant Properties

The scavenging ability of each extract (0–0.75 mg extract/mL for TP and wheat flour; 0–3.6 mg extract/mL for the steamed breads) for DPPH radicals was determined on the basis of the method of Shimada et al. [32]. Reducing power of each extract (0–3 mg extract/mL for TP and wheat flour; 0–15 mg extract/mL for the steamed breads) was determined according to the method of Oyaizu [33]. The $EC_{50}$ value (mg extract/mL) is the effective concentration at which the DPPH radicals were scavenged by 50%, and the absorbance was 0.5 for reducing power. The $EC_{50}$ values of TP and tamarillo steamed breads were obtained by the interpolation method of linear regression analysis, while the $EC_{50}$ values of flour and control steamed bread were obtained by the extrapolation method of linear regression analysis.

### 2.7. Determination of Starch Hydrolysis and Predicted Glycemic Index (pGI)

Starch hydrolysis of the cooked steamed breads was measured according to the confinement (dialysis) system method of Germaine et al. [34]. Briefly, the cooked steamed breads (50–60 °C) containing 1 g of available carbohydrate were weighed into tubes containing 5 mL of 0.05 M sodium potassium phosphate buffer (pH 6.9) and were determined for the starch hydrolysis with the restricted system. At 0, 15, 30, 45, 60, 90, 120, and 180 min, aliquots (2 mL) of the dialysate were withdrawn, analyzed for reducing sugar content using the 3,5-dinitrosalicylic acid method, and compared with a maltose standard curve. The area under the curve (AUC) value for each starch hydrolysis curve was calculated using Sigmaplot 10.0 (Systat Software, San Jose, CA, USA). The hydrolysis index (HI) was obtained by dividing the area under the hydrolysis curve of the sample by the area of white bread. The pGI of starch hydrolysis value at 180 min was calculated using the formula (pGI = 39.71 + (0.549 × HI)) developed by Goñi et al. [35]. Each analysis was carried out in triplicate.

### 2.8. Sensory Evaluation

The SB samples (50–60 °C) were subjected to sensory evaluation using the hedonic test. Ninety untrained consumers (32 males and 58 females, aged 18–60 years) who were interested in the SB samples were recruited from the students and staff at Hungkuang University. Each SB sample was placed on a white plate and identified with a random three-digit number, then presented to consumers in a balanced and random order. Consumers were asked to clean their mouths with warm water between sampling sessions to reduce the effect of sample carryover. The sensory quality characteristics of the SB samples was assessed using a nine-point hedonic scale questionnaire, including crust color, crumb color, sweetness, aroma, flavor, texture, and overall preference (1 = extremely disliked, 5 = neither liked nor disliked, 9 = extremely liked).

### 2.9. Statistical Analysis

With the exception of weight ($n$ = 10), dimension ($n$ = 10), texture profile analysis ($n$ = 10), color value ($n$ = 10), and hedonic testing ($n$ = 90), each measurement was performed in triplicate. The experimental data were subjected to analysis of variance using a statistical analysis system (SAS Institute, Cary, NC, USA). Duncan's multiple range tests were used to determine the difference among means at the 0.05 level.

## 3. Results and Discussion

### 3.1. Physicochemical Qualities of Tamarillo Powder and Wheat Flour

The TP used in this study was rich in total dietary fiber (TDF) and crude ash but lower in moisture, carbohydrate, and water activity than wheat flour (Table 2). The presence of a large amount of ash correlated with the quantity of minerals present in the TP [11]. TP has a higher water holding capacity (WHC) than wheat flour. Robertson and Eastwood [36] reported dietary fiber has water holding capacity. Therefore, this result should be attributed to the abundance of TDF in TP.

**Table 2.** Physicochemical quality characteristics of tamarillo powder and wheat flour.

| | TP [1] | MPF | LPF |
|---|---|---|---|
| Moisture (g/100 g) | 1.28 ± 0.04 [c2] | 6.94 ± 0.09 [a] | 6.20 ± 0.09 [b] |
| Crude fat (g/100 g) | 0.94 ± 0.03 [a] | 0.93 ± 0.02 [a] | 0.94 ± 0.02 [a] |
| Crude protein (g/100 g) | 10.51 ± 0.21 [b] | 11.07 ± 0.08 [a] | 8.15 ± 0.08 [c] |
| Crude ash (g/100 g) | 6.91 ± 0.01 [a] | 0.38 ± 0.01 [c] | 0.44 ± 0.01 [b] |
| Carbohydrate (g/100 g) | 80.36 ± 0.21 [c] | 80.68 ± 0.18 [b] | 84.27 ± 0.02 [a] |
| Total dietary fiber (g/100 g) | 36.45 ± 9.21 [a] | 1.93 ± 0.73 [b] | 2.07 ± 0.89 [b] |
| Water activity | 0.254 ± 0.002 [c] | 0.641 ± 0.005 [a] | 0.593 ± 0.006 [b] |
| Water holding capacity ($g\ H_2O$ absorbed/g sample) | 2.97 ± 0.01 [a] | 0.81 ± 0.01 [b] | 0.59 ± 0.01 [c] |
| | | Color property | |
| $L^*$ | 53.50 ± 0.12 [c] | 93.11 ± 0.07 [b] | 93.98 ± 0.08 [a] |
| $a^*$ | 19.16 ± 0.12 [a] | 0.19 ± 0.04 [b] | 0.02 ± 0.03 [b] |
| $b^*$ | 21.91 ± 0.08 [a] | 8.61 ± 0.06 [b] | 7.26 ± 0.04 [c] |
| $WI^*$ | 44.99 ± 0.17 [c] | 88.97 ± 0.01 [b] | 90.57 ± 0.06 [a] |
| $c^*$ | 29.23 ± 0.12 [a] | 8.55 ± 0.06 [b] | 7.30 ± 0.04 [c] |
| $h^*$ (°) | 48.83 ± 0.01 [c] | 88.74 ± 0.01 [b] | 89.84 ± 0.01 [a] |
| $\Delta E$ | | 45.99 ± 0.13 [b] | 47.23 ± 0.02 [a] |

[1] TP: Tamarillo powder. MPF: Middle protein flour. LPF: Low-protein flour. [2] Each value is expressed as mean ± standard deviation ($n$ = 3). Means with different lowercase letters within a row differ significantly ($p < 0.05$).

The $a^*$, $b^*$, and $c^*$ values of TP were significantly higher than those of wheat flour, but the $L^*$ value showed the reverse (Table 2), which means that the color of TP is redder, yellower, darker, and has higher color intensity than that of wheat flour. The 0°, 60°, 120°, 180°, 240°, and 300° of the $h^*$ are red, yellow, green, cyan, blue, and magenta, respectively [37]. The $h^*$ value of TP is 48.83° (Table 2), which is between red and yellow, consistent with the color of tamarillo powder in Figure 1. Interestingly, wheat flour is a white powder and cannot be suitably expressed by $h^*$. Instead, it should be expressed by whiteness index. After calculation, the whiteness index of middle protein flour and low-protein flour were 88.97 and 90.57, respectively. As for TP, it is not suitable to be represented by whiteness index. The $\Delta E$ value of TP and wheat flour was >3. Song et al. [38] reported that the $\Delta E$ value can be explained as follows: $\Delta E < 1$, the color difference was not obvious to the human eye; $1 < \Delta E < 3$, the color has no obvious difference from the human eye; $\Delta E > 3$, the color difference was obvious to the human eye. The $\Delta E$ value of middle protein flour and low-protein flour was 1.24, which means that there was no obvious difference in the color of the two flours when viewed by the human eye.

TP and wheat flour were extracted with 50% aqueous ethanol. The extraction yield of the freeze-dried extract of TP was significantly higher than that of wheat flour extract (Table 3). Compared with wheat flour, this showed that TP has many components soluble in 50% ethanol. As expected, the TP in this study was rich in total phenols, ascorbic acid, total anthocyanins, and total carotenoids (Table 3). This result was similar to other studies [10–15]. Free radicals have been reported to be involved in the progression of various human diseases such as diabetes and atherosclerosis [39]. Naturally occurring antioxidants have been reported to delay the progression of many chronic human diseases by scavenging free radicals [40]. Studies have pointed out that TP has good antioxidant properties [10,12,18,19]. This should be related to the phenolic compounds and ascorbic acid [41], anthocyanins [42], and carotenoids [43] having good antioxidant properties. The results of the antioxidant capacity of the extracts in this study were expressed as $EC_{50}$ values, which were negatively correlated with the antioxidant capacity. The scavenging ability on DPPH radicals and reducing power of TP extract were significantly higher than those of wheat flour extract, which should be related to the fact that TP was rich in phenolic compounds, ascorbic acid, anthocyanins, and carotenoids (Table 3), especially phenolic compounds and ascorbic acid [41]. Previous studies showed that the $EC_{50}$ values of ascorbic acid for DPPH radical scavenging ability and reducing power were 15.49 μg/mL and 19.92 μg/mL, respectively [41]. Clearly, TP is a rich source of dietary fiber and bioactive components, and substituting TP for wheat flour in traditional steamed bread formula would result in a functional product.

**Table 3.** The yield, bioactive components, and antioxidant property of tamarillo powder and wheat flour extracts.

| | TP [1] | MPF | LPF |
|---|---|---|---|
| Yield (g extract/100 g powder) | 52.69 ± 0.86 [a3] | 7.16 ± 0.16 [b] | 6.31 ± 0.38 [b] |
| Total phenols (mg GAE/g extract) [1] | 28.90 ± 0.20 [a] | 13.27 ± 0.24 [b] | 12.31 ± 0.51 [c] |
| Ascorbic acid (mg/g extract) | 3.25 ± 0.05 [a] | 0.32 ± 0.01 [b] | 0.31 ± 0.01 [b] |
| Total anthocyanins (μg C3GE/g extract) [1] | 316.35 ± 0.02 | nd [4] | nd |
| Total carotenoids (μg βCE/g extract) [1] | 12.68 ± 1.33 | nd | nd |
| $EC_{50}$ value of antioxidant property (mg extract/mL) [2] | | | |
| Scavenging ability of DPPH radicals | 0.29 ± 0.01 [c] | 3.84 ± 0.05 [b] | 4.13 ± 0.02 [a] |
| Reducing power | 0.95 ± 0.05 [b] | 46.27 ± 0.32 [a] | 46.35 ± 0.72 [a] |

[1] TP: Tamarillo powder. MPF: Middle protein flour. LPF: Low-protein flour. GAE: gallic acid equivalent. C3GE: cyanidin-3-glucoside equivalent. βCE: β-carotene equivalent. [2] $EC_{50}$ value: The effective concentration at which DPPH radicals were scavenged by 50%, and the absorbance was 0.5 for reducing power. The $EC_{50}$ value of TP was obtained by the interpolation method of linear regression analysis, while the $EC_{50}$ value of flour was obtained by the extrapolation method of linear regression analysis. [3] Each value is expressed as mean ± standard deviation ($n$ = 3). Means with different lowercase letters within a row differ significantly ($p < 0.05$). [4] nd: not detected.

### 3.2. Physical Characteristics of the Steamed Bread

There was no significant difference in the weight of steamed breads (Table 4). However, replacing wheat flour with TP will reduce the dimensions of steamed breads, including length, width, height, spread ratio, volume, and specific volume. The volume (mL) of the steamed bread samples decreased significantly from 78.5 (control) to 44.0 (T20) as the level of TP in the formulae increased. The decreased volume (%) of T5, T10, T15, and T20 was 1.91%, 11.5%, 26.8%, and 43.95%, respectively, compared to the control. This could be attributed to the gluten dilution effect, since TP does not contain gluten-type proteins. The gliadins and glutenins in flour are the gluten-forming proteins that provide the desired texture and volume of bread [44] In addition, the dietary fiber disrupted the gluten matrix of wheat flour in the steamed bread, which reduces the gas holding capacity of bread and the mechanical strength of dough during proofing, and steaming is also one of the reasons [45,46]. This change due to gluten dilution has been previously observed in steamed breads fortified with other functional ingredients, such as wholegrain flour, quinoa flour, high-amylose corn starch, and purple sweet potato flour [7,46–48].

**Table 4.** Physical quality characteristics of tamarillo steamed breads.

| | Control | T5 | T10 | T15 | T20 |
|---|---|---|---|---|---|
| Weight (g) | 25.25 ± 0.10 [a1] | 25.19 ± 0.19 [a] | 25.11 ± 0.10 [a] | 25.49 ± 0.34 [a] | 25.40 ± 0.11 [a] |
| | | | Dimension | | |
| Length (mm) | 59.15 ± 1.45 [a] | 58.90 ± 0.50 [a] | 45.90 ± 0.10 [b] | 43.20 ± 0.30 [c] | 43.70 ± 1.10 [c] |
| Width (mm) | 51.65 ± 0.65 [a] | 48.48 ± 0.60 [b] | 45.40 ± 0.30 [c] | 42.76 ± 0.25 [d] | 37.90 ± 0.6 [e] |
| Height (mm) | 32.65 ± 0.35 [a] | 32.11 ± 0.10 [b] | 31.25 ± 0.15 [c] | 31.05 ± 0.15 [c] | 28.75 ± 0.30 [d] |
| Spread ratio | 1.582 ± 0.004 [a] | 1.510 ± 0.034 [b] | 1.453 ± 0.023 [c] | 1.377 ± 0.021 [d] | 1.318 ± 0.023 [e] |
| Volume (mL) | 78.5 ± 0.7 [a] | 76.0 ± 0.1 [b] | 69.5 ± 0.7 [c] | 57.5 ± 0.7 [d] | 44.0 ± 1.4 [e] |
| Specific volume (mL/g) | 3.11 ± 0.03 [a] | 3.02 ± 0.02 [a] | 2.77 ± 0.02 [b] | 2.26 ± 0.05 [c] | 1.73 ± 0.05 [d] |
| | | | Crust color property [2] | | |
| $L^*$ | 82.83 ± 0.49 [a] | 76.12 ± 1.30 [b] | 68.41 ± 0.79 [c] | 64.09 ± 0.62 [d] | 59.15 ± 0.66 [e] |
| $a^*$ | −0.18 ± 0.08 [e] | 5.71 ± 0.14 [d] | 10.98 ± 0.25 [c] | 15.01 ± 0.44 [b] | 17.45 ± 0.59 [a] |
| $b^*$ | 11.15 ± 0.32 [d] | 25.77 ± 0.44 [c] | 34.18 ± 0.52 [a] | 34.95 ± 0.81 [a] | 31.36 ± 0.74 [b] |
| $c^*$ | 11.15 ± 0.32 [d] | 26.39 ± 0.43 [c] | 35.90 ± 0.48 [b] | 38.04 ± 0.75 [a] | 35.88 ± 0.88 [b] |
| $h^*$ (°) | 90.92 ± 0.26 [a] | 77.51 ± 0.20 [b] | 72.19 ± 0.26 [c] | 66.76 ± 0.67 [d] | 60.91 ± 0.29 [e] |
| $\Delta E$ | | 17.12 ± 0.37 [d] | 29.38 ± 0.39 [c] | 33.89 ± 0.51 [b] | 35.77 ± 0.60 [a] |
| | | | Crumb color property | | |
| $L^*$ | 79.82 ± 1.08 [a] | 75.14 ± 0.86 [b] | 68.74 ± 0.98 [c] | 67.34 ± 0.79 [d] | 60.62 ± 0.74 [e] |
| $a^*$ | −0.14 ± 0.01 [e] | 4.61 ± 0.15 [d] | 8.79 ± 0.17 [c] | 11.34 ± 0.19 [b] | 14.14 ± 0.40 [a] |
| $b^*$ | 12.17 ± 0.44 [d] | 24.62 ± 0.88 [c] | 30.83 ± 0.31 [b] | 31.94 ± 0.47 [a] | 30.77 ± 0.43 [b] |
| $c^*$ | 12.17 ± 0.44 [d] | 25.05 ± 0.88 [c] | 32.06 ± 0.31 [b] | 33.89 ± 0.47 [a] | 33.86 ± 0.47 [a] |
| $h^*$ (°) | 90.66 ± 0.16 [a] | 79.39 ± 0.26 [b] | 74.09 ± 0.11 [c] | 70.45 ± 0.31 [d] | 65.32 ± 0.21 [e] |
| $\Delta E$ | | 14.12 ± 0.96 [d] | 23.47 ± 0.89 [c] | 26.14 ± 0.79 [b] | 30.33 ± 0.74 [a] |
| | | | Texture profile analysis | | |
| Hardness (N) | 4.86 ± 0.12 [e] | 6.18 ± 0.30 [d] | 7.61 ± 0.20 [c] | 10.08 ± 0.23 [b] | 11.65 ± 0.41 [a] |
| Cohesiveness | 0.81 ± 0.01 [a] | 0.74 ± 0.01 [b] | 0.69 ± 0.01 [c] | 0.64 ± 0.01 [d] | 0.61 ± 0.02 [e] |
| Springiness | 0.99 ± 0.02 [a] | 0.93 ± 0.02 [a] | 0.86 ± 0.03 [b] | 0.80 ± 0.04 [c] | 0.76 ± 0.05 [c] |
| Gumminess (N) | 3.94 ± 0.08 [e] | 4.57 ± 0.41 [d] | 5.25 ± 0.03 [c] | 6.45 ± 0.16 [b] | 7.11 ± 0.46 [a] |
| Chewiness (N) | 3.89 ± 0.11 [e] | 4.25 ± 0.37 [d] | 4.52 ± 0.14 [c] | 5.16 ± 0.35 [b] | 5.40 ± 1.44 [a] |
| Resilience | 0.43 ± 0.01 [a] | 0.29 ± 0.01 [b] | 0.27 ± 0.02 [c] | 0.23 ± 0.01 [d] | 0.21 ± 0.02 [d] |

[1] Each value is expressed as mean ± standard deviation ($n$ = 10). Means with different lowercase letters within a row differ significantly ($p < 0.05$). [2] $L^*$, $a^*$, $b^*$, $WI^*$, $c^*$, $h^*$, and $\Delta E$ values corresponding to lightness, redness, yellowness, whiteness index, chroma, hue angle, and total color difference, respectively.

Color is one of the important quality characteristics of food. The significant difference on the crust and crumb color of the steamed breads can be seen in Figure 2. The color of the steamed breads was greatly affected by the replacement of flour with TP (Table 4). For crust and crumb color, $L^*$ values decreased with the increase in TP levels, but the $a^*$, $b^*$, and $c^*$ values showed the reverse. This result indicated that the color of steamed breads was darker, redder, yellower, and had higher color intensity with the increase in TP levels. The color change of the steamed breads should be related to the pigments in TP, including anthocyanins and carotenoids. From the $h^*$ results of T5–T20, it was observed that the progression from yellow to red in crust color (77.51–60.91°) and crumb color (79.39–65.32°) correlated with increased TP substitution (Table 4). Compared with the control, the $\Delta E$ of the crust and crumb color of the tamarillo steamed bread samples ranged from 17.12 to 35.77 and 14.12 to 30.33, respectively (Table 4). This result indicated that the color difference between T5–T20 and control could be obviously distinguished by the human eye [38]. The crust color ($\Delta E$, 1.88) of T15 and T20 and the crumb color ($\Delta E$, 2.67) of T10 and T15 were difficult to distinguish. The $\Delta E$ values of other tamarillo steamed breads were greater than 3, which should be easy to distinguish with the naked eye [38].

**Figure 2.** The appearance and color of tamarillo steamed breads.

Hardness is an indicator commonly used to evaluate the texture quality of food. In texture profile analysis, the order of hardness of the steamed breads was T20 > T15 > T10 > T5 > control, indicating that the samples became firmer with the increase in TP levels in the mixed formulae (Table 4). In this study, there was no significant difference in the weight of any of the steamed breads. Therefore, the increase in hardness is mainly related to the volume or specific volume of these samples. The increase in firmness of the food might be attributed to the density of the tested sample, which is inversely proportional to its specific volume. The internal resistance of the steamed breads' structure can be expressed by cohesiveness; that is, the steamed breads with high cohesiveness form clumps rather than disintegrating during chewing [49]. The cohesiveness of the steamed breads decreased with the increase in TP level. Springiness is measured by the degree of recovery between the first and second compressions. Resilience is the percentage at which energy can be recovered when the first compression is released. The results of TPA showed that the springiness and resilience of steamed breads were in the order control > T5 > T10 > T15 > T20. This might be due to the higher proportion of TP replacing wheat flour in the formula of steamed breads, resulting in lower gluten content, smaller volume, and higher density. The energy required to break down food for swallowing can be expressed in terms of chewiness. Chewiness is calculated by multiplying gumminess by springiness, while gumminess is calculated by multiplying hardness by cohesiveness. Gumminess and chewiness of the SB increased with the increase in TP levels. Overall, as the level of TP in the formulae increased, the hardness, gumminess, and chewiness of the steamed breads increased, while their cohesiveness, springiness, and resilience decreased. Changes in the texture data of steamed breads could be used as quality indicators for subsequent consumer evaluation results.

### 3.3. *Proximate Composition, Total Dietary Fiber, and Water Activity of the Steamed Bread*

The crude ash, crude protein, and TDF contents of the steamed breads increased with the increase in levels of added TP (Table 5). This might be attributed to the fact that ash, protein, and TDF content of TP were higher than those of low-protein flour (Table 2). The moisture content of control and T5 did not differ significantly but was higher than that of T10, T15, and T20 (Table 5). This might be attributed to the fact that moisture content of low-protein flour was higher than that of TP (Table 2). The water activity of the steamed breads also decreased with the increase in TP levels, which could be attributed to the fact that TP has a higher WHC than low-protein flour (Table 2). The American Diabetes Association recommends that diabetes mellitus patients or the general population should consume 20 to 35 g of fiber from raw vegetables and unprocessed grains (or about 14 g of fiber per 1000 kcal ingested) per day [50]. According to the TDF data in Table 5, the tamarillo steamed breads in this study should have potential, especially T20. However, further research is required.

**Table 5.** Proximate compositions, total dietary fiber, and water activity of tamarillo steamed breads.

| | Control | T5 | T10 | T15 | T20 |
|---|---|---|---|---|---|
| Moisture (%) | 37.58 ± 0.07 [ab1] | 37.66 ± 0.07 [a] | 37.48 ± 0.16 [b] | 37.25 ± 0.09 [c] | 37.02 ± 0.07 [d] |
| Crude ash (%) | 0.41 ± 0.01 [e] | 0.54 ± 0.01 [d] | 0.67 ± 0.03 [c] | 0.84 ± 0.01 [b] | 1.05 ± 0.01 [a] |
| Crude fat (%) | 0.61 ± 0.01 [a] | 0.61 ± 0.01 [a] | 0.61 ± 0.01 [a] | 0.62 ± 0.01 [a] | 0.62 ± 0.01 [a] |
| Crude protein (%) | 7.63 ± 0.04 [c] | 7.63 ± 0.05 [c] | 7.82 ± 0.11 [b] | 7.90 ± 0.17 [b] | 8.09 ± 0.04 [a] |
| Carbohydrate (%) | 53.77 ± 0.07 [a] | 53.56 ± 0.01 [ab] | 53.42 ± 0.171 [b] | 53.39 ± 0.184 [b] | 53.22 ± 0.07 [b] |
| Total dietary fiber (%) | 1.19 ± 0.47 [e] | 2.01 ± 0.53 [d] | 3.15 ± 0.091 c | 4.25 ± 0.073 [b] | 5.11 ± 0.602 [a] |
| Water activity | 0.984 ± 0.004 [a] | 0.982 ± 0.062 [a] | 0.975 ± 0.058 [a] | 0.965 ± 0.053 [b] | 0.964 ± 0.058 [b] |

[1] Each value is expressed as mean ± standard deviation ($n$ = 3). Means with different lowercase letters within a row differ significantly ($p < 0.05$). Mean values of color and texture from 10 replicates, and mean values of other quality characteristics from three replicates.

### 3.4. Bioactive Components and Antioxidant Activity of the Steamed Bread Extract

The steamed bread used 50% aqueous ethanol as the extraction solvent, and the extraction yields (15.72–25.43%) increased as the amount of low-protein flour replaced by TP increased (Table 6). The total phenols, ascorbic acid, total anthocyanins, and total carotenoids content of the lyophilized extracts also increased significantly with the increase in TP levels. This should be attributed to the higher bioactive content of the TP than that of low-protein flour (Table 3). When the extraction yields of the steamed bread were taken into consideration, the contents of total phenols, ascorbic acid, total anthocyanins, and total carotenoids of the steamed bread also increased significantly with the increase in TP levels. The DPPH radical scavenging efficiency and reducing ability of the steamed bread extracts were in the order of T20 > T15 > T10 > T5 > control (Table 6). The results showed that the substitution of TP for low-protein flour enhanced the antioxidant activity of the steamed bread. The improved antioxidant properties of the steamed bread could be attributed to the bioactive components of TP, including phenolic compounds, ascorbic acid, and carotenoids. Therefore, compared with traditional steamed bread, tamarillo steamed bread retained a large number of bioactive components and stronger antioxidant properties after cooking, which is beneficial to consumers and provides them with so-called physiological properties.

**Table 6.** The yield, bioactive components, and antioxidant property of tamarillo steamed breads extracts.

| | Control | T5 | T10 | T15 | T20 |
|---|---|---|---|---|---|
| Yield (g extract/100 g powder) | 15.72 ± 0.11 [e3] | 18.48 ± 0.08 [d] | 20.63 ± 0.93 [c] | 24.09 ± 0.06 [b] | 25.43 ± 0.12 [a] |
| Total phenols (mg GAE/g extract) [1] | 10.58 ± 0.25 [e] | 12.19 ± 0.54 [d] | 14.74 ± 0.05 [c] | 15.22 ± 0.08 [b] | 15.92 ± 0.06 [a] |
| Ascorbic acid (mg/g extract) | 0.241 ± 0.008 [d] | 0.384 ± 0.008 [c] | 0.546 ± 0.003 [b] | 0.621 ± 0.029 [a] | 0.627 ± 0.004 [a] |
| Total anthocyanins (μg C3GE/g extract) [1] | nd [4] | 243.1 ± 2.1 [d] | 265.5 ± 1.3 [c] | 357.1 ± 3.5 [b] | 389.3 ± 2.3 [a] |
| Total carotenoids (μg βCE/g extract) [1] | nd | 25.39 ± 0.48 [c] | 27.25 ± 0.23 [b] | 27.14 ± 0.16 [b] | 30.34 ± 0.23 [a] |
| | $EC_{50}$ value of antioxidant property (mg extract/mL) [2] | | | | |
| Scavenging ability of DPPH radicals | 12.93 ± 0.06 [a] | 2.14 ± 0.10 [b] | 1.34 ± 0.12 [b] | 0.97 ± 0.04 [b] | 0.88 ± 0.05 [b] |
| Reducing power | 42.29 ± 0.38 [a] | 6.39 ± 0.21 [b] | 4.22 ± 0.18 [c] | 2.98 ± 0.72 [d] | 2.35 ± 0.04 [e] |

[1] GAE: gallic acid equivalent. C3GE: cyanidin-3-glucoside equivalent. βCE: β-carotene equivalent. [2] $EC_{50}$ values: The effective concentration at which DPPH radicals were scavenged by 50%, and the absorbance was 0.5 for reducing power. The $EC_{50}$ values of tamarillo SB were obtained by the interpolation method of linear regression analysis, while the $EC_{50}$ value of control was obtained by the extrapolation method of linear regression analysis. [3] Each value is expressed as mean ± standard deviation ($n$ = 3). Means with different lowercase letters within a row differ significantly ($p < 0.05$). [4] nd: not detected.

### 3.5. Starch Hydrolysis and pGI of the SB

The in vivo measurement of GI requires the use of human subjects to assess glycemic responses 2 h after food intake; thus, it is resource-intensive and relatively time-consuming. The in vitro enzymatic assays to estimate the possible rates of starch digestion and glucose absorption in the small intestine are a faster and more cost-effective method than in vivo GI measurements [34,35]. The sensitivity of starch to the action of digestive enzymes is an important factor in both in vivo and in vitro responses; thus, the in vitro method allows the estimation of biological responses. It is particularly suitable for screening starch-rich products for GI during product development before selection for in vivo GI testing [35]. Therefore, the in vitro starch hydrolysis was used to estimate the pGI of the steamed breads in this study. As shown in Figure 3, when using white bread as a reference, the percentage of starch hydrolysis reached 48.01% at 180 min in the restricted system. The results were consistent with those of Granfeldt et al. [51] (nearly 50%), Germaine et al. [34] (about 46%), and Lai et al. [52] (about 42%) when using a restricted system. If an unrestricted system was used, the percentage of starch hydrolysis of white bread was approximately 70–80% after 180 min of incubation [34,35]. The starch hydrolysis curves of the tested steamed breads using the dialysis system followed a linear model. At the end of the 180 min incubation period, the percentage of starch hydrolysis was significantly lower for T5 (43.82%), T10 (41.57%), T15 (37.41%), and T20 (35.63%) than for the control (49.80%). The results suggest that the steamed breads made with TP instead of low-protein flour may be more beneficial to the health of general consumers and type 2 diabetes patients. The calculated pGI values for the steamed breads estimated from the ratio between the area under the hydrolysis curve for a given steamed breads and white bread are given in Table 7. The pGI values of control, T5, T10, T15, and T20 ranged from 97.28 to 80.02 when using white bread as the reference food. Traditional steamed bread is generally classified as a high-GI food [2–4]. Among the tested steamed breads, T20 had the lowest pGI value and was a medium-GI food (Table 7), which could be attributed to it having the highest content of TDF and bioactive components (Tables 2 and 3).

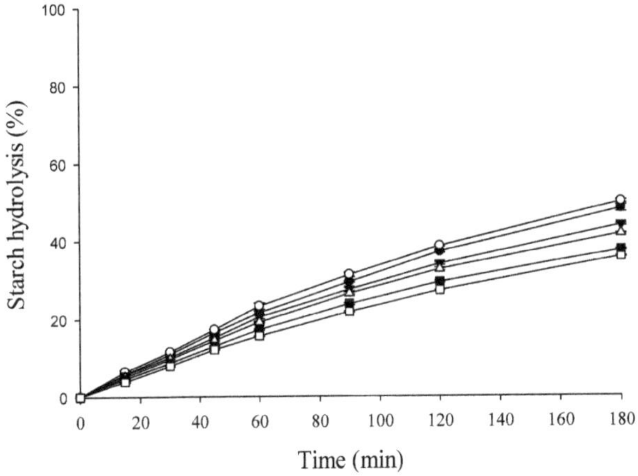


**Figure 3.** Starch hydrolysis (%) of white bread and tamarillo steamed breads. Each value is expressed as mean ± standard deviation ($n$ = 3). White bread (•), control (○), T5 (▼), T10 (△), T15 (■), and T20 (□).

**Table 7.** Predicted glycemic index (pGI) of tamarillo steamed breads in the in vitro test.

| Sample | pGI | GI Category [1] |
|---|---|---|
| Control | 97.28 ± 0.47 [A2] | High |
| T5 | 94.82 ± 0.38 [B] | High |
| T10 | 89.72 ± 0.15 [C] | High |
| T15 | 86.37 ± 0.44 [D] | High |
| T20 | 80.02 ± 0.13 [E] | Medium |

[1] GI categories, with the reference food of white bread are as follows: high GI value, GI > 85; medium GI value, 85 > GI > 60; and low GI value, GI < 60. [2] Each value is expressed as mean ± SD ($n$ = 3). Means with different capital letters within a column differ significantly ($p < 0.05$).

### 3.6. Sensory Evaluation

The nine-point hedonic scale could be used as a measure of preference for products and provided reliable and valid results [53]. No significant differences were observed with regard to the color of the crust and crumb, sweetness, aroma, flavor, texture, and overall preference scores between the control and T5 (Figure 4). Results for all hedonic quality characteristics scores were 6.8–7.8 for control and T5, indicating that both samples were moderately to very much liked by consumption. The sweetness (5.1–6.6), aroma (5.3–6.6), flavor (5.1–6.6), texture (5.0–5.9), and comprehensive (5.1–6.4) scores of T10, T15, and T20 were significantly lower than those of the control. Consumption preferences ranged from "neither liked nor disliked" to "moderately liked", although the hedonic quality characteristics scores of T15 and T20 were lower than those of other steamed breads. From the opinions of consumers in the questionnaire, if the moistness and texture of T15 and T20 could be improved, the consumers would also like T15 and T20, which are rich in bioactive components, TDF, and antioxidant capacity.

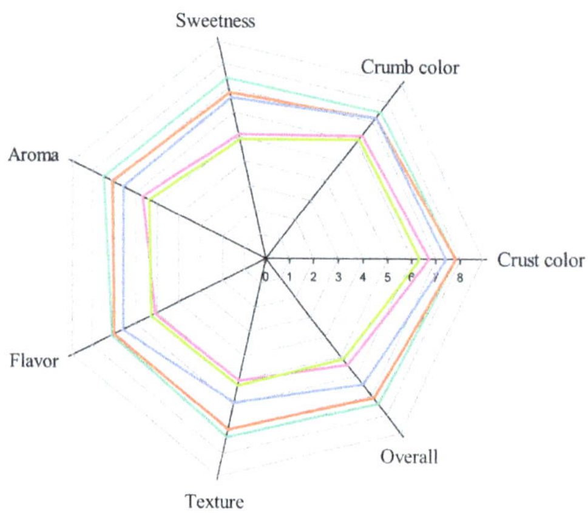


**Figure 4.** The radar plot of hedonic quality characteristics results of tamarillo steamed breads. Each value is expressed as mean ± standard deviation ($n = 90$). Control (——), T5 (——), T10 (——), T15 (——), T20 (——). Values of each hedonic scale were averaged from 90 replicates. Nine-point hedonic scale with 1, 5, and 9 representing extremely disliked, neither liked nor disliked, and extremely liked, respectively.

### 3.7. Effect of Added Water Amount on the Physical and Sensory Properties of T20

Since T20 contains more bioactive components, TDF, and better antioxidant properties than T15, it had the lowest pGI value. Consequently, T20 was used as the test sample in this study to investigate the effects of various water additions on the physical and sensory properties of T20. The addition of various amounts of water to the T20 formula is shown in Table 1. The appearance and color of T20 with various amounts of water added are shown in Figure 5. There was no significant difference in weight among the steamed breads (Table 8). However, the moisture content and water activity of T20 with various amounts of water added increased as the amount of water in the formula increased. The volume and specific volume of T20-15 and T20-20 were the highest among T20 samples, indicating that the formula of T20 with an extra 15–20% water added could improve the volume and specific volume of T20. This might be due to the fact that TP has higher total dietary fiber content (36.45%) and water holding capacity (2.97 g $H_2O$ absorbed/g sample) than wheat flour (Table 2), which enabled the steamed breads to retain more water vapor during steaming. However, when an extra 25–30% of water was added to the T20 formula, the texture structure of the sample became wet and sticky and collapsed, resulting in a decrease in product volume. The crust color, crumb color, aroma, flavor, texture, and overall preference scores of T20-15 and T20-20 were higher than those of other T20 samples, and there was no significant difference from the control (Table 8).

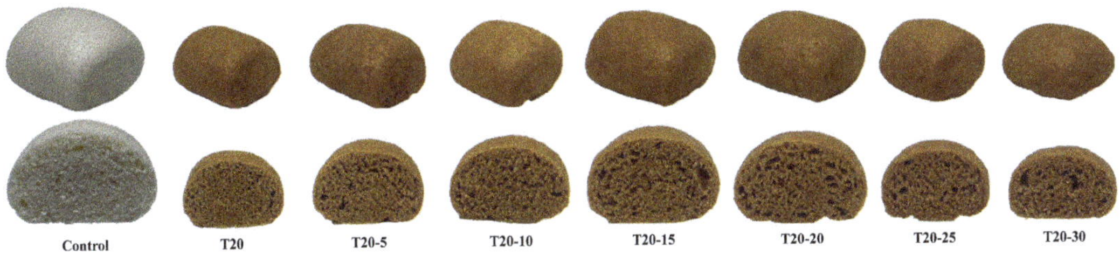

Figure 5. The appearance and color of control and T20 steamed breads with various amounts of water.

**Table 8.** Physical quality characteristics and hedonic test results of control and T20 with various amounts of water added.

| | Control | T20 | T20-05 | T20-10 | T20-15 | T20-20 | T20-25 | T20-30 |
|---|---|---|---|---|---|---|---|---|
| Weight (g) | 25.25 ± 0.10 [a1] | 25.40 ± 0.63 [a] | 25.25 ± 0.33 [a] | 24.95 ± 0.26 [a] | 25.31 ± 0.24 [a] | 25.15 ± 0.32 [a] | 25.29 ± 0.48 [a] | 25.06 ± 0.40 [a] |
| Moisture (%) | 38.72 ± 0.20 [d] | 38.46 ± 0.57 [d] | 39.49 ± 0.14 [c] | 40.10 ± 0.82 [c] | 43.18 ± 0.39 [b] | 43.81 ± 0.13 [b] | 45.46 ± 0.33 [a] | 45.67 ± 0.17 [a] |
| Water activity | 0.961 ± 0.006 [d] | 0.950 ± 0.003 [e] | 0.954 ± 0.001 [e] | 0.960 ± 0.001 [d] | 0.963 ± 0.003 [cd] | 0.967 ± 0.001 [bc] | 0.970 ± 0.001 [ab] | 0.972 ± 0.002 [a] |
| Volume (mL) | 78.50 ± 0.50 [a] | 44.00 ± 1.00 [e] | 62.77 ± 0.58 [d] | 65.33 ± 0.58 [c] | 70.67 ± 0.29 [b] | 68.63 ± 0.29 [b] | 65.50 ± 0.5 [c] | 64.00 ± 0.05 [cd] |
| Specific volume (mL/g) | 3.11 ± 0.03 [a] | 1.73 ± 0.05 [d] | 2.49 ± 0.06 [c] | 2.62 ± 0.02 [bc] | 2.79 ± 0.04 [b] | 2.73 ± 0.11 [b] | 2.59 ± 0.07 [bc] | 2.55 ± 0.06 [bc] |
| | | | | Hedonic test | | | | |
| Crust color | 7.8 ± 1.1 [a] | 6.3 ± 1.3 [c] | 6.3 ± 1.3 [c] | 6.3 ± 1.3 [c] | 7.6 ± 1.2 [a] | 7.3 ± 1.1 [ab] | 6.8 ± 1.3 [bc] | 6.5 ± 1.2 [c] |
| Crumb color | 7.3 ± 1.3 [a] | 6.0 ± 1.2 [b] | 6.0 ± 1.4 [b] | 6.2 ± 0.8 [b] | 7.0 ± 1.5 [a] | 6.9 ± 1.5 [a] | 6.2 ± 0.9 [b] | 6.1 ± 1.1 [b] |
| Sweetness | 7.2 ± 1.6 [a] | 5.5 ± 1.2 [c] | 5.8 ± 0.9 [bc] | 5.9 ± 1.1 [bc] | 6.0 ± 1.0 [bc] | 6.0 ± 1.1 [bc] | 6.6 ± 1.5 [abc] | 6.8 ± 1.6 [ab] |
| Aroma | 7.0 ± 1.5 [a] | 5.4 ± 1.2 [b] | 5.5 ± 1.2 [b] | 6.5 ± 1.0 [a] | 6.8 ± 1.4 [a] | 6.9 ± 1.3 [a] | 6.4 ± 1.0 [a] | 6.5 ± 1.1 [a] |
| Flavor | 7.1 ± 1.3 [a] | 5.4 ± 1.2 [c] | 5.7 ± 0.8 [c] | 5.9 ± 0.9 [c] | 6.9 ± 1.0 [a] | 6.9 ± 1.5 [a] | 6.8 ± 1.4 [ab] | 6.1 ± 1.3 [bc] |
| Texture | 7.6 ± 0.9 [a] | 5.4 ± 1.1 [b] | 6.3 ± 0.5 [b] | 6.6 ± 1.4 [b] | 7.6 ± 1.2 [a] | 7.8 ± 1.0 [a] | 7.8 ± 1.1 [a] | 7.6 ± 1.4 [a] |
| Overall | 7.6 ± 0.9 [a] | 5.6 ± 0.8 [e] | 6.8 ± 1.2 [cd] | 6.9 ± 0.9 [bcd] | 7.5 ± 0.7 [ab] | 7.4 ± 1.1 [abc] | 6.7 ± 1.2 [d] | 6.5 ± 1.0 [d] |

[1] Means ± standard deviation ($n$ = 90 for hedonic scale test and $n$ = 3 for others) with different lowercase letters within a row differ significantly ($p < 0.05$). Nine-point hedonic scale with 1, 5, and 9 representing extremely disliked, neither liked nor disliked and extremely liked, respectively.

## 4. Conclusions

This study successfully developed a new formula of tamarillo steamed bread. Tamarillo is rich in dietary fiber and bioactive components and has good antioxidant capacity. Compared with traditional steamed breads prepared with 100% wheat flour, steamed breads prepared with up to 20% tamarillo powder partial replacement of wheat flour are rich in dietary fiber and bioactive components while having lower GI value and pleasant tamarillo flavor. Overall, tamarillo powder could be added to steamed breads to provide tamarillo steamed breads with more functional components and more effective antioxidant properties. Future studies will focus on an in vivo model to reflect the potential benefits of TP.

**Author Contributions:** Conceptualization, S.-D.L.; data curation, P.-C.S., Q.-F.Z. and S.-D.L.; formal analysis, P.-C.S., Q.-F.Z. and S.-D.L.; investigation, P.-C.S. and S.-D.L.; methodology, P.-C.S., Q.-F.Z. and S.-D.L.; project administration, S.-D.L.; resources, S.-D.L.; supervision, S.-D.L.; validation, S.-D.L.; writing—original draft, S.-D.L.; writing—review and editing, S.-D.L. All authors have read and agreed to the published version of the manuscript.

**Funding:** This work was supported by the National Science and Technology Council, Taiwan (MOST 107-2221-E-241-006).

**Institutional Review Board Statement:** Not applicable.

**Informed Consent Statement:** Not applicable.

**Data Availability Statement:** Data are contained within the article.

**Acknowledgments:** Thanks to the National Science and Technology Council, Taiwan (MOST 107-2221-E-241-006) and all of the co-authors who participated in the work on this study.

**Conflicts of Interest:** The authors declare no conflict of interest.

## References


1. Zhu, F. Influence of ingredients and chemical components on the quality of Chinese steamed bread. *Food Chem.* **2014**, *163*, 154–162. [CrossRef] [PubMed]
2. Jiao, Z.S.; Yu, M.; Xia, C.; Meng, N.; Wang, B.T.; Mo, J.X.; Che, S.P. Analysis on glycemic indexes of the staple foods in Tianjin City. *J. Tianjin Med. Univ.* **2003**, *9*, 152–155.
3. Lan, S.Q.; Meng, Y.N.; Li, X.P.; Zhang, Y.L.; Song, G.Y.; Ma, H.J. Effect of consumption of micronutrient enriched wheat steamed bread on postprandial plasma glucose in healthy and type 2 diabetic subjects. *Nutr. J.* **2013**, *12*, 64.
4. Yang, Y.X.; Wang, H.W.; Cui, H.M.; Wang, Y.; Yu, L.D.; Xiang, S.X.; Zhou, S.Y. Glycemic index of cereal and tubers produced in China. *World J. Gastroenterol.* **2006**, *12*, 3430–3433. [CrossRef]
5. Ma, X.Y.; Liu, J.P.; Song, Z.Y. Glycemic load, glycemic index and risk of cardiovascular diseases: Meta-analyses of prospective studies. *Atherosclerosis* **2012**, *223*, 491–496. [CrossRef]
6. Zhu, F. Glycemic control in Chinese steamed bread: Strategies and opportunities. *Trends Food Sci. Technol.* **2019**, *86*, 252–259. [CrossRef]
7. Hsieh, P.H.; Weng, Y.M.; Yu, Z.R.; Wang, B.J. Substitution of wheat flour with wholegrain flours affects physical properties, sensory acceptance, and starch digestion of Chinese steam bread (Mantou). *LWT Food Sci. Technol.* **2017**, *86*, 571–576. [CrossRef]
8. Zhu, F.; Chan, C. Effect of chia seed on glycemic response, texture, and sensory properties of Chinese steamed bread. *LWT Food Sci. Technol.* **2018**, *98*, 77–84. [CrossRef]
9. Guo, D.; Yin, X.; Cheng, H.; Chen, J.; Ye, X. Fortification of Chinese steamed bread with *Glycyrrhiza uralensis* polysaccharides and evaluation of its quality and performance attributes. *Foods* **2022**, *11*, 2253. [CrossRef]
10. Espin, S.; Gonzalez-Manzano, S.; Taco, V.; Poveda, C.; Ayuda-Duran, B.; Gonzalez-Paramas, A.M.; Santos-Buelga, C. Phenolic Composition and Antioxidant Capacity of Yellow and Purple-red Ecuadorian Cultivars of Tree Tomato (*Solanum betaceum* Cav.). *Food Chem.* **2016**, *194*, 1073–1080. [CrossRef]
11. Abdul Mutalib, M.; Rahmat, A.; Ali, F.; Othman, F.; Ramasamy, R. Nutritional compositions and antiproliferative activities of different solvent fractions from ethanol extract of *Cyphomandra betacea* (tamarillo) fruit. *Malays J. Med. Sci.* **2017**, *24*, 19–32. [CrossRef]
12. Muliarta, M.; Tirtayasa, K.; Prabawa, P.Y.; Wiryadana, K.A. Tamarillo consumption associated with increased acetylcholinesterase activity and improved oxidative stress markers in farmers exposed to daily pesticide-related activities in Baturiti, Bali, Indonesia. *Open Access Maced. J. Med. Sci.* **2020**, *8*, 244–250. [CrossRef]
13. Wang, S.; Zhu, F. Tamarillo (*Solanum betaceum*): Chemical composition, biological properties, and product innovation. *Trends Food Sci. Technol.* **2020**, *95*, 45–58. [CrossRef]
14. Diep, T.T.; Rush, E.C.; Yoo, M.J.Y. Tamarillo (*Solanum betaceum* Cav.): A review of physicochemical and bioactive properties and potential applications. *Food Rev. Int.* **2022**, *38*, 1343–1367. [CrossRef]
15. Hu, C.; Gao, X.; Dou, K.; Zhu, C.; Zhou, Y.; Hu, Z. Physiological and Metabolic Changes in Tamarillo (*Solanum betaceum*) during Fruit Ripening. *Molecules* **2023**, *28*, 1800. [CrossRef]
16. Abdul Kadir, N.A.A.; Rahmat, A.; Jaafar, H.Z.E. Protective effects of tamarillo (*Cyphomandra betacea*) extract against high fat diet induced obesity in sprague-dawley rats. *J. Obes.* **2015**, *2015*, 846041. [CrossRef] [PubMed]
17. Ordóñez, R.M.; Ordóñez, A.A.L.; Sayago, J.E.; Nieva Moreno, M.I.; Isla, M.I. Antimicrobial activity of glycosidase inhibitory protein isolated from *Cyphomandra betacea* Sendt. fruit. *Peptides* **2006**, *27*, 1187–1191. [CrossRef] [PubMed]
18. Senthil Kumar, C.; Sivakumar, M.; Ruckmani, K. Microwave-assisted extraction of polysaccharides from *Cyphomandra betacea* and its biological activities. *Int. J. Biol. Macromol.* **2016**, *92*, 682–693.
19. Orqueda, M.E.; Rivas, M.; Zampini, I.C.; Alberto, M.R.; Torres, S.; Cuello, S.; Sayago, J.; Thomas-Valdes, D.; Jiménez-Aspee, F.; Schmeda-Hirschmann, G.; et al. Chemical and functional characterization of seed, pulp and skin powder from chilto (*Solanum betaceum*), an Argentine native fruit. Phenolic fractions affect key enzymes involved in metabolic syndrome and oxidative stress. *Food Chem.* **2017**, *216*, 70–79. [CrossRef]
20. Suárez-Montenegro, Z.J.; Ballesteros-Vivas, D.; Gallego, R.; Valdés, A.; Sánchez-Martínez, J.D.; Parada-Alfonso, F.; Ibáñez, E.; Cifuentes, A. Neuroprotective potential of tamarillo (*Cyphomandra betacea*) epicarp extracts obtained by sustainable extraction process. *Front. Nutr.* **2021**, *8*, 769617. [CrossRef] [PubMed]
21. Ramakrishnan, Y.; Adzahan, N.M.; Yusof, Y.A.; Muhammad, K. Effect of wall materials on the spray drying efficiency, powder properties and stability of bioactive compounds in tamarillo juice microencapsulation. *Powder Technol.* **2018**, *328*, 406–414. [CrossRef]
22. Fernandino, C.M.; Nepomuceno, A.T.; Fonseca, H.C.; Bastos, R.A.; Lima, J.P. Physicochemical properties of tamarillo pulp (*Solanum betaceum*) and its applicability in the production of ice cream. *Braz. J. Food Technol.* **2021**, *24*, e2020090. [CrossRef]
23. Weickert, M.O.; Pfeiffer, A.F.H. Impact of dietary fiber consumption on insulin resistance and the prevention of type 2 diabetes. *J. Nutr.* **2018**, *148*, 7–12. [CrossRef] [PubMed]
24. Zhou, Q.-Y.; Huang, Y.-X.; Lin, Z.-E. *White Mantou (Knife Cut)—Making of Chinese Traditional Wheat Food—Ferment Dough*; China Grain Products Research & Development Institute: New Taipei, Taiwan, 2007; Chapter 6, 6-1-1; pp. 31–33.

25. AACC International. *Approved Methods of the AACC*, 10th ed.; American Association of Cereal Chemists: St. Paul, MN, USA, 2000.
26. AOAC. *Official Methods of Analysis. Method 985.29. Total Dietary Fiber in Foods, Enzymatic-Gravimetric Method*, 17th ed.; Association of Official Analytical Chemists International: Gaithersburg, MD, USA, 2003.
27. Mau, J.L.; Lee, C.C.; Yang, C.W.; Chen, R.W.; Zhang, Q.F.; Lin, S.D. Physicochemical, antioxidant and sensory characteristics of bread partially substituted with aerial parts of sweet potato. *LWT Food Sci. Technol.* **2020**, *117*, 108602. [CrossRef]
28. Lu, T.M.; Lee, C.C.; Mau, J.L.; Lin, S.D. Quality and antioxidant property of green tea sponge cake. *Food Chem.* **2010**, *119*, 1090–1095. [CrossRef]
29. Mau, J.L.; Lee, C.C.; Chen, Y.P.; Lin, S.D. Physicochemical, antioxidant and sensory characteristics of chiffon cake prepared with black rice as replacement for wheat flour. *LWT Food Sci. Technol.* **2017**, *75*, 434–439. [CrossRef]
30. Knockaert, G.; Lemmens, L.; Van Buggenhout, S.; Hendrickx, M.; Van Loey, A. Changes in β-carotene bioaccessibility and concentration during processing of carrot puree. *Food Chem.* **2012**, *133*, 60–67. [CrossRef]
31. Attia, T.Z. Simultaneous determination of rutin and ascorbic acid mixture in their pure forms and combined dosage form. *Spectrochim. Acta A Mol. Biomol. Spectrosc.* **2016**, *169*, 82–86. [CrossRef]
32. Shimada, K.; Fujikawa, K.; Yahara, K.; Nakamura, T. Antioxidative properties of xanthan on the autoxidation of soybean oil in cyclodextrin emulsion. *J. Agric. Food Chem.* **1992**, *40*, 945–948. [CrossRef]
33. Oyaizu, M. Studies on products of browning reactions: Antioxidative activities of products of browning reaction prepared from glucosamine. *Jpn. J. Nutr.* **1986**, *44*, 307–315. [CrossRef]
34. Germaine, K.A.; Samman, S.; Fryirs, C.G.; Griffiths, P.J.; Johnson, S.K.; Quail, K.J. Comparison of in vitro starch digestibility methods for predicting the glycaemic index of grain foods. *J. Sci. Food Agric.* **2008**, *88*, 652–658. [CrossRef]
35. Goñi, I.; Garcia-Alonso, A.; Saura-Calixto, F. A starch hydrolysis procedure to estimate glycemic index. *Nutr. Res.* **1997**, *17*, 427–437. [CrossRef]
36. Robertson, J.A.; Eastwood, M.A. An examination of factors which may affect the water holding capacity of dietary fibre. *Br. J. Nutr.* **1981**, *45*, 83–88. [CrossRef]
37. Smith, A.R. Color gamut transform pairs. *ACM Siggraph Comput. Graph.* **1978**, *12*, 12–19. [CrossRef]
38. Song, K.Y.; Joung, K.Y.; Shin, S.Y.; Kim, Y.S. Effects of basil (*Ocimum basilicum* L.) seed mucilage substituted for fat source in sponge cake: Physicochemical, structural, and retrogradation properties. *Ital. J. Food Sci.* **2017**, *29*, 681–696.
39. Halliwell, B. Free radicals and antioxidants: Updating a personal view. *Nutr. Rev.* **2012**, *70*, 257–265. [CrossRef] [PubMed]
40. Prior, R.L. Fruits and vegetables in the prevention of cellular oxidative damage. *Am. J. Clin. Nutr.* **2003**, *78*, 570S–578S. [CrossRef] [PubMed]
41. Chiou, S.Y.; Sung, J.M.; Huang, P.W.; Lin, S.D. Antioxidant, antidiabetic, and antihypertensive properties of *Echinacea purpurea* flower extract and caffeic acid derivatives using in vitro models. *J. Med. Food* **2017**, *20*, 171–179. [CrossRef]
42. Kähkönen, M.; Heinonen, M. Antioxidant Activity of Anthocyanins and Their Aglycons. *J. Agric. Food Chem.* **2003**, *51*, 628–633. [CrossRef]
43. Nuray, K.; Feryal, K. Changes of bioactive compounds and anti-oxidant activity during cold storage of carrot. *Int. J. Food Sci. Technol.* **2008**, *43*, 2019–2025.
44. Goesaert, H.; Brijs, K.; Veraverbeke, W.S.; Courtin, C.M.; Gebruers, K.; Delcour, J.A. Wheat flour constituents: How they impact bread quality, and how to impact their functionality. *Trends Food Sci. Technol.* **2005**, *16*, 12–30. [CrossRef]
45. Ahmed, J.; Almusallam, A.S.; Al-Salman, F.; AbdulRahman, M.H.; Al-Salem, E. Rheological properties of water insoluble date fiber incorporated wheat flour dough. *LWT Food Sci. Technol.* **2013**, *51*, 409–416. [CrossRef]
46. Zhu, F.; Sun, J. Physicochemical and sensory properties of steamed bread fortified with purple sweet potato flour. *Food Biosci.* **2019**, *30*, 100411. [CrossRef]
47. Wang, S.; Opassathavorn, A.; Zhu, F. Influence of quinoa flour on quality characteristics of cookie, bread, and Chinese steamed bread. *J. Texture Stud.* **2015**, *46*, 281–292. [CrossRef]
48. Wang, S.; Khamchanxana, P.; Zhu, F.; Zhu, C.; Pan, J. Textural and sensory properties of steamed bread fortified with high-amylose maize starch. *J. Texture Stud.* **2017**, *48*, 3–8. [CrossRef]
49. Matos, M.E.; Rosell, C.M. Relationship between instrumental parameters and sensory characteristics in gluten-free breads. *Eur. Food Res. Technol.* **2012**, *235*, 107–117. [CrossRef]
50. American Diabetes Association. Standards of Medical Care in Diabetes—2019. *Diabetes Care* **2019**, *42* (Suppl. 1), S46–S60. [CrossRef] [PubMed]
51. Granfeldt, Y.; Björck, I.; Drews, A.; Tovar, J. An in vitro procedure based on chewing to predict metabolic response to starch in cereal and legume products. *Eur. J. Clin. Nutr.* **1992**, *46*, 649–660. [CrossRef] [PubMed]
52. Lai, M.H.; Liu, K.L.; Chen, P.Y.; Ke, N.J.; Chen, J.J.; Sung, J.M.; Wu, Y.L.; Lin, S.D. Predicted glycemic index and glycemic index of rice varieties grown in Taiwan. *Cereal Chem.* **2016**, *93*, 150–155. [CrossRef]
53. Stone, H.; Sidel, J.L. *Sensory Evaluation Practices*, 2nd ed.; Academic Press Inc.: San Diego, CA, USA, 1993; p. 234.

*foods*

MDPI

*Article*

# Effect of 3D Food Printing Processing on Polyphenol System of Loaded *Aronia melanocarpa* and Post-Processing Evaluation of 3D Printing Products

**Quancheng Zhou [1,2,†], Xijun Nan [2,†], Shucheng Zhang [3], Liang Zhang [4], Jian Chen [5], Jiayi Li [2], Honglei Wang [2] and Zheng Ruan [1,*]**

1 State Key Laboratory of Food Science and Technology, School of Food Science and Technology, Nanchang University, Nanchang 330047, China

2 Department of Food Science, School of Agricultural Engineering and Food Science, Shandong University of Technology, Zibo 255049, China; a985354232@163.com (X.N.)

3 Yantai Shuangta Food Co., Ltd., Yantai 265404, China

4 Arkhum Health Technology Co., Ltd., Zibo 255035, China

5 Shandong Jiucifang Biotechnology Co., Ltd., Zibo 256102, China

* Correspondence: ruanzheng@ncu.edu.cn

† These authors contributed equally to this work.

**Abstract:** *Aronia melanocarpa* polyphenols (AMP) have good nutritional values and functions. This study aimed to explore the printability and storage properties of AM gels in 3D food printing (3DFP). Therefore, 3DFP was performed on a loaded AMP gel system to determine its textural properties, rheological properties, microstructure, swelling degree and storage performance. The results revealed that the best loading AMP gel system to meet the printability requirements of 3DFP processing was AM fruit pulp:methylcellulose:pea albumin: hyaluronic acid = 100:14:1:1. Compared with other ratios and before 3DFP processing, the best loading AMP gel system processed by 3DFP exhibited the lowest deviation of 4.19%, the highest hardness, the highest elasticity, the least adhesion, a compact structure, uniform porosity, difficulty in collapsing, good support, a high degree of crosslinking, and good water retention. Additionally, they could be stored for 14 d at 4 °C. After post-processing, the AMP gel had a favorable AMP release rate and good sustained release effect in gastrointestinal digestion, which conformed to the Ritger–Peppas equation model. The results revealed that the gel system had good printability and applicability for 3D printing; as well, 3DFP products had good storage properties. These conclusions provide a theoretical basis for the application of 3D printing using fruit pulp as a raw material.

**Keywords:** 3D food printing; *Aronia melanocarpa* polyphenol; gel; printability; applicability; storage

## 1. Introduction

Three-dimensional food printing (3DFP) technology is a type of processing technology that can form 3D structures by stacking edible materials [1]. Recently, 3DFP has been increasingly favored by researchers owing to its personalization and novelty. The composition and processability of 3DFP ink, particularly the rich ink in fruits and vegetables, have been key to the development of the 3DFP technology [2]. However, fruits may be considered among the hardest materials for 3DFP because of their low viscosity [3]. A common method used in research involves converting fruits and vegetables into fruit and vegetable powders. For example, Chen et al. [4,5]. studied the printing characteristics and sensory evaluation properties of five different animal and vegetable proteins as a matrix and fruit powder composite printing ink. Nevertheless, these processing methods cannot retain the original nutrients in the fruit. Furthermore, few studies have been conducted on the application of 3D printing in the direct processing of fruits and vegetables [3]. Eating more nutritious fruit can effectively regulate body health and reduce the risk of diseases [6]. Consequently, research on the printability and applicability of fruits as a raw material is of great significance for theoretical and practical applications in the field of 3D printing. Moreover, the fruit of functional components can also be well preserved and play their functions, the formula can be adjusted according to its nutritional characteristics and physical properties, and the model can be designed and printed to meet the precise nutritional and personalized needs of people. The fruit of *Aronia melanocarpa* (AM) is rich in polyphenols. AM polyphenols (AMP) have antioxidants, and lipid-lowering, anti-inflammatory, and anticancer effects [7]. Additionally, AMP can improve the intestinal flora and assist in lowering blood glucose and pressure [4]. The demand for AM is gradually increasing owing to the good nutritional value and function of AM. However, the common AM processing method is dry, whereas other AM processing methods and applications have been studied less extensively.

Methyl cellulose (MC) is a safe, non-toxic food additive with good biocompatibility and gelling properties; it has been used as a cytoskeleton in disease treatment [8]. MC has good water solubility and film-forming properties, and it can be used as an edible membrane material [9]. Additionally, it can be made into gels [10], which can be used as 3D printing materials, in theory; nonetheless, studies in the field of 3DFP processing are limited. Hyaluronic acid (HA) is a straight-chain polysaccharide widely found in various human and vertebrate tissues. It has good biocompatibility and physiological functions, including water retention, thickening, and delayed aging [11,12]. Some studies have added HA to nutritional preparations to improve their rheological properties [13]; however, HA as a functional factor, added to the 3DFP printing system, has not been reported. Pea albumin (PA) has good gel properties, solubility, and emulsification, as well as certain functional properties, such as improving obesity and mediating metabolism [14]. Nonetheless, studies on PA and its scope of application are limited.

Accordingly, this paper aimed to build an AM/MC gel system suitable for 3D printing processing. By discussing the influence of 3D printing processing on the physicochemical properties, structure, and physical properties of the gel system, the physical property requirements of AM for realizing 3D printing were revealed, and PA and HA were added to the printing system. Hence, the objective of this study was to explore the effects of PA and HA on the physicochemical properties, structure, and physical properties of the gel system.

## 2. Material and Methods

### 2.1. Preparation of AM 3D Printing Inks and Printing

In this experiment, 100 g of Aronia melanocarpa (AM) fruit pulp was mixed with methylcellulose (MC), pea albumin (PA), and hyaluronic acid (HA) in specific proportions (refer to Table 1 for the composition of each ingredient) [13,15]. The fruit pulp was sourced from Fukangyuan no. 1, Zibo Forestry Protection and Development Center, China. The mixture was then prepared as an ink for the extrusion-based 3D food printer (FOODBOT-S, Hangzhou Printing Technology Co., Ltd., Beijing, China).

The 3D printer used the provided "letter D" model to print the samples. The printing parameters included a 1.20 mm inner diameter for the printing needle, a 3 mm printing height, a 30 mm/s moving rate, and a 1.20 mm layer height.

Before and after the printing process, the samples were placed in disposable petri dishes, sealed with plastic wrap, and tested for various indicators within 1 h. Furthermore, the AM gel powders were obtained after lyophilization (−50 °C, vacuum degree is 22) by a lyophilizer (Beijing Boyikang Experimental Instrument Co., LTD., FD-1A-50, Beijing, China) and stored in −20 °C refrigerator for future use.

**Table 1.** Formability of 3DFP material with different mass ratios.

| Formulation | MC/PA/HA | Print Span (mm) | Print Deviation (%) |
|---|---|---|---|
| 1 | 12% MC | 7.00 ± 0.32 | 39.97 ± 6.43 |
| 2 | 13% MC | 6.26 ± 0.55 | 25.21 ± 11.05 |
| 3 | 16% MC/1% PA | 5.28 ± 0.20 | 5.67 ± 4.09 |
| 4 | 17% MC/1% PA | 5.24 ± 0.18 | 4.86 ± 3.50 |
| 5 | 16% MC/1% HA | 6.12 ± 0.17 | 22.48 ± 3.42 |
| 6 | 17% MC/1% HA | 5.52 ± 0.38 | 10.41 ± 7.50 |
| 7 | 13% MC/1% PA/1% HA | 5.50 ± 0.22 | 10.01 ± 4.32 |
| 8 | 14% MC/1% PA/1% HA | 5.21 ± 0.23 | 4.19 ± 4.60 |
| 9 | 14% MC/1% PA/2% HA | 6.78 ± 0.34 | 35.60 ± 6.80 |
| 10 | 14% MC/2% PA/1% HA | / | / |
| 11 | 14% MC | / | / |
| 12 | 14% MC/1% PA | / | / |
| 13 | 14% MC/1% HA | / | / |
| 14 | 15% MC/1% PA/1% HA | / | / |

Note: AM fruit pulp was 100%; MC, PA, and HA were added according to the percentages in the table. The print width of the model design value was 5.00 mm.

### 2.2. Study on the Influence of 3D Printing Materials on Formability

The formability of the 3D printed food products was evaluated through visual analysis, focusing on the smoothness of the 3D printing process, the printed shape, and the forming effect of the printed objects. This qualitative assessment aimed to identify any defects or inconsistencies in the printed structures and to determine whether the chosen materials and parameters resulted in successful printing outcomes.

### 2.3. Determination of Printing Deviation of 3DFP Products

A micrometer caliper (VolScan Profiler; Stable Micro Systems, Surrey, UK) was used to determine the maximum width of the 3DFP products. This was compared with the model design width and calculated width deviation. Width deviation was calculated using the following equation [16].

$$\text{Departure (\%)} = \frac{\text{3DFP products measurement width} - \text{Model design width}}{\text{Model design width}} \times 100\%. \quad (1)$$

### 2.4. Texture Analysis

A texture analyzer (TA-XT 2, Stable Micro Systems, Ltd., Surrey, UK) measured the texture of the 3DFP materials and products (hereinafter referred to as samples) at 25 ± 0.2 °C using a disc type probe (P/75); additionally, the TPA mode was used to measure the hardness, springiness, chewiness, and adhesiveness of samples. The probe compressed the samples with the strain set to 70%. The initial distance between the plate and probe was set to 3.0 cm, and the down, compression, and return speeds were set to 5, 1, and 1 mm/s, respectively. The starting point induction force was 5 g, and the interval between the two compression times was 5 s. Each sample was tested in triplicate with different formulations [1].

### 2.5. PCA (Principal Component Analysis)

In order to obtain AM printing products with good physical properties, principal component analysis and a membership comprehensive scoring method were used to determine the comprehensive score of AM printing products [17]. According to the importance of the physical performance indicators of AM printing products, multiple indicators are simplified into a single indicator, and the weight of the importance of each indicator is determined according to the principal component analysis. In the practical application process, the greater the hardness, elasticity, and chewability, the better, with the membership degree calculated according to Formula (2); the smaller the adhesion, the better, with its membership calculated according to Equation (3) [18].

$$P = \frac{A_i - A_{min}}{A_{max} - A_{min}} \times 100\% \quad (2)$$

$$P = \frac{A_{max} - A_i}{A_{max} - A_{min}} \times 100\% \quad (3)$$

where P is the index membership degree; $A_i$ is the index value; $A_{min}$ is the minimum value of the same index; and $A_{max}$ indicates the maximum value of the same indicator. The comprehensive physical properties of the film are divided into S and calculated according to Formula (4).

$$S = a\,P_1 + b\,P_2 + c\,P_3 + d\,P_4 \quad (4)$$

In the formula, $P_1$, $P_2$, $P_3$, and $P_4$ are the membership degrees of four indexes, namely deviation, hardness, elasticity, chewability, and adhesion. a, b, c, and d are the weights of the four indicators, respectively. The weights obtained are substituted into the formula for calculation.

### 2.6. Characterization of the Rheological Properties of Samples

The rheological properties of the samples were characterized using a Kinexus Pro rheometer (Malvern Instruments Ltd., Worcestershire, UK) operated with a serrated parallel plate geometry with a diameter of 40 mm and gap of 1 mm. Subsequently, the frequency sweep test was performed at 25 °C in the range of 0.10–10.00 Hz at a strain of 0.20%. Each test was performed in triplicate [19].

### 2.7. Fourier Transform Infrared (FTIR) Spectroscopy

The method of Xiao [20] was slightly modified. Before and after printing, 2 mg of the sample powder was accurately weighed and thoroughly mixed with 100 mg of potassium bromide powder. The powder was evenly ground using a spherical grinding machine. After pressing, the samples were loaded into a Fourier transform infrared (FTIR) spectrometer (Thermo Fisher Scientific, Inc., Nicolet5700, Madison, WI, USA) for detection.

### 2.8. Scanning Electron Microscopy (SEM)

Scanning electron microscopy (SEM) (American FEI Company, Tecnai G2F 20, Hillsboro, OR, USA) was used to obtain micrographs of the samples before and after printing, at an accelerated voltage of 20 kV. Before the measurement, the printed material was freeze-dried, fixed on the sample holder, and sprayed with gold. The material was enlarged from different directions of the cross and longitudinal sections, and SEM images with a magnification of 500 times were selected to observe the microstructure of the samples. The most representative micrographs were selected for sample analysis [21].

### 2.9. Swelling Tests

The lyophilized samples (0.1 g) were accurately weighed (the mass was denoted as $M_0$) and then placed in a 50 mL centrifuge (Shanghai Anting Scientific Instrument Factory, TDL-40B, Shanghai, China) tube. The masses of the samples and centrifuge tubes were denoted as $M_1$. Subsequently, 30 mL of distilled water was accurately added with a pipette, and the samples were shaken in a water bath at 30 °C for 1 h and placed at 25 ± 2 °C for 1 h. The supernatant was removed by high-speed centrifugation at 4000 rpm for 30 min. The water on the gel surface was dried with a filter paper to a constant weight. The sediment and centrifuge tube at this time were weighed and recorded as $M_2$. The parallel experiments were repeated three times. The swelling (%) was quantified using Equation (2) [22].

$$\text{Swelling} = \frac{(M_2 - M_1)}{M_0}. \quad (5)$$

### 2.10. Determination of Storage Properties

After printing, the 3DFP products were divided into several groups and stored in beakers sealed with plastic films at 4 °C and 25 °C in dry environments. The storage period was terminated when the surface of the 3DFP products deteriorated. During the storage period, the polyphenol content was measured every 2 d, and each treatment was repeated three times [23].

### 2.11. Post-Processing of 3DFP Products

We created the best system for the 3D printing and cured it. The post-processing method was slightly modified according to the one described by [24]. Three post-processing methods are used: boiling, steaming, and roasting. The specific methods are as follows.

Boiling: add water in an induction cooker according to a 1:8 ratio of material to water, and set the temperature to 100 °C for 15 min.

Steaming: steam the sheet on the shelf of an induction cooker at 100 °C for 15 min.

Roasting: place the 3DFP products in a microwave oven at 160 °C for 12 min. Flip sides once in a while.

### 2.12. Texture Properties

After post-processing, the product was cut into pieces using scissors to simulate chewing with teeth. The hardness and chewability of the 3DFP products after post-processing were measured using a texture meter (TA-XT 2, Stable Micro Systems, Ltd., Surrey, UK); the test method is the same as in method Section 2.4. The remaining 3DFP products were stored for future use.

### 2.13. Total Color Change Rate

The printed AM/MC gel was cut into round cakes with a radius of 1.50 cm, and the luminance (L*), red and green (a*), and yellow and blue (b*) of the 3DFP products were measured and recorded using a WSD-3C whiteness meter (Background Kangguang Instrument Co., LTD, WSD-3C, Beijing, China). After post-processing of the gel, the $L_0$*, $a_0$*, and $b_0$* values of the gel were detected again and recorded. The total color change rate ΔE was calculated according to the following formula [25]:

$$\Delta E = \sqrt{(L^* - L_0{}^*)^2 + (a^* - a_0{}^*)^2 + (b^* - b_0{}^*)^2}, \quad (6)$$

### 2.14. In Vitro Digestion Simulation

The in vitro digestion simulation was based on an international consensus [26], with slight modifications.

#### 2.14.1. Preparation of Simulated Digestive Fluid Reserve

Simulated saliva, gastric juice, and intestinal juice were prepared according to the method of Minekus [26].

#### 2.14.2. Simulated Oral Digestion In Vitro

A 0.50 g sample of the cured 3DFP products was placed in 50 mL centrifuge tubes. In addition, 3.50 mL simulated saliva, 0.50 mL amylase (enzyme activity of 1500 U/mL), 25 μL 0.30 M $CaCl_2$, and 975 μL distilled water were added to the centrifuge tubes and mixed thoroughly. The cured 3DFP products were digested in a 37 °C constant temperature oscillator (Changzhou Wanfeng Instrument Manufacturing Co., Ltd., WHY-2, Changzhou, China) for 2 min. After digestion, the reaction was terminated with liquid nitrogen and stored for later use.

2.14.3. Simulated Gastric Digestion In Vitro

The residual 3DFP products after simulated oral digestion were added to 50 mL centrifuge tubes in the gastric digestion group, where 7.50 mL of simulated gastric juice, 1.60 mL of 25,000 U/mL pig pepsin, 5 μL of 0.30 M $CaCl_2$, 702 μL of 5 M HCl, and 193 μL of distilled water were added into the centrifuge tubes. Simulated gastric juice was used as a blank control. The 3DFP products were digested in a constant temperature oscillator at 37 °C for 2 h and sampled at 0, 24, 48, 72, 96, and 120 min. The reaction was terminated immediately after the end of the reaction in liquid nitrogen.

2.14.4. In Vitro Simulated Intestinal Digestion

The residual 3DFP products, after the gastric digestion simulation, were added to 50 mL centrifuge tubes of the intestinal digestion and blank control groups. Intestinal digestion group: 11 mL simulated intestinal fluid, 5 mL 800 U/mL trypsin, 2.50 mL 4% pig bile, 40 μL 0.30 M $CaCl_2$, 460 μL 5 M NaOH, and 1 mL distilled water. Simulated intestinal fluid was used as a blank control. The 3DFP products were digested in a constant temperature oscillator at 37 °C for 2 h and sampled at 0, 24, 48, 72, 96, and 120 min. Finally, the centrifuge tubes were quickly placed in liquid nitrogen to terminate the reaction and were subsequently set aside.

The release solution was centrifuged at 3000 rpm for 10 min, the supernatant was collected, and the absorbance of the release solution at 765 nm was measured. According to the change in polyphenol concentration in the release system, the cumulative release rate (Q, %) of polyphenols from the 3DFP products at different times (t, min) was calculated. It is calculated as follows [27]:

$$Q\ (\%) = \frac{M_t}{M} \times 100\%, \tag{7}$$

where $M_t$ is the cumulative AMP mass released at time t and M is the total AMP mass initially embedded in the gel.

### *2.15. Statistical Analysis*

SPSS 17.0 and Origin 2018 software were used to analyze the test results. The recorded data are presented as the mean ± standard error (SE) for each analysis. To investigate the significant differences between the data points, analysis of variance (ANOVA) was used ($p \leq 0.05$).

## 3. Results and Analysis

### *3.1. Effects of MC, PA, and HA Concentrations on the 3DFP Formability and Deviation*

In the experiment, varying concentrations of the components (MC, PA, and HA) in the AM/MC gel were tested to determine the optimal formulation for 3D food printing (3DFP). As shown in Figure 1 and Table 1, a 13% MC concentration resulted in poor molding performance and low printing accuracy. When 14% MC was used without adding PA or HA, the mixture became too thick for printing.

However, when 14% MC was combined with 1% PA and 1% HA, the mixture could be printed and formed effectively. The best printing outcome was achieved with this combination (Figures 1–8), with a width deviation of only 4.19%, making it more accurate than the yam starch gel used in a study by Wu et al. [28].

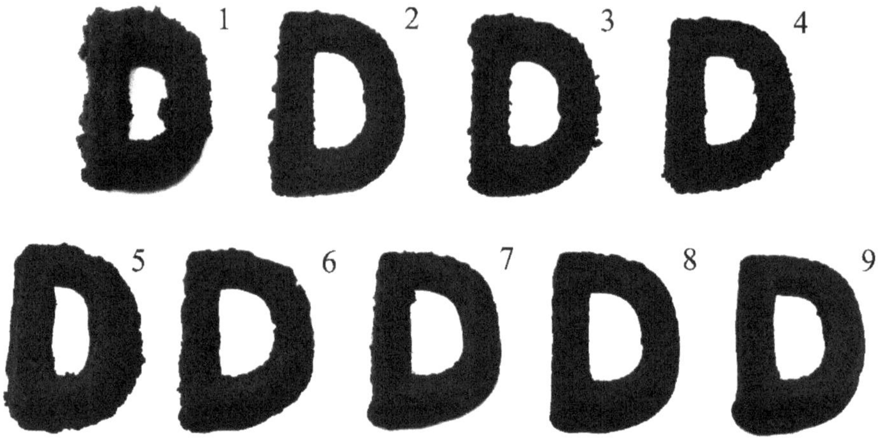


**Figure 1.** Images of 3DFP products with different mass ratios. Note: Numbers 1–9 in the chart indicate that AM pulp was added according to the mass ratio of MC, PA, and HA in Table 1.

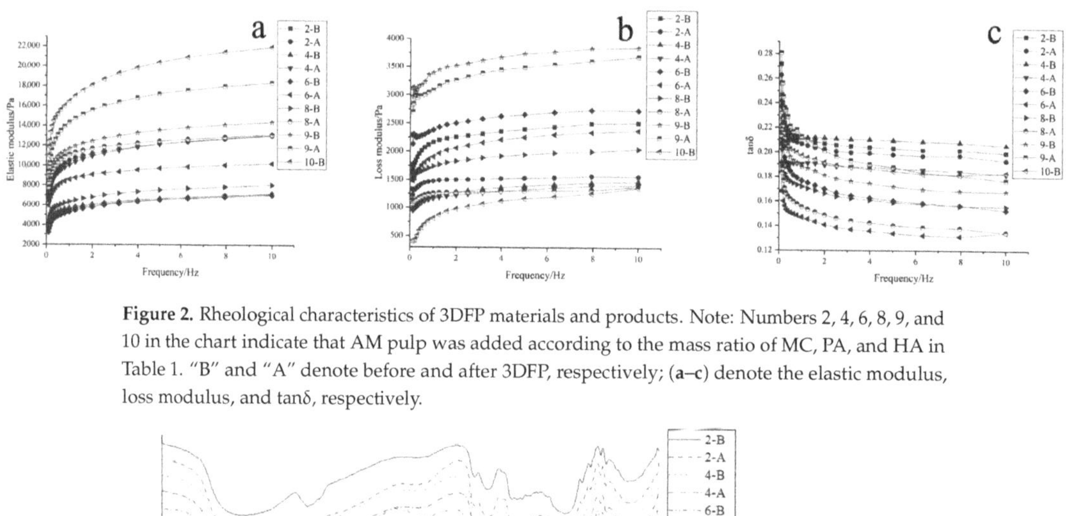


**Figure 2.** Rheological characteristics of 3DFP materials and products. Note: Numbers 2, 4, 6, 8, 9, and 10 in the chart indicate that AM pulp was added according to the mass ratio of MC, PA, and HA in Table 1. "B" and "A" denote before and after 3DFP, respectively; (**a–c**) denote the elastic modulus, loss modulus, and tanδ, respectively.

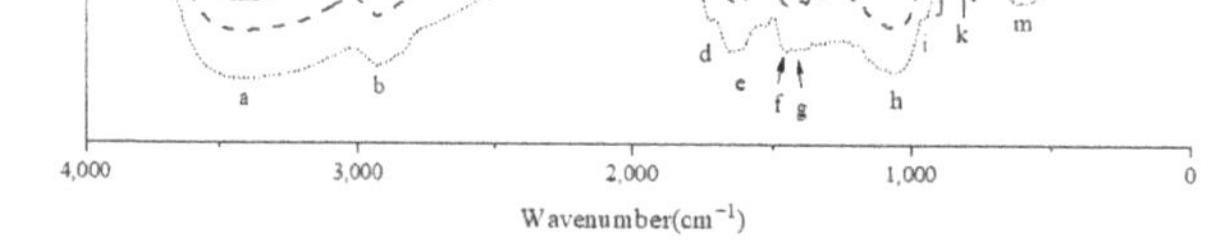


**Figure 3.** FTIR micrograph of 3DFP materials and 3DFP products.

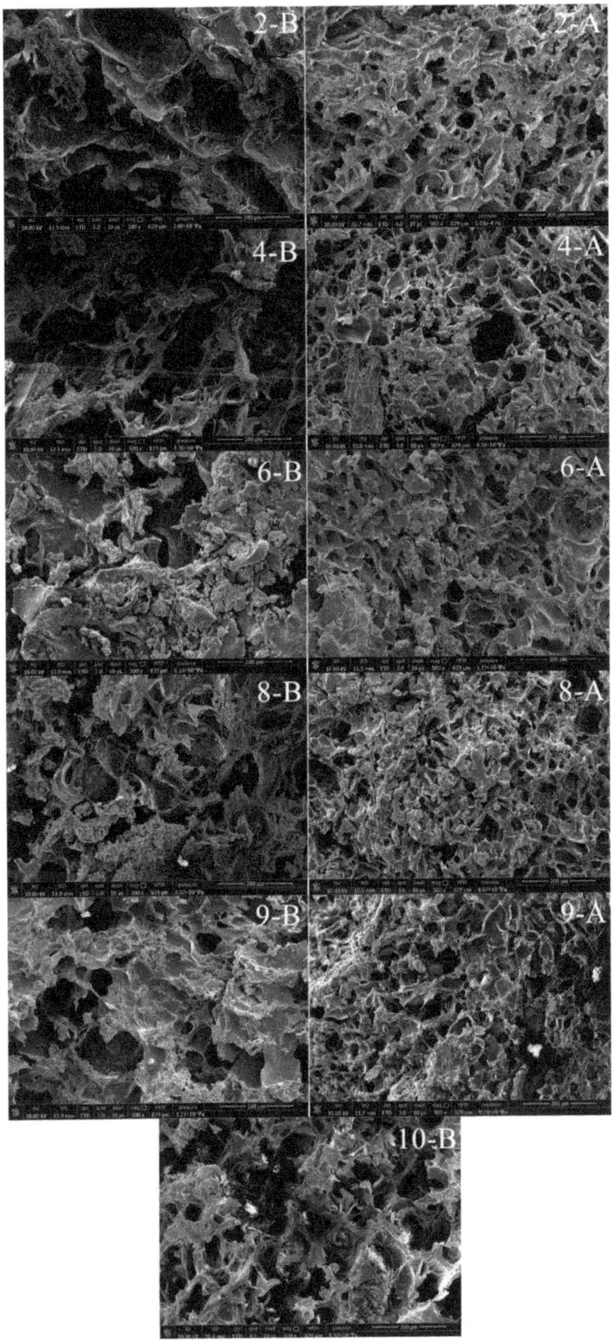


**Figure 4.** SEM micrograph of 3DFP materials and products.

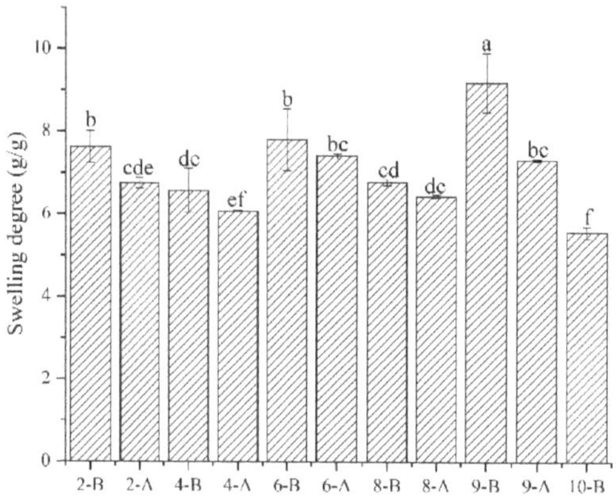


**Figure 5.** Swelling degree of 3D printing material and 3D printed products. Note: (a–f) Different letters indicate significant difference ($p < 0.05$).

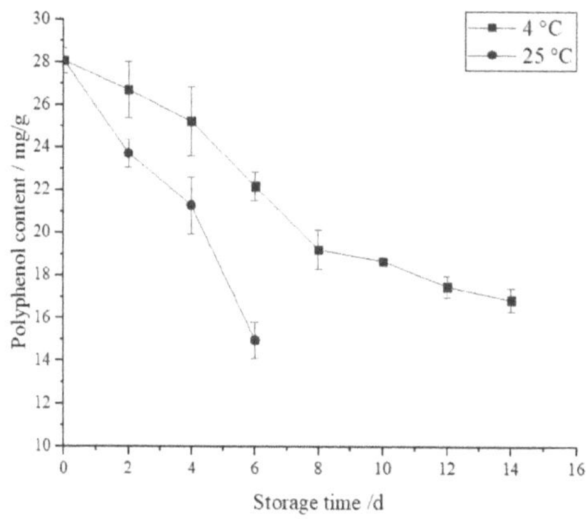


**Figure 6.** Storage time on AMP content of 3DFP products in optimal system.

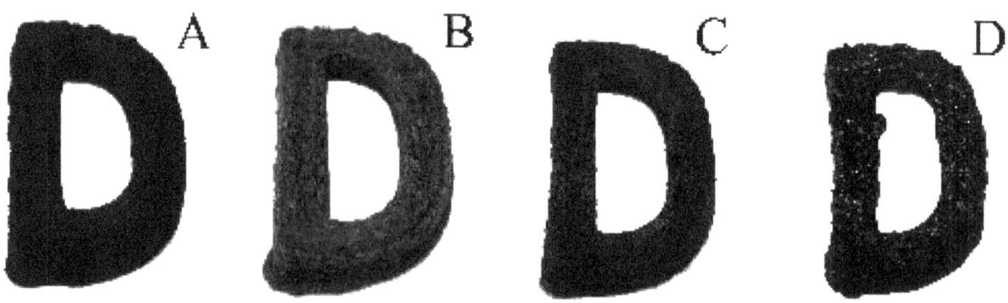


**Figure 7.** Effects of different post-processing methods on the appearance of 3D products. Note: (**A–D**) represent uncured 3DFP products, and 3DFP products after boiling, steaming, and roasting, respectively.

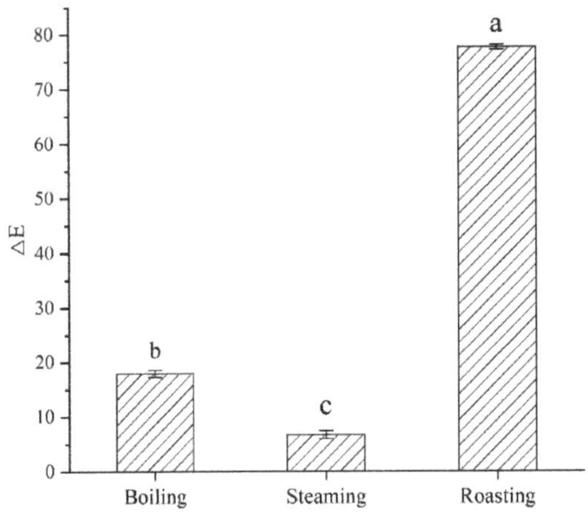


**Figure 8.** Chromatic aberration of 3DFP products by different post-processing methods. Note: (a–c) Different letters indicate significant difference ($p < 0.05$).

The success of the formulation can be attributed to the interaction between MC, PA, and HA. While MC plays a significant role in printability, the wetting effect of HA [29] and the synergistic interaction between PA and HA [30] improve printing accuracy and reduce deviation.

Increasing the concentration of MC, PA, or HA beyond the successful formulation led to poor molding performance due to excessive thickness or poor product support. The results indicate that the optimal formulation for 3DFP using the AM fruit pulp-based hydrogel consists of 14% MC, 1% PA, and 1% HA, displaying good balance in printability, formability, and accuracy.

### 3.2. *Effects of MC, PA, and HA Concentrations on Texture Characteristics of Samples*

In texture analysis, several factors are used to characterize the properties of 3D food printed samples: hardness—the force needed to cause a shape change to the sample; elasticity—the degree to which the sample can recover its shape between the first and second compressions; adhesion—the force required to separate the sample surface from other attached objects (e.g., tongue, teeth, or mouth); and degree of mastication (chewiness)—the energy required to chew a solid sample to a state sufficient for swallowing [31].

As shown in Tables 2 and 3, the different formulations impact these properties: Increasing MC content led to higher hardness, elasticity, and chewiness, while adhesion decreased. Adding PA improved hardness, elasticity, and chewiness, and reduced adhesion due to its good gel properties, enhancing the appearance and support ability of the 3DFP products. Adding HA slightly decreased hardness, elasticity, and chewiness, but increased adhesion due to its lubricating and moisturizing qualities.

The optimal formulation—consisting of 14% MC, 1% PA, and 1% HA—demonstrated the highest hardness and elasticity, the lowest adhesion, and the best molding during printing. This combination allowed for successful 3DFP, while providing a balance between texture properties and nutritional value. Overall, the texture analysis reveals how the different components (MC, PA, and HA) influence the performance of the resulting 3DFP products, highlighting the importance of selecting the proper formulation for achieving the desired printing outcomes and product quality.

**Table 2.** Texture characteristics of 3DFP materials with different mass ratios.

| Formulation Codes | Hardness (g) | Springiness | Chewiness (g·mm) | Adhesiveness (g·s) |
|---|---|---|---|---|
| 1-B | 526.59 ± 42.11 $^{e}$ | 0.1821 ± 0.0291 $^{bc}$ | 15.09 ± 1.35 $^{de}$ | −135.78 ± 8.01 $^{de}$ |
| 2-B | 755.23 ± 62.44 $^{c}$ | 0.1893 ± 0.0410 $^{b}$ | 24.87 ± 1.63 $^{ab}$ | −103.4 ± 11.16 $^{bd}$ |
| 3-B | 646.27 ± 19.55 $^{e}$ | 0.1535 ± 0.0369 $^{bc}$ | 16.03 ± 2.67 $^{d}$ | −114.78 ± 20.65 $^{cd}$ |
| 4-B | 895.45 ± 45.49 $^{b}$ | 0.1442 ± 0.0164 $^{bc}$ | 20.47 ± 1.92 $^{c}$ | −81.99 ± 14.19 $^{b}$ |
| 5-B | 519.58 ± 63.22 $^{e}$ | 0.1245 ± 0.0281 $^{c}$ | 12.07 ± 2.33 $^{e}$ | −150.72 ± 9.23 $^{e}$ |
| 6-B | 711.39 ± 45.38 $^{cd}$ | 0.1660 ± 0.0361 $^{bc}$ | 27.01 ± 2.29 $^{a}$ | −120.04 ± 16.89 $^{cd}$ |
| 7-B | 663.42 ± 36.54 $^{d}$ | 0.1941 ± 0.0329 $^{b}$ | 22.95 ± 1.23 $^{bc}$ | −107.69 ± 6.79 $^{c}$ |
| 8-B | 926.14 ± 34.99 $^{b}$ | 0.2753 ± 0.0180 $^{a}$ | 23.45 ± 0.99 $^{bc}$ | −83.85 ± 10.33 $^{b}$ |
| 9-B | 807.90 ± 71.32 $^{bc}$ | 0.1473 ± 0.0193 $^{bc}$ | 25.36 ± 2.59 $^{ab}$ | −80.30 ± 13.00 $^{b}$ |
| 10-B | 1065.72 ± 58.94 $^{a}$ | 0.2533 ± 0.0432 $^{a}$ | 32.28 ± 2.66 $^{a}$ | −23.06 ± 14.67 $^{a}$ |

Note: Numbers 1–10 in the table indicate that AM pulp was added according to the mass ratio of MC, PA, and HA in Table 1. "B" indicates before 3DFP. Different letters in the same column indicate significant difference ($p < 0.05$).

**Table 3.** Texture characteristics of 3DFP products with different mass ratios.

| Formulation Codes | Hardness (g) | Springiness | Chewiness (g·mm) | Adhesiveness (g·s) |
|---|---|---|---|---|
| 1-A | 686.47 ± 69.54 $^{d}$ | 0.1968 ± 0.0350 $^{cd}$ | 24.66 ± 2.46 $^{b}$ | −100.80 ± 15.39 $^{ab}$ |
| 2-A | 890.92 ± 95.66 $^{ab}$ | 0.2055 ± 0.0109 $^{bc}$ | 30.68 ± 3.44 $^{a}$ | −81.30 ± 20.47 $^{a}$ |
| 3-A | 728.55 ± 58.47 $^{cd}$ | 0.2340 ± 0.0412 $^{abc}$ | 23.90 ± 0.99 $^{b}$ | −99.22 ± 32.43 $^{ab}$ |
| 4-A | 976.32 ± 73.55 $^{ab}$ | 0.2430 ± 0.0262 $^{abc}$ | 31.65 ± 2.86 $^{a}$ | −71.11 ± 12.77 $^{a}$ |
| 5-A | 654.15 ± 77.23 $^{d}$ | 0.1228 ± 0.0333 $^{e}$ | 18.95 ± 1.93 $^{c}$ | −133.71 ± 9.64 $^{b}$ |
| 6-A | 865.59 ± 78.01 $^{bc}$ | 0.1496 ± 0.0291 $^{de}$ | 33.57 ± 2.34 $^{a}$ | −96.52 ± 23.67 $^{ab}$ |
| 7-A | 744.64 ± 92.13 $^{cd}$ | 0.2574 ± 0.0294 $^{ab}$ | 29.97 ± 2.01 $^{a}$ | −87.96 ± 32.16 $^{a}$ |
| 8-A | 1029.98 ± 86.49 $^{a}$ | 0.2867 ± 0.0160 $^{a}$ | 33.34 ± 1.67 $^{a}$ | −65.92 ± 17.58 $^{a}$ |
| 9-A | 913.08 ± 68.40 $^{a}$ | 0.2010 ± 0.0538 $^{bc}$ | 31.34 ± 3.23 $^{a}$ | −88.16 ± 17.47 $^{a}$ |

Note: Numbers 1–9 in the table refer to the mass ratios of AM pulp, MC, PA, and HA in Table 1. "A" indicates after 3DFP. Different letters in the same column indicate significant difference ($p < 0.05$).

### 3.3. PCA

The physical properties of each AM printing product, including hardness, elasticity, chewability, and adhesion, were taken as the analysis object, and the principal component analysis of nine kinds of different AM printing products was carried out by using SPSS software [17].

The software analysis results are shown in Table 4. It can be seen from Table 4 that there is one principal component whose eigenvalue is greater than 1. When extracting the first principal component, its variance contribution rate is 79.769%, close to 80%, indicating that this principal component can basically reflect the information of all indicators and can replace the four selected indicators, successfully achieving the purpose of dimension reduction.

**Table 4.** Results of contribution rate by principal component analysis.

| Components | Characteristic Root | Variance Contribution Rate (%) | Cumulative Variance Contribution Rate (%) |
|---|---|---|---|
| 1 | 3.1910 | 79.7690 | 79.7690 |
| 2 | 0.6320 | 15.8050 | 95.5740 |
| 3 | 0.1490 | 3.7330 | 99.3070 |
| 4 | 0.0280 | 0.6930 | 100 |

The factor loading in Table 5 reflects the contribution rate of each index to the principal component. From the table, it can be observed that hardness, elasticity, chewability, and adhesiveness had higher loadings on the first principal component. This indicates that the first principal component mainly reflects the information of these four indices, with adhesiveness having the largest loading.

**Table 5.** Component matrix of principal component analysis.

| Components | Hardness | Springiness | Chewiness | Adhesiveness |
|---|---|---|---|---|
| 1 | 0.908 | 0.767 | 0.898 | 0.985 |

Combining Tables 4 and 5, the weights of the four indices for evaluating the performance of AM printing products were determined, and are presented in Table 6. The weight of hardness is 0.2552, the weight of elasticity is 0.2156, the weight of chewability is 0.2524, and the weight of adhesion is 0.2768. Using Formula (4), the sum score of AM printing products can be obtained from Formula (8):

$$S = 0.2552P_1 + 0.2156P_2 + 0.2524P_3 + 0.2768P_4 \quad (8)$$

**Table 6.** Normalized data.

| Indexes | Normalized Data |
|---|---|
| Hardness | 0.2552 |
| Springiness | 0.2156 |
| Chewiness | 0.2524 |
| Adhesiveness | 0.2768 |

By combining Equations (2) and (3), membership values of the 9 different AM printing products were calculated (Table 7). The comprehensive score of physical properties of AM printing products can be obtained by using Formula (4), which is shown in Table 8. Among them, 8-A (AM gel system with 14% MC, 1% PA, and 1% HA added) achieved the highest score of 0.997. This indicates that the 8-A AM printing product exhibited the best overall performance in terms of its physical properties.

**Table 7.** Membership degree value.

| | $P_1$ (Hardness Positive Effect) | $P_2$ (Springiness Positive Effect) | $P_3$ (Chewiness Positive Effect) | $P_4$ (Adhesiveness Negative Effect) |
|---|---|---|---|---|
| 1-A | 0.0860 | 0.4515 | 0.3906 | 0.4855 |
| 2-A | 0.6300 | 0.5046 | 0.8023 | 0.7731 |
| 3-A | 0.1980 | 0.6785 | 0.3386 | 0.5088 |
| 4-A | 0.8572 | 0.7334 | 0.8687 | 0.9234 |
| 5-A | 0.0000 | 0.0000 | 0.0000 | 0.0000 |
| 6-A | 0.5626 | 0.1635 | 1.0000 | 0.5486 |
| 7-A | 0.2408 | 0.8212 | 0.7538 | 0.6749 |
| 8-A | 1.0000 | 1.0000 | 0.9843 | 1.0000 |
| 9-A | 0.6890 | 0.4771 | 0.8475 | 0.6719 |

**Table 8.** Comprehensive score sheet.

| Formulation Codes | 1-A | 2-A | 3-A | 4-A | 5-A | 6-A | 7-A | 8-A | 9-A |
|---|---|---|---|---|---|---|---|---|---|
| Synthesis score | 0.370 | 0.670 | 0.456 | 0.842 | 0.000 | 0.532 | 0.641 | 0.997 | 0.655 |

To further investigate the printability and applicability of the different formulations, six printing systems with high scores and varying levels of MC, PA, and HA were selected for study. These systems included four groups with better printability and texture properties and two groups featuring different combinations of PA and HA. The mass ratios of AM pulp, MC, PA, and HA in these groups were as follows: 100:13:0:0 (Table 1, No. 2, 0.670 points); 100:17:1:0 (Table 1, No. 4, 0.842 points); 100:17:0:1 (Table 1, No. 6, 0.532 points); 100:14:1:1 (Table 1, No. 8, 0.997); 100:14:1:2 (Table 1, No. 9, 0.655); and 100:14:2:1 (Table 1, No. 10, too thick for printing).

Each group was evaluated before and after 3D food printing to assess its printability and applicability. By comparing the performance of these printing systems, the study aimed to identify the best formulation for achieving optimal printing results, texture properties, and overall quality in the final 3DFP products.

### 3.4. Effects of MC, PA, and HA Concentrations on Rheological Characteristics of Samples

The dynamic viscoelasticity of samples is crucial for their practical application properties in 3D food printing. There are three main parameters in dynamic viscoelasticity: energy storage modulus (G′)—represents the elasticity and rigidity of the printed material. Higher G′ values indicate better support and reduced likelihood of collapse [32]; loss modulus (G″)—reflects the viscosity of the print [33]; and tanδ—represents fluidity and helps to differentiate between elastic and viscous behaviors.

Both G′ and G″ showed some dependence on frequency, indicating a relationship between frequency and the viscoelastic properties of the gel system. As the frequency increased, the G′ and G″ values for the AM gel systems also showed varying degrees of increase. In all samples, G′ was higher than G″, which demonstrates that the prints were primarily elastic within the scanned frequency range and had a reticular gel structure with a certain level of strength.

Understanding the dynamic viscoelasticity of the samples is important for optimizing the formulation and printing parameters in 3D food printing. By considering the energy storage modulus, loss modulus, and fluidity, researchers can better control and predict the printed products' performance and texture properties.

Figure 2 shows the differences in the energy storage modulus (G′) and loss modulus (G″) for the 3DFP products before and after printing. Some key observations from this data include the following: G′ values increased and G″ values decreased after 3D printing extrusion, indicating an increase in hardness and elasticity, and a decrease in adhesion for the gel system. The addition of PA increased G′ (elasticity) and decreased G″ (adhesion), due to its good gel characteristics and its interactions with AM pulp and MC that reduce gaps in the system. The addition of HA led to decreased elasticity and increased adhesion because it functions as a lubricant. When PA and HA were added simultaneously, the AM gel system exhibited maximum G′, while the difference between G′ and G″ also reached its peak. This resulted in a stable gel structure with good formability, making the material more suitable for printing.

Increasing the amount of PA led to higher elasticity, lower adhesion, and improved gel strength but reached a point where it could not be printed. On the other hand, increasing HA resulted in lowered elasticity and increased adhesion, negatively impacting the printing effect.

The trend of tanδ decreasing with frequency suggests a reduction in the mobility and viscoelasticity of the 3DFP products, which leads to an increase in their solid-like properties.

In conclusion, to achieve the desired product type after processing, different composition ratios should be carefully selected based on the required solid and liquid states of the 3D printing products. By considering the dynamic viscoelastic properties and subsequent processing requirements, 3D printed products can be optimized for both performance and quality.

### 3.5. Effects of MC, PA, and HA Concentrations on FTIR of Samples

Fourier Transform Infrared Spectroscopy (FTIR) is an effective method for studying the structures of copolymers, as it can measure the vibration of functional groups and polar bonds in samples, allowing the analysis of changes in their chemical bonds [20]. In this study, FTIR was used to investigate the effects of 3DFP on the ink components and the interactions between them before and after adding PA and HA, including the potential formation of covalent bonds.

As shown in Figure 3, both 3DFP materials and 3DFP products exhibit a strong, wide absorption band at 3200–3500 $cm^{-1}$ (peak a), generated by the stretching vibration of phenolic hydroxyl O-H in AM, and a strong absorption band at 1200–1000 $cm^{-1}$ (peak h) produced by the tensile vibration of phenolic C-O in AM. The addition of PA and HA had minimal effects on these peaks, indicating that they did not significantly impact AMP.

Note: (a–m) denote 3386.87, 2931.58, 2063.57, 1724.28, 1645.03, 1449.52, 1409, 1075.95, 947.47, 890.45, 818.49, 776.52, 615.10 $cm^{-1}$, respectively.

Upon adding PA, the peak at 1724.28 $cm^{-1}$ (peak d) became more pronounced, suggesting an enhancement of the C=O stretch of ketone in the 3DFP products. However, no new peaks formed, meaning PA and AM binding was weak. The addition of HA led to the formation of a new peak at 2063.57 $cm^{-1}$ (peak c), implying a new covalent bond was formed between HA and the AM/MC gel.

When PA and HA were added simultaneously, a new absorption peak emerged at 1409 $cm^{-1}$ (peak g), indicating a covalent bond formed between PA and HA. This bond was stronger with the addition of 1% PA and 1% HA. Comparing before and after printing, the bond strength of the 3DFP products increased, possibly due to 3DFP making the sample structure more compact.

### 3.6. Effects of MC, PA, and HA Concentrations on the SEM Images of the Samples

The microstructural analysis of the 3DFP materials and products is illustrated in Figure 4. When PA or HA is added to the AM gel containing MC, the pores become closer due to the gelling and thickening effects of PA and HA, respectively. However, HA's wetting effect leads to unevenly distributed pores and a relatively loose structure, resulting in lower printing accuracy.

In contrast, the system containing both PA and HA has more uniform and smaller pores. The 3DFP products with 1% PA and 1% HA exhibit the most compact and well-proportioned pore structure, allowing the printed gel to maintain its shape without collapsing. This stability might be attributed to the covalent bonds formed between PA and HA.

When more PA is added to the system, there are fewer pores, but it becomes unprintable due to the high gel performance, which interrupts the material during the printing process. On the other hand, increasing the HA content leads to larger porosity and reduced uniformity in the 3DFP products, possibly due to its high wettability causing uneven discharge, severe deformation, and poor support. These observations are consistent with the results reported by Wang et al. [16].

Additionally, after 3DFP, the pores in all materials significantly reduced, and the structure became more compact, which aligns with previous findings in the study.

### 3.7. Effects of MC, PA, and HA Concentrations on the Swelling Degree of the Samples

Swelling refers to the phenomenon where a colloidal gel dissolves in a solution, and the volume increases as the colloid absorbs the solution. It is an important property of gels, and the degree of swelling can be used as a measure of this property. The swelling characteristics of different gels are related to the degree of crosslinking; the higher the degree of crosslinking, the lower the swelling degree of the gel under the same conditions [34].

As shown in Figure 5, after adding PA, the swelling degree of the printing material decreased because it has a gelling effect that makes the sample structure more compact. When HA was added, the swelling degree of the samples increased due to HA's wettability. Compared to other systems, the swelling degree of the printing material decreased after simultaneously adding 1% PA and 1% HA. This may be attributed to the formation of covalent bonds between PA and HA, resulting in increased gel crosslinking.

Additionally, the water retention of all materials after printing was better than before printing, likely because the 3DFP process made the sample structure more compact.

In conclusion, the gel system containing 14% MC, 1% PA, and 1% HA had the highest score (0.997), the most stable structure, the best printing effect, the best indicators, and the ideal printability and applicability. Therefore, this proportion of printing material was selected for the storage, post-processing, and intestinal digestion simulations.

### 3.8. Effect of Storage Time on AMP Content of 3DFP Products in Optimal System

The quality of the best 3DFP products during storage was investigated. As shown in Figure 6, the 3DFP product of the optimal system could be stored at 4 °C and 25 °C for 14 d and 6 d, respectively. Because of the high nutrient content of the printing material, the storage time was relatively short. During the storage period, the AMP content of the prints supplemented with 14% MC, 1% PA, and 1% HA gradually decreased at 4 °C and 25 °C. At the end of storage, AMP content decreased by 39.90% and 46.63%, respectively.

### 3.9. Effect of Different Post-Processing Methods on Chromatic Aberration and Texture Properties of 3DFP Products of Optimal System

Figures 7 and 8 show the 3DFP product of the optimal system after post-processing and its chromatic aberration before post-processing, respectively. The results indicate that steaming had a negligible effect on color. However, under boiling treatment, the color was lighter than before post-processing due to the solubility of anthocyanins and other color substances in water.

Compared with the 3DFP products before post-processing, the color difference and color intensity of the 3DFP products after roasting were greater, possibly due to the Maillard reaction, which imparts a pleasing flavor and color. Investigating the biological activities and other effects of roasting requires further research.

Table 9 lists the textural properties of 3DFP in the best post-processing system. Baked 3DFP products had high chewiness and hardness, potentially because the water loss in the baked system was the highest. Cooking chewiness and hardness were lower, likely due to the water-based treatment resulting in less water loss for the 3DFP product.

**Table 9.** Texture properties of 3D printed products treated by different post-processing methods.

| Post-Processing Method | Hardness (g) | Chewiness (g·mm) |
|---|---|---|
| Boiling | 5359.91 ± 59.13 $^{c}$ | 1877.61 ± 544.99 $^{c}$ |
| Steaming | 8521.74 ± 988.99 $^{b}$ | 3327.78 ± 925.57 $^{b}$ |
| Roasting | 33,988.44 ± 383.69 $^{a}$ | 11,399.75 ± 320.79 $^{a}$ |

Note: Different letters in the same column indicate significant difference ($p < 0.05$).

These findings demonstrate that different post-processing methods have varied effects on the color, hardness, and chewiness of the 3DFP products. The appropriate post-processing method can be selected based on specific requirements and desired product characteristics.

### 3.10. In Vitro Gastrointestinal Digestion Simulation of 3DFP Products of Optimal System

Figure 9 shows that during the 120 min gastric digestion and 120 min intestinal digestion, the cumulative release rate of AMP content under each treatment gradually increased. This indicates that the 3DFP products have sustained release characteristics in simulated gastric juice and simulated intestinal juice.

During gastric digestion, the cumulative release rates of boiling, steaming, and roasting were 8.04%, 15.12%, and 24.79%, respectively, while during intestinal digestion, they were 73.87%, 71.64%, and 72.25%, respectively. This suggests that AMP release characteristics were better in the intestinal simulation solution after sample post-processing, similar to the findings of Bermúdez-Soto et al. [35].

After 120 min of gastric digestion and 120 min of intestinal digestion, the sum of the cumulative release rates under boiling, steaming, and roasting treatments reached 81.91%, 86.76%, and 97.04%, respectively. The roasting treatment had a significantly higher sum of cumulative release rates than other groups, potentially due to the lower temperature causing less damage and less influence on AMP content. Compared to the roasting treatment, the boiling treatment temperature was the same, but part of the water-soluble AMP was lost due to being dissolved in water.

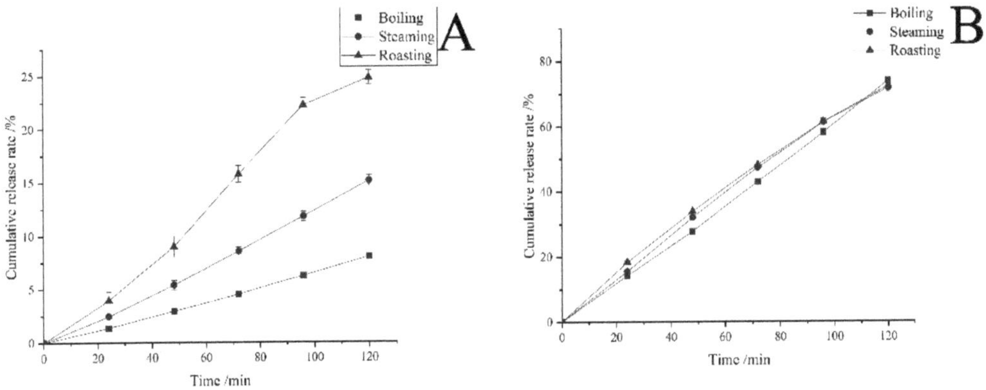


**Figure 9.** AMP release rate of the best 3DFP products. Note: (**A**) denotes gastric digestion and (**B**) denotes intestinal digestion.

The results show that the gastric environment has a promotional effect on the release of AMP, possibly because AMP contains abundant weakly acidic phenolic hydroxyl groups [36] and is relatively stable under acidic conditions. The acidic environment of gastric juice may cause polyphenol hydrolysis of the bound state in AM, resulting in AMP dissociation [37]. Additionally, some polyphenols undergo hydrolysis under acidic conditions, and part of the glycoside compounds are converted into aglycones, increasing the release rate of AMP [38]. These results are consistent with [39].

AMP is better retained in the intestine, and its release in intestinal digestion simulation is greater than its degradation. This could be due to the hydrolyzing action of pancreatic juice and bile on the chemical bonds that react with AMP, and the stability of some neutral polyphenols in the intestinal simulation solution at pH 7, allowing AMP to be better retained. These findings align with those reported by Rao et al. [39].

In vitro release of AMP content is commonly fitted using several kinetic models, such as the zero-order kinetic model, first-order kinetic model, Niebergall square root equation, and Ritger–Peppas equation model [40,41]. In this study, Origin software was employed to fit four models of samples after 3D printing and curing post-processing to establish the kinetic release equation of AMP content in simulated gastric juice and simulated intestinal juice; the results are presented in Table 10.

**Table 10.** The kinetic model fitting of AMP release.

| Post-Processing Method | Types of Digestion | Model | Fitted Equation | $R^2$ |
|---|---|---|---|---|
| Boiling | Peptic digestion | Zero-order dynamics | $Q = 0.06714t - 0.1773$ | 0.9973 |
| | | First-order dynamics | $\ln(100 - Q) = (-6.9938 \times 10^{-4})t + 4.6074$ | 0.9957 |
| | | Niebergall square root equation | $(1 - Q)^{1/2} = -0.0034t + 10.0101$ | 0.8589 |
| | | Ritger–Peppas equation | $\ln Q = 1.0922\ln t - 3.1529$ | 0.9998 |
| | Intestinal digestion | Zero-order dynamics | $Q = 0.6141t - 0.7201$ | 0.9991 |
| | | First-order dynamics | $\ln(100 - Q) = -0.0108t + 4.7129$ | 0.9410 |
| | | Niebergall square root equation | $(1 - Q)^{1/2} = -0.0401t + 10.2275$ | 0.8756 |
| | | Ritger–Peppas equation | $\ln Q = 1.0248\ln t - 0.6182$ | 0.9991 |
| Steaming | Peptic digestion | Zero-order dynamics | $Q = 0.1272t - 0.4115$ | 0.9969 |
| | | First-order dynamics | $\ln(100 - Q) = -0.0014t + 4.6112$ | 0.9936 |
| | | Niebergall square root equation | $(1 - Q)^{1/2} = -0.0066t + 10.0252$ | 0.8538 |
| | | Ritger–Peppas equation | $\ln Q = 1.1323\ln t - 2.7$ | 0.9999 |
| | Intestinal digestion | Zero-order dynamics | $Q = 0.6076t + 1.4759$ | 0.9940 |
| | | First-order dynamics | $\ln(100 - Q) = -0.0106t + 4.6732$ | 0.9844 |
| | | Niebergall square root equation | $(1 - Q)^{1/2} = -0.0396t + 10.0834$ | 0.9179 |
| | | Ritger–Peppas equation | $\ln Q = 0.9598\ln t - 0.2788$ | 0.9959 |

**Table 10.** *Cont.*

| Post-Processing Method | Types of Digestion | Model | Fitted Equation | $R^2$ |
|---|---|---|---|---|
| Roasting | Peptic digestion | Zero-order dynamics | $Q = 0.2209t - 0.6349$ | 0.9849 |
| | | First-order dynamics | $\ln(100 - Q) = -0.0025t + 4.6173$ | 0.9828 |
| | | Niebergall square root equation | $(1 - Q)^{1/2} = -0.0118t + 10.0453$ | 0.8560 |
| | | Ritger–Peppas equation | $\ln Q = 1.1872\ln t - 2.3824$ | 0.9901 |
| | Intestinal digestion | Zero-order dynamics | $Q = 0.6002t + 2.9779$ | 0.9911 |
| | | First-order dynamics | $\ln(100 - Q) = -0.0106 + 4.6560$ | 0.9872 |
| | | Niebergall square root equation | $(1 - Q)^{1/2} = -0.0393t + 10.0016$ | 0.9390 |
| | | Ritger–Peppas equation | $\ln Q = 0.8561\ln t - 0.2002$ | 0.9991 |

According to the correlation coefficient $R^2$ results, the in vitro release of AMP from 3DFP products of the best system closely adhered to the Ritger–Peppas equation model fitting. The coefficient n of $\ln Q = n\ln t + k$ was greater than 0.89, with the exception of the intestinal digestion simulation of the roasting treatment, indicating that the release mechanism was mainly skeleton dissolution. However, the coefficient of intestinal digestion in the roasting treatment fell between 0.45 and 0.89, suggesting that the release mechanism was primarily a combination of diffusion and skeleton dissolution [42].

## 4. Conclusions

In this study, a gel system capable of loading AMP and suitable for 3D printing was developed using AM pulp as the main material, MC as the auxiliary material, and PA and HA as functional additives. The optimal ratio was determined to be AM pulp:MC:PA:HA = 100:14:1:1. This system demonstrated high printing accuracy, excellent physical and rheological properties, and achieved ideal 3D printing suitability compared to other proportions and pre-printed samples.

The system featured a compact structure, uniform pores, good support, high degree of cross-linking, and strong water retention. It could also be stored at 4 °C for 14 days, highlighting its potential applicability for 3D printing. After post-processing, the 3DFP products of this system exhibited a favorable AMP release rate and sustained release effect during gastrointestinal digestion, which is consistent with the Ritger–Peppas equation model.

These findings demonstrated that 3D printing positively influenced the printability and applicability of the material, providing a theoretical foundation for the application of 3D printing using fruit pulp as a raw material.

In future research, relevant technical means can be used to modify the materials and verify and broaden the application scope of 3D printing technology in the field of food to meet the market demand for personalized food. Additionally, the use of methylcellulose as a gel-forming thickener to embed other functional substances requires further research. Moreover, methylcellulose can be added to other fruit pulps to form fruit pulp gel, which can be applied to 3D printing. Their printability and applicability in future food value chain applications require further research.

**Author Contributions:** Q.Z. and X.N. prepared the AM gel, carried out the formal analysis, and wrote the manuscript. S.Z. and L.Z. studied the structure of the AM gel and carried the formal analysis. J.C. studied the properties of the AM gel and carried out the formal analysis. J.L. and H.W. reviewed the manuscript. Z.R. guided the experiments and reviewed the manuscript. All authors have read and agreed to the published version of the manuscript.

**Funding:** This research and the APC were funded by Research Project of the State Key Laboratory of Food Science and Technology (Project No. SKLF-ZZB-202125).

**Data Availability Statement:** Data sharing is not applicable—no new data were generated, or the article describes entirely theoretical research.

**Conflicts of Interest:** The authors declare no conflict of interest.

## References


1. Montoya, J.; Medina, J.; Molina, A.; Gutiérrez, J.; Rodríguez, B.; Marín, R. Impact of viscoelastic and structural properties from starch-mango and starch-arabinoxylans hydrocolloids in 3D food printing. *Addit. Manuf.* **2021**, *39*, 101891. [CrossRef]
2. Agarwal, T.; Costantini, M.; Maiti, T.K. Extrusion 3D printing with Pectin-based ink formulations: Recent trends in tissue engineering and food manufacturing. *Biomed. Eng. Adv.* **2021**, *2*, 100018. [CrossRef]
3. Ricci, I.; Derossi, A.; Severini, C. 3D Printed Food From Fruits and Vegetables. *Fundam. 3d Food Print. Appl.* **2019**, *5*, 117–149.
4. Chen, X.H.; Sheng, G.H.; Nan, X.J.; Yin, C.P.; Shi, Y.K.; Zhang, J.; Zhou, Q.C. Preliminary Studies on the Antihypertensive Effect of Polyphenols from *Aronia melanocarpa* Based on Intestinal Flora. *J. Chin. Inst. Food Sci. Technol.* **2021**, *21*, 88–97.
5. Chen, Y.; Min, Z.; Pattarapon, P. 3D printing of protein-based composite fruit and vegetable gel system. *LWT—Food Sci. Technol.* **2021**, *141*, 110978. [CrossRef]
6. Aljohani, N.E. Effects of a Nutrition Education Intervention on Fruit and Vegetables Intakes in College Students. *Am. J. Food Nutr.* **2021**, *9*, 112–116. [CrossRef]
7. Sun, W.T.; Wang, S.; Wang, M.; Xu, H.T.; Zhou, Q.C. Research Progress of Extraction, Purification and Function of Overseas *Aronia melanocarpa* in Recent Years. *Food Ind.* **2019**, *40*, 262–265.
8. Li, K.; Yu, T.; Chao, Q.; Xia, Z. Experimental Study on Simvastatin/Methylcellulose Gel Promoting Tendon-bone Healing after Acute Rotator Cuff Repair in Rabbits. *Chin. J. Sports Med.* **2019**, *38*, 46–53.
9. Azlin, M.; Baba, N.B. Characterization of Solid Polymer Electrolyte Based on Methylcellulose Doped Ammonium Bromide. *J. Phys. Conf. Ser.* **2021**, *1874*, 012072. [CrossRef]
10. He, Z.J.; Zang, H.X.; Zhu, L.; Huang, K.; Yi, T.L. An anti-inflammatory peptide and brain-derived neurotrophic factor-modified hyaluronan-methylcellulose hydrogel promotes nerve regeneration in rats with spinal cord injury. *Int. J. Nanomed.* **2019**, *14*, 721–732. [CrossRef]
11. Cai, Z.; Zhang, F.; Wei, Y.; Zhang, H. Freeze–Thaw-Induced Gelation of Hyaluronan: Physical Cryostructuration Correlated with Intermolecular Associations and Molecular Conformation. *Macromolecules* **2017**, *50*, 6647–6658. [CrossRef]
12. Xu, Q.; Torres, J.E.; Hakim, M.; Babiak, P.M.; Pal, P.; Battistoni, C.M.; Nguyen, M.; Panitch, A.; Solorio, L.; Liu, J.C. Collagen-and hyaluronic acid-based hydrogels and their biomedical applications. *Mater. Sci. Eng. R Rep.* **2021**, *146*, 100641. [CrossRef] [PubMed]
13. Wei, Y.Y.; Zhang, H.B.; Ma, A.Q.; Gao, S. Rheological Properties of Hyaluronan Thickened Enteral Nutritional Preparations for Dysphagia Management. *Food Sci.* **2019**, *40*, 50–55.
14. Liu, N.; Zhang, S.C.; Wang, P.J.; Yang, J.J.; Zang, J.Q.; Wu, S.M.; Li, Y.X.; Wang, R.; Liu, S.Y. *Application of Pea Albumin in Preparation of Composition for Improving Obesity and Metabolic Syndrome*; State Intellectual Property Office of the People's Republic of China: Beijing, China, 2021; Published Application Number CN113546152A; pp. 1–10.
15. Lin, Y.; Jiang, G.Q.; Zan, J.; Ding, F.X. Physical and in vitro release properties of methylcellulose-based thermo-sensitive hydrogels. *J. Tsinghua Univ. (Nat. Sci. Ed.)* **2006**, *46*, 836–838.
16. Wang, H.; Tan, C.; Chen, J.; Xu, J.X.; Li, D.N.; Meng, X.J. Effect of Konjac Gum on 3D Printing Properties of Blueberry Gel System. *Food Sci.* **2019**, *40*, 104–110.
17. Chen, S.S.; Tao, H.J.; Wang, Y.J.; Ma, Z.S. Process optimization of soy protein isolate-based edible films containing nanocrystalline cellulose from sunflower seed hull and chitosan. *Trans. Chin. Soc. Agric. Eng.* **2016**, *32*, 306–314.
18. Sun, H.T.; Shao, X.R.; Jiang, R.P.; Wang, S.; Su, D.J.; Ma, Z.S. Preparation and Physical Properties of Corn Distarch Phosphate/Corn Straw Cellulose Edible Film. *Food Sci.* **2016**, *37*, 21–28.
19. Yi, S.M.; Liang, F.; Yang, L.; Li, X.P.; Li, J.R.; Mi, H.B.; Xie, J.; Yu, J.Y.; Wang, L.M.; Zhou, X.M. Effect of cassava starch on 3D printing of Nemipterus virgatus surimi. *J. Fish. China* **2021**, *7*, 1111–1120.
20. Xiao, Z.; Zhang, Q.; Chen, G.; Shan, X.; Wang, L.; Tao, L.; Yang, Q. Study on the regulation mechanism of sugar beet pectin on starch gel. *Food Sci. Technol.* **2019**, *44*, 283–288.
21. Liu, Y.Y.; Xu, L.N.; Zhang, H.; Zhang, H.J.; Chi, Y.J.; Zhang, L.L.; Liu, X.X. Preparation and Properties of Modified Nano-$TiO_2$/SPI Antimicrobial Film. *Food Sci.* **2018**, *39*, 302–310.
22. Bhat, M.A.; Rather, R.A.; Yaseen, Z.; Shalla, A.H. Viscoelastic and smart swelling disposition of Carboxymethylcellulose based hydrogels substantiated by Gemini surfactant and in-vitro encapsulation and controlled release of Quercetin. *Int. J. Biol. Macromol.* **2022**, *207*, 374–386. [CrossRef] [PubMed]
23. Li, J.; Wang, H.R.; Lin, Z.B.; Huang, K.X.; Li, X.D. Effects of Storage and Heat Treatment on Polyphenols Content in Mulberry Wine. *Food Ind.* **2020**, *41*, 30–34.
24. Oyim, I.R.; Anyango, J.O.; Omwamba, M.N. Effect of Pre-Gelatinization Conditions on the Total Oxalate Content and Techno-Functional Properties of Taro (*Colocasia esculenta*) Flour. *Food Nutr. Sci.* **2022**, *13*, 511–525. [CrossRef]
25. Pelit, H.; Alkan, A.; Yalcin, M. Decay Resistance and Color Change of Pine and Beech Wood Impregnated with R. Luteum and R. Ponticum Plant Extracts. *Fac. For.* **2021**, *72*, 353–363. [CrossRef]
26. Minekus, M.; Alminger, M.; Alvito, P.; Ballance, S.; Bohn, T.; Bourlieu, C.; Carrière, F.; Boutrou, R.; Corredig, M.; Dupont, D.; et al. A standardised static in vitro digestion method suitable for food—An international consensus. *Food Funct.* **2014**, *5*, 1113–1124. [CrossRef] [PubMed]
27. Chen, J.; Mei, L.Y. Preparation and Properties of Polylactic Acid/Tea Polyphenol Composite Packaging Film. *Packag. Eng.* **2021**, *42*, 100–108.

28. Wu, J.; Feng, L.; Song, J.; Li, D.; Li, M.; Zhang, Z.; Zhang, P.; Liu, C. 3D Printing Characteristics of Chinese Yam Starch Gel. *Mod. Food Sci. Technol.* **2021**, *37*, 211–216, 256.
29. Dallari, D.; Sabbioni, G.; Del Piccolo, N.; Carubbi, C.; Veronesi, F.; Torricelli, P.; Fini, M. Efficacy of intra-articular polynucleotides associated with hyaluronic acid versus hyaluronic acid alone in the treatment of knee osteoarthritis: A randomized, double-blind, controlled clinical trial. *Clin. J. Sport Med.* **2020**, *30*, 1–7. [CrossRef] [PubMed]
30. Hao, L.; Sun, J.; Pei, M.; Zhang, G.; Li, C.; Li, C.; Ma, X.; He, S.; Liu, L. Impact of non-covalent bound polyphenols on conformational, functional properties and in vitro digestibility of pea protein. *Food Chem.* **2022**, *383*, 132623. [CrossRef]
31. Liu, A.; Bai, Y.; Li, M.; Chen, X.; Wang, Q.; Zhang, Y. Process optimization, quality research and product development of extrusion on barley powder. *Food Res. Dev.* **2019**, *40*, 118–123.
32. Figueiredo, T.; Jing, J.; Jeacomine, I.; Olsson, J.; Gerfaud, T.; Boiteau, J.G.; Rome, C.; Harris, C.; Auzély-Velty, R. Injectable Self-Healing Hydrogels Based on Boronate Ester Formation between Hyaluronic Acid Partners Modified with Benzoxaborin Derivatives and Saccharides. *Biomacromolecules* **2019**, *21*, 230–239. [CrossRef]
33. Lei, M.; Huang, J.; Tian, X.; Zhou, P.; Zhu, Q.; Li, L.; Li, L.; Ma, S.; Wang, X. Effects of insoluble dietary fiber from wheat bran on noodle quality. *Grain Oil Sci. Technol.* **2021**, *4*, 1–9. [CrossRef]
34. Asy-Syifa, N.; Kusjuriansah Waresindo, W.X.; Edikresnha, D.; Suciati, T.; Khairurrijal, K. The Study of the Swelling Degree of the PVA Hydrogel with varying concentrations of PVA. *J. Phys. Conf. Ser.* **2022**, *2243*, 012053. [CrossRef]
35. Bermúdez-Soto, M.J.; Tomás-Barberán, F.A.; García-Conesa, M.T. Stability of polyphenols in chokeberry (*Aronia melanocarpa*) subjected to in vitro gastric and pancreatic digestion. *Food Chem.* **2007**, *102*, 865–874. [CrossRef]
36. Gao, N.X.; Li, B.; Du, S.S.; Meng, X.J.; Zhang, Q.; Jiao, X.Y.; Chen, S.F. Stability of Polyphenols from *Aronia melanocarpa* Fruits. *Food Sci.* **2016**, *37*, 20–24.
37. Lu, J.; Dun, H.Y.; Xiang, X.Z.; Fu, C.Y.; Mo, K.D.; Zeng, X. Effect of in Vitro Simulated Gastrointestinal Digestion on Bioactive Components and Antioxidant Activities of Six Kinds of Black Foods. *Food Sci.* **2018**, *39*, 47–56.
38. Amaregouda, Y.; Kamanna, K.; Gasti, T. Fabrication of intelligent/active films based on chitosan/polyvinyl alcohol matrices containing Jacaranda cuspidifolia anthocyanin for real-time monitoring of fish freshness. *Int. J. Biol. Macromol.* **2022**, *218*, 799–815. [CrossRef]
39. Rao, X.T.; Zeng, X.A.; Lin, S.Y.; Liu, Z.H.; Sun, J.J. Changing Trends in the Phenolic Substances and Antioxidant Activities of Chokeberry (*Aronia melanocarpa*) Subjected to in Vitro Simulated Digestion. *Mod. Food Sci. Technol.* **2020**, *36*, 77–83.
40. Farahmand, A.; Emadzadeh, B.; Ghorani, B.; Poncelet, D. Droplet-based millifluidic technique for encapsulation of cinnamon essential oil: Optimization of the process and physicochemical characterization. *Food Hydrocoll.* **2022**, *129*, 107609. [CrossRef]
41. Xu, X.F.; Zhong, H.Q.; Liu, W.; Xia, W.X.; Yang, J.G. A Kinetic Study on the In Vitro Simulated Digestion of Gummy Vitamin C and Calcium Candies. *Mod. Food Sci. Technol.* **2018**, *34*, 83–88, 274.
42. Zhou, J.M.; Hu, Y.; Fang, P.C.; Jiang, B.; Yuan, Z.X. Study on the preparation and release in vitro of Wujiwan gastric-mucoadhesive tablets. *Pract. Pharm. Clin. Remedies* **2019**, *22*, 1182–1185.

*foods*

MDPI

*Article*

# Phytochemical-Rich Colored Noodles Fortified with an Aqueous Extract of *Clitoria ternatea* Flowers

Sy-Yu Shiau [1,2,*], Yanli Yu [1], Jing Li [1], Wenbo Huang [1] and Haixia Feng [1]

1 Department of Food Nutrition and Safety, Sanda University, Shanghai 201209, China
2 Department of Food Science and Technology, Tajen University, Pingtung County 90741, Taiwan
* Correspondence: syshiau@sandau.edu.cn; Tel.: +86-13926739052; Fax: +86-57384239300

**Abstract:** *Clitoria ternatea* (CT) flowers are rich in phytochemicals. An innovative approach was taken to utilize CT flower extract (CTFE) as a functional ingredient with natural pigment by incorporating it into noodles. The aim of this study was to examine the effect of the CTFE amount (0–30%) on the color, texture, phytochemicals, and sensory quality of both dried and cooked noodles. Dried noodles with 30% CTFE had the highest total anthocyanins (9.48 μg/g), polyphenols (612 μg/g), DPPH radical scavenging capacity (165 μg TE/g), and reducing power (2203 μg TE/g). Cooking resulted in a significant decrease in the anthocyanin levels and blue color, while also increasing the greenness of the noodle. Both dried and cooked noodles with 20–30% CTFE showed a significantly higher color preference compared to the control sample. Despite a significant reduction in the cutting force, tensile strength, and extensibility of cooked noodles with 20–30% CTFE, the sensory attributes such as flavor, texture, and overall preferences were similar to those of noodles with 0–30% CTFE. Blue noodles with high phytochemicals, antioxidant activities, and desirable sensory qualities can be produced by the incorporation of 20–30% CTFE.

**Keywords:** natural pigment; butterfly pea; anthocyanin; antioxidant; noodle; color

## 1. Introduction

Noodle products are a pleasant staple food in Asia. Noodle quality, such as color, texture, and nutrition, is affected by the ingredients and processing methods [1,2]. Since humans continuously seek a healthy, long life, and colorful food products with good quality, there is a growing worldwide trend toward the manufacturing of food products with colored functional ingredients such as grape pomace [3], black rice bran [4], and pitaya peel [5].

The perennial leguminous herb *Clitoria ternatea* (CT) (or butterfly pea) is widely cultivated in tropical regions, and its flower can be utilized in the food industry as a natural pigment [6]. Since CT flowers and extracts are rich in bioactive substances, they have antioxidation, anti-inflammation, antibacterial, antidiabetic, and antitumor activities [7–9]

Anthocyanin is a water-soluble natural pigment that mainly occurs in the flowers, leaves, fruits, and seeds of plants. The structure of anthocyanins consists of an anthocyanidin (aglycon), sugars, and/or organic acids [7,10]. The main anthocyanidins that occur in plants are pelargonidin, cyanidin, peonidin, delphinidin, petunidin, and malvidin. The CT flower is rich in blue polyacylated anthocyanins, which are known as ternatins and are derivatives of delphinidin 3, 3′, 5′-triglucoside [6,11]. Currently, researchers have identified about eighteen delphinidin and cyanidin derivatives from CT flower [9,12,13]. Anthocyanins can be used in intelligent films to monitor the freshness of food [14] and offer many health-promoting benefits such as anticancer, antidiabetic, and vascular protective effects [15].

The stability of anthocyanins relates to the chemical structure, temperature, pH, light, oxygen, and water activity [6,16,17]. Voss et al. [18] indicated that 10-catechyl-pyranoanthocyanins, formed by reacting anthocyanins with caffeic acid, displayed an excellent thermal stability. Anthocyanins were generally more stable at a lower temperature, pH, and water activity without oxygen and light. The extract of CT anthocyanin was very stable at pH 4.0–8.0, and with an increase in the pH from 0.5 to 13, the color of the extract changed from purple, red, and blue to green and yellow [19]. Anthocyanins extracted from CT flowers demonstrated stability at temperatures ranging from 60 to 70 °C within a pH range of 3.6 to 5.4. However, it was observed that the rate of thermal degradation raised significantly beyond 70 °C [6].

Some articles concerning the effects of anthocyanins on noodle and pasta properties have been reported. Pasta quality including antioxidant activity, anthocyanin, and fiber contents could be significatly improved by purple wheat (bran) and black rice bran [4,20,21]. An addition of up to 5% of extracts from the coat of black soybean could enhance the nutritional values of noodles with a satisfactory textural quality [2]. CT flowers are rich in blue anthocyanins and other bioactive compounds. Therefore, incorporating the flowers or their extracts into noodle products has the potential to enhance the color, nutritional value, and antioxidant activity of the final product. Yet, reports on the physicochemical properties of CT flower-containing noodles and on the stability of anthocyanins during the cooking process of noodles and pasta are rarely few. The aims of this study were as follows: (1) to investigate the color, texture, phytochemicals, antioxidative, and sensory properties of both dried and cooked noodles fortified with various CTFE amounts, and (2) to investigate the thermal stability of phytochemicals during the cooking process of noodles.

## 2. Materials and Methods

### 2.1. Materials

Wheat flour with 11.00% crude protein and medium gluten strength was purchased from Taichung Chia-Fha Enterprise Co. (Taichung, Taiwan) Fresh flowers of *Clitoria ternatea* (CT) were bought from a local market. All the chemicals and reagents used were of analytical grade.

### 2.2. Preparation of CTFE

Dry CT sample was obtained by drying (50 °C for 12 h) fresh flowers. Factors affecting the extraction yield of anthocyanins from CT flowers were solvent type, the ratio of substrate and solvent, extraction temperature, and time. Water heated to between 40 and 100 °C has been effectively used to extract anthocyanins from the flowers of butterfly pea [6], due to the economy, convenience, and non-toxicity. In the preparatory study for extracting anthocyanin from dried CT flowers, the total anthocyanin content of the aqueous extracts gradually increased from 4.79 to 24.55 mg/L when the extraction temperature increased from 30 to 90 °C, while maintaining a fixed extraction time of 30 min. To prepare aqueous extracts, 6 g of dried CT flowers were mixed with 300 mL of distilled water (90 °C) in this study. The mixture was then stirred on a hot plate with a magnetic stirrer for 30 min at 90 °C to extract bioactive compounds. After cooling, the extract was passed through a filter using a Buchner funnel and vacuum suction. Then, the filtered extract was mixed with distilled water until the final volume reached 300 mL.

### 2.3. Noodle Preparation

Wheat doughs with a 35.00% moisture content (wet basis) were produced according to the method [5]. Basic wheat dough was prepared by mixing 500 g of wheat flour (11.97% moisture) and distilled water (177.2 g). For making the dough fortified with 10%, 20%, and 30% CTPE, the amounts of CTFE and distilled water added were 50 and 127.9 g, 100 and 78.6 g, as well as 150 and 29.3 g, respectively.

After being allowed to rest at room temperature (approximately 26 °C) for 10 min, the dough was kneaded by hand to develop a cohesive mass. Next, the noodle dough was rolled into a sheet using a dough sheeter (Model HC1-A, Dean Food Equipment Co., Kaohsiung, Taiwan). The dough sheet underwent a gradual reduction in thickness by passing through roll gaps of 9.0, 7.0, 5.5, 4.2, 2.9, 1.5, and 0.8 mm until it reached a final thickness of approximately 1.6 mm. Subsequently, the dough sheet was sliced into raw noodles using a noodle cutter with a width of 3 mm. The noodle cutter used for this process was the DRB-36 model, manufactured by the Dean Food Service Equipment Co., Ltd. Dried noodles with about 11.5% of moisture content were prepared by air drying overnight in an air-conditioned room (26 °C, 60% RH).

To prepare cooked noodle samples, around 30 g of the dried noodles were boiled with 400 mL of water until the white core was no longer visible. After cooking, the noodles were transferred to a container filled with cold water at a temperature of 25 °C and left to soak for 20 s. Once cooled, the cooked noodles were promptly placed into sealable plastic bags and analyzed as soon as possible. Dry and cooked noodles with CTFE concentrations ranging from 0 to 30% were prepared by three replications for analysis of their physicochemical properties.

### 2.4. Total Anthocyanin Content (TAC)

TACs of CTFEs and noodles were determined by pH differential methods [22,23]. Firstly, both dried and cooked noodle samples (2 g) were extracted with 10 mL of 80% ethanol for 30 min, and then were centrifuged for 10 min at 6000× *g*. To determine TAC, the supernatant was collected and the extraction process was repeated. The combined supernatant and CTFE were analyzed using spectrophotometry. The extract sample (1 mL) was mixed thoroughly with 4 mL of pH 1.0 buffer containing 0.025 M KCl (adjusted by HCl), and another 4 mL of pH 4.5 buffer of 0.4 M sodium acetate. The resulting solutions were then measured for absorbance at 520 and 700 nm. The TAC values of both the CTFE and noodles were calculated using the following equation:

$$\text{TAC}\ (\mu g/g,\ \text{DB}) = (A \times F \times MW \times 1000 \times V)/(\varepsilon \times x \times W) \quad (1)$$

where $A = [(A_{520nm} - A_{700nm})_{pH\,1.0} - (A_{520nm} - A_{700nm})_{pH4.5}]$; F: dilution factor; MW: 449.2 g/mol of cyanidin-3-glucoside; V: volume of extract (mL); ε: 26,900 L/mol.cm, molar extinction coefficient of cyanidin-3-glucoside; x: path length of cuvette (cm); W: sample weight (g); and DB: dry weight basis.

### 2.5. Free Polyphenolic Content (FPC)

The method specified in reference [5] was employed to repeatedly extract the free form of total polyphenols from the noodles using 80% ethanol. The combined supernatant obtained from the extraction was used to determine the FPC and antioxidant activities. For FPC determination, the Folin–Ciocalteau method, as described in reference [24], was utilized. The absorbance was measured at 725 nm, and a standard curve was generated using ferulic acid. The FPC value was expressed as μg of ferulic acid equivalent (FAE) per gram of noodle (on a dry basis).

### 2.6. Antioxidant Activity

The radical scavenging activity of dried and cooked noodles' ethanolic extracts was assessed using the DPPH method, with Trolox as the standard. The spectrophotometric analysis was carried out at 517 nm [25], and the results were expressed in terms of Trolox equivalent (TE) per gram of noodles on a dry basis (DB). Furthermore, the reducing power of the ethanolic extract was determined using the ferric chloride and potassium ferricyanide method [26], and the development of a ferric–ferrous complex was observed by spectrophotometry at 700 nm. The results were also expressed in terms of TE per gram of noodles (DB).

### 2.7. Color of Noodle

The color of dried and cooked noodles was measured using a ColorFlex colorimeter from Hunter Associates Laboratory, Inc., located in Reston, VA, USA. Each noodle sample underwent six color measurements. The L* value, ranging from 0 to 100, indicates the level of whiteness or blackness. The a* and b* values indicate the degree of redness and yellowness or greenness and blueness, respectively, with positive values indicating the former and negative values indicating the latter. To calculate the white index (WI), the following equation was used:

$$WI = 100 - \sqrt{(100 - L^*)^2 + a^{*2} + b^{*2}} \quad (2)$$

*2.8. Breaking and Cutting Forces of Noodle*

Both breaking and cutting forces of the noodle samples were measured using the method [5]. To determine the breaking force (N), a single dried noodle measuring 8 cm in length was placed horizontally on the platform of a three-point bend rig (HDP/3PB) attached to a texture analyzer (TA-XT2i, Stable Micro Systems, Godalming, UK). The test was conducted at a speed of 1.0 mm/s, and the maximum force on the compression curve was recorded as the breaking force (mN). Cutting force of five cooked noodle strands was determined using a knife probe (A/LKB-F). Test speed was set to 0.5 mm/s and compression depth was set to 1.0 mm. The cutting force (mN) of the cooked noodles was read from the maximum force on compression curve. The measurements of both breaking and cutting forces of noodle samples were conducted in six repetitions for each treatment.

*2.9. Tensile Strength and Extensibility of Noodle*

Tensile properties of one cooked noodle were determined by the texture analyzer operated in tension mode and equipped with a spaghetti/noodle tensile rig. Both pretest and test speeds were set to 2.0 mm/s. Tensile strength (TS) and extensibility (E) of the cooked noodle can be measured by reading the maximum force and the distance at extension limit points during a tensile test. Tensile measurement was performed in 6 replications for each treatment.

*2.10. Stress Relaxation of Noodle*

According to the method of Pan et al. [27], three repetitions of stress relaxation testing were performed on cooked noodles for each treatment using the texture analyzer rigged with a P20 probe. Over a period of 240 s, the force remaining in the sample during testing at a 20% strain was recorded continuously as a function of time. Percent stress relaxation (%SR), a simple parameter for food viscoelasticity, can be obtained from the following equation [28]:

$$\%SR = \frac{F_0 - F_{t=20}}{F_0} \times 100 \quad (3)$$

where $F_0$ is initial force, $F_{t=20}$ is the force at 20 s after the initial strain was achieved. Furthermore, the data of stress relaxation were non-linearly regressed by using Peleg–Normand model [29], Equation (4).

$$\frac{F_0 t}{F_0 - F(t)} = k_1 + k_2 t \quad (4)$$

where $F_0$ is initial force, F(t) is the momentary force at time t, and both $k_1$ and $k_2$ values are constants.

*2.11. Cooking Quality of Noodle*

Cooking quality of noodle was executed according to the method [5]. As mentioned earlier, the dried noodle samples were cooked until the white core was no longer visible, and the appropriate cooking time was determined to be 10 min. The weights of the dried and cooked noodles, as well as the solids in the cooking water, were measured. The moisture content and cooking loss of cooked noodles could be determined using the gravimetric method, with three replications conducted for each treatment of noodle sample.

### 2.12. Sensory Evaluation

For consumers' sensory evaluation of the noodles enriched with various CTFE levels, thirty untrained panelists (16 females and 14 males, within the age range of 19 to 60 years) were invited by students and employees at the university. Each panelist was given four noodle samples to assess at the same time. Each panelist was served four samples of dried or cooked noodles, with each sample weighing approximately 30 or 60 g and placed on a white plate labeled with a three-digit code. The samples were presented to the panelists simultaneously. They were then asked to rate the noodles using a 7-point hedonic scale based on color, flavor, texture, and overall acceptability. The highest value represents the highest degree of acceptance (7 = extremely like, 1 = extremely dislike).

### 2.13. Statistical Analysis

The data from triplicate samples of the same type of noodles but different treatments were analyzed using SPSS software version 20.0 (SPSS, Chicago, IL, USA). To ascertain the statistical significance ($p < 0.05$) of the variations observed among the values, one-way ANOVA and Duncan's new multiple range tests were used.

## 3. Results and Discussion

### 3.1. Total Anthocyanins of Noodle

The TAC of the CTFE prepared in this study was 24.55 mg/L or 2668 mg/kg (DB). It was similar to the study [30] in that the extracts of the CT flower via an ultrasound-assisted extraction with water for 30–60 min at 40–80 °C contained 1050–2190 mg cyanidin-3-glucoside/kg (DW). The TACs of dried and cooked noodles were highly augmented by the levels (0–30%) of the CTFE (Figure 1), which ranged from 0 to 6.37–9.48 µg/g noodle (DB) for the dried noodle and ranged from 0 to 3.31–5.18 µg/g noodle (DB) for the cooked form, respectively. Thus, the TAC of dried noodles was reduced by about 45.34–49.00% during the cooking process. The decrease in the TAC of the noodle may be attributed to thermal degradation [6,31] and leaching of anthocyanins, as evidenced by the mild blue color of the cooking water when the noodle fortified with CTFE was cooked for 10 min. The use of black soybean coat, black rice bran, and purple wheat could have similar enriching effects on the anthocyanin content of both noodles and pasta [2,4,20].

Although the thermal degradation mechanisms of anthocyanins are not yet fully understood, it is known that their chemical structure plays a crucial role in their stability. Research has shown that polyacylated and polyglycosylated anthocyanins are more resistant to changes in heat and pH [31]. For example, tenatins, which are the main polyacylated anthocyanins found in CT flowers, are relatively stable under acidic and neutral conditions [11]. However, anthocyanins can undergo degradation during heat treatment as a result of deacylation, deglycosylation, oxidation, and the cleavage of covalent bonds [6,19]. This study also found evidence of thermal degradation of anthocyanins, as the dried CTFE-fortified noodle had a higher TAC than the cooked noodle.

In a study by Eliasova et al. [16], it was found that 45–51% of anthocyanins were lost during the baking stage of blue-grain wheat bread. Similarly, only 41.8% of the initial anthocyanin content was retained when cupcakes enriched with anthocyanin extract from CT flowers were baked at 170 °C for 20 min [32]. Additionally, the total anthocyanin content (TAC) of steamed bread with CT flower extract was reduced by approximately 36–40% after steaming [23]. These findings suggest that the stability of anthocyanins in food systems is influenced by factors such as the intensity, duration, and pattern of heating during thermal processing. However, there are limited reports on the thermal degradation of anthocyanins in noodle and pasta products, making it difficult to compare the results of this study with the other related literature on wheat-based products.

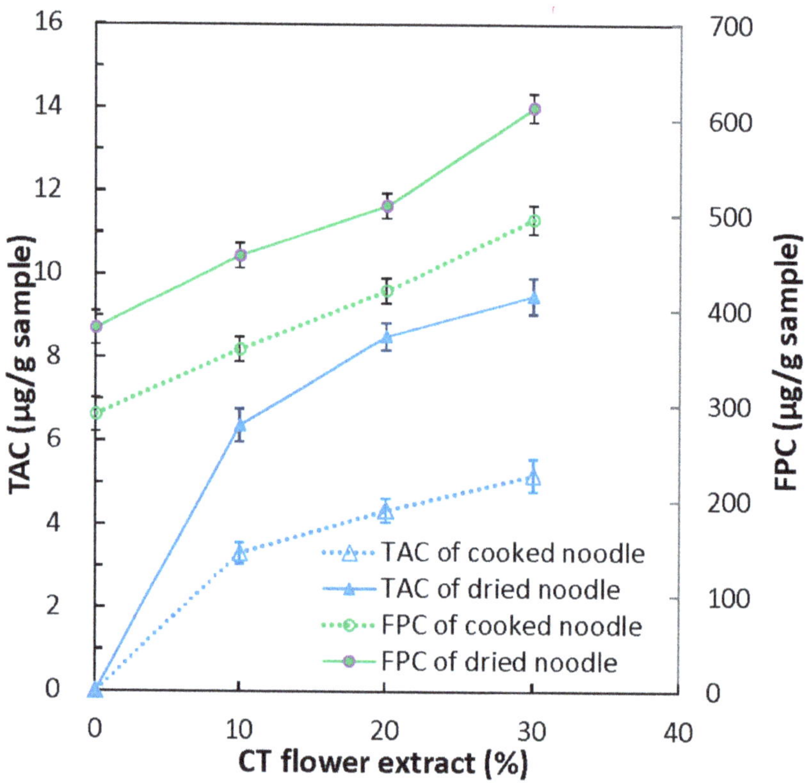


**Figure 1.** Total anthocyanin content (TAC) and free phenolic content (FPC) of dried and cooked noodles with various CTFE levels.

### *3.2. FPC and Antioxidant Capacity of Noodle*

CT flowers contained a high level of phytochemicals (such as anthocyanins, phenolic acids, and flavanols), which enhanced the antioxidant activity of enriched foods [6,33]. The CTFE used in this study had 65,401 μg FAE/g for the FPC, 15,080 μg TE/g for the DPPH scavenging activity, and 168,476 μg TE/g (DB) for the reducing power.

The results in Figures 1 and 2 showed that the FPC and antioxidant activities of both dried and cooked noodles obviously enhanced by increasing the amount of CTFE. Increasing the amount of CTFE in the dried noodles from 0 to 30% resulted in significant improvements in their FPC, DPPH radical scavenging capacity, and reducing power. The FPC increased by 20.1–61.0%, the DPPH radical scavenging capacity by 18.3–34.8%, and the reducing power by 3.0–10.1%. For the cooked noodles, FPC and two antioxidant activities improved by 23.6–70.8%, 15.2–32.3, and 5.6–20.5%, respectively. Moreover, phytochemicals (TAC and FPC) of the noodles positively correlated with the antioxidant activities (DPPH radical scavenging capacity and reducing power) ($r > 0.87$, $p < 0.05$). The findings of this study are consistent with previous research demonstrating that bread [34], sponge cake [35], steamed bread [23], and breakfast cereals [33] fortified with CT flower extract had significantly higher levels of total polyphenols and antioxidant capacity compared to the control products.

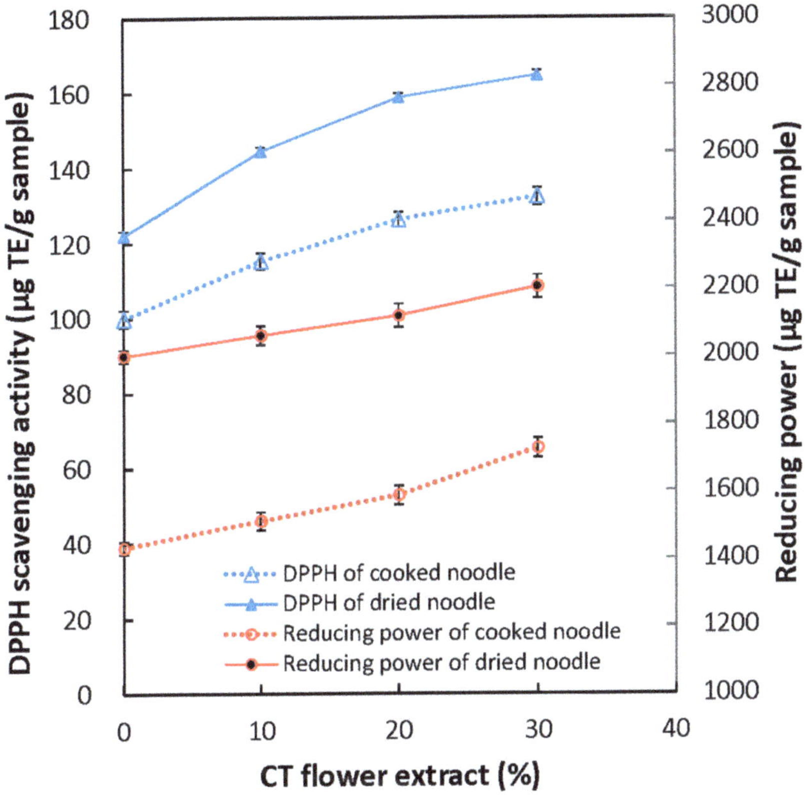


**Figure 2.** DPPH scavenging capacity and reducing power of both dried and cooked noodles fortified with various CTFE amounts.

### *3.3. Color of Noodle*

Food color is one of the important attributes in evaluating food quality. Due to the rich ternatins, the petals of CT and CTFE appeared navy blue. The control noodle displayed a white appearance in Figure 3, whereas the dried and cooked noodles containing 10–30% CTFE exhibited color variations ranging from light blue to blue. The effects of the CTFE levels (0–30%) on the color parameters of dried and cooked noodles are listed in Table 1. As the CTPE level increased, both Hunter L* and WI values of dried and cooked noodles significantly decreased, but the greenness (-a*) and blueness (-b*) of the noodles appeared to be remarkably increased. After the cooking process, the cooked CTFE-fortified noodle had an obviously higher brightness, WI, and greenness than the uncooked noodle. It was observed that the cooked noodles fortified with CTFE had lower blueness levels compared to their dried counterparts, which was consistent with the decrease in TAC shown in Figure 1. This could be attributed to the thermal degradation of ternatins from their blue neutral quinonoid base to the yellow chalcone form [19], as well as the release of anthocyanins into the cooking water of the noodle. Marpaung et al. [36] indicated that the color degradation of CT anthocyanins at pH 7 occurred through the deacylation, while at pH 4, it occurred through the unfolding of the hydrophobic interaction.

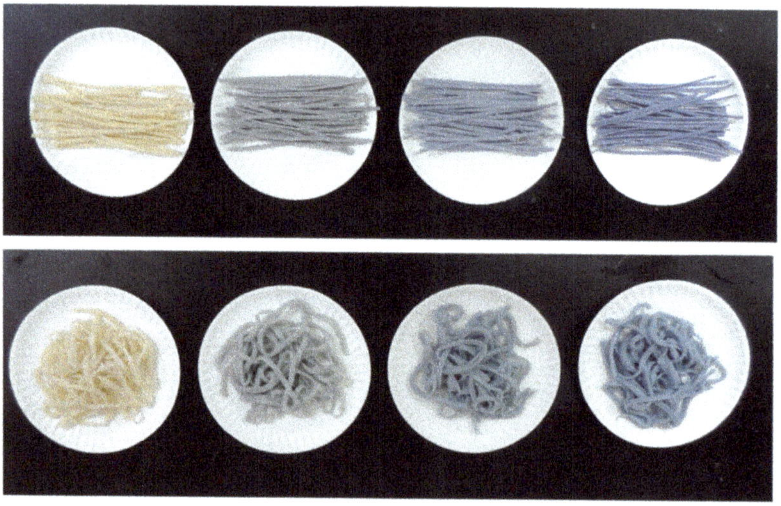

**Figure 3.** Appearances of dried noodle (**upper**) and cooked noodle (**lower**) fortified with 0, 10, 20, and 30% CTFE, respectively (from **left** to **right**).

**Table 1.** The color parameters of noodles fortified with different CTFE levels.

| CTFE (%) | Kind | L* | a* | b* | WI |
|---|---|---|---|---|---|
| 0 | Dried noodle | 70.72 ± 0.74 [b] | 1.24 ± 0.22 [a] | 17.69 ± 0.56 [a] | 65.77 ± 0.90 [b] |
| 10 | Dried noodle | 58.40 ± 0.40 [e] | −3.35 ± 0.18 [c] | −1.30 ± 0.25 [d] | 58.25 ± 0.42 [e] |
| 20 | Dried noodle | 54.97 ± 0.95 [g] | −3.81 ± 0.23 [d] | −7.11 ± 0.22 [f] | 54.25 ± 0.96 [g] |
| 30 | Dried noodle | 46.73 ± 0.59 [h] | −3.95 ± 0.18 [d] | −11.00 ± 0.19 [h] | 45.46 ± 0.59 [h] |
| 0 | Cooked noodle | 73.70 ± 0.41 [a] | −1.54 ± 0.14 [b] | 8.04 ± 0.25 [b] | 72.46 ± 0.36 [a] |
| 10 | Cooked noodle | 64.50 ± 0.67 [c] | −4.84 ± 0.27 [e] | −0.41 ± 0.18 [c] | 64.17 ± 0.67 [c] |
| 20 | Cooked noodle | 60.38 ± 0.60 [d] | −5.95 ± 0.19 [f] | −6.42 ± 0.17 [e] | 59.42 ± 0.60 [d] |
| 30 | Cooked noodle | 56.85 ± 0.31 [f] | −6.67 ± 0.13 [g] | −9.20 ± 0.12 [g] | 55.38 ± 0.32 [f] |

The reported values represent means ± standard deviations ($n = 6$), and any means within the same column that have different letters are considered to be significantly different ($p < 0.05$). CTFE: *Clitoria ternatea* flower extract; WI: white index.

During processing, the crumb and crust of the sponge cake containing 5–20% CT dried extract exhibited a greenish and brownish color, which can be attributed to the pH change [35]. Additionally, the bread crumb of anthocyanin-rich bread containing 30% blue-seeded maize flour changed color to a range of red shades [37]. The addition of citrus juice to CT ice cream led to color changes from blue to purple as its pH value decreased from 6 to 4 [38]. In this study, both dried and cooked noodles enriched with CTFE retained their blue color even after undergoing various processing stages such as mixing, sheeting, drying, and cooking. Previous research by Gamage et al. [6] highlighted the potential of blue anthocyanins from CT flowers to serve as a natural blue food colorant in neutral and mildly acidic food products. These observations suggest that factors such as the heating method, pH, and chemical structure of anthocyanins are crucial in determining the color stability of foods that are fortified with anthocyanin-rich ingredients.

### *3.4. Textural, Rheological and Cooking Properties of Noodle*

The texture and viscoelasticity of noodle products are chiefly influenced by the integrity of the gluten network. Table 2 lists the effect of various CTFE levels on the texture and cooking properties of noodles. Although the breaking force of dried noodles was not significantly affected by the CTFE level, cooked noodles fortified with high CTFE level (20% or 30%) had a significantly lower cutting force and tensile properties than the control noodles. Hence, the addition of a high CTFE level might result in a weaker and less extensible cooked noodle, due to the dilution of the gluten protein and protein–anthocyanin interaction [11,39,40].

**Table 2.** Influence of CTFE amount on texture and cooking properties of noodles.

| CTFE (%) | Breaking Force (mN) | Cutting Force (mN) | Tensile Strength (mN) | Extensibility (mm) | Moisture Content (%) | Cooking Loss (%) |
|---|---|---|---|---|---|---|
| 0 | 1779 ± 85 [a] | 1732 ± 67 [a] | 399.7 ± 24.5 [a] | 55.21 ± 3.07 [a] | 64.77 ± 0.47 [a] | 4.31 ± 0.13 [a] |
| 10 | 1760 ± 98 [a] | 1687 ± 66 [ab] | 393.3 ± 10.7 [a] | 52.49 ± 2.68 [a] | 64.39 ± 0.59 [a] | 4.43 ± 0.11 [a] |
| 20 | 1674 ± 82 [a] | 1595 ± 70 [bc] | 385.8 ± 16.6 [a] | 46.41 ± 3.02 [b] | 64.93 ± 0.51 [a] | 4.28 ± 0.12 [a] |
| 30 | 1712 ± 73 [a] | 1550 ± 57 [c] | 323.8 ± 10.1 [b] | 39.90 ± 2.67 [c] | 65.58 ± 0.54 [a] | 4.42 ± 0.15 [a] |

The reported values represent means ± standard deviations (n = 3), and any means within the same column that have different letters are considered to be significantly different ($p < 0.05$).

The hardness, elasticity, and cohesiveness of cooked noodles with 2.5–5% of vacuum-dried anthocyanins-rich extract from black soybean coat were not significantly changed ($p > 0.05$) [2]. However, cooked noodles fortified with higher levels of extract (7.5% or 10%) showed a significant decrease in the hardness, elasticity, and cohesiveness, and the continuous network structure of gluten was destroyed at the higher extract levels [2]. The incorporation of anthocyanin-rich black rice bran to the pasta led to a shorter cooking time, pasta firmness, and stickiness with higher cooking losses, since the bran containing high fiber disrupted the protein–starch matrix [4]. Boiled pasta supplemented with purple sweet potato powder (2.5–10%) had a lower hardness, but its elasticity was insignificantly ($p > 0.05$) influenced by the powder amount [41].

Food viscoelasticity can be determined by the stress relaxation experiment. The stress relaxation data of cooked noodles can be fitted well by the Peleg–Normand model (Equation (4)), with an $R^2 > 0.999$. Cooked noodles with 0–30% CTFE had similar %SR (53.44–54.39), $k_1$ (10.02–10.79 s), and $k_2$ (1.499–1.506) values. This indicated that the elasticity of cooked noodles was not influenced by the CTFE level. In Table 2, different CTFE levels do not significantly affect the moisture content of the cooked noodle (64.39–65.58%) and the cooking loss (4.28–4.43%), which is low and acceptable.

### 3.5. Sensory Evaluation of Noodle

The quality of the cooked noodles determined by different analytical instruments in Tables 1 and 2 was significantly or insignificantly affected by various CTFE levels. An objective instrumental measurement generally has a better sensitivity than a subjective sensory evaluation. The sensory scores of consumers for dried and cooked noodles fortified with different levels of CTFE are presented in Table 3. The addition of various levels of CTFE did not have a significant impact on the flavor acceptability of dried noodles ($p > 0.05$). However, dried noodles fortified with 20–30% CTFE had a significantly higher color acceptability compared to the control sample. Similarly, cooked noodles fortified with 20–30% CTFE also had a significantly higher color acceptability than the control sample. In terms of the flavor, texture, and overall acceptability, there were no significant differences ($p > 0.05$) observed in the cooked noodles fortified with 0–30% CTFE. These findings suggest that the addition of 20–30% CTFE can be a viable option to produce anthocyanin-rich colorful noodles, which can help increase the phytochemical intake without compromising the sensory acceptability. Santiago et al. [41] found that pasta supplemented with 7.5% and 10% of purple sweet potato powder had a similar sensory quality to the control pasta. No differences were found in the overall acceptability between sponge cake with 5–20% of CT extract and the control [35]. However, cupcakes fortified with the anthocyanin extract of the CT flower exhibited a higher color, flavor, and overall acceptability when compared to the control sample [32].

**Table 3.** Sensory evaluation of dried or cooked noodle fortified with various CTFE levels.

| | Dried Noodle | | Cooked Noodle | | | |
|---|---|---|---|---|---|---|
| CTFE (%) | Color | Flavor | Color | Flavor | Texture | Overall |
| 0 | 4.45 ± 1.26 [b] | 4.46 ± 1.04 [a] | 4.31 ± 1.13 [b] | 4.26 ± 1.09 [a] | 5.17 ± 1.03 [a] | 5.21 ± 1.02 [a] |
| 10 | 4.35 ± 1.17 [b] | 4.50 ± 1.38 [a] | 4.37 ± 1.26 [b] | 4.43 ± 0.95 [a] | 4.92 ± 1.29 [a] | 4.93 ± 1.16 [a] |
| 20 | 5.56 ± 1.35 [a] | 4.43 ± 0.87 [a] | 5.25 ± 1.30 [a] | 4.96 ± 1.12 [a] | 5.16 ± 1.45 [a] | 5.25 ± 1.23 [a] |
| 30 | 5.93 ± 0.92 [a] | 4.70 ± 0.92 [a] | 5.72 ± 1.05 [a] | 4.83 ± 1.36 [a] | 5.03 ± 1.67 [a] | 5.36 ± 1.15 [a] |

The reported values represent means ± standard deviations (n = 30), and any means within the same column that have different letters are considered to be significantly different ($p$ <0.05).

## 4. Conclusions

Due to their rich blue pigments and bioactive compounds, *Clitoria ternatea* (CT) flowers have the potential to be used as a natural colorant and functional ingredient in noodle products. The CTFE level significantly affected the color, antioxidant activity, texture, and sensory attributes of both dried and cooked noodles. An addition of 10–30% CTFE resulted in a significant improvement in the TAC, FPC, radical DPPH scavenging capacity, and reducing power of the noodles, which exhibited a light blue–blue coloration. However, during the cooking process, there was a noticeable reduction in the TAC and blueness of the noodle, along with an increase in the greenness, due to the thermal degradation and leaching of anthocyanins. Despite the significant reduction in the cutting force and tensile properties of the cooked noodle with 20–30% CTFE, the CTFE levels did not significantly affect the sensory flavor, texture, and overall acceptability of the noodle. This study demonstrates that blue noodles containing high bioactive compounds and antioxidant capacities with acceptable sensory qualities can be produced by the incorporation of 20–30% CTFE.

**Author Contributions:** The five authors have participated in the preparation of the article jointly. Supervision: S.-Y.S.; Data curation: Y.Y. and J.L.; Validation: W.H.; Visualization: H.F.; Writing—Original draft preparation: S.-Y.S., Y.Y., J.L., W.H. and H.F. All authors have read and agreed to the published version of the manuscript.

**Funding:** This work was supported by the project (MOST-107-2221-E-127-002) of the Ministry of Science and Technology, Taiwan, and the Foundation Key project (2022ZD04) and Educational Reform project (2021025) of Shanghai Sanda University.

**Institutional Review Board Statement:** Not applicable.

**Informed Consent Statement:** Not applicable.

**Data Availability Statement:** Data are contained within the article.

**Acknowledgments:** Thanks to the financial supports from the projects of the Ministry of Science and Technology, Taiwan, and Shanghai Sanda University.

**Conflicts of Interest:** The authors declare no conflict of interest.

## References

1. Cai, M.; Shen, C.; Li, Y.; Xiong, S.; Li, F. The Quality Characteristics Comparison of Stone-Milled Dried Whole Wheat Noodles, Dried Wheat Noodles, and Commercially Dried Whole Wheat Noodles. *Foods* **2022**, *12*, 55. [CrossRef] [PubMed]
2. Ma, M.; Xie, Y.; Wang, C. Effect of anthocyanin-rich extract from black soybean coat on wheat dough rheology and noodle texture. *J. Food Process. Preserv.* **2020**, *45*, 15007. [CrossRef]
3. Boff, J.M.; Strasburg, V.J.; Ferrari, G.T.; Schmidt, H.D.O.; Manfroi, V.; de Oliveira, V.R. Chemical, Technological, and Sensory Quality of Pasta and Bakery Products Made with the Addition of Grape Pomace Flour. *Foods* **2022**, *11*, 3812. [CrossRef] [PubMed]
4. Sethi, S.; Nanda, S.K.; Bala, M. Quality assessment of pasta enriched with anthocyanin-rich black rice bran. *J. Food Process. Preserv.* **2020**, *44*, 14952. [CrossRef]
5. Shiau, S.; Li, G.; Pan, W.; Xiong, C. Effect of pitaya peel powder addition on the phytochemical and textural properties and sensory acceptability of dried and cooked noodles. *J. Food Process. Preserv.* **2020**, *44*, 14491. [CrossRef]
6. Gamage, G.C.V.; Lim, Y.Y.; Choo, W.S. Anthocyanins from *Clitoria ternatea* Flower: Biosynthesis, Extraction, Stability, Antioxidant Activity, and Applications. *Front. Plant Sci.* **2021**, *12*, 792303. [CrossRef]

7. Escher, G.B.; Wen, M.; Zhang, L.; Rosso, N.D.; Granato, D. Phenolic composition by UHPLC-Q-TOF-MS/MS and stability of anthocyanins from *Clitoria ternatea* L. (butterfly pea) blue petals. *Food Chem.* **2020**, *331*, 127341. [CrossRef]
8. Li, C.; Tang, W.; Chen, S.; He, J.; Li, X.; Zhu, X.; Li, H.; Peng, Y. Phytochemical Properties and In Vitro Biological Activities of Phenolic Compounds from Flower of *Clitoria ternatea* L. *Molecules* **2022**, *27*, 6336. [CrossRef]
9. Nair, V.; Bang, W.Y.; Schreckinger, E.; Andarwulan, N.; Cisneros-Zevallos, L. Protective Role of Ternatin Anthocyanins and Quercetin Glycosides from Butterfly Pea (*Clitoria ternatea* Leguminosae) Blue Flower Petals against Lipopolysaccharide (LPS)-Induced Inflammation in Macrophage Cells. *J. Agric. Food Chem.* **2015**, *63*, 6355–6365. [CrossRef]
10. Jokioja, J.; Yang, B.; Linderborg, K.M. Acylated anthocyanins: A review on their bioavailability and effects on postprandial carbohydrate metabolism and inflammation. *Compr. Rev. Food Sci. Food Saf.* **2021**, *20*, 5570–5615. [CrossRef]
11. Vuong, T.T.; Hongsprabhas, P. Influences of pH on binding mechanisms of anthocyanins from butterfly pea flower (*Clitoria ternatea*) with whey powder and whey protein isolate. *Cogent Food Agric.* **2021**, *7*, 1889098. [CrossRef]
12. Terahara, N.; Toki, K.; Saito, N.; Honda, T.; Matsui, T.; Osajima, Y. Eight new anthocyanins, ternatins C1-C5 and D3 and preternatins A3 and C4 from young *Clitoria ternatea* flowers. *J. Nat. Prod.* **1998**, *61*, 1361–1367. [CrossRef]
13. Thuy, N.M.; Minh, V.Q.; Ben, T.C.; Nguyen, M.T.T.; Ha, H.T.N.; Tai, N.V. Identification of anthocyanin compounds in butterfly pea flowers (*Clitoria ternatea* L.) by ultra performance liquid chromatography/ultraviolet coupled to mass spectrometry. *Molecules* **2021**, *26*, 4539. [CrossRef] [PubMed]
14. Sohany, M.; Tawakkal, I.S.M.A.; Ariffin, S.H.; Shah, N.N.A.K.; Yusof, Y.A. Characterization of Anthocyanin Associated Purple Sweet Potato Starch and Peel-Based pH Indicator Films. *Foods* **2021**, *10*, 2005. [CrossRef]
15. Blesso, C.N. Dietary Anthocyanins and Human Health. *Nutrients* **2019**, *11*, 2107. [CrossRef]
16. Eliášová, M.; Kotíková, Z.; Lachman, J.; Orsák, M.; Martinek, P. Influence of baking on anthocyanin content in coloured-grain wheat bread. *Plant Soil Environ.* **2020**, *66*, 381–386. [CrossRef]
17. Guo, Y.; Zhang, H.; Shao, S.; Sun, S.; Yang, D.; Lv, S. Anthocyanin: A review of plant sources, extraction, stability, content determination and modifications. *Int. J. Food Sci. Technol.* **2022**, *57*, 7573–7591. [CrossRef]
18. Voss, D.M.; Miyagusuku-Cruzado, G.; Giusti, M.M. Thermal stability comparison between 10-catechyl-pyranoanthocyanins and anthocyanins derived from pelargonidin, cyanidin, and malvidin. *Food Chem.* **2023**, *403*, 134305. [CrossRef]
19. Fu, X.Y.; Wu, Q.; Wang, J.; Chen, Y.L.; Zhu, G.P.; Zhu, Z.X. Spectral characteristic, storage stability and antioxidant properties of anthocyanin extracts from flowers of butterfly pea (*Clitoria ternatea* L.). *Molecules* **2021**, *26*, 7000. [CrossRef]
20. Verma, J.; Rani, V.; Sangwan, V.; Karnika, K. Physical, sensory and nutritional quality of anthocyanins rich pasta prepared using biofortified purple wheat. *Ann. Phytomedicine Int. J.* **2022**, *11*, 78–85. [CrossRef]
21. Zanoletti, M.; Parizad, P.A.; Lavelli, V.; Cecchini, C.; Menesatti, P.; Marti, A.; Pagani, M.A. Debranning of purple wheat: Recovery of anthocyanin-rich fractions and their use in pasta production. *LWT* **2017**, *75*, 663–669. [CrossRef]
22. Lee, J.; Durst, R.W.; Wrolstad, R.E.; Wrolstad, R.E. Determination of Total Monomeric Anthocyanin Pigment Content of Fruit Juices, Beverages, Natural Colorants, and Wines by the pH Differential Method: Collaborative Study. *J. AOAC Int.* **2005**, *88*, 1269–1278. [CrossRef]
23. Shiau, S.; Yu, Y.; Pan, W.; Li, G. Colorful and health improving Chinese steamed bread fortified by anthocyanin-rich extract of butterfly pea flower. *J. Food Process. Preserv.* **2022**, *46*, 16925. [CrossRef]
24. Buçukoglu, T.Z.; Albayrak, S.; Halici, M.G.; Tay, T. Antimicrobial and Antioxidant Activities of Extracts and Lichen Acids Obtained from Some *Umbilicaria* Species from Central Anatolia, Turkey. *J. Food Process. Preserv.* **2013**, *37*, 1103–1110. [CrossRef]
25. Liyana-Pathlrana, C.M.; Shahidi, F. Antioxidant and free radical scavenging activities of whole wheat and milling fractions. *Food Chem.* **2007**, *101*, 1151–1157. [CrossRef]
26. Oyaizu, M. Studies on products of browning reaction. Antioxidative activities of products of browning reaction prepared from glucosamine. *Jpn. J. Nutr. Diet.* **1986**, *44*, 307–315. [CrossRef]
27. Pan, W.C.; Liu, Y.M.; Shiau, S.Y. Effect of okara and vital gluten on physico-chemical properties of noodle. *Czech J. Food Sci.* **2018**, *36*, 301–306. [CrossRef]
28. Bellido, G.; Hatcher, D. Asian noodles: Revisiting Peleg's analysis for presenting stress relaxation data in soft solid foods. *J. Food Eng.* **2009**, *92*, 29–36. [CrossRef]
29. Peleg, M.; Normand, M.D. Comparison of two methods for stress relaxation data presentation of solid foods. *Rheol. Acta* **1983**, *22*, 108–113. [CrossRef]
30. Salacheep, S.; Kasemsiri, P.; Pongsa, U.; Okhawilai, M.; Chindaprasirt, P.; Hiziroglu, S. Optimization of ultrasound-assisted extraction of anthocyanins and bioactive compounds from butterfly pea petals using Taguchi method and Grey relational analysis. *J. Food Sci. Technol.* **2020**, *57*, 3720–3730. [CrossRef] [PubMed]
31. Patras, A.; Brunton, N.P.; O'Donnell, C.; Tiwari, B. Effect of thermal processing on anthocyanin stability in foods; mechanisms and kinetics of degradation. *Trends Food Sci. Technol.* **2010**, *21*, 3–11. [CrossRef]
32. Thanh, V.T.; Tran, N.Y.T.; Linh, N.T.V.; Vy, T.A.; Truc, T.T. Application of anthocyanin natural colors from Butterfly Pea (*Clitoria ternatea* L.) extracts to cupcake. *IOP Conf. Ser. Mater. Sci. Eng.* **2020**, *736*, 062014. [CrossRef]
33. Singh, R.; Yu, C.-C.; Chen, G.-W.; Chen, C.-H.; Sinaki, N.Y.; Lin, J.; Koksel, F. Butterfly Pea Flower as a Novel Ingredient to Produce Antioxidant-Enriched Yellow Pea-Based Breakfast Cereals. *Foods* **2022**, *11*, 3447. [CrossRef] [PubMed]
34. Chusak, C.; Henry, C.J.; Chantarasinlapin, P.; Techasukthavorn, V.; Adisakwattana, S. Influence of *Clitoria ternatea* Flower Extract on the In Vitro Enzymatic Digestibility of Starch and Its Application in Bread. *Foods* **2018**, *7*, 102. [CrossRef] [PubMed]

35. Porntip, P.; Theerapat, P.; Nontapat, S.; Chonlada, C.; Piyawan, W.; Sirichai, A.; Sathaporn, N. Physicochemical, antioxidant and sensory characteristics of sponge cakes fortified with *Clitoria ternatea* extract. *J. Food Sci. Technol.* **2018**, *55*, 2881–2889.
36. Marpaung, A.M.; Andarwulan, N.; Hariyadi, P.; Faridah, D.N. The Difference in Colour Shifting of *Clitoria ternatea* L. Flower Extract at pH 1, 4, and 7 During Storage. *Curr. Nutr. Food Sci.* **2019**, *15*, 694–699. [CrossRef]
37. Simić, M.; Žilić, S.; Šimuruna, O.; Filipčev, B.; Škrobot, D.; Vančetović, J. Effects of Anthocyanin-Rich Popping Maize Flour on the Phenolic Profile and the Antioxidant Capacity of Mix-Bread and Its Physical and Sensory Properties. *Pol. J. Food Nutr. Sci.* **2018**, *68*, 299–308. [CrossRef]
38. Rashid, N.F.M.; Zulkifli, N.A.; Anwar, N.Z.R.; Gani, H.S.M.; Jusoh, T. Physicochemical properties, antioxidant activities and sensory evaluation of butterfly pea (*Clitoria ternatea* L.) ice cream flavoured with citrus juice. *Biosci. Rep.* **2022**, *19*, 73–81.
39. Li, J.; Wang, B.; He, Y.; Wen, L.; Nan, H.; Zheng, F.; Liu, H.; Lu, S.; Wu, M.; Zhang, H. A review of the interaction between anthocyanins and proteins. *Food Sci. Technol. Int.* **2021**, *27*, 470–482. [CrossRef]
40. Rakhesh, N.; Fellows, C.; Sissons, M. Evaluation of the technological and sensory properties of durum wheat spaghetti enriched with different dietary fibres. *J. Sci. Food Agric.* **2015**, *95*, 2–11. [CrossRef]
41. Santiago, D.M.; Kawashima, Y.; Matsushita, K.; Noda, T.; Pelpolage, S.; Tsuboi, K.; Kawakami, S.; Koaze, H.; Yamauchi, H. Noodle Qualities of Fresh Pasta Supplemented with Various Amounts of Purple Sweet Potato Powder. *Food Sci. Technol. Res.* **2016**, *22*, 307–316. [CrossRef]

MDPI AG
Grosspeteranlage 5
4052 Basel
Switzerland
Tel.: +41 61 683 77 34

*Foods* Editorial Office
E-mail: foods@mdpi.com
www.mdpi.com/journal/foods

www.ingramcontent.com/pod-product-compliance
Lightning Source LLC
Chambersburg PA
CBHW061829210926
51850CB00006B/842
*9783725845538*